TECHNOLOGY OF WOOD BONDING
Principles in Practice

Alan A. Marra

Professor Emeritus
University of Massachusetts, Department of Forestry and Wildlife
University of Michigan, College of Architecture

VNR VAN NOSTRAND REINHOLD
_____ New York

Copyright © 1992 by Van Nostrand Reinhold

Library of Congress Catalog Card Number 91–45707
ISBN 0-442-00797-3

Printed in the United States of America.

Van Nostrand Reinhold
115 Fifth Avenue
New York, New York 10003

Chapman and Hall
2-6 Boundary Row
London, SE1 8HN, England

Thomas Nelson Australia
102 Dodds Street
South Melbourne 3205
Victoria, Australia

Nelson Canada
1120 Birchmount Road
Scarborough, Ontario MIK 5G4, Canada

16 15 14 13 12 11 10 9 8 7 6 5 4 3 2 1

Library of Congress Cataloging-in-Publication Data

Marra, Alan A.
    Technology of wood bonding: principles in practice/by Alan A.
  Marra.
       p.   cm.
    Includes bibliographical references and index.
    ISBN 0-442-00797-3
    1. Wood—Bonding. 2. Adhesives. 3. Gluing. I. Title.
  TS857.M37   1992
    668´.3—dc20
                                                                91–45707
                                                                CIP

To George, whose total dedication to the field of forest products inspired this book, and whose collaboration would have made it more comprehensive and more definitive.

Contents

Foreword

Interest in renewable resources is increasing worldwide because of concern about depleting resources and energy shortages. Wood is the only construction material that is readily renewable, and it is energy efficient in production, processing, and use. In the past, wood was used largely as it came from the forest, its properties assumed to be beyond the control of man and determined simply by identifying the tree from which it came. Quality was more a process of selection than a part of the manufacturing process. Then wood scientists and wood technologists found ways to improve the properties of wood by the application of various kinds of treatments. Today, new wood materials can be created that have properties more precisely tailored to use requirements. Control of wood properties is now the dominant theme, extending not only to processing, but also to genetics and forest management as well.

The materials science approach to wood products requires that a relation be developed between properties and their composition, structure, and processing. The operational term that derives from such knowledge is *synthesis,* in which a desired property of a material may be achieved through knowledgeable combination of ingredients and control of processing variables. This is not a futuristic goal, but one now in current practice. Over 50 percent of all wood used in the United States is first taken apart and then reconstituted in some form such as paper, fiberboard, particleboard, and plywood. In most such products the final properties are dependent more on the nature of the processing than on the properties of the original material.

The greatest gain in synthesis capability beyond laminated lumber and plywood occurred with the invention and use of a number of small wood elements such as fibers, particles, strands, flakes, and wafers shown in Figure 1. These basic wood elements, each with its own distinctive geometry, together with controllable processing variables, such as size of element, type and amount of adhesive, moisture content, pressure, pressure sequence, mix of elements, alignment of elements and layered construction, provide a virtually unlimited field for the exercise of the principle of synthesis. The fact that this goal can be achieved with wood that is otherwise unsuitable for the more conventional lumber and plywood applications adds greatly to the technical triumph. The net effect will be a vastly increased supply of wood materials for the world's growing population.

The practice of synthesis in wood materials accomplishes three important benefits for a society striving to maintain a high-technology standard of living beyond the year 2000:

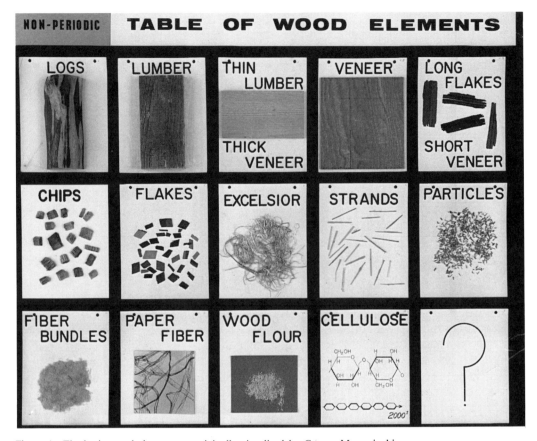

Figure 1   The basic wood elements as originally visualized by George Marra in his lectures on the materials science of wood.

1. It provides a greatly increased produc output from the limited annual harvest of forest resources.
2. It provides wood materials with improved properties for an expanded range of uses.
3. It provides a safe alternative to rapid depletion of metallic and polymeric (petrochemical) resources for fabrication and construction purposes.

It is obvious that, in addition to a knowledge of solid wood as produced by nature, professionals in this discipline will have to understand the mechanisms by which individual wood elements interact to confer properties to the larger body. These wood elements vary not only in species, but also in size, geometry, and surface quality. The contribution of each element ultimately depends on its orientation and on the quality of bonding between it and adjacent elements. Since bonding is a molecular phenomenon involving dissimilar materials, this area of knowledge can accommodate a wide range of fundamental scientific research.

A second consideration is the manipulation of variables in the manufacturing process, which in large measure controls the properties of the resulting material. The selection of a bonding system is one of the most critical decisions, affecting both properties and economics. The amount of resin, its distribution, and its rate of cure are all controllable variables with pronounced influence on final material properties. Equally potent are the layering and orientation of elements during board formation. Heat delivery, heat transfer, pressure, and pressure sequence add further levels of control. Throughout the manufacturing process, moisture content hangs as a pernicious factor ready to spoil properties by slight variations at any step. With precise knowledge of these factors still minimal, it is amazing that a wayward factor at one point in the process can often be compensated by judicious modification of another factor at another point. In this sense, the manufacturing process becomes a delicately balanced ecosystem, and its control requires fundamental knowledge of many "environmental" factors.

A third consideration is the properties of the material and their relation to use requirements. Much of what now constitutes design criteria for wood structures is based on knowledge of solid wood properties. The properties of engineered wood materials differ from those of solid wood in a number of ways, two of which are worthy of note here: (1) the controlled variation in properties in the three orthogonal directions; and (2) the uniformity of properties from piece to piece. These two factors greatly expand the design freedom of the architect and engineer, but new design criteria must be developed in order to take advantage of the full potential of these new materials. The net result of improved properties and new design criteria is likely to be a reduction in the amount of wood used in structures, together with increased reliability in terms of performance.

It seems safe to say that the age of wood lies ahead of us, not behind us, and that it rests in the realm of materials engineering in the truest sense of this term.

This foreword has been adapted from the speeches and writings of GEORGE G. MARRA, late Deputy Director, USDA Forest Products Laboratory, Madison, Wisconsin; formerly Professor of Materials Science and Engineering, and Assistant Dean for Research, College of Engineering, Washington State University, Pullman, Washington.

The presentations include the following:

"A Materials Science Approach to Wood Science," in the Proceedings of IXth International Symposia on Forest Science, National Academy of Sciences, 1981 Seoul, Korea.

"The Age of Wood Materials Engineering," keynote address at the World Consultation on Wood-Based Panels, New Delhi, India, 1975.

"The Coming Age of Wood Materials Engineering," *The Idaho Forester,* January 1975.

"A Vision of the Future of Particleboard," proceedings of the Third Washington State University Symposium on Particleboard, 1969.

Preface

Because it is one of the most abundant and widely used materials in the world, because its warmth and natural ambiance are universally appealling, and because its strength and easy workability endear it to crafts and industry people alike, wood touches virtually everyone. A high percentage of this wood, 50 percent or more by some estimates, is glued. The gluing of wood is therefore also widespread, and performed, it may be presumed, by a wide range of variously trained people.

The forests that produce wood represent a far-flung biomass interwoven into an ecosystem that sustains a seemingly endless array of life forms, many in critical balance. It has become increasingly obvious that forests are essential to the welfare of not only humans but also these other forms of life, which in turn contribute, in still largely unknown ways, to our own existence. We are all part of the same ecosystem, in a mutually dependent but fragile relationship. Out of sheer self-interest if nothing else, it would seem prudent to support the growth of forests anywhere they will grow.

Forests are a prime and practical example of the ability to eat your cake and have it too. They are renewable. However, in order to perpetuate a forest and enjoy its wood, it is necessary to harvest judiciously. In the past, and still to a large extent in the present, the utilization of timber has resulted in preferences for certain species, sizes, and forms of trees. Since the resulting selective harvesting tends to warp the composition of forests by removing only certain trees, any technology that helps maintain diversity would benefit the system. Gluing is one such technology. It permits homogenizing different qualities of trees into products of wide utility.

Gluing requires a level of technology beyond that of the hammer and saw. More factors need to be controlled, conditions adjusted, judgments made, responsibilities assigned, trouble assessed, and performance predicted. Because the outcome is never certain except as experience and attention to detail favors good results, many who use adhesives feel some degree of trepidation as clamps close on an assembled product. Even after the clamps open there may be uncertainty whether an adequate bond has been formed. The operator is then faced with a Hobson's choice of destroying the bond to verify its quality, or in the hope that everything has been done correctly, taking a chance that it is satisfactory. In any case it would be mentally and economically comforting if there were someone to turn to for advice and suggestions, especially if something new had been added to the procedure. This book tries to be such a "someone."

With these thoughts in mind, it seems obvious that more widespread practical knowledge of wood gluing is essential not only for the benefit of more and better forestry, but also to support a vital segment of wood-based industry. While wood gluing has been practiced for centuries, it is not well enough understood on the shop floor, the work bench, or even in the laboratory, to deal confidently on a daily basis with the many factors in the process. This book is intended to provide a measure of understanding to anyone using wood adhesives, developing them, researching their performance, or learning about them for the first time.

A substantial but not overwhelming number of books have been written in this field, covering subjects such as wood adhesives, gluing processes, equipment, and glued products including their economics, their chemistry, their mechanics, and their importance. It seems appropriate that one be written emphasizing wood and its interactions with the adhesive. Since wood is the main ingredient in glued constructions, it dominates the gluing processes in ways that are often not well accounted.

The field of wood gluing draws on so many disciplines, such as physics, chemistry, mechanics, engineering, rheology, physical chemistry, polymer chemistry, wood anatomy, and wood properties, that it sometimes does not seem to have a science and technology of its own. Contributions from various disciplines have been illuminating the wood gluing scene in ever-increasing volume since World War II. While each added significantly to specific points of interest, together they did little to unify the subject. Progress nevertheless has been rapid, too rapid perhaps for it to have gelled into a discrete field of knowledge with neatly regimented theories and a formalized line of study. This book assumes a unifying role in integrating disparate islands of knowledge into a more comprehensive discipline, covering as many facets as possible that have a bearing on the practice of wood gluing.

The science and technology of wood gluing include so many fields that it exceeds the working knowledge of any one individual. Because I am only one individual, bounded by specialized training and experience, I have had to borrow freely in order to support some basic tenets on which this book is constructed. And in order to overcome similar bounds that readers may have, a concept of wood gluing is developed that obviates to a certain degree the need to know intricate details of these ancillary fields. This was done by vesting all glues with certain activities essential to bond formation, and by focusing on the space between two pieces of wood. This is the scene and the environment where wood and glue interact to form bonds, and maintain them. A number of principles are formulated which cover most of the events that occur in the glue line, and simplify the reasoning that must apply. The approach allows thousands of glued products and processes to be reduced to their distinctive essentials and categorized into a relatively few processes.

In a broad and subtle sense, the philosophy is a rationale for dealing with ignorance. The importance of this approach lies in the evolving nature of knowledge and new products. Continuous evolution virtually guarantees that any one individual, at any point in time, will lack certain information. The power of principles, generalized as proposed herein, is that they override or bridge gaps in knowledge, and are always ready for application. In the early 1940s for example, the introduction of urea resins as an improvement over then current glues, created many problems resulting from lack of knowledge concerning use factors that affect performance of these adhesives. The principles in this book, together with a minimum of new information, might have mediated most of the unfavorable consequences of using that new product.

The wide diversity in knowledge underlying the field suggests that wood gluing has been developed in a piecemeal manner, supported largely by phenomenological observation and trial-and-error experimentation, and less by rigorous scientific synthesis and extrapolation of theories. It further suggests that understanding the art and controlling the process depends on powers of observation and a sense of how factors of diverse origin interact to affect performance. This premise further supports the approach followed in this book. Wherever possible, observable characteristcs are woven into a logic that leads to simplification and confidence.

In carrying out the objectives of the book, pertinent knowledge of wood gluing was drawn together, systematized where appropriate, and reduced to principles when facts appeared to represent laws of nature. Such a treatment, it was hoped, would accelerate the further development of the field, as well as ensure the correct application of present knowledge. The reader will note that tables often contain information that is not only itemized but also sequentialized. Sequences carry meanings of their own. Besides aiding the learning process, they provide logic in a form that can be interpolated and extrapolated into a great deal of additional information.

This book is not exhaustive. Its main objective is to develop the essential parameters of wood gluing in a way that they make projectible sense and ensure predictable results. Considerable interesting, even relevant, material has been left to other sources so as to concentrate on principles essential to an understanding of proper gluing technique. For the same reason, specific mechanical details of many gluing processes have been omitted except where they illustrate principles applicable to all.

An exception has been made in the case of wood properties and its preparation, which I feel need more emphasis than is traditionally accorded. Many books on adhesives and wood composites do not pay sufficient attention to the substrate. Wood is wood, the thing being glued, they seem to believe, and one piece is the same as another. Actually, no two pieces of wood are alike. Sometimes the differences are profound, sometimes trivial. But since wood is the major constituent in wood composites, even the trivial, such as a few percent moisture content, degrees of grain angle, or slight change in pH, can overwhelm the properties of a thin glue line.

The principles, and the factors they represent, are universal; they apply no matter what adhesive is being used, in what construction, or in what process. They are based on what a glue has to do to form a bond, and to maintain it. The rest is chemistry, mechanics, wood properties, environment, all multiplying into an infinity of processes. Consequently, details of the latter, beyond the operational level, have been largely shorn away in a considered attempt to make the principles stand out. Operators are thus relieved of "status quoism," and are free to innovate their way to alternative ways of gluing wood.

In organizing this book, two options needed to be resolved: either treating the subject as a series of unit operations, or treating individual glued products and their separate methods. The latter would have been helpful for those interested in a single glued product, but it would have required repeating information common to several. On the other hand, covering the subject in terms of unit operations would make it better pedagogically, but would require a practitioner to leaf through various parts of the book to piece together the information relevant, for example, to edge-gluing of lumber. I have tried to combine the two approaches, separating common information from the specific, and tying the whole together with threads of principles. This should be particularly appealing considering the fact that problems can have their origins in more than one part of the process. Operators, knowing how their work affects or is affected by other operators, will better appreciate the cautions and constraints of their particular part of the process. As a consequence, readers will find the same principle repeated several times, like the refrain from a song, and may sense duplication. In most cases, however, many circumstances are different. The universality of the concepts on which the principles are based is thus emphasized.

The book should be useful for practitioners and students alike. Practitioners of the art will see here many thought processes and logic they have used in the day-to-day course of their work. For some it will be "old hat." For others, the reduction to principles should make reasoning more logical and the art less mystifying. As for individual craftspeople, their art, particularly new art, tends to percolate to them from the large-scale industrial operations which are the prime motivators of new developments. What percolates then blends with the old, sometimes unobtrusively, sometimes jarringly. In any case, the new learning that must take place should be eased by the principles discussed herein.

Wood gluing is regarded as a very complex subject, mainly because of the many factors involved. Since most of the factors interact in some way, the subject seems to have no logical entry point to ease the learning process. Factors come piling in from the beginning, requiring explanation based on knowledge of other factors yet unexplained. There is no recourse but to begin somewhere and simplify, oversimplify if necessary, and weave it together into an understandable whole by the end. I have oversimplified liberally, with some sacrifice of purity and scientific elegance. But if the result is greater understanding, and greater utility, I hope the reader will consider it a small price to pay.

The approach taken has been to formulate an *Equation of Performance* (Table 1-1) as a guide to unifying the field by defining the various parts and inferring their summative nature in establishing the quality of a glue bond. Ultimately, better-trained scientists will discover ways of quantifying the effects of the diverse factors that make up the equation of performance, and will develop computer logic to predict results. In the meantime, the equation may serve as both a context and a road map to the world of wood gluing, providing bite-size pieces of information for operators concerned with forming a bond between pieces of wood.

However, before engaging the equation of performance, it seemed appropriate to discuss in Chapter 1 some background about the field, and the philosophy behind the equation. This is followed in Chapter 2 by a cursory look at how wood gluing is practiced

in its various aspects, the aim being to provide a mental picture or outline on which later detail can more easily be attached. The four basic gluing processes—lumber gluing, veneer gluing, flake gluing, and fiber gluing—are sketched here, together with a brief survey of other kinds of gluing associated with the production of final shapes. These processes embody much of the basic technology to be discussed.

Chapter 3 presents the fundamental knowledge underlying the principles. This chapter is a generalization and systematization of information that applies to all glues and all gluing processes. It isolates factors, defines functions, and explains how bonds form in wood. Bond formation is couched in dynamic terms suggestive of lifelike actions which occur in stages, sequences, and magnitudes, and which, fortunately, leave identifiable clues to their operation. This chapter is the heart of the book. The principles are used in subsequent chapters to extract information from the equation of performance, to explain how the thousands of variations in glued products and processes might affect a result, and, most important, to suggest corrective measures when results are not up to expectations. In other words, the principles are the keys to understanding, and the paths to success in wood gluing operations.

Chapters 4 and 5 describe factors in the composition of adhesives and the composition and structure of wood that are pertinent to the gluing operation. In this effort at simplification, properties are described only as they bear on the principles and the actions that must occur in the glue line. More information is available and should be pursued in the many texts and reference books that deal specifically with the subject.

The basic characteristics inherent in wood and adhesives are inevitably modified or influenced in some way by the processes of preparing or applying them. These, together with their effects, are discussed in Chapters 6 and 8. Again the effects are strained through the web of bond-forming actions to demonstrate how problems might originate and evidence themselves.

Wood Geometry Factors, Chapter 7, represent the chief distinction among different gluing processes, influencing choice of adhesive, production system, and performance of the product. This chapter is an attempt to systematize the factor of geometry in terms of wood elements having certain sizes and producing certain properties in the glued product. Important relationships of both general and technological interest can also be drawn between geometry and the quality of the resource from which the wood came, the degree of mechanization that may be employed, the conversion energy required, and the economic consequences that may apply.

All the factors discussed in the previous chapters come together in various gluing processes, consummating the bond and fixing its performance. In Chapter 9 we see how these factors integrate and interact in specific gluing processes. We see how differences develop as a result of interactions throughout the Equation of Performance, what new situations occur, and what new cautions apply.

After a bond is formed, its performance depends on a number of factors associated with the environment in which it is placed. The environment normally includes two main influences that may have degradative effects: stress and the surrounding atmosphere. The composition and magnitude of each, over time, determines how well glued wood will remain glued and able to deliver the functions of the product. In Chapter 10 the degradative factors are discussed and related to various products and adhesives. This chapter also covers the methods that have been developed to assess the quality of a given adhesive bond. Most have been standardized by agency actions as a means of defining the performance to be expected from manufactured products. Time has shown them to be sufficiently reliable and repeatable for use in contractual agreements, as well as in research, and in quality control programs. A procedure is proposed which reveals bond-forming potential of an unknown adhesive in a quantitative manner.

In using adhesives to bond pieces of wood together, it is natural to assume that proper choice of adhesive ordains success. However, it has been ruefully learned by many that something besides what is in the can is necessary to assure good bonding. That something is control of the factors that inhabit all gluing operations. Because it is impossible to know for certain that a given bond is a good one without breaking it, one is left with the necessity of making it right in the first place. Chapter 11 offers procedures for determining that stipulations in use instructions have been met.

In a final acknowledgment of the fallibility of instructions and operators, there is included in Chapter 12 a guide to troubleshooting, a glue line doctor as it were, as a source of self-help when the unexpected happens.

Some aids to the reader who may feel buffeted by scientific, technical, and practical advice are included to ease the search for information. A glossary defines terms that may be new or unclear to the reader, and an index, heavily cross-listed, should help the reader find information spread through more than one heading. When information that is too detailed, too specific, or too voluminous to be included in this book is needed, the reader is encouraged to consult the works listed in the bibliography.

The pursuit of brevity in trying to confine the subject matter to one book may have led to a potential disappointment for some readers who may be searching for complete solutions to their particular problems. There are simply too many permutations of variables in any one situation to draw together the ones that may be acting, and at what level, in all possible combinations. Specific instructions on how to conduct an operation, and how to run and maintain different types of equipment, would have turned this book into an encyclopedia. Each area has its own arena of technology and its own specialists, whose literature can provide the information one needs for specific job training. I have therefore merely set up guideposts for rationalizing general solutions. Readers are encouraged to pick up the train of thought and extrapolate it through their own set of variables to the solution of the problem at hand. In this way, a few generalizations can lead to thousands of specifics.

As mentioned earlier, no one individual can be sufficiently versed in the many fields bearing on the science and technology of wood gluing to cover the subject with his own knowledge. This especially includes me, whose basic training was in wood technology. Consequently I am deeply indebted to many people for the presumptions of knowledge offered in this book. A great number have contributed materially, and I hope I have properly credited all of them. To any I may have missed, my sincere apologies, and my promise to rectify at the first opportunity.

The staff of the U.S. Forest Products Laboratory (USFPL) at Madison, Wisconsin, were unstinting in providing charts and photographs that illuminate the book. I am also particularly grateful to many others who contributed either directly or indirectly. They include: the manufacturers of equipment and supplies and the manufacturers of adhesives, who are the chief innovators modernizing the art of gluing; the users of adhesives, who represent the practice and in a real sense define the status of adhesive technology; the many scientists and technologists who have done and are doing research that explains and makes progress possible; and everlastingly the hundreds of former students who endured the honing of the principles, and confirmed their utility.

An (un)fair amount of enduring was also incurred by my patient wife, Maxine, and my three sons, Dan, Nick, and Frank, all of whom provided impetus and encouragement that kept me focused—and glued to the word processor.

All of this only produced 25 pounds of highly comminuted manuscript. For consolidating it into the form of a book I am indebted to the staff of Van Nostrand Reinhold; to Brian Rivers of the U.S. Forest Products Laboratory, whose sharp technical editing resolved kinks and inaccuracies; and to Cheryl and Paul Wolfe of Hardwick, Massachusetts, whose skill with computer-aided design transformed rough sketches into printable art.

Alan A. Marra

1

Introduction

It has been stated that the tonnage use of wood exceeds that of all other materials combined. A large part of this wood is used by breaking it down and reconstituting it to form composites such as laminated beams, plywood, flakeboard, and fiberboard, which then become materials for fabricating buildings and furnishings. These products are part of a modern approach to wood utilization which, together with nondestructive testing, is helping to establish a "material science of wood." With this science, wood properties are enhanced, made more uniform, and engineered to specification.

This new materials science of wood in turn owes much of its power to the use of modern adhesives. New products and processes are spawned almost daily from the technologies they permit. The products themselves owe their ultimate market acceptance to craftspeople who assemble them into useful structures. The assembly process, though often less sophisticated, depends on various fastening devices to create desired shapes, using adhesives when it is feasible to do so. Adhesives, and their proper use, therefore represent keys to the success of this entire segment of the forest products industry. The fact that millions of tons of adhesives are consumed annually further points to the importance of wood gluing in converting trees to useful products.

Moreover, the importance of glued wood products seems destined to persist far into the future as the renewability and the relatively benign convertability of forest resources become increasingly attractive alternatives to the utilization of exhaustible and environment-threatening mineral resources. Wood gluing will be a key not only to the efficient utilization of forests as a source of raw material, but also to the efficient manufacture of wood products. Ensuring efficiency in converting trees to useful products will help to assure the continual husbanding of forest resources, and thus perpetuate this universal benefits system.

The role of adhesives in the perpetuation of forests extends into the entire ecosystem the forests represent. By permitting the utilization of trees regardless of their form, size, or species, gluing allows forests to be managed for other objectives besides production of conventional timber values. Species, in this cornucopic forest, can be selected for maximum erosion control and climate control, as well as water runoff control to protect rivers and reservoirs, and the fish they contain. Diversity in forest composition, different species and different ages particularly, is known to enhance resistance to catastrophic destruction by fire, disease, and insects, and it supports more forms of wildlife. Moreover, heterogenous forests may be easier to maintain than monogenous forests, because they tend to regenerate in response to

1

the natural selection process, though at a slower pace, and may not need as much human intervention. The knowledge that trees will have economic value at almost any stage in their life cycle should encourage greater development of amenity values. Economic value in a highly diverse forest may well depend on the ability to take wood apart (i.e., reduce it to small pieces, such that the original size and form of the tree becomes less critical), reorganize it, and glue it back together again into useful structures.

In order for this philosophy to have widespread and steadfast credibility, it must be anchored in a high level of integrity with regard to the performance of the products. Since performance of glued products is inexorably tied to the quality of the glue bonds that hold pieces of wood together, the entire concept is dependent on the ability to use glue properly, and to be able to assure others that it has been used properly. History has already confirmed the general utility of adhesives in remaking wood to better serve the needs of people.

Artifacts of glued wood recovered from the tombs of Egypt suggest an early expertise and a surprising degree of durability. The glues used then were of starch or protein origin (developed, one might surmise, from difficulty in cleaning pots used to prepare food). Today, glues of similar origin are still being used, but an impressive array of synthetic glues has been added. These new resins have greatly improved the durability of adhesive bonds, to the point where they exceed the durability of wood itself.

The process of gluing wood, however, has remained basically unchanged in the intervening years: The wood is prepared by cutting to size and surfacing; glue is mixed and applied; and the surfaces are brought together and held firmly until the glue hardens.

Two major developments, in addition to new adhesive materials, have slowly evolved over the years to give wood gluing its modern touch. One is the mechanization and automation that has occurred in gluing processes. The other is the great diversity in the sizes of the pieces of wood being glued (from

thick lumber to parts of fibers), multiplied by many kinds of structures or composites into which they are glued.

It should be noted at the outset that knowledge about adhesive bonding of materials has a somewhat reverse genesis compared to that in current high-technology industries. Wood gluing is an art, an ancient art, and it is well established as a technology among modern methods of production. The science that normally underlies such a technology, however, is still evolving, trying to explain what we already know how to do in a reasonably acceptable fashion. As it evolves, new knowledge on adhesion and cohesion is continually being applied to wood and to adhesive compounds, and more and more of it participates directly in the development of new bonding processes.

The development of new polymers, and the expanded scale of gluing operations, have combined to produce a continuous source of challenges and problems as new situations compound their way into the glue line. The need for extremely high production efficiency and low unit costs further complicates the gluing process. In many cases it is these latter aspects of the operation that make modern gluing a sensitive process— and quality control a vital ingredient of success. The reason for sensitivity in such high-speed processes is that many factors are pushed to extremes, leaving little room for unexpected variation, and setting the stage for costly failures. In order to control any gluing operation, large or small, a rather large number of factors from many disciplines need to be considered. These tend to shroud the field in a blanket of complexity, which may undermine the confidence of many practicioners. Hence some form of simplification seems to be in order.

Although all gluing processes are dictated by the needs of the glue in forming a bond, what the glue needs is itself dictated by the wood being glued, its species, its properties, its preparation, its shape, its size, its surface, and the space between the adjoining surfaces. The variabilities of wood, added to the variabilities of the adhesive, and the variabilities of machines and operators, produce a situation that requires constant monitoring and adjusting. Understanding the interactions offers the chief means of sustaining successful day-to-day operations.

One escape from detailed knowledge is achieved by withdrawing into the glue line and centering attention primarily on what goes on there. What goes on there is then further simplified by visualizing adhesives as having lifelike characteristics. This endows them with functions which allow them to perform as living organisms preordained to create and support life. One can then begin "thinking" like an adhesive, and develop a sense for its needs, knowing only how it goes about its job of forming and maintaining a bond. In many cases, there is no more need to know compositional details of a glue than to know the composition of a human being in order to stop a nose bleed, or to learn how to dance. Controlling the functioning of a glue on the glue line has similar, though more complicated, aspects.

In all gluing processes, there are many factors operating, any one or all of which can vary controllably or uncontrollably, and they may affect the gluing process either directly or indirectly. They present a formidable jungle of effects that seemingly defy understanding. Some separation of effects would aid in bringing the field of wood gluing into more easily manageable pieces. One dissection separates the technology into two parts, each defining a rather specific area of interest: *bond formation* and *bond performance*. In this analysis, *bond formation* involves the fluid properties of adhesives and their conversion to solids, while *bond performance* involves the solid properties of adhesives and how they respond to stress and degradative influences. Unfortunately, they are not independent, since the qualities produced in bond formation read into those observed in bond performance. The distinction, however, aids in emphasizing that they represent different functions of the glue, and that therefore they respond to different factors in the gluing operation, and require different means of observation.

It is sometimes desirable to visualize a third part, *product performance,* in recognition of the possibility that performance of the final glued product may differ from that characterized by the quality of the bond. In other words, it is possible that even though the bond may perform satisfactorily, the product it is in may not. For example, warping, loss of smoothness, or dimensional changes can affect the quality of a product even though the bond remains intact. This is because the total amount and organization of wood in the final product now dominates the main interaction with the environment. Stresses are created and transferred in and across all glue lines, usually in a nonuniform manner, concentrating some or diffusing some as the structure works with or against the glue line.

THE EQUATION OF PERFORMANCE

The above dissections serve to categorize effects, but do not identify the factors that create those effects. A further sectioning is needed to organize factors into some sense of commonality. Factors with common origins seems a logical approach. If one visualizes the gluing process as a sequence of events, factors can be identified that group themselves into different materials and operations. Each group can be looked upon as representing a distinct area of activity, all contributing accumulatively to the ultimate bond performance. Since the contributions can be considered to be summative, the groups of factors naturally organize into the Equation of Performance (Table 1-1). The first four groups in the equation affect bond formation, the last one affects bond performance. The fifth group is strongly pivotal, affecting both ends of the process.

The equation of performance serves to sequester factors into distinct groups, but without further apparent utility. However, the summative nature of the factors suggest that there ought to be a numerical result. While this is not readily done with the present state of knowledge, a means of extracting useful information employs observable

Table 1–1   The Equation of Performance

GLUED PRODUCT PERFORMANCE	= 70,000 PSI ± Σ	ADHESIVE COMPOSITION FACTORS ± Σ	WOOD PROPERTY FACTORS ± Σ	WOOD PREPARATION FACTORS ± Σ	ADHESIVE APPLICATION FACTORS ± Σ	WOOD GEOMETRY FACTORS ± Σ	PRODUCT SERVICE FACTORS
Strength	Potential	Viscosity	Species	Cutting to size	Storage	Element	*Stress*
Durability	adhesion	Hardening	Strength	Surface	Weighing	Thickness	External
Appearance	forces	Mechanism	Density	Smoothness	Mixing	Width	Shear
		Rate	Stability	Trueness	Mix age	Length	Tension
		Degree	Sapwood	Damage	Application	Grain	Cleavage
		Strength	Heartwood	Plane of cut	Method	Parallel	Duration
		Durability	Springwood	Contamination	Speed	Cross	Short
		pH	Summerwood	Inactivation	Amount	Number of plies	Long
		Solvents	Permeability	Grain angle	Distribution	Organization	Impact
		Diluents	pH	Moisture	Assembly time	Aligned	Cyclic
		Fillers	Age	Content	Open	Random	Internal
		Extenders	Extractives	Distribution	Closed	Structure	M.C. change
		Fortifiers	Hygroscopicity	Age of surface	Semiclosed	Homogenous	Continuous
		Tackifiers	Reaction wood	Impregnants	Ambient	Heterogenous	Cyclic
		Catalysts/hardeners	Cellulose	Temperature	Temperature	Stratified	Temp. change
		Solids content	Lignin	Honeycombed	Humidity	Graded	Continuous
		Resin solids	Anisotropy	Casehardened	Circulation	Mat formed	Cyclic
		Shrinkage	Decay	Checked	Prepress	Extruded	Stress conc.
		Rheology	Knots	Burnished	Pressure	Architecture	Creep
		Molecular or particulate	Conductivity	Dust	Temperature	Size	Relaxation
		Size	Dielectric properties	Warp	Press	Linear	*Environment*
		Design	Creep	Supply	Glue line	Planar	Heat
		Distribution	Relaxation	Variability	Rate	Volumnar	Continuous
		Hybrids	Stability		Press time	Curved	Cyclic
		Surface tension	Variability		Conditioning	Variability	Moisture
					Variability		Continuous
							Cyclic
							Organisms
							Radiation
							Chemicals
							Controlled
							Uncontrolled

4

functions derived from the basic imperatives of all adhesives. A virtual common denominator emerges despite a wide diversity of direct and indirect effects and differing units of measure as yet unresolved to make them truly additive.

A closer look at the parts of the equation will illustrate its usefulness. The equation of performance offers a broad view of the field, displaying more than 100 variables that affect wood gluing operations. The variables in each group have common origins, though unfortunately not common effects. Although there seems to be no logical entry point to the subject of wood gluing, the process of understanding has a beginning with recognition of what factors are involved and how, where, and when they operate. Categorizing the bonding factors into groups is a further step in simplification because it permits factors to be discussed in their own isolated terms, as distinct areas of concern, knowing they ultimately must have their specific contribution to the total bond.

The equation of performance begins with the assumption that if all the forces of adhesion theoretically residing in the surface regions of a piece of wood could be engaged, they would approach an estimated 70,000 psi. This may seem a preposterous figure considering that wood, in the across-the-grain direction involved in gluing, is one to two orders of magnitude weaker. The discrepancy is primarily due to the fact that wood is both a material (e.g., an alloy of cellulose and lignin) and a structure (e.g., a system of tubes). The figure of 70,000 psi comes from the material. The estimate is conservative considering that it is based on the observation that water adsorbed on wood substance experiences an attractive force sufficient to compress it to a specific gravity of 1.3. This suggests that there is a very large pool of attractive forces that might participate in bond formation. For reasons that will be discussed in greater detail later, a high percentage of these forces are prevented from becoming part of the bond because they are preempted by other substances or blocked out of reach.

The first consideration in planning a glued product is the glue to be used. The first category in the equation of performance, adhesive composition factors, includes the factors involved in this choice. Glues should be chosen on the basis of how and where the glued product is to perform. This determines the level of strength and durability that must be met. Although it is usually not necessary to know the exact chemical makeup of the adhesive, some knowledge of gross composition is helpful in determining limits of expectation and special procedures that may apply. Strength and durability are determined largely by the principal chemical constituent. Other constituents are incorporated to enhance particular properties, such as viscosity, but the characteristic that distinguishes one type of glue from another is the "backbone" chemical entity, usually one, but sometimes more than one molecular species.

A normal expectation is that the bond will be as strong as the wood. The plus and minus signs in front of the adhesive category (as well as the others) in the equation suggests that under a combination of certain conditions, a bond can be stronger than or weaker than the wood. Most adhesives designed for use with wood can form a bond at least equal to the strength of the wood. However, if they are modified, diluted, overextended with fillers, or misused, they may be weaker. Bonds stronger than the wood can of course be made but they could never be measured beyond "equal to," since the point at which wood fails represents the maximum strength that could be measured.

The second category in the equation of performance, wood property factors, represents primarily the target strength of the bond to be formed. However, factors that originate in the wood sometimes inhibit and sometimes favor the attainment of that strength. Moreover, whether they inhibit or favor depends on the type of glue and how it is being used. The properties of wood fall into four distinct classifications: anatomical properties, mechanical properties, chemical properties, and physical properties. Each includes factors that bear on bond qualities. For example, it has been lamented that the stronger the wood, the more difficult it is to bond.

Exploring this further, if two pieces of aspen and two pieces of oak are bonded with the same glue, one might observe on testing that the oak bonded system could carry twice as much load as the aspen bonded system. Is the oak bond therefore superior to the aspen bond? It is mechanically stronger, but is it better? Maybe. Observing the fractured surfaces, one might note that the aspen bond had failed entirely in the wood whereas the oak bond had failed entirely in the glue line. The conclusion of "stronger than the wood" or "weaker than the wood" therefore depends upon the strength of the wood.

In this example the physical properties of both woods also participate in the final determination. Cycling the bonds through high and low relative humidities could destroy the oak bond but not the aspen bond. Shrinking and swelling properties of the two woods thus add further effects in determining the adequacy of this adhesive bond. Anatomical properties such as cell cavity size and chemical properties such as pH also play subtle roles in formation of the bond and its subsequent performance.

Unfortunately, the wood confronting the adhesive is not only that described as the inherent properties of wood, but also those that are changed by processing into the wood elements to be glued. These changes are so profound that they occupy their own category in the equation. In some respects, preparation of wood confers more important properties than the wood originally had. Seasoning of the wood, for example, a prerequisite to all gluing, controls much of the behavior before gluing, in how it machines, and how it maintains shape and size; during gluing in how it interacts with the adhesive; and after gluing, in how it may cause internal stress as the wood shrinks and swells against restraints that glue lines impose. The seasoning process also alters mechanical properties, increasing the strength dramatically. Surface properties also incur changes. These can cut into the 70,000 psi of avail-

able forces the wood may have started with, all but destroying them in some cases.

An insidious change occurs below the surface of the wood when the process of producing the surface damages the fibers beneath. This is almost a universal occurrence, causing strong minus repercussions in bond performance of bonds. Overriding all other factors in wood preparation is the necessity that the surfaces be cut straight and uniform so that they mate with a high degree of proximity and require a minimum amount of pressure. These and other factors in preparing the wood can propagate throughout the gluing operation to defeat the most well-chosen glue.

Like the inherent properties of wood, the inherent properties of glue are modified by the manner in which they are prepared for use and applied to the wood. Like wood preparation, glue preparation and application thus also merits its own grouping of factors. The best glue in the world will be useless if it is not handled properly. Each glue has a preferred regime or treatment from the time it is purchased to the time it must solidify between two pieces of wood. In addition to preparation and application, the most important factors in the process are the many time intervals that need to be managed along the way to a completed bond, such as storage time, mix time, assembly time, heating time, pressing time, and conditioning time. Some glues are more tolerant of irregularities in procedure, and these tend to be favored over those that might be stronger and more durable, but more prone to dysfunction.

The category of wood geometry factors includes, perhaps surprisingly, some of the more decisive factors in choice of gluing process and performance of glued products. One senses, for example, the difference between a laminated lumber beam and a sheet of Masonite: high strength versus smoothness; strength provided by the original grain and therefore the original strength of the wood versus smoothness provided by reducing the wood to fiber and reconstituting under high heat and pressure. The size of the wood element also controls the amount of stress that the glue line might be called upon to sustain, and therefore the quality of the glue line needed to sustain it. For example, the glue line in a laminated lumber beam needs to be very strong, because large stresses will be imposed through the wood. On the other hand, the glue line between fibers can be composed of atomized spots of glue, because the stress imposed from fiber to fiber is considerably less.

When the completed bond is ready to perform, the final category of factors comes into play, those that might degrade or destroy the bond. These include the type and magnitude of stress that might be imposed on the bond, and the type and amount of debilitating agents such as heat, water, and solvents that may be part of the service environment. Obviously, glues of different compositions will withstand different amounts of both mechanical and environmental stress. Testing and evaluation of bonds involve a high degree of fracture mechanics combined with experience to sort out causes and effects.

Summing the Equation of Performance

The factors in the equation of performance can be assumed to be summative by treating them as if each one affects performance in positive or negative ways. Putting the factors in equation form helps emphasize at the outset that every glue bond is the sum of many plus and minus effects. The point should be made, however, that a single factor such as pressure, acting strongly in a negative fashion, can alone negate all positive factors and prevent a bond from forming.

There is some intellectual danger lurking behind the simple plus and minus signs, because the factors can have ramifications over a wide range, and any one can have a plus or a minus effect depending on its level or the level of some other factor. Many direct effects as well as complex, cancelling, or compounding interactions occur between factors. Some are known, some are unknown; some are measurable, some are unmeasurable. Of those that are known and measurable, few have common units that allow summation in a mathematical sense.

The conglomeration of effects thus seems to comprise an unfathomable web of obscure relationships, precluding any single definitive solution for a particular combination of conditions. Nevertheless, as mentioned before, it does have important utility as a vehicle for understanding the total process.

The Common Denominator

The key to using the equation of performance is a rationalized common denominator which allows information to flow up and down the equation. Partial summations or solutions can then be put together to reveal the actual status of a bond or bonding process, and, most important, to diagnose problems and suggest corrections.

The common denominator is derived from an understanding of how adhesives function. With only a slight stretch of logic, all adhesives can be construed as functioning through their rheological properties in forming a bond. (Rheological properties are those that control how a material flows, spreads, creeps, or otherwise deforms physically in response to external forces.) This reasoning allows one to consider both bond formation and bond performance in terms of motion. Becuse all the factors in the equation of performance can rather easily be reasoned to affect the motions an adhesive must perform to produce a bond, a very useful *mobility parameter* emerges.

Mobility thus becomes the common denominator which indicates how the adhesive responds to a factor or factors. Although observations of mobility are largely qualitative, and usually reflect the operation of more than one factor, a good deal of information can be deduced from them. The power of the mobility function is that no matter how many factors are operating, observations always reduce to just three, usually visible, situations:

1. Too much mobility
2. Too little mobility
3. Optimum mobility

The mobility parameter allows most of the major factors in the equation of performance to be further collapsed into two groups: those that increase mobility, and those that decrease mobility. How to affect or change the mobility of the adhesive through manipulation of factors in either group then becomes the only additional piece of information needed to change a given result in controlling a gluing operation. A basis for a quality control program that can respond immediately to visible problems is thus possible.

A Framework for Understanding

The equation of performance also serves a pedagogic function by providing a general format for discussing each factor and its influence on the bonding process. However, since the discussion depends on the mobility parameter, it is necessary first to establish a general working knowledge of how adhesives function. The different parts of the equation can then be approached in order, beginning with the pertinent wood adhesives and the essential features of each. Wood need be described only in general terms of the properties that affect adhesive performance. These include: (1) those that are inherent, (2) those conferred in processing, (3) those due to geometry, and (4) those due to organization in the glued structure.

Commonality of Adhesive Action

Though there are thousands of different adhesives and gluing processes, the actions of adhesives in forming a bond are remarkably similar. Thus it is possible to generalize and eventually isolate different adhesive actions during bond formation. This permits another round of simplifications in the study of wood gluing: a *commonality* of adhesive action. The infinite differences associated with the production of glued articles reduce to the same few basic actions that are common to all adhesives. It is this commonality that allows an attack on the equation of performance, at the nose-bleed level, extracting information from it and thereby controlling the gluing process. Commonality also produces what may seem as repetitious descrip-

tions of similar events throughout the book. This is because the same event may have its origins in many facets of the gluing process, some that begin in the tree and pick up compounding or cancelling effects as each successive operation impacts upon bond formation and performance.

When adhesive meets wood, the bonding action begins, bringing together variables from three sources: wood, adhesive, and process. The summation ritual, orchestrated by unseen chemistry and unseen mechanics, weaves the best bond the various contributors permit. There is only one "go" at it, as the hardening of the glue "freezes" the bonding actions at whatever point of completion they may be at. Thus the final bond contains a solidified history of how it was formed. Our objective is to control the history before it happens, or to read it after it happens. As a means of providing a map of the glue line, it can be visualized as a series of links, each defining a region where certain adhesive actions take place.

PRINCIPLES OF WOOD GLUING

The following *principles of wood gluing* summarize the logic developed throughout this book. They are fairly universal, springing as they do from a common point: the region between two pieces of wood and the environment created there in which the adhesive does its work. The application of these principles comprises the essence of the wood gluing art as it is or will be practiced no matter what innovations may come. Knowing what an adhesive is supposed to do, and knowing how it is affected by the environment to which it is subjected, is the foundation for understanding the many factors which affect all gluing operations, be they large manufacturing processes or small corner-block gluing at the craftsman's bench.

- *Principle 1.* Wood gluing is divided into two distinct but connected areas: *bond formation* and *bond performance.*
- *Principle 2. Bond formation* invokes the

liquid properties of glues (interacting with the manner in which they are used, and the wood on which they are applied).

- *Principle 3. Bond performance* invokes the solid properties of glues (interacting with the wood with which they are used, and the environment to which they are exposed).
- *Principle 4.* The total performance of every glue bond is the sum of the plus or minus effects of many factors (isolated and grouped into an *equation of performance*).
- *Principle 5. Adhesives* (glues) are defined functionally as liquids that can be converted to solids (in the glue line).
- *Principle 6.* The *anatomy* of an adhesive bond can be characterized as nine links in a chain, each link being the locus of a particular adhesive action.
- *Principle 7.* In forming a bond, adhesives execute five separate actions or *motions:*

  1. Flow
  2. Transfer
  3. Penetrate
  4. Wet
  5. Solidify

- *Principle 8.* The motions of an adhesive in forming a bond occur in *sequence,* beginning with "flow" and ending with "solidify."
- *Principle 9.* The motions of an adhesive in forming a bond vary in *magnitude* (as bonding conditions vary).
- *Principle 10.* The magnitude of adhesive motions is the chief determinant of bond quality (for a given adhesive application).
- *Principle 11.* The magnitude of the most decisive adhesive motions can be assessed by visual characteristics.
- *Principle 12.* As the magnitude of adhesive motions varies on the glue line, one of four distinct, observable *glue line conditions* will develop:

  1. Starved
  2. Bonded
  3. Unanchored
  4. Prehardened

- *Principle 13.* The motions of an adhesive in forming a bond occur at definite *locations* (links) in the joint.
- *Principle 14.* Sequence, magnitude, and location of adhesive motions provide the main diagnostic characteristics for monitoring the behavior of an adhesive on the glue line.
- *Principle 15.* The motions of an adhesive in bond formation are controlled by factors primarily in the first categories of the equation of performance (working with or against composition factors).
- *Principle 16.* The motions of an adhesive can be associated with the factors that control them by means of a *mobility parameter,* an intuitive index of the degree of adhesive motion.
- *Principle 17.* The mobility parameter functions as a *common denominator* that permits integrating to useful solutions the effects of diverse factors in the equation of performance.
- *Principle 18. Bond formation* and *bond performance* are basically dependent on the rheological properties of adhesives.
- *Principle 19.* The *rheological properties* of adhesives can be construed basically as:

  1. How fluid they are (viscosity)
  2. How they solidify (chemically and physically)
  3. How fast they solidify (curing speed)
  4. How solid they get (strength)
  5. How well they maintain their solidity (durability)

- *Principle 20. Bond formation* is influenced primarily by the first three rheological properties.
- *Principle 21. Bond performance* is influenced primarily by the last two rheological properties.
- *Principle 22.* Glues can be intuitively organized into a declining order of *durability* based on overall resistance to environmental conditions, combining the effects of stress, heat, moisture, and organisms (the SHMO factor).
- *Principle 23.* Glues can be intuitively or-

ganized into a declining order of *ease of use* based on tolerance to irregularities in the gluing process, i.e., resistance to misuse (the SHMO subverse: combining the negative effects of variations in mixing, spread rate, assembly time, wood moisture, species, temperature, pressure, press time, gaps, etc.).

- *Principle 24.* As adhesives harden on the glue line, they may achieve different solid states depending on conditions prevailing during the pressure period:

  1. Undercured
  2. Overcured
  3. Chalky
  4. Grainy
  5. Filtered
  6. Crazed
  7. Frothy
  8. Solid

- *Principle 25.* Bond performance of a well-made bond is ordained by the composition of the adhesive, mediated by the wood on both sides of the glue line.
- *Principle 26.* The less wood per glue line, the less stress per glue line.
- *Principle 27.* The higher the density of the wood, the greater the stress on the glue line.
- *Principle 28.* Properties of the glued product can be engineered and anticipated through the laws of *diminishing dimensions,* a sequence of wood elements ordered by decreasing size of one or more dimensions.
- *Principle 29.* The behavior of wood varies widely and is strongly consequential, affecting the behavior of adhesives variously, even within individual pieces.
- *Principle 30.* The essentials of wood behavior that have a bearing on adhesive performance include the following factors:

  1. Wood is 20 times stronger along the grain than across the grain.
  2. Wood is hygroscopic.
  3. Moisture content affects strength.
  4. Moisture content affects dimensions.
  5. Dimension changes differ with direction.
  6. Density varies between species, within species, and within the same piece.
  7. Density influences strength and dimensional stability.
  8. Porosity and permeability vary with species, pieces, and direction.
  9. No two pieces of wood are alike, therefore:
  10. No two pieces will behave alike;
  11. Nor glue alike.
  12. Hence, adhesives must cope with disparate wood differences in bond formation and bond performance.

2

Overview

Before delving into the detail posed by the equation of performance, an overview of the field of wood gluing seems appropriate to provide a broad perspective of the products and processes that are addressed by the science and technology to be discussed. It provides also a skeleton upon which can be hung specific details of wood gluing with some prior sense of what adhesives are expected to accomplish.

A METAPHORICAL APPROACH

The approach to the glue world has a metaphor in the approach to a new city. Cities vary primarily in size and number of buildings, their architecture and organization. They have a common element, people; and while the buildings are the most prominent and visible, what the people do, provides the most important feature characterizing them. Similarly, glued products vary in size and number of wood pieces they contain. They too have a common and less visible element, glue; and what the glue does provides the most important feature characterizing their performance.

In this overview the sizes of the wood pieces generally glued provide an easily visualized territory of products. They also provide a basis for discussing glues and gluing procedures since they derive some of their primary differences from the geometry, i.e.,

the size and shape of the wood. For example, lumber, the largest wood element in the wood gluing field, is glued into massive beams and arches; veneer, the next in size, is glued into different kinds of plywood; flakes into flakeboard; particles into particleboard, and fiber into fiberboards. In each case, a different glue and a different procedure would be used.

It may ease entry into this territory of wood and glue if we took a brief overflight to sense its outline and identify some of its distinctions. From above, the process of converting trees to consumer products can be seen as proceeding down a one-way street. A chain of events ensues more or less in stages, with some activity in the side streets.

THE SIX STAGES IN THE CONVERSION CHAIN

At the beginning of the street, trees are cut up into manageable pieces, logically into the largest size having marketable value, or greatest convenience. See the left column of Figure 2–1a. This part of the process, particularly the production of the larger pieces at the top of the column, is usually carried out in the forest. The smaller pieces in the lower part of the column are normally cut at some collection point, though increasingly they are produced in the forest with portable machinery.

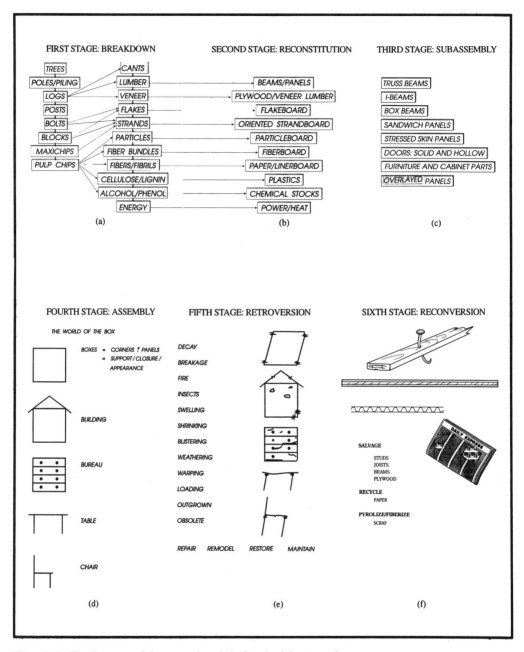

Figure 2-1   The six stages of the conversion chain for glued forest products.

At the mill, the pieces are further broken down, subdivided, into more discrete pieces of wood having rather precise dimensions and surface characteristics. (Figure 2-1a, right column). These two columns represent Stage One in the chain of events in what are known as "breakdown" operations, primary breakdown—left column, secondary breakdown—right column. The products of secondary breakdown are referred to as "wood elements." It is these wood elements that form the basis of a materials science of wood, since they can be reconstituted in endless combinations into products having predictable properties and assured qualities.

Reconstitution (Figure 2–1b) may be considered as Stage Two in proceeding down the street. This is the first meeting of wood and glue. The products are usually flat panels or sheets, essentially planar with some molding to shape or heavy structural members, which are mostly linear in aspect.

As an aside at this point, it should be noted that the arrows connecting the products in the three columns suggest a very important relationship between product properties and the qualities of the forest resource that may produce them. One can sense that the elements in the upper portions require more perfect trees in terms of size and shape, and those in the lower portions can be derived from virtually any quality tree. A crucial deduction with implications in the utilization of low-grade timber is the possibility of producing plywood properties, flakeboard, from trees that might otherwise yield only stove wood.

At this time, it is also useful to note that the wood elements are arranged in a series in which their dimensions are progressively smaller. Placing the elements in such a series further emphasizes the connection between tree quality and end product. It suggests that any element can be derived from the ones above merely by additional cutting, and thus presenting options for the utilization of waste inevitably generated under production conditions: It also shows that low quality timber can produce high quality products. More importantly, as will be discussed in Chapter 7, the size sequence of the wood elements provides a measure of predictability with respect to properties of consolidated products.

After reconstitution, many products go through a second gluing operation for the purpose of adding overlays, edge bandings, or further aggregating into subassemblies by gluing to a frame or to a core of another material. Subassembly then is the third stage in the chain of events leading toward a final product. Some representative products of the subassembly stage are listed in Figure 2–1c.

The fourth stage is the development of shape, an assembly process. The conversion to shape produces ultimate utility for consumers, and permits final expression of material properties engineered by earlier processing. Various forms of mechanical fastening and gluing that provide fast assembly are essential at this stage.

It is worthy of special note that the shape most generally demanded is that of a box, or parts of a box (i.e., rectangular rather than rounded structures) as shown in Figure 2–1d. A box is basically a system of panels and corners. What commands our interest here is that the making of corners is the work of craftspeople, in contrast to the making of panels which is a machine-dependent process. Further, the making of corners is the most costly of all operations in the conversion of wood and it confers almost the total worth (quite independently of all the previous values added) of the final consumer product. Moreover, along with the total worth, this assembly step confers a major share of the performance expected of the article, since loosening or breaking of corners usually results in virtual failure of the product. Because of this dependence on performance and cost, the making of corners can be regarded as the most important stage in the production of wood products.

After glued wood objects are finished and placed in service, they encounter the rigors of the environment in which they are used, and a certain amount of wear, breakage, delamination, or obsolesence sets in. While proper engineering, design, quality control, and testing should minimize such adverse performance, there is a continuing need for *retroversion:* the repairing, reworking, or remodeling of glued products. This fifth stage in the conversion chain is illustrated in Figure 2–1e. Very special gluing processes are needed to accomplish this fifth stage, which is generally the most difficult of all gluing operations from the standpoint of pressure application and the generally poor condition of the joining surfaces.

Finally, in the sixth stage, as shown in Figure 2–1f, wood articles that have reached the end of their service life need to be *reconverted* to serve a new use. Usually this means taking advantage of one of the two major characteristics of wood: biodegradability and combustibility. However, the application of modern technologies should permit more and more discarded wood to be salvaged, either directly as individual pieces, or recycled by comminution and reconstitution. The major impediments to recycling are the gathering and transportation costs, contaminants such as heavy metals as well as nails and plaster, mixed sizes and species, unassured continuity of supply, and the usually dry condition of the wood. The latter limits the kinds of wood elements that can be efficiently produced. These impediments seem formidable, but they are primarily engineering problems rather than technological problems, and many of them have already been solved.

The above flyover has shown the broad scope of the wood gluing field from tree to glued product and final disposition after the product has served its purpose. A closer look will show the broad scope from another angle: the wide spectrum of glued products made possible by the processes of comminution and reconstitution.

AN EXPLOSION OF PRODUCTS

The kinds of products listed in Figure 2–1b are typically produced by aggregating and consolidating wood elements in a gluing operation. This list however contains only products that can be called "homogenous," that is, composed entirely of one type of element. A rather dramatic explosion of reconstituted product possibilities emerges when one considers the combination of two and three element products that might be engineered from this series of wood elements. Figure 2–2, a three-dimensional array of 10 wood elements, suggests that over 1000 combinations are possible. Some of these may

not be viable, and some may be redundant. Many combinations, however, also offer the opportunity to produce different organizational modifications such as orienting or stratifying to enhance a particular property (e.g., strength, stiffness, smoothness, stability, or density). Moreover, each combination can be reorganized to reverse core and face elements, further increasing the number of products that can be produced from these 10 elements.

The above combinations are rationalized on the basis of type of element and organization of elements. However, within type of element there is a range of sizes, as indicated in Table 7–1. If each element is considered to have only four different sizes, the theoretical number of combinations increases accordingly. Moreover, some of these combinations can be further modified by incorporating other materials, such as fiberglass, plastics, and metals, as overlays, as stratified layers, or as random mixtures, raising the total product possibilities again. The introduction of different species expands the possibilities even further. Since each species contributes its own particular properties, this is tantamount to introducing different materials. A new product emerges when, for example, the high strength of oak is combined with the light weight of aspen. Also, different qualities of wood from the same species can be judiciously organized; in a beam, for example, higher-quality wood may be placed in the higher-stress areas of the product, and lower quality in low-stress areas. The potential for creating new products by reconstituting is virtually endless. Current technology has so far not invoked more than 2 percent of this potential.

GLUING PARAMETERS

The gluing process actually begins retroactively in all cases, with a tree standing in the forest. Some of the most important factors affecting gluing are set while the tree is growing. Anatomical, physical, chemical, and mechanical properties, the basic characteristics of wood, become fixed as the cell

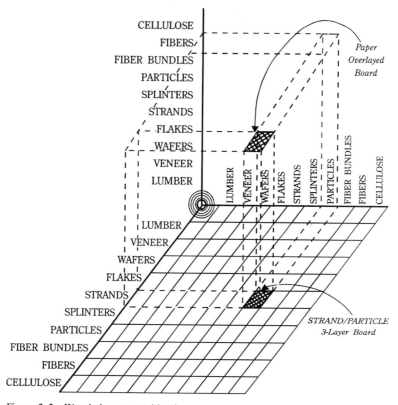

Figure 2-2 Wood element combinations.

structure being formed responds to the dictates of species, genetics, and environment. Variations in properties, one of the more obstinate sources of problems in wood gluing, also arise in the forest as a result of differences in growth conditions—the weather, the soil, the competition. While differences between species are readily accepted as the work of nature, the variabilities that exist within species, within different parts of individual trees, and even within the same piece of wood, are sometimes exasperating to contend with on a daily basis.

The properties of wood as produced by the tree are invariably modified in some way by the processing it receives in preparation for gluing. Cutting methods influence the quality of the surface; seasoning operations influence the strength, the moisture content, the distribution of moisture throughout the piece, the stability of shape and size; treatment with preservatives, fire retardants, and

dimensional stabilizers influence gluability (usually negatively); and, most inconveniently, time alone can change some physical properties, sometimes also adversely.

Within this context of wood characteristics, the actual gluing process includes most of the following operations, essentially in the order given:

1. Cutting to size
2. Seasoning
3. Surfacing
4. Preparation of the glue
5. Application of the glue to the wood surface
6. Assembly of the parts
7. Application of pressure
8. Application of heat (if required)
9. Conditioning
10. Trimming
11. Sanding
12. Inspection

Gluing processes differ primarily according to the dictates of size of the wood element being glued. The differences consist mainly in the manner in which the above operations are executed. There is also an important difference, often overlooked, in the order of the first three operations, which strongly affects the quality of the wood surface presented to the glue. For example, in those processes where surfacing precedes seasoning, as in products made from veneer or flakes, the surface the glue sees is very different from those where seasoning precedes surfacing, as in most lumber gluing.

Most of these operations involve technologies that go far beyond the mechanics of carrying them out. Some of the technologies that affect the gluing process directly are discussed in Chapters 6 and 8. Here, however, we will take a brief look at several representative products, in order to get an idea of some of the grosser differences that exist.

Before doing this, however, it is necessary to deal with a little local jargon, some terms that have crept into the art and persist in the technology. Processes are often referred to as either "hot press" or "cold press," depending on whether heat is used in solidifying the glue or whether ambient conditions are used. (The word "cold" is really a misnomer in this usage because it implies refrigeration or winter conditions, but in wood gluing it means room temperature, and often no lower than 70°F (21°C) except as specifically allowed by the manufacturer of the glue.) Lumber is normally glued in a cold press operation, although sometimes elevated temperatures are used to speed the hardening. Veneer is reconstituted mainly in hot press operations. Flakes and smaller elements, with a few notable exceptions, are also reconstituted at high temperature. Each of the wood elements has its own relatively specific gluing process, which revolves around the temperature required to solidify the glue. The role of temperature in gluing these wood elements will become evident in subsequent discussion.

Gluing processes may also be referred to, for example, as "lumber gluing," "plywood gluing," or "phenol gluing," emphasizing, respectively, the wood element being glued, the glued product, or the glue used. All these terms are useful for those who understand their implications. Other terms will be explained as they are needed.

THE THREE KINDS OF GLUING PROCESSES

One can also look at wood gluing from another angle, that of the stages in generating the all-important shape of the final consumer product. The first gluing of the wood elements after they have been cut from the tree is termed *primary gluing*. When additional gluing is performed on the panel or sheet products of primary gluing, the processes are usually referred to as *secondary gluing*. Secondary gluing conventionally includes all other gluing that might be done to make a finished article, but it is often desirable to further separate secondary gluing operations into two categories because of the distinctly different operations they involve. Processes that produce subassemblies or components by bonding products of primary gluing to something else, such as overlays on particleboard, edgebanding on plywood, fiberboard on lumber frames, and the like, may be considered the secondary gluing processes, since they involve a second gluing to products previously reconstituted by gluing. These products are still relatively flat, or in some cases curved or vaulted, and in a sense are precursors of the final shaped article. Glues that harden either with or without heat are used in these operations depending on the size of the operation and the rate of production demanded.

The products of secondary gluing as defined above are still merely components. They require assembly into shapes of various configurations and sizes to achieve utilitarian value in the marketplace. This last assembly process uses different kinds of fasteners to achieve and maintain the shapes into which panels are converted. Nails, screws, brackets, or plates are commonly used in the shape-generating process. When glues are used to create shapes from component materials, the process is *tertiary gluing*. Because these processes are geared to the speed of driving nails, the glues used must offer equivalent speed. Speed is so important in these operations that durability plays a lesser role in choice of adhesive. In addition to speed, practitioners of the art also demand ease of use; gap filling, no mixing, tolerance to ambient conditions, etc. A common characteristic of secondary and tertiary gluing is that surfaces may not be fresh nor well-mated.

Combinations of primary and secondary gluing exist where overlays, for example, are applied at the same time as the reconstituted panel is formed. Bonding of phenolic-impregnated paper to plywood is a typical example of such a process. Combining operations saves a pass through a hot press. Of greater potential are combinations of primary and tertiary gluing where shapes are created directly from wood elements by molding processes using veneer, flakes, or fibers. Such processes, though they may have limited versatility, address the cost of assembly fastening steps.

The cost of the gluing operation relative to the value of the glued product bears consideration. In primary gluing, except for the gluing of some lumber and veneer products, the cost of the gluing operation is the single most expensive item in the cost of manufacturing. The cost of the adhesive alone in some cases exceeds the cost of the wood. In any case, the products of primary gluing, being still in the early stages of conversion, are usually commodities with narrow profit margins. They are often caught in a cost squeeze where the cost of gluing can spell the difference between success and failure.

In secondary gluing, the surfaces to be bonded are smaller, and the value of the product is greater. Hence, the ratio of glues and gluing costs to product value is lower, and a greater margin for cost variability ensues. This margin reaches its lowest point in tertiary gluing, where only small amounts of glue are sufficient to assemble an object of relatively great worth, such as a cabinet or chest of drawers. The amount of the glue is so small, in fact, that a higher-priced glue can be used to gain efficiencies in this otherwise expensive process. In the combination processes of making components or final shaped products directly from wood elements, the ratio of glue costs to product value are intermediate.

SURVEY OF GLUED PRODUCTS

A brief survey of the major types of glued products should provide a fuller picture of gluing processes as well as a point of focus for later amplification. We begin with primary glued products and end with illustrations of secondary, tertiary, and combination glued products. Only the conceptual outlines are presented at this time, leaving details for later addition after the background for the different technologies has been developed.

As we discuss the primary glued products, it is not too early to begin noting several progressions. These derive their order and logic from the geometries of the wood elements and their orientation in the reconstituted product. First is the progression that begins with single-grain direction, preserving the maximum strength of wood in one direction, as in glued lumber products. This proceeds to two-grain direction, which produces the bidirectional properties characteristic of plywood; then to multidirectional grain, and the multidirectional strength of flakeboard; and finally to no grain, and the multidirectional strengths of fiberboard and particleboard. In short, the progression goes from complete grain dominance of properties, which reflect the primal cell structure of solid wood, to no-grain effects, in which the original wood structure is no longer present. One of the marvels of newer technologies is the creation of one-directional strength with the smaller wood elements by aligning them prior to consolidation.

Along with the progression in grain-dependent properties is a progression of decreasing tree-quality dependence, ending finally in any wood capable of being ground to particle or fiber size. Other progressions will be seen in the sizes and shapes of wood elements, which suggest incremental or decremental differences in such factors as surface smoothness, panel stability, product strength, moldability, capital/labor intensiveness, energy consumption, forest diversity, and product uniformity. These progressions are discussed more fully in Chapter 7.

Glued Lumber Products

The gluing of lumber takes three distinct paths depending on which surfaces are being glued—the edges, the faces, or the ends (see Figure 2–3). In all cases the objective is to make the along-the-grain properties of wood

available in sizes that are wider, longer, and/or thicker than can be obtained from trees. (Combinations of end grain to face grain or edge grain to face grain are special arrangements associated with assembly operations to create final shape, and are discussed in Chapter 7.)

Each of the three gluing surfaces has its own technology and its own pressing system. The operations are dictated largely by the surface being glued and are described more fully in Chapter 9. After the wood has been seasoned, the gluing process begins by creating the gluing surfaces—planing in the case of face gluing, jointing in the case of edge gluing, and scarfing or finger-jointing in the case of end gluing. In each case, the primary

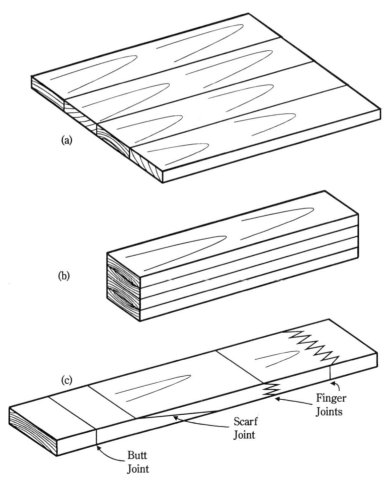

Figure 2–3 Lumber is glued in one of three configurations: (a) edge to edge; (b) face to face; or (c) end to end.

requirements are straightness, smoothness, trueness, surface integrity, and, of course, cleanliness (see Chapter 6). Selection of the glue is made on the basis of the performance desired.

Lumber is glued face to face to produce thick members for such products as beams, arches, posts, or turnings. Several advantages accrue in the case of members for structural uses. The process produces beams of greater strength than solid wood by allowing the strategic location of each board so that the strongest pieces are in the most highly stressed regions of the beam. Greater strength also derives from better seasoning, and less splitting. There are also bonuses in reduced warpage, more uniform strength, and more design options.

Laminated beams and arches provide some of the most spectacular uses of wood by allowing the fabrication of large buildings and bridges with long, clear spans (Figure 2-4d). Since these are load-carrying members, in which safety is the most important factor, their integrity is a first-order consideration in manufacture. Consequently, strict procedures must be followed during every step of the process. Promulgated by the U.S. Department of Commerce, and effectuated by the American Institute of Timber Construction, the procedures stipulate how the lumber is to be selected and prepared, and

(a)

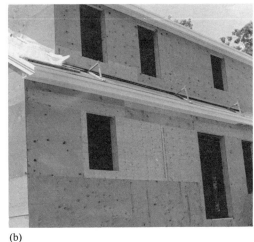

(b)

(c)

(d)

Figure 2-4 Glued products: (a) beams and arches; (b) plywood sheathing; (c) furniture paneling; and (d) bowl.

how the gluing process is to be carried out. Procedures for monitoring, testing, grading, and certifying are also detailed.

The gluing of lumber edge to edge produces lumber in panel form for use primarily in solid wood furniture, tables, dressers, etc., or as core stock in lumber-core veneered panels. After preparation of the wood to control moisture content and machining the edges, the glue is applied, and the preselected boards are placed in a clamping device until the glue hardens. Edge-glued lumber provides panels that tend to remain flatter than solid lumber of the same dimensions and that are technically the same in strength. The art has a long history in furniture making. Modernization of the processes has mostly mechanized and automated them, but the basic operations and cautions remain the same.

The end gluing of lumber is one of the most difficult wood gluing processes, mainly because of the end grain presented to the glue, which not only represents a very porous surface, susceptible to overpenetration, but also has to carry the maximum stresses that wood can sustain. Square butt joints, regardless of the glue used, cannot begin to have sufficient strength, and are therefore never used where strength is a factor. To improve both bond formation and bond performance, the end of the board is cut on an angle, or *scarfed*. This reduces the end-grain effect while at the same time increasing the bond area. A bond in such a scarf joint can bear loads up to 90 percent of that of solid wood. The finger joint is similar in effect to the scarf joint except that it is cut as a stacked series of sloping areas resembling fingers (see Figure 2–3c). It wastes less wood, is somewhat easier and faster to produce, and only slightly less strong. The finger joint, when well cut, can be jammed together and will hold without maintaining pressure. It therefore is amenable to small shop production. The process allows for continuous production of an endless strip of lumber. One of its major uses is in salvaging short scraps of lumber and converting them to a high-value product.

Veneer Products

The slicing of wood into thin sheets makes possible the deliberate distribution of wood properties in more than one, commonly two, directions. In so doing, the strength is made more equal in two directions, shrinking and swelling is markedly reduced in the two directions of the face plane, splitting is virtually eliminated, the product is more sheetlike than boardlike, and natural grain patterns are greatly enhanced. The product, plywood, represented the first, and perhaps the most important, improvement in the properties and uses of wood in all history. While veneering had been practiced on a small decorative scale for centuries, plywood attained commodity status in the mid-1930s with the advent of durable phenolic resins and the hot presses in which to cure them. There are two basic forms of plywood: all veneer (Figure 2–5a) and lumber core (Figure 2–5b).

Plywood is assembled with veneer arranged so that the grain of alternating layers are at 90° to each other. Typically, an odd number of layers are used to produce a balanced (later defined) construction. Therefore the minimum number of layers is three, and the increment from there is two layers at a time to whatever thickness is desired. Each layer has a name that represents its location and/or its function. The *face* and *back* are located and function as their names imply. Usually the face has the best veneer from an appearance standpoint, and the back is of lesser quality; the grain of both run the long direction of the panel. The middle ply in all constructions is called the *core* or *center ply*. The core is usually the thickest ply in a panel. Sometimes, in order to save cutting and stocking a separate thickness for the core, two regular-thickness plies will be assembled with the grain parallel to form a core. This gives the appearance of a four-ply panel, but the center two function as one. In a five-ply construction, the two plies next to the face and back are called *crossbands,* and their grain runs the short direction of the panel. In a seven-ply panel, two more layers are added next to the core with their grain

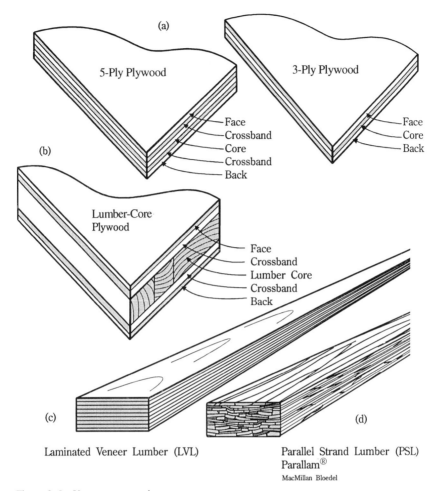

Figure 2–5 Veneer construction.

parallel to the face. They are called simply *inner bands* or *layers*.

In a comparatively recent development, the veneer is assembled with the grain all in one direction, producing what is called *laminated veneer lumber* (LVL). In this construction, as shown in Figure 2–5c, the overall strength of the product is increased to the upper levels of the strength range for the species by judicious selection and allocation of each veneer in the assembly. Equally important, the normal board-to-board variation found in mill-run lumber is greatly reduced by the multiplicity of pieces all summing toward the same average. This gluing of veneer to produce more uniform strength properties for a lumber product is identical

in concept to the gluing of lumber to produce more uniform strength properties in beams, but with a smaller wood element. A product having similar properties as LVL, but assembled in a more random manner using veneer cut to narrow strips, designated as Parallel Strand Lumber (PSL) Parallam® (Figure 2–5d) is a most recent development of Macmillan Bloedel, Vancouver, B.C.

The market for plywood consists of two distinct kinds of plywood: softwood, used mainly for structural purposes (Figure 2–4b) and hardwood, used mainly for decorative purposes (Figure 2–4c). Within each category, particular species, grades of veneer, and thicknesses of cut offer specific strength or decorative properties. Hence, there are many types and grades of plywood that define or assure properties for specific uses, such as sheathing and cabinetry for softwood plywood. Originally, in the United States, structural grades of softwood plywood were made exclusively of Douglas fir. Later, when second-growth Southern pine came of sufficient size, this species was brought into use. Local hardwoods have also been introduced as inner plies in softwood plywood to extend the available resources. By judicious accounting of properties, mixtures of species in softwood plywood panels have maintained the performance requirements specified for various purposes. Figure 2–6, developed by the plywood industry, lists species that produce equivalent properties.

Hardwood and softwood plywood production are sufficiently different that one might almost consider them to be different industries. Hardwood and softwood producers are located in different regions, use different glues, serve a different market, sometimes cut veneer differently, and often have different plant sizes and capacities. Veneer gluing is conducted by two different processes, one using cold pressing, and one using hot pressing (see Chapter 9).

The softwood plywood industry produces two types of plywood: *exterior plywood,* for use where it will be exposed to the weather or any condition that wood itself can withstand; and *interior plywood,* usable only where it will be protected from the weather.

The distinction between them resides mainly in the glues used. However, each type is also made in various grades, depending on the quality of the veneer, both on the faces and in the inner plies. Exterior plywood has clearly defined performance criteria supported by severe accelerated test methods. Interior plywood is characterized more by appearance features than by durability. Hardwood plywood, although used mostly in locations that are protected from the effects of water and weather, also includes types that range from fully waterproof to limited resistance to moisture.

The process for making the different plywoods begins with the production of the veneer, of which there are two mechanically different methods (see Chapter 6). One method, *rotary cutting,* the essential of which is lathe action where a long knife, set parallel to the log, peels the veneer from the log in a continuous ribbon—like unwinding a roll of paper—with the grain running crosswise to the length. In the other method, *slicing,* the blade is held in the same relative position with respect to the log, but in this case is made to travel across the log rather than around it as in rotary cutting. This produces a slice of wood that is equal in size to the length and width of the log.

Rotary cutting, because it is cheaper and faster, is generally used for producing veneer to be incorporated into building panels. Because softwoods are the species of choice for

Group 1	Group 2		Group 3	Group 4	Group 5
Apitong	Cedar, Port Orford	Maple, Black	Alder, Red	Aspen	Basswood
Beech, American	Cypress	Mengkulang	Birch, Paper	Bigtooth	Poplar, Balsam
Birch	Douglas Fir 2	Meranti, Red	Cedar, Alaska	Quaking	
Sweet	Fir	Mersawa	Fir, Subalpine	Cativo	
Yellow	Balsam	Pine	Hemlock, Eastern	Cedar	
Douglas Fir 1	California Red	Pond	Maple, Bigleaf	Incense	
Kapur	Grand	Red	Pine	Western Red	
Keruing	Noble	Virginia	Jack	Cottonwood	
Larch, Western	Pacific Silver	Western White	Lodgepole	Eastern	
Maple, Sugar	White	Spruce	Ponderosa	Black (Western Poplar)	
Pine	Hemlock, Western	Black	Spruce	Pine	
Caribbean	Lauan	Red	Redwood	Eastern White	
Ocote	Almon	Sitka	Spruce	Sugar	
Pine, Southern	Bagtikan	Sweetgum	Engelmann		
Loblolly	Mayapis	Tamarack	White		
Longleaf	Red Lauan	Yellow Poplar			
Shortleaf	Tangile				
Slash	White Lauan				
Tanoak					

Figure 2–6 Classification of species: each group contributes similar strength properties to softwood plywood and are therefore interchangeable within group. (U.S. Product Standard PS1-83)

construction purposes (high strength to weight ratio), rotary cutting is practiced throughout the softwood plywood industry.

Slicing is used mainly for producing hardwood veneer for decorative panels. The log is first sawn into *cants* of various sizes and angles. This allows a choice of decorative features to appear on the face. Because decorative properties are most likely to occur in hardwoods, slicing is practiced mostly in hardwood plywood manufacture. Slicing permits cutting thinner veneer and hence yields more product from expensive logs. However, since many hardwoods have decorative features that are exhibited on rotary-cut veneer, this lower-cost method of cutting is chosen when the wood and the product permit. Moreover, since inner plies do not need decorative features, rotary cutting offers lower-cost processing for a majority of the plies in a hardwood panel. A unique feature of slicing operations is that, after cutting, the veneer sheets are always kept together in exactly the same order they came from in the log. Such a bundle of veneer is called a *flitch*. The flitch permits one of the more artistic advantages of gluing: the matching of veneers to form designed patterns of elegant beauty (Figure 2–4c).

For reasons peculiar to each industry, hardwood veneer is normally cut much thinner than softwood veneer. For softwood plywood the wood cost is lower but the panel thicknesses are greater. In order to reduce the number of costly glue lines, plies are made thicker, in the range of $\frac{1}{10}$ to $\frac{1}{4}$ in. For hardwood plywood, panels tend to be thinner, but since a minimum number of plies are needed to make a balanced panel, the plies must also be thinner, Also, since the wood costs considerably more, particularly for face plies, the veneer is cut as thin as practicable, consistent with subsequent finishing operations. Face veneers are typically cut $\frac{1}{45}$ to $\frac{1}{32}$ in. thick for face veneer and $\frac{1}{16}$ to $\frac{1}{8}$ in. thick for inner plies. Veneer as thin as $\frac{1}{100}$ in. can be cut when warranted to stretch the supply. The thickness of the veneer is also a matter of some importance to bond formation and bond performance, because

the quality of the surface varies with the thickness. This factor is discussed at greater length in Chapter 6.

Different qualities of veneer develop from the surface of the log inward as more knots, decay, and heartwood are encountered. The quality of the veneer determines whether a sheet becomes the face of a panel, the back, or some inner ply. Different thicknesses are cut depending on whether the veneer will end up as an outer lamination or an inner lamination. Clipping out defects, drying, patching knot holes, and splicing to produce panel-size sheets prepare the veneer for gluing.

Pressure, and heat if required, is applied by any of a genera of presses until the glue has hardened (see Chapter 9). After pressing, the panel is allowed to remain in a stack for a period of time that depends on the degree of hardening achieved in the press. At the scheduled time the panels are trimmed, removing about 2 in. from all edges, sanded, inspected, patched again if necessary, stamped with the appropriate grade and other information, and delivered for further processing.

Flakeboards

Flakeboards represent the technological answer to the shortage of large, straight trees necessary for cutting veneer to make plywood. Flakes are, in essence, small pieces of veneer, 2 to 3 in. square, and usually somewhat thinner than veneer. Whereas veneer is assembled by hand to provide a definite 90° orientation in adjacent—relatively few—plies, flakes are deposited randomly in great numbers, such that the equivalent of a 90° angle occurs statistically from point to point throughout the length and thickness of a panel. When properly bonded, the flakes restrain and complement each other in the same way as the cross plies in plywood (see Figure 2–7).

The process is similar to plywood only from the press to the finished product. From the forest to the press, the process is entirely different. Small, crooked trees and residue

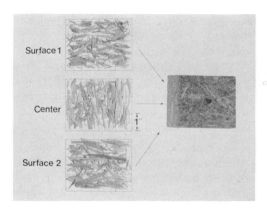

Figure 2-7 Composition of flakeboard or waferboard. [Courtesy of United States Forest Products Laboratory (USFPL).]

from conventional logging operations that can be reduced to bolt-wood size are acceptable. A *flaker,* with small knives mounted on the surface of a drum or disk, addresses the bolts with knives in the same aspect as in veneering.

Flakes are produced at high speed and in great volume, stored in bins, and transported on belts or blown through tubes. They are dried by tumbling in large cylindrical dryers propelled by very hot air and turning vanes. Resin is sprayed on while the flakes are tumbled in a *blender.* They then proceed to a *felter,* which deposits them randomly but uniformly at a controlled rate onto a *caul,* a thin metal pan like a cookie sheet that will later carry the *mat* into the press. The caul is carried on a moving belt. The thick mat of resin-coated flakes that is formed on the caul contains a quantity of flakes which when pressed to a certain thickness will produce a panel with predesigned density.

Since panel size is not limited by log or lathe length as it is in plywood, flakeboards can be made in multiples of 4 ft by 8 ft. Two advantages immediately ensue: Longer lengths are available to the market; and in cutting to the standard 4 by 8 ft, two edges will incur no trim losses. Trim losses of 2 to 4 in. are sometimes needed to remove poorly consolidated material around the mat edges. After panels emerge from the press, they are placed in closed stacks ("hot-stacked"), so

that the residual heat can be used to drive the final cure of the resin.

With the above as the basic embodiment of the art of reconstituting small pieces of wood, several important differentiations can be discerned. One derives mostly from the thickness of the flakes. If the flakes are 0.050 in. or so in thickness, the product is known as *waferboard.* Pioneered by Dr. James d'Acre more than 35 years ago, this product was one of the first successful flake-type panels to be manufactured and marketed. If the flakes are 0.025 in. or less in thickness, the resulting product has come to be known as the generic form of *flakeboard.* The performance of flakeboard and waferboard is essentially the same, though waferboard may be slightly rougher in texture.

Another derivative of greater structural consequence is the reduction in width of flakes to what are known as *strands.* Optimally 0.5 in. wide and 1.5 to 2 in. long, strands acquire by their geometry the ability to be deposited in prescribed directions. Simply by dropping them through narrow slots, strands can be aligned in the direction of the slots. By aligning one set of slots in one direction, another set perpendicular to the first set, and a third set parallel to the first, a three-layer mat can be formed on a passing caul which, when consolidated, has directional properties similar to plywood (see Figure 2-8). This product, developed by Elmendorf Research Lab, is appropriately called *oriented strandboard* (OSB). OSB resembles

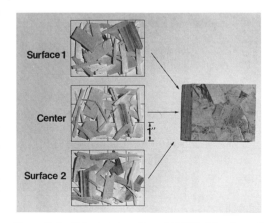

Figure 2-8 Oriented strandboard (OSB). (Courtesy of USFPL.)

plywood even more than flakeboard because it has the grain attributes of veneer on the face and back. Thus it is stiffer than flakeboard when applied across floor and roof joists. Strands possess a production advantage over flakes in that they can be produced from an even lower quality of tree. Wood that can be reduced to rather large chips, 1 to 3 in. long, can be processed into strands in a *ring flaker.* The advantage accrues only to the extent that chips are easier to handle

mechanically than bolts. The strands are of slightly lower quality and contain a higher percentage of smaller wood fragments.

A derivative of oriented strandboard involves a thin, veneerlike board in which the strands all run in the same direction. This OSB is used like veneer as the core in otherwise conventional plywood production where the face and back are solid veneer. The product resembles plywood, and is readily accepted as such (Figure 2–9a). A similar

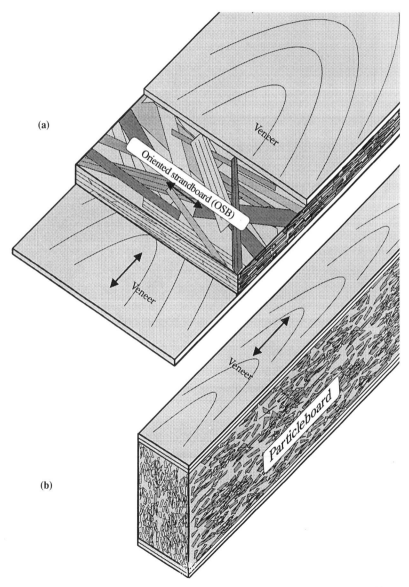

Figure 2–9 Composite products: (a) veneer face and back and OSB core for sheathing; (b) veneer edges on particleboard core for structural purposes.

concept but with different organization produces framing members, 2 × 4's, 2 × 6's, etc. In this case the core is particleboard 1.5 in. thick cut to 3-in. widths. Veneer is glued to the edges to produce an I-beam effect (Figure 2–9b).

Fiberboards and Particleboards

The products discussed so far are all made with wood pieces derived from the log entirely by knife action. In fiberboard and particleboard (Figure 2–10) the wood elements are made by grinding. Two different methods are used. Fibers for fiberboard are ground in an attrition mill, similar to those used to produce cereal flours, by grinding the elements between rotating disks bearing teeth or ridges. The particles for particleboard, on the other hand, are made by grinding in a hammermill, where flailing hammers smash the wood to bits. In either case, the wood is in chip or shaving form before the grinding.

The fibers are actually clumps of whole cells, and are more properly called "fiber bundles" (as opposed to fibers used in papermaking, which are parts of cell walls shredded to microscopic ribbons). They are essentially linear in aspect. Particles are dis-

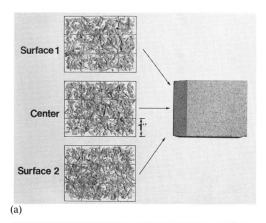

(a)

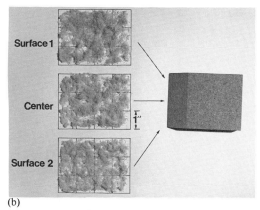

(b)

(c)

(d)

Figure 2–10   (a) Particleboard and (b) fiberboard composition. (Courtesy of USFPL.) (c) and (d) Representative uses: doorskin and siding. (Courtesy of Masonite Corporation.)

tinct from fiber bundles in being actually small pieces of solid wood, larger, relatively blunt, and with less linearity. (The word *particle* is sometimes used as a generic term for any comminuted wood, including flakes, fibers, and granulated wood. However, it was coined during the time when boards were first being reconstituted, and the only material considered for use then was hammer-milled waste wood, "particles" of wood.) Current usage of the term is more accurately restricted to small, rather cubical pieces of wood, with a major direction not more than two or three times the smaller dimensions. The word *element* is a more general term referring to any distinct piece of wood that is used in a reconstituting process. When the elements are small, en masse, and ready for consolidation, they often are referred to as the *furnish*.

The felting operation—i.e., making the mat—provides a point of distinction for classifying several types of products made from particles or fibers. Three processes are recognized: wet process, dry process, and semidry process.

The wet process for forming mats derives its technology from papermaking. Here the fibers are suspended in water and flooded onto a moving screen from which the water drains or is sucked out by vacuum, leaving a wet mat. The mat is transferred to another screen which carries it into the hot press, where it serves to allow water and steam to be expressed from the mat while it is being consolidated. The screen leaves a characteristic impression on the back side of the finished panel.

In dry-formed mats the elements are either suspended in air and allowed to float or drift toward the caul below, or cascaded over rolls to drop freely onto the caul. The apparatus that handles the furnish during its deposition on the caul is called a *forming head*. Usually a minimum of three forming heads in series are used. Such an arrangement not only produces a more uniform mat, but most important, allows the formation of distinct layers in the board, because each head can deliver a different furnish. One example might be to put fine particles in the first head, coarse particles in the second head, and fine particles again in the third head. This would produce a panel with smooth outside surfaces. Other differences, such as moisture content and resin content, can be stratified into the panel through the three heads. Dry-formed panels, being formed on smooth cauls, do not have a screen impression on the back but are smooth on both sides.

The semidry process combines features of both the wet and dry processes. The fiber furnish is not actually "wet," but it is very high in moisture content. It therefore needs a screen caul to allow moisture to escape during pressing. The fibers are felted out of an air suspension, thereby avoiding the handling of masses of water. As in wet-process panels, a screen impression remains on the back of semidry formed panels.

Fiber mat forming is increasingly done by the dry process, progressing from wet, to semidry, to dry, in the pursuit of production efficiency. However, there are some notable exceptions. The success of the original Masonite process was due in part to the use of lignin remaining on the fiber as the binder instead of resin. In order to make lignin function as an adhesive, high heat, high pressure, high moisture, and a screen backing are needed, all of which present their own production necessities.

A range of fiber products is defined by density, of which there are three classes: (1) low density, less than 37 pounds per cubic foot, (2) medium density, 37 to 50 pounds per cubic foot, and (3) high density, over 50 pounds per cubic foot. The original Masonite and similar products are classed as high-density fiberboards. Low-density fiberboards include the insulating boards used as sheathing in house construction, and ceiling panels with decorative surface treatments. These are made by wet-process forming followed by drying in a heated tunnel without pressure. Medium-density fiberboards are relative newcomers to the fiberboard scene, pioneered by S. Hunter Brooks in the early 1960s. These fiberboards were a response to the need for a panel material to be used in furniture construction, where smooth surfaces for thin overlays, better machining properties, and lower weight represented an improvement over current practice. Medium-density fiberboard also finds acceptance as siding on houses because of its smoothness (for ease of painting) and its greater dimensional stability. The outstanding property of the higher-density products is smoothness of surface. This allows the addition of overlays by direct printing, application of decorative films, or gluing on of veneers. The smoothness must be permanent to avoid later "telegraphing" of roughness through the overlay.

Secondary Gluing

After a panel has been made from elemental wood pieces by primary gluing, value is added by subsequent gluing procedures,

combining them with other elements or other materials. For example, plywood webs can be glued to the top and bottom chords of a truss system, either singly to form an I-beam, or doubly to form a box beam (Figure 2–11a). The application of Formica on particleboard to make countertops is another familiar example (Figure 2–11b). Others include the production of stressed-skin structures by gluing plywood to a framework of studs (Figure 2–11c); the edging of thin decorative panels with lumber to rigidize them and make them more fastenable (Figure 2–11d); and the making of hollow-core doors, where plywood or fiberboard is glued to a frame of lumber incorporating a honeycomb core (Figure 2–11e). Variations of these processes are seen in house construction, where subflooring panels are nail-glued to floor joists to increase

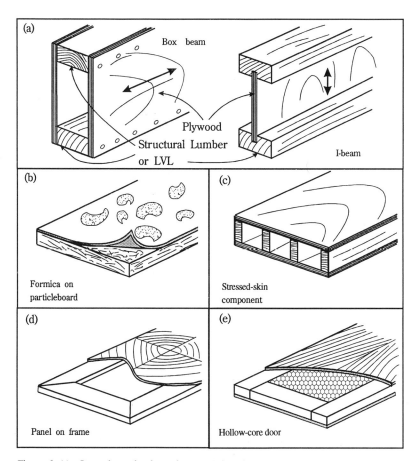

Figure 2–11   Secondary glued products: (a) box beam and I-beam; (b) Formica on particleboard; (c) stressed-skin panel; (d) panel on frame; (e) hollow-core door.

stiffness of the floor, and to eliminate squeaks. The lumber-core plywood of Figure 2–5, and the composite products of Figure 2–9, are also examples of secondary gluing. All these examples fall into the third stage of converting trees to consumer articles, the subassembly or componentizing stage.

Different glues and different means of applying pressure are associated with secondary gluing, in response to the special situations that exist at this stage. For panel products, wood surfaces generally are given no preparation except sanding. Because surfaces are taken as they come from prior operations, good surface contact or cleanliness is not assured. Hence, in applications of gluing to structural lumber, some tolerance to gaps and surface roughness must be exhibited by the glue. Speed is also essential; parts should not have to remain in presses for long periods of time. The parts are sometimes held together with nails, as in gluing subflooring to joists or making box beams. Because the glue line often is too far from the outside surface for heat to be delivered to it, fast, cold-setting glues are preferred. Typical glue choices might be casein or resorcinol for nail-gluing box beams and for stressed-skin panels, urea and hot or cold press for gluing veneer on particleboard, urea and dielectric press for gluing panels to lumber frames, and casein or polyvinylacetate for hollow-core doors. In all cases, it should be noted, glue costs are not as important a factor as they are in primary gluing.

Tertiary Gluing

Up to this point, the objective in gluing has been linear or planar products. The final gluing stage (the fourth stage in the conversion chain), is the assembly of those products into cubical or volumetric shapes, e.g., houses or furniture. At this stage, speed of fastening is the driving imperative. Mechanical fasteners such as nails, screws, and bolts provide the speed, and are often the first choice when they are appropriate. However, several disadvantages accompany their use: They provide point fastening, with resultant concentration of stresses; they "slip" and are therefore less rigid; and they allow racking (deformation of shape). Mechanical fasteners also provide no sealing capacity, and they often protrude above the surface. Adhesive fasteners partially or entirely overcome these disadvantages. A picture frame, for example, that is held together by brads or corrugated fasteners can rather easily be deformed, or racked out of square. When it is glued at the corners, however, particularly with a spline insert, the frame has greater ridigity.

The assembly of a piece of furniture, such as a chest of drawers presents choices in fastening systems depending on the quality of the piece. A high-quality chest would use a number of wood-to-wood glue-joining methods to ensure a rigid structure. These might include the liberal use of dowels to supplement side grain-to-side grain joints, mortise-and-tenon joints for end grain-to-side grain joints, dovetail joints for end grain-to-end grain joints at corners, and corner blocks to provide filet action at high-stress corners. Such methods are discussed further in Chapter 7.

Since fast production is a primary concern in tertiary gluing, glues are chosen for their speed of hardening. All of the available glues of reasonable cost are at the lower end of the durability scale. While this is not a severe limitation because most of the uses are in protected environments, high humidities (and occasional wetting), which cause softening of the glue and swelling stresses between wood pieces, suggest the use of more durable glues when severe environments are anticipated.

Combinations of Primary and Secondary Gluing

The secondary gluing process of making components by gluing a second element to a previously formed composite such as veneer on particleboard, or paper on plywood, means that the product must enter a press twice. To avoid double pressing, it is sometimes advantageous to do all the bonding at one time. In the case of veneer on particle-board, a mat of resin-coated particles would be formed on a sheet of veneer, and then another sheet of veneer would be placed on top before inserting the piece into the press. The two elements thus would be consolidated simultaneously.

The apparent simplicity of such a process hides some difficulties. Two different bonding processes must be integrated. The amount of resin applied for particle-to-particle bonding is not enough for veneer-to-particle bonding. Some means of increasing the amount of adhesive at the interface is therefore necessary. The shuffling of veneer under and over the mat may slow the process to an inefficient status. Increased breakage of valuable face veneer can occur, both from handling and from the shifting of particles during compaction. Trim allowances will be more costly. While these difficulties yield to engineering solutions, they still must respond to economies of scale.

Similar problems accompany the manufacture of paper-overlayed plywood in one step, though in this case the problems are easier to overcome. Since defects that occur during pressing of plywood cannot be easily repaired when there is paper on the surface, the problem reduces mainly to one of producing defect-free plywood in the first place. Also, since the plywood cannot be sanded for smoothness and thickness with the paper on it, face veneers must be sanded in advance, and other thickness controls must be instituted. Problems tend to simplify as the elements to be combined become smaller and there is a greater degree of uniformity between them.

Combinations of Primary and Tertiary Gluing

Basically a process of producing shaped articles from the original wood elements, the combination of primary and tertiary gluing offers distinct advantages. It eliminates the most costly operation associated with producing the final consumer article: the craftsmanship of fitting and fastening pieces. At one extreme of the process of forming shapes, one can visualize the building of a box or a pallet with lumber, nailing in place one piece at a time. At the opposite extreme would be a pallet molded in one piece using flakes or particles, as shown in Figure 2-12. By using a combination of primary and tertiary gluing, veneers can be molded into such disparate products as salad bowls and airplane fuselages. The principles of the composites that underlie such molded shapes is discussed in Chapter 7; here we will discuss only the process of glue application.

The process involves applying adhesive to the wood elements, and charging them into a mold either as a preformed mat, or filled directly into the mold cavity. In the case of flakes, particles, or fibers, the operations up to pressing are similar to those for flat panels. For veneer, an entirely different technology is required to make large structural products. Because of its anisotropic properties, sheets of veneer and larger elements cannot be bent in more than one direction without breaking. To avoid breaking, the veneer is first cut into narrow strips, 2 to 3 in. wide. This narrow width allows the twisting and bending needed for conforming to compound curves.

The system seems to embody the ultimate

Figure 2-12 Combination of primary and tertiary gluing: a molded pallet. (Courtesy of USFPL.)

in processing to produce useful wood articles. Unfortunately, while the technology is readily available and reasonably applicable, the economics are sometimes difficult to bring into line. The main difficulty is the molds, which have to be heated to high temperature, and have to deliver high pressure. Such molds are costly, and produce only one item per press cycle. Thus, many molds and presses are needed for large production runs, but only large runs can justify the cost of the molds. Moreover, a different mold is required for each size and shape of product manufactured.

Despite the problems and costs of molding processes, they offer a means of overcoming one of the greater negative aspects of working with wood: the fabrication of corners, and their resulting performance. Corners represent a discontinuity in a stress-transmitting system. They inevitably invoke the weak side-grain properties of wood somewhere around the corner. In molding a corner, the material properties are essentially the same all the way around. A chair part, for example, that is molded in the form of a right angle with parallel grain veneer, is able to transmit all its stresses from the seat to the legs along the grain without involving end-grain weaknesses. Compare Figures 2–13a and 2–13b.

Extruded Products

Extrusion can be considered a molding process in which two dimensions are kept constant, and the third is linear and infinite. Plastic moldings, tubes, channels, and rods are familiar examples of extruded objects. Extrusion of comminuted wood has a long history. First practiced in the making of fireplace logs and briquettes, it was developed into a panel process in Germany in the 1940s by the necessities of war and scarcity.

In making fireplace logs, a round plunger or screw drives a charge of particles into a round barrel with sufficient force and heat to consolidate the mass without binders. In making panels by extrusion, the plunger and

(a)

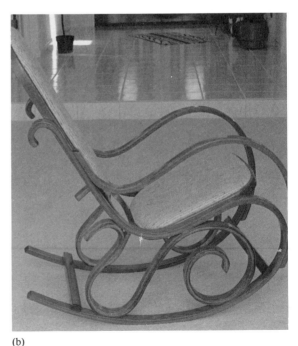

(b)

Figure 2-13 Corners in chairs; (a) mortise and tenons; (b) molded parallel laminated veneer.

the barrel have the cross-sectional shape of the panel, rectangular. Pressure is therefore applied to the edge rather than to the face of the panel, as is the case in platen-pressed panels. The panel is extruded in a continuous ribbon, sometimes with holes lengthwise through the center to reduce weight (Figure 2–14a).

The extrusion process has several distinctive features: (1) Panels can be as long as desired; (2) stability in the thickness direction is better than for flat-pressed panels; (3) stability and strength in the length direction is relatively poor; (4) compared to other panel processes, the process is very low in capital and operating costs. While extrusion seems ideal for small operations where the main concern is to use up internally generated waste wood, the industry places extruded panels in a separate classification from platen-pressed panels because of their lower strength. These kinds of panel materials therefore are never used in their primary glued form, but always go through a second-

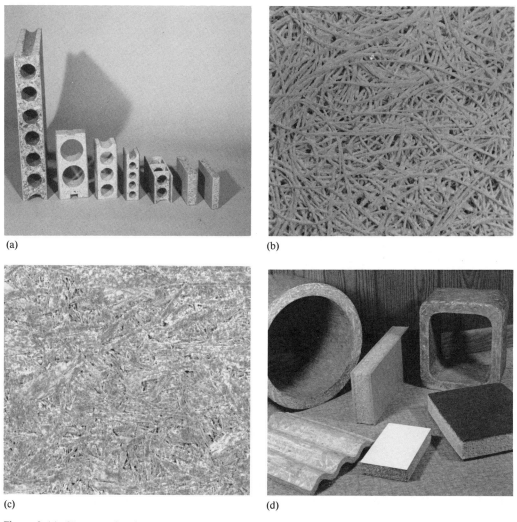

(a)

(b)

(c)

(d)

Figure 2–14 Unconventional products: (a) extruded panels; (b) excelsior/gypsum; (c) flake/Portland cement; (d) fiber/foam.

ary gluing operation to add overlays which both improve strength and provide decorative features.

Cold-Pressed Composites

The processes of comminuting wood and reconstituting it through various stages of manufacture to a final product involves, in many cases, knife cutting, application of a small amount of adhesive, and pressing at high heat and pressure. To a disproportionate degree, the adhesive controls the process: the choice of species, the need for low moisture content, the need for high heat and pressure, the need for high capitalization, the need for economies of scale, and the need to accept high density in the product. Some of these factors are mutually restrictive.

Boards and blocks made of comminuted wood and Portland cement, gypsum, or magnesite (Figure 2–14b&c) remove the constraints imposed by the necessity to heat, and thereby greatly simplify the consolidation steps. Although this process is not widely used in the United States, it is appealing for producing building materials that are resistant to fire, vermin, and weather in a process that utilizes low-cost materials and equipment, and it is comparatively forgiving of many variables encountered in previously discussed products.

A composite of foamed resin and a spe-cially configured wood element takes another step toward reducing or eliminating some of the above constraints (Figure 2–14d). The wood element, fibrous in nature, can be formed from green wood of most species, including the high-density hardwoods. When the element is made into a mat and suffused with a self-curing foaming resin, consolidation occurs spontaneously, and no pressure is needed except to obtain the desired density, which can range from 8 to 48 pounds per cubic foot. With heat, and to some degree pressure, removed from the operation, the process can transcend the entire spectrum of gluing from primary to the production of molded shapes in continuous or batch operations. The simplicity, low equipment cost, and versatility of the process allow scale-up or scale-down to reconcile markets, material resources, financial resources, and human resources over a wide range from farmer to industrialist. Although resin costs are high, the economies in processing, and the high-value products that can be made, are offsetting.

A concluding thought that emerges from consideration of these composites is that adhesives having the proper properties can be the servants of the bonding process by making it easier and more versatile, rather than the bonding process being the servant of the adhesive, so that complicated procedures must be followed.

3

Fundamentals of Bond Formation

Gluing technology, as previously mentioned, is divided into two parts: *bond formation* and *bond performance*. In this chapter, attention is centered on bond formation with a discussion of the different parts of a bond, and what the adhesive has to do to form them. The bond is described in rather precise detail, depicting it as an anatomic entity with clearly defined regions where specific adhesive activities take place. Adhesive activities are then evolved as an organic system, using terms descriptive of the different functions an adhesive must perform in consummating a bond.

In simplest terms, adhesives are liquids that convert to solids. Liquid properties are needed for bond formation to achieve the proximity essential for adhesion to take place, and solid properties are needed for bond performance to deliver the strength required in the glued product. The concept of proximity is so crucial to bond formation, it pervades most of the factors which affect the gluing process.

Bond formation requires the adhesive to attain the closest possible conformation to surface irregularities. If surfaces could be made smooth enough so that, when brought together, molecular attractive forces would come into operation, a distance of a few angstroms, no adhesive would be necessary. A sense of this possibility is evident in Johansen blocks, blocks of steel, ground and

polished to precise dimensions, and used for gauging thickness. When they are very clean, these blocks tend to stick together. In this case, it should be noted that though they appear by common forms of measurement to be absolutely smooth, at the molecular level, there still are too many gaps beyond the reach of molecular forces. Hence the blocks can be separated with relatively little additional force.

In the case of wood, its inherent porosity creates an initial surface roughness that is not only inescapable but also highly variable due to anatomical differences from point to point on a surface of some species. When the roughness created by machining is added to that due to porosity, there is very little opportunity for molecular distances to exist between any two wood surfaces. Nevertheless, with well prepared surfaces it is sometimes possible to perceive a Johansen block effect especially with even-textured woods like maple. How well wood surfaces come together can be sensed by simply sliding one piece over the other and noting the friction or "drag" that exists. The more friction or drag there is, the smoother the surfaces are. This will be more evident with small pieces than large ones, because the latter are more likely to contain deviations from plane that add to the gap.

The primary function of an adhesive, therefore, is to compensate for the lack of closeness between the molecules on both surfaces of a joint. This means that the process of adhesion deals with molecular distances. Only liquids (and, of course, gasses) can accomplish such proximity. This explains why all adhesives must have a liquid or at least a semiliquid phase if they are to accomplish their task. It should be noted, however, that adhesion produces only a bond between the adhesive and the wood. To form a bond between wood and wood, it is necessary to create a bridge to span any gaps.

The actions of an adhesive in bond formation are associated with liquid properties and the process of conversion to a solid. It is important to bear in mind that the actions and the conversion occur in the environment *between* pieces of wood. This environment is composed of a space or volume to be filled, and an atmosphere that includes temperature, pressure, and sorption variables, all of which affect the bond-forming activities of the adhesive.

A METAPHOR: THE BOND AS A LIVING ENTITY

The term "adhesive activities" in reference to bond formation is symbolic of the entire process of bond formation because it suggests that the adhesive has to undergo certain actions or motions to do its job. This infers that bond formation is a dynamic process. In fact, the bond can almost be described in living terms: It is born, it is malformed or well formed, it mates, it matures, it lives, it dies. Certain things happen at each stage. Knowing what happens or is supposed to happen provides the basis for understanding and doing what is proper to help the adhesive. The metaphor of life provides a context for related terms in bond development.

1. Just as time is an important factor in real life, determining when some phase, such as reproduction, should occur, *time* plays an important role in bond formation, in controlling the occurrence of certain actions.

2. Just as growing phases occur in sequence in real life, adhesive growth actions occur in *sequence* in bond formation.

3. Just as growing actions occur at certain places in real life, bond-forming actions occur at definite *locations* in bond formation.

4. Just as growing actions occur to different degrees in real life, adhesive actions occur to different degrees, or, as we shall say, have *magnitude,* in bond formation.

5. Just as in real life, wherein the amount of an action has consequences, so too in bond formation, the amount (magnitude) of an action has a consequence. Consequences result in definite glue line conditions. The conditions are identifiable, and serve as indicators of what the glue did, i.e., actions it performed or failed to perform.

Thus the events in bond formation are conceived in terms of:

1. Actions
2. Time
3. Sequence
4. Location
5. Magnitude
6. Consequence

These terms represent the framework within which a concept of bond formation is developed. Of great significance to the pertinent logic is the premise that running through all the above terms is a common operator: the mobility of the adhesive *on the glue line.* Thus a mobility parameter becomes the single most important concept in this book, because through it we will be able to relate what happens on the glue line to what caused it and what would change it. It will make possible the practical solution of the equation of performance, and allow the operation of responsive and reliable quality control programs.

In this chapter, a general concept of bond

formation is drawn by weaving together the mobility-related terms mentioned above. First, however, it is appropriate to dissect an adhesive bond to learn the various parts and their locations, in order to develop a sense of the structure the adhesive is trying to produce.

THE ANATOMY OF AN ADHESIVE BOND

A visual representation of a bond as a system of links, Figure 3–1 suggests that the bond is made up of distinct parts in specific locations. One can infer that each part or location requires a different action by the

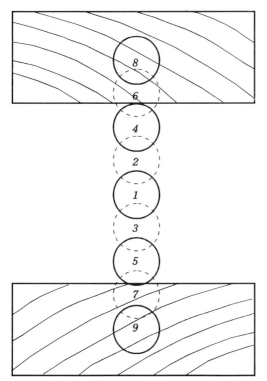

LINK 1	The Adhesive Film
LINKS 2 and 3	Intraadhesive Boundary Layer
LINKS 4 and 5	Adhesive-Adherend Interface
LINKS 6 and 7	Adherend Subsurface
LINKS 8 and 9	Adherend Proper

Figure 3–1 Links in an adhesive bond (dashed links are most vulnerable to malformation).

adhesive in order to form that link. A considerable amount of information can thus be associated with each of the links by recognizing what must happen, did happen, or did not happen, action-wise, at each location.

There are nine distinct—but not always identifiable—links in any bonded system. In Figure 3–1 they appear to be mirror images about the center link. This is seldom true in actual practice because, for example, species may be different, adhesive application may vary, and surface qualities may differ. However in this general discussion they may be viewed as mirror images.

At one time, not too long ago, a bond could be characterized with only three links, the two wood pieces and the adhesive in between. As knowledge grew, this was expanded to five links, the five bold links in Figure 3–1, in recognition of the basic attachment mechanism between wood and glue, links 4 and 5. Two more links were added to recognize the contribution of the condition of the wood immediately under the surface, links 6 and 7 (dashed circles to indicate obscurity and potential weakness). A final two links were added to emphasize the possibility that the cured glue line may not remain homogenous due to the differential influence of wood properties on some of the ingredients, links 2 and 3 (also dashed circles suggestive of obscurity and potential weakness). These four new links represent vulnerabilities that beset many glue lines. The latter two are particularly subject to malformation with glues that harden by chemical action because the first few molecules of the adhesive can be overpowered by the chemistry of the wood surface.

A beginning point in considering these links is the realization that the main objective in wood gluing, as in most joining operation, is to form a *stress transfer mechanism* between two pieces of wood. Two kinds of stress transfer mechanisms may be recognized: one that operates through the bulk of the material, such as screws, bolts, dowels, or plates; and another, represented by adhesives, that operates by surface attachment. A crucial difference arises from the fact that the condition of the surfaces to be bonded to a large degree controls the success in wood gluing because six of the nine links in the bond (2, 3, 4, 5, 6, and 7) are dependent upon this surface region.

In Figure 3–1, links 8 and 9 represent the two wood pieces in a joint. It is the strength of these two pieces that the other links are expected to match. However, these links also represent many other factors that affect both bond formation and bond performance. In bond formation, for example, the porosity and moisture content of the wood pieces affect the movement of the adhesive into the wood, and their thickness affects the amount of solvent they can absorb from the glue. Thickness also affects the ability to apply heat in curing the glue. Later, in bond performance, the size and properties of the wood pieces affect the amount of stress they can generate or transfer to the bond.

Links 6 and 7 represent the region in the wood just under the surface. This subsurface region contains damaged wood. All methods of surfacing wood, with the possible exception of hand planing, rupture the fiber structure for some depth below the surface. Unless the glue penetrates this region and repairs the damage, the final bond can be no stronger than this ruptured wood. These links thus form another critical region, the location where subtle changes in procedure or properties can invade the operation and produce major changes in performance.

Links 4 and 5 represent *adhesion,* that is, the basic attraction, molecule to molecule, that must be established between the adhesive and the wood in order to achieve good bond performance. Three different mechanisms might operate here:

1. A strictly mechanical bonding, in which tendrils of glue penetrate the pore and interstitial structure of the wood and interlock in a manner similar to plaster on a lath wall.
2. A chemical reaction between adhesive molecules and wood molecules to form a new compound. While it is known to be possible for some adhesives to form chemical bonds, experimental proof of such reactions is limited. However, a bond formed by chemical reaction of surface molecules would be the preferred bond because of the greater strength and durability it could deliver.
3. A physical bond, in which residual attractive forces, which exist on all molecules as a result of electron distribution, draw molecules together electrostatically. These attractive forces vary greatly in degree, depending on the molecule, resulting in what are termed polar materials with a high degree of attractiveness, and, at the opposite extreme, nonpolar materials having little attractiveness. Water, paper towels, and wood, for example, are highly polar materials while gasoline, plastic films, and wax are nonpolar.

Physical adhesion is generally considered to be the main mechanism by which adhesives attach to wood. However, mechanical or interlocking actions always occur and contribute their measure of strength. In fact, it is not unreasonable to state that the attraction between molecules can also result in a maximum of interlocking and coincidentally a maximum of strength.

Links 2 and 3 represent the portion of the glue film closest to the wood but inside the first layer of molecules which have participated in the adhesion action. Their location implies that they may be subject to some influence arising in the wood that may affect how they react with each other and therefore influence the quality of their cured or hardened state, or the development of cohesion. Such factors as the pH of the wood, its buffering potential, or its ability to differentially absorb certain ingredients (e.g., solvents and catalysts) can produce a localized weakness in the glue film.

Link 1 is the glue line proper, which is expected to hold the two pieces of wood together through whatever stress they will experience. In addition to its original composition, its performance depends overwhelmingly on the conditions of its conversion from a liquid to a solid. Most wood glues, when properly applied and cured, attain sufficient strength in link 1 for the intended usage. Link 1 must be regarded as a layer of solid material. As such, it has thickness, and it has bulk properties as well as mechanical, physical, and chemical properties of its own. The fact that it is bonded on both sides to wood changes some of its properties from what might develop when hardened as a free film. This is covered in greater detail in a later section.

ADHESIVE MOTIONS

Motion is the essence of bond formation. It arises from the actions that the adhesive takes to form a bond, and it leads to concepts of sequence and consequence, culminating finally in a cause-and-effect system of observation.

Some actions begin the instant the adhesive is placed on the wood surface. Others occur after the surfaces are brought together. These take place in sufficient sequence to warrant close attention and understanding since they become part of the useful logic of the process.

Adhesives undergo five distinct actions on the glue line in forming a bond. These are:

1. *Flowing* in the plane of the glue line in response to pressure and to high and low spots on the surface
2. *Transferring* to the opposite surface in the manner of a printing operation
3. *Penetrating* the pore and interstitial structure of the wood surface
4. *Wetting* the wood substance at available surfaces to produce adhesion
5. *Solidifying* to produce strength

These actions represent the physical and chemical means by which an adhesive transforms itself into a mechanism for holding two pieces of wood together. Each action requires a great deal of specific knowledge to design and to produce. Even more knowledge is needed to harmonize them into a formulation for a particular operation where species of wood, temperatures, pressures, and other factors impose constraints on effectiveness.

From a user's point of view, less knowledge is needed if the actions are considered to be motions, each with a different degree of mobility. This takes the emphasis off "composition" and places it on "process," which is what the user is doing. By attending to the motions an adhesive must perform, the user can gauge what is happening and assure successful bonding.

A sense of motion is developed below with respect to each of the actions. It is particularly important to understand that motions differ in degree, from coarse mass movements to very fine molecular movements, and even finer molecular orientations.

In Figure 3–2 the adhesive is seen first as a series of drops or beads on the surface of one of the blocks. (For practical purposes and for simplification, this is considered to be the standard procedure in wood gluing. Except in special cases, the adhesive is usually applied to only one surface.) In dealing with adhesive motion, it is helpful to imagine it as originating from a definite pattern, a starting point. This greatly improves the observation system, because it facilitates estimation of degree or magnitude of motions. The device is similar to the use of a precise geometric pattern in TV transmission to help detect sharpness of the image on the screen.

Flow

When the opposing block or surface is brought into contact, a very definite and easily observed motion occurs. The adhesive moves laterally in the plane of the surface, flowing into the grosser irregularities and forming a continuous film. This first motion is the easiest to perform because it involves only coarse movement of the adhesive en masse. Nevertheless, it can occur only if (1) there is sufficient adhesive, (2) the adhesive is sufficiently fluid, and (3) there is contact pressure with the opposing surface. (Adhe-

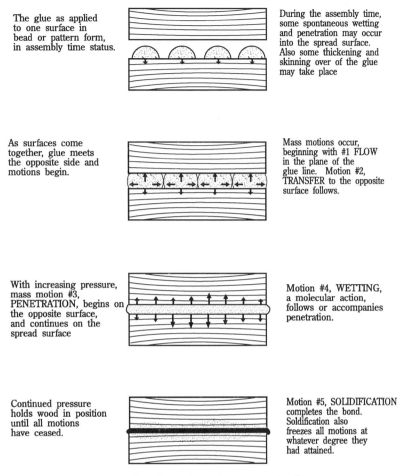

The glue as applied to one surface in bead or pattern form, in assembly time status.

During the assembly time, some spontaneous wetting and penetration may occur into the spread surface. Also some thickening and skinning over of the glue may take place

As surfaces come together, glue meets the opposite side and motions begin.

Mass motions occur, beginning with #1 FLOW in the plane of the glue line. Motion #2, TRANSFER to the opposite surface follows.

With increasing pressure, mass motion #3, PENETRATION, begins on the opposite surface, and continues on the spread surface

Motion #4, WETTING, a molecular action, follows or accompanies penetration.

Continued pressure holds wood in position until all motions have ceased.

Motion #5, SOLIDIFICATION completes the bond. Solidification also freezes all motions at whatever degree they had attained.

Figure 3–2   The sequence of motions of an adhesive on a glue line.

sives that flow without pressure require special attention to avoid starving the glue lines.) These three conditions with all their degrees of variation are the basis of many cause-and-effect observations utilized in troubleshooting (see Chapter 12).

Transfer

The second motion of the adhesive is transfer to the opposite surface. This may occur simultaneously with flow, but in some instances flow will occur without transfer. This leads to an important inference: Transfer is achieved with slightly more difficulty than flow and as a consequence is more easily inhibited. While still a coarse motion of the adhesive en masse, the adhesive must have

an additional "something" to allow this motion. That something is an extra degree of mobility, beyond that needed for flow alone. The extra mobility may be visualized as "tack," as in a printing process. (The reverse, in which transfer occurs but not flow, seldom happens, and only in special, easily observed, cases where excessive pressure was used to try to overcome effects of excessive assembly time.)

The applicability of the mobility parameter can already be discerned in understanding the difference between flow and transfer motions. Later we shall see what factors are responsible for the change in mobility in this and other cases. One can perceive, for instance, how low viscosity or high pressure induces high mobility. Because all factors operate through mobility effects, relationships will become more and more obvious as we proceed through the equation of performance.

The concept of transfer as distinct from flow is important mainly under extreme conditions, where the two can actually be distinguished as separate motions. Factors and combinations of factors that produce extremes are part of the technology to be learned. Also, it may be noted that when the adhesive is applied to both surfaces, the motion of transfer does not need to occur since the adhesive is already in place on both surfaces. Other methods of aiding the transfer motion are part of the technology that emanates from the mobility parameter.

Penetration

The third motion, penetration, brings the adhesive into the coarse capillary structure of wood and into checks and ruptures created in processing. Like flow and transfer, this is also a motion of the adhesive en masse. However, because of the small openings into which the adhesive must move, a still higher degree of mobility is necessary for this motion to take place. Penetration is commonly thought to be a motion of capillary action of a liquid, and to some extent it is in this case. However, as will be explained in the following section, wood adhesives are formulated to modify or control such spontaneous action, and therefore it becomes more dependent upon clamping pressure to aid the motion.

Wetting

The fourth motion, wetting, is basically a spontaneous motion, driven by the attraction of the polar molecules of the wood surface for the polar molecules of the adhesive. A similar and more familiar action occurs between towels and water, or between ink and blotting paper. An opposite action (repulsion) occurs between water and wax, or between water and a vinyl film. Instances of repulsion or lack of wetting also occur in wood gluing. When lack of wetting occurs, it raises very specific questions about the quality of the wood surface, or the conduct of the gluing operation.

Wetting is not a mass motion like the previous ones but a motion of the mobile molecules in the adhesive achieving final proximity and physical attachment with the fixed molecules of the wood substance. In this sense molecular mobility represents a finer degree of movement than liquid mobility en masse. The implications are many, and some of them seem counter to the accepted laws of physics governing wetting, spreading, capillary action, and adhesion, in which these movements are driven by residual forces residing in the molecules. The contradiction arises from the formulation of wood adhesives. They are not pure liquids like those on which the physical laws were observed. Rather, they are elaborate mixtures designed to control all the motions of adhesives in the face of great variability in the way they are used, in the way wood grows, and the way it is processed.

In pure liquids on pure solids, molecular forces largely control movement on surfaces and into capillaries; wetting and adhesion then occur by the natural tendency of matter to reduce its state of potential energy. Attractive forces between molecules can be thought of as free energy. When brought together, forces are neutralized, and both experience a lower and therefore a more stable energy level.

The same tendencies exist to a high degree, and necessarily so, in any adhesive. In a wood adhesive, however, they must be restrained or controlled so they do not dominate adhesive mobility, causing excess movement and loss of film-forming capability. The restraint is introduced primarily through viscosity modifiers and fillers, which unfortunately reduce both molecular mobility as well as mass mobility. Hence the natural tendency of adhesives to wet and adhere to wood surfaces must be prompted by an external force such as pressure.

This trade-off of wetting power for mobility control has a wide operational range.

Under some circumstances wetting can be allowed to maximize, while in other circumstances some wetting must be sacrificed so that mobility can be reduced. For example, when it is necessary to encourage deep penetration, particularly into cell walls, only molecular attractive forces coupled with complete molecular mobility can succeed. On the other hand, when the adhesive must function on a very porous surface, its en-masse mobility must be reduced.

In most cases, wetting is a time-dependent function. Because of the way adhesives must be formulated for wood bonding, they must be forced or otherwise brought to the site where wetting is to occur. That is one reason why penetration precedes wetting in the sequence of adhesive motions on wood, instead of vice versa as physical laws would predict. Another reason is that since a finer (higher) degree of mobility (i.e., at the molecular level) is required, it is also more difficult to achieve (and more easily thwarted).

The amount of wetting that occurs depends on:

1. The molecular nature of the adhesive
2. The quality of the wood surface
3. The molecular mobility in the adhesive
4. The pressure on the adhesive
5. The time available before hardening of the glue arrests mobility

Solidification

The final adhesive motion is solidification, which stops all motions and consummates the bond-forming actions that have preceded. Solidification is a property of the adhesive designed in by the manufacturer. Four basic mechanisms for hardening are available:

1. Loss of solvent or liquid carrier
2. Loss of heat
3. Chemical reaction
4. Combinations

In all cases the solidifying action deals with motions and interactions between molecules or particles within the adhesive system as influenced by the environment in the glue line. Those adhesives that solidify by chemical reaction are particularly sensitive to their environment because they involve the finest degree of mobility between molecules. They must not only achieve very close proximity with each other for reactions to occur, but they must be able to orient themselves with respect to each other in order to properly align the chemical bonds being formed. Factors that affect the mobility of molecules and particles in a mixture include their size, their configuration, their concentration, the viscosity, and the type and amount of solvent.

Depending on the composition of the adhesive and the environment in the glue line, a number of motions or lack of motions at the molecular or particulate level can occur *during solidification*. These produce different states of solidification which have a bearing on solid properties of link 1. Some of them can adversely affect strength of the bond (Figure 3–3). They include:

1. *Undercure.* Solidification is interrupted or terminated before being fully accomplished, as might happen with a hot-press glue that received too short a press cycle, or a room-temperature-curing glue applied to cold wood. Undercuring produces no clearly visible signs and is therefore not illustrated in Figure 3–3, but it yields a bond with low strength and reduced durability.
2. *Overcure.* Solidification is carried out beyond the full cure stage, so that the glue begins to break down. This happens mostly with hot-press glues that are heated too long, and applies mostly to urea resins. Again, overcuring shows no visible characteristics at the glue line and is therefore not illustrated, but it produces a bond with low strength.
3. *Grainy.* The appearance of graininess in a glue line (Figure 3–3a) indicates use of too much filler, or too rapid dry-out so that particulates in the adhesive do not have time to assimilate fully into the glue film.

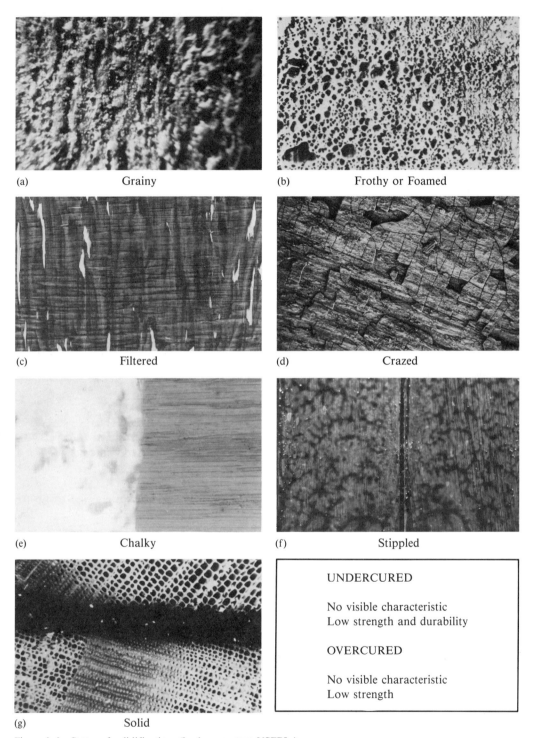

Figure 3-3   States of solidification. (b, d, g courtesy USFPL.)

4. *Frothy.* Sometimes bubbles or foam are formed in mixing or spreading that persist in the glue line, or they may be formed by the action of heat, especially dielectric or microwave, which volatilize the glue water or solvent while the adhesive is curing in the glue line (see Figure 3-3b).

5. *Filtered.* A filtered or "wash-out" effect results from differential absorption of adhesive ingredients, in which the wood acts as a filter and allows the resin fraction to penetrate but holds fillers and extenders on the surface. The glue line then becomes deficient in resin, and less resistant to forces of destruction (Figure 3-3c).

6. *Crazed.* Random breaking up of a glue line, usually occurring in thick glue lines due to shrinkage of the hardened adhesive occasioned by loss of solvent from the film (Figure 3-3d). Tightening of the adhesive mass as a result of further curing also creates stresses that cause the film to break up. This is not a frequent occurrence, and it can usually be avoided by proper formulation.

7. *Chalky.* This is a condition that occurs chiefly with polyvinylacetate adhesives applied to wood that is too cold. The low temperature causes the resin globules to precipitate in the glue line rather than coalesce as they would on wood at room temperature. Instead of the clear glue line that would otherwise occur, on cold wood the glue line appears opaque and clouded, or chalky (Figure 3-3e).

8. *Stippled.* Stippling develops when surfaces spread with glue are brought into contact so that flow and transfer occur, but then the surfaces are allowed to spring apart before hardening can take place. Adhesive appears on both surfaces of the joint, with elevations or projections reaching toward the opposite surface, sometimes dentate in nature, sometimes filamentous, depending on the amount of tack in the adhesive (Figure 3-3f). In thick gap situations some adhesives may draw themselves into a system of columns between the surfaces when loss of the liquid carrier causes shrinkage in glue volume.

9. *Solid.* The desired condition of a glue line is solid: All ingredients have functioned optimally and remained uniformly distributed throughout the glue line (Figure 3-3g). The condition has one distinctive feature: It is usually obscured by wood on both sides. Generally it cannot be viewed except by a very low angle cut from the wood on one side of the bond line to the wood on the other side.

Some of these conditions are obvious; some cannot be revealed except by destructive tests. All, however, represent factors that operate in the glue line and cause disruption of the hardening process. We will examine them again in discussing effects of operational factors on the quality of the bond.

LOCATION OF ADHESIVE MOTIONS

Having established the nature of adhesive motions that contribute to a concept of mobility, it is appropriate to relate the motions with particular locations in the joint, and thus establish observation points for each action.

The representation of an adhesive joint as a series of links (Figure 3-1) can be used as a map showing where some of the actions or motions of adhesives occur.

Flow, by determining the distribution of the adhesive in the bond line, has a direct bearing on the existence and uniformity of link 1. As the first level of mobility, it also influences the motions that produce links 4, 5, 6, and 7.

Transfer operates at link 4 or link 5, whichever surface did not receive glue by applicator. Link 6 or link 7 of the same surface is also in need of the transfer motion. These two links, dependent as they are on operational factors of the gluing process, can be considered the most vulnerable in the entire chain, and are therefore the focus of observation in assaying bond quality by visual characteristics.

Penetration is associated with links 6 and 7, the subsurface region of the wood. It is this region that is most often damaged during surfacing and must therefore be repaired by the adhesive. Porosity and grain angle, as well as surface checks, affect the amount of penetration that might occur at a given instant with a given pressure.

Wetting involves links 4 and 5, where the crucial physical attachment of adhesive molecules to wood molecules occurs.

Solidification forms link 1, the key stress transfer mechanism that is ultimately responsible for the strength of the joint.

Links 2 and 3 are affected by a special kind and combination of motions. These involve disturbances in the solidification action occasioned by contact with wood and its physical-chemical properties, chiefly its pH, its extractives, and its sorption effects.

SEQUENCE OF ADHESIVE MOTIONS

The motions of an adhesive on the glue line occur in sequence. This is a direct result of the solidification motion, which normally occurs not instantaneously, but over a period of time in most cases. Hardening of the adhesive essentially begins the moment it is applied to the surface, although for some, hardening begins when they are first mixed for use or when first compounded by the manufacturer. (For this reason many adhesives carry a designated shelf or storage time and temperature beyond which they may be unusable.) The consequent progressive reduction in mobility produces the sequential effect on the motions of the adhesive on the glue line because each motion has a *mobility limit*. As hardening progresses, the first motion to reach its limit and be inhibited is the one that requires the greatest mobility, wetting, because it has to have the delicate mobility to achieve molecular contact. The inhibition point for wetting, however, still permits penetration, transfer, and flow to

occur. (It could be argued on the basis of previous reasoning, that solidification by chemical reaction, involving as it does molecular movement and orientations, and requiring the greatest mobility would inhibit its own motions first. This argument is engaged in greater detail in the chapter on adhesive composition.)

Further hardening can inhibit penetration but not transfer or flow, which are less demanding of mobility effects. More hardening would inhibit transfer but not flow. Finally, sufficient hardening will inhibit even flow, thwarting all bond-forming motions, and leaving the glue line much as it was first deposited.

Looking at this sequence from the other direction, if at a certain stage of hardening, flow cannot occur, none of the other motions—transfer, penetration, or wetting—can either. Continuing down the sequence, it is possible that sufficient mobility exist for flow to take place but not enough for transfer. If transfer cannot take place, neither can penetration or wetting because they require even more mobility. A stage of hardening could produce a situation where flow and transfer occur but not penetration, in which case wetting, a finer motion, would be absent also. Finally it is possible for flow, transfer, and penetration—the coarsest motions—to take place but not wetting, a situation that produces a most insidious bond, because it can have enough strength to hold parts together during processing, but not enough to resist service conditions.

This differential effect of mobility changes on the motions in the glue line not only establishes a sequence but also provides an important key to relating causes and effects among different gluing factors. These relationships then become the intelligence and the means by which adjustments can be made in bond formation.

It is important to note that some of the mobility effects occur before the glue is applied to the wood, and some between application and clamping. Shortly after pressure is applied, the coarser motions occur, and only the finer motions remain to be executed as the adhesive continues the hardening process. In being the last motions to occur during hardening, the finer motions are thus in greatest peril of being inhibited.

MAGNITUDE AND CONSEQUENCE OF ADHESIVE MOTIONS

In addition to sequence, adhesive motions also exhibit magnitude. Motions can be massive or nonexistent, depending upon a number of compositional or operational factors. The important determinants are the solidification action and the fluidity of the adhesive, both of which control motions on the glue line. Although these properties are basically ordained by the composition of the adhesive, the actual amount of motion that ensues in a given instance also depends upon the conditions of use. The conditions of use include most of the factors in the first five categories of the equation of performance, where the pervasiveness of the mobility parameter is most evident. These will be drawn into the logic as various factors are discussed.

For the present, it is only necessary to sense the basic manifestations of mobility effects on the glue line. Interest is centered on the condition of the adhesive at the time

pressure is applied because that is the precise point at which the most important bond forming actions occur. For instance, if the adhesive is very fluid (for whatever reason) at the time pressure is applied, flow will be excessive and the adhesive will be squeezed out of the joint. This effect is, in fact, called *squeeze-out* (Figure 3–4a). Also with some species, penetration can be excessive, resulting in the adhesive disappearing into the pore structure of the wood. If the adhesive penetrates through the wood piece and out the other side, the effect is called *bleed-through* (Figure 3–4b).

In these examples flow and penetration (and undoubtedly also wetting) have been maximized, and links 4, 5, 6, and 7 have been well established. However, the end result of either squeeze-out or bleed-through may be that there is insufficient adhesive left on the glue line to form link 1. This condition, which is often observable with the unaided eye, is appropriately called *starved* (Figure 3–5a). A frequent characteristic of the starved condition, particularly in hot-press work, is the void structure of the wood surface remaining unfilled. In some cases, the glue pattern as applied will have failed to merge even though pressure and flow appear to have been favorable. The question then reduces to whether enough adhesive was ap-

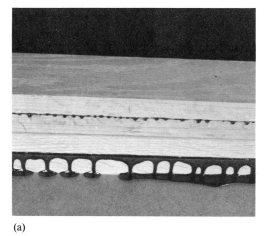

(a)

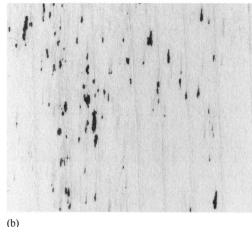

(b)

Figure 3–4   Evidence of excessive mobility: (a) squeeze-out in the lower glue line of a lumber assembly (the upper glue line shows a well-formed bead of glue); (b) bleed-through in plywood.

| Starved | (b) | Bonded |

| (c) | Unanchored | (d) | Prehardened |

Figure 3-5   Four consequences of the magnitude of adhesive motions on the glue line.

plied or whether it disappeared into the wood because of too much mobility. Assuring in advance that enough was applied allows the answer to be searched in its disappearance.

At the opposite extreme, if the adhesive is dried or partially cured when pressure is applied, no flow will occur. If flow cannot occur because of insufficient mobility, neither can transfer, penetration, or wetting. This condition is called by various descriptive names: *dried, precured,* or *prehardened.* The latter is preferred because it is not specific as to the mechanism involved. This condition is also easily observed with the unaided eye, the most revealing characteristic being that

the glue line looks much as it did when it was applied, since no movement has occurred. In this case link 1 has solidified, but none of the other links have been formed (Figure 3-5d).

Between these two extremes of too much and too little motion lies the optimum amount of motion for each action, and this produces the strongest bond when the adhesive fully hardens. The result of optimum motion is not observable directly, because any attempt to pry into this well-made glue line always produces a wood surface. The condition is therefore called simply *bonded* (Figure 3-5b).

Another condition that can be observed is

caused by a rather specific amount of motion occurring on the glue line. In this condition, which falls between bonded and prehardened, there is sufficient mobility for flow, some transfer, and penetration, but not enough for wetting. In this case one usually sees a replica of one surface or the other impressed in the glue line. The replica will include not only the irregularities of the surface, but sometimes even the pore and pit structure of the wood (Figure 3–5c).

When wetting does not occur, links 4 and 5 do not form. As mentioned previously, this results in a joint that may have deceptive initial strength, but that will fail at a later time. The bond is primarily of the interlocking mechanical type, and it can loosen over time as a result of cyclic environmental conditions. Because the bond lacks links 4 and 5, the condition is called *unanchored* (Figure 3–5c).

The condition of no wetting, it must be noted, is due not only to mobility failure, but also to what is termed *surface inactivation*. This is a surface property of wood in which it acts like a nonpolar material, and is otherwise unreceptive to the adhesive. This lack of receptivity has many origins, some between species, some within species, some within the same piece, and some contributed while the wood is being processed. This subject is covered in greater detail in Chapters 5 and 6.

Although the mobility changes of present interest occur in the glue line, it is important to remember that a base or inherent mobility is a property of the adhesive. Unfortunately, the property is not constant. It often changes (some more than others) from the moment it is "born" in a vat to its introduction into a joint. Sometimes the change is beneficial, though usually it is not.

Four time periods in the life of an adhesive have a bearing on its mobility:

1. The period from the date of manufacture to the date of use, called the *shelf life* or *storage life*. This period is always stated on the package along with conditions of humidity and temperature that affect it.
2. The period between mixing and application to the surface, called the *mix time, mix age, working life,* or *pot life*. This period applies only to adhesives that have to be mixed or otherwise prepared by the user.
3. The period between application of adhesive to the surface and application of pressure to close the joint. This period is divided into two important parts: *Open assembly time* is the time between applying the adhesive and placing the two surfaces of the joint together; *closed assembly time* is the time between placing the glue-spread surfaces together and the application of pressure.
4. The period during which the glue joint is under pressure, called the *press time*.

RHEOLOGICAL EFFECTS OF ADHESIVES ON BOND FORMATION

The motions and deformations of adhesives in forming and performing in a bond depend upon their *rheological* properties, that is, how the materials move or deform in response to force. Since these properties change as solidification takes place during the course of bond formation, all of the factors that affect these changes also affect the actions or motions the adhesive must undergo on the glue line. Five distinct areas of concern, involving behavioral properties of adhesives can be identified. The terminology used is intended to be more descriptive and helpful in learning rather than rigorously scientific notation. The terms also take on a semblance of sequence in bond formation and bond performance. The effects of interest are due to:

1. Liquid properties
2. Mechanism of solidification
3. Speed of solidification
4. Degree of solidification
5. Integrity of solidification

The first three of these properties affect bond formation and interact with all the factors affecting mobility on the glue line. The last two properties relate to bond performance and interact with factors of stress and exposure.

These rheological properties are reflected to some degree in all adhesives. They can be used to explain most of the motions to which adhesive bonds are subject. More important, they provide the connection for understanding the effect of adhesive functions on wood and process factors, and conversely, the effects of wood and process factors on adhesive functions. Rheological effects related to bond formation and the physical nature of the reactions are discussed in this section. Bond performance consequences of rheological properties are more related to the composition of the adhesive mixture and are discussed in Chapter 4.

Effects Due to Liquid Properties

Two properties of liquids are of special significance to bond formation: *fluidity,* the root of the coarser motions of the adhesive on the glue line; and *adhesion potential,* the basic driving force in creating the adhesive/wood interfacial bond. Fluidity has a direct bearing on how well a glue conforms to the surface (i.e., flow and penetration), achieves proximity, and responds to clamp pressure. The proximity of interest is measured in atomic diameters, the distances within which attractive forces operate between molecules. By being able to move freely and somewhat independently, the molecules of a liquid are able to come atomically close to the molecules of a solid. This is one of the crucial functions of the liquid state simply defined. From this point adhesion actions can take place. Adhesion (i.e., transfer and wetting), although born of the chemical constituents of the adhesive, responds mostly in a physical manner on the glue line. The pH of adhesives transcends all rheological properties in affecting the mechanisms of solidification. Although it too is born of chemical constituents, its impact begins very early in the process, and is briefly discussed in this section also.

Fluidity.   Despite its common use, *fluidity* is an ambiguous word. It connotes the ability to flow but is usually measured as a resistance to flow, or a resistance to shear. Although fluidity is simple to conceptualize, its scientific basis is complex, even in pure liquids. The phenomenon is a consequence of the comparatively loose attraction of molecules for each other in a liquid. In the case of liquids like adhesives, which are mixtures of solvents, solidifying molecules and particles, catalysts, fillers, extenders, and fortifiers, all either dissolved, suspended, or emulsified, the net attraction as measured requires considerable interpretation.

Liquid properties are achieved and modified in many ways. Some materials are solids at room temperature but melt at elevated temperatures. Certain solid synthetic materials, for instance, melt at temperatures of 300° F or more. These materials can be used to make fast-setting adhesives.

Another means of producing a liquid adhesive is to dissolve the adhesive material in a solvent. For example, a solution of rubber dissolved in benzene, or cellulose acetate dissolved in acetone, has adhesive properties. Synthetic resins of low molecular weight or natural polymers that can be reduced to low molecular weight are dissolved in water.

Molecular materials that form strong films but do not dissolve readily are made into liquids by emulsifying them so they can be suspended in a liquid. Rubber, for example, is made into a water-base adhesive by adding emulsifiers that keep it in suspension.

Finally, some polymeric systems are liquid even without being dissolved in a solvent. These are referred to as 100% solids adhesives. They occupy a special niche in adhesive technology because they provide a high level of adhesion and cohesion, and they do not shrink appreciably upon hardening. They are therefore good gap fillers.

Hence some sense of liquids, pertinent to

their function as adhesives, is in order. Fluidity characteristics are generally described in terms of viscosity, a scientifically derived and measured quality. The unit of measurement is the *poise,* usually expressed in centipoises. Although measured in various ways, the poise is a function of the internal friction, or resistance to shear of the liquid. It is precisely described as the force necessary to shear one surface of a 1-cm cube of liquid through a displacement of 1 cm in 1 s.

Water has a viscosity of 1.002 cp at a temperature of 20°C. Most adhesives have viscosities ranging from 4 cp up to 75,000 cp for some roller-applied adhesives. Adhesives such as mastics, which require troweling or extrusion from a caulking gun, may be much higher in viscosity, and are actually more semisolid than liquid.

Liquids differ in response to rate of shear. Some appear to have a higher viscosity as the rate of shear (e.g., stirring) increases. This is known as *dilatancy.* A good example is a mixture of corn starch in water. At rest, or when stirred slowly, it appears to be very fluid. When stirred fast, it appears to have a high viscosity. Some of the newer latex paints, on the other hand, appear to be thick and jellylike at rest or when stirred slowly. At high stirring speed, however, the viscosity appears to be much lower, an advantage in brushing since paint flows easily while being stroked, but will not drip afterward. This behavior is called *thixotropy.* There are also liquids, water being one, that show no change in viscosity with changes in rate of shear. These latter liquids are known as Newtonian liquids; the others are called non-Newtonian liquids (see Figure 3–6).

Most wood adhesives are thixotropic, which means that they have no fixed viscosity—their viscosity can be defined only in

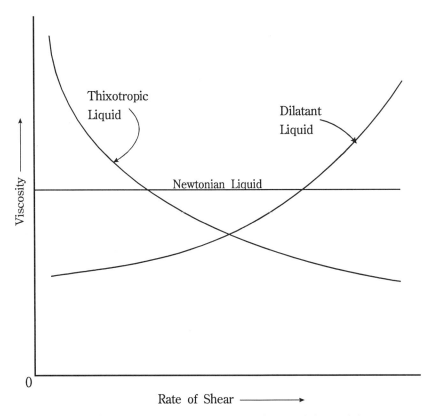

Figure 3–6   Viscosity behavior of liquids with changes in rate of shear (stirring).

terms of a series of measurements at different speeds. For this reason, any one reading is more correctly termed "apparent viscosity" or, to avoid the issue altogether, "consistency." The word "fluidity" connotes motion and therefore is more descriptive of the adhesive's role in bond formation. In any case, it is accepted practice to qualify an adhesive viscosity statistic with both the method used to measure it and the speed at which the reading was made. Because temperature has a profound effect on viscosity, this factor must also be recorded with the reading.

Despite the ambiguities inherent in viscosity measurements of wood adhesives, they provide some of the most important information in the development, production, and use of adhesives. For example, they indicate the course of polymerization in producing resin adhesives; the stability of adhesives against time, temperature, and humidity (e.g., storage life and mix life); the effects of solvents, fillers, and additives; and the potential handling properties for different applications. In general, it can be said that viscosity is used for monitoring the physical and chemical status of an adhesive from its origin to its application on a wood surface. After the adhesive meets wood, the viscosity effects are not readily quantifiable and hence the need for the observation system based on mobility discussed in the previous section.

It is precisely at this crucial point, when the adhesive engages wood and is about to perform its all-important function of creating a bond, that the process goes blind. On the glue line, viscosity, and therefore mobility, changes rapidly, and adhesive functions that depend on a certain degree of mobility may be inhibited or bypassed. Unfortunately, since measurements cannot be made during the delicate hardening stage to determine how well the adhesive is doing, all determinations must be made through hindsight and postmortems. Liquid-stage functions are therefore traditionally assessed through solid-state conditions. This is normally done by means of strength and durability tests and judgments are made with respect to expectations or specifications. The observation system described earlier adds corroborative detail to the strength tests, and may even make a strength reading unnecessary for on-the-spot quality control.

Viscosity measurements reflect the fluidity of the adhesive as a uniform liquid mass. On the glue line, although some of the motions were described earlier as being mass actions, this is true mostly in a gross sense. Actually, more useful deductions can be made if every ingredient in the adhesive mix is assumed to have its own independent mobility as well as a combined mobility. Certainly the solvent can be expected to move by diffusion or evaporation, as well as by flow. Other ingredients, especially those in solution, can behave independently in contact with wood. Particulate material in suspension or in emulsion will have the least mobility, being roughly related to molecular or particulate size, or configuration. The differential behavior of adhesive ingredients remains one of the major unknowns in bond formation with wood.

Adhesion Potential.   By far the most important factor in forming strong bonds is the achievement of a high level of adhesion between adhesive molecules and wood molecules (links 4 and 5). Without adhesion, the bond is reduced to whatever mechanical interlocking can take place as a result of adhesive flowing into the larger interstices at the surface. With adhesion, bond formation can take advantage of some of the same forces that hold all materials together, forces that produce cohesion. Adhesion may be defined as the attraction between one material and another. Cohesion, on the other hand, is the attraction that exists within a material. In other words, cohesion is a force that is expressed internally, while adhesion is a force that is expressed externally.

Adhesives have the ability to achieve both adhesion, and cohesion. These forces are of various kinds but of the same source, all arising within atoms and their organization into molecules. Since adhesion and cohesion forces arise from the same source, it is logical to assume that they have the potential of being equally strong, a reassuring thought from the adhesive standpoint.

Bond forces bear names which relate to their source or discoverer, e.g., ionic, covalent, metallic, dipole, hydrogen, induced, London dispersion forces. They generate through spatial distribution of electrons and atomic nucleii in a system, and vary in strength, and the distances over which they are effective. Only one of these, the covalent bond, is the result of a chemical reaction between adhesive molecules and substrate molecules, and is considered to be the strongest and most durable (and, unfortunately, the least likely to occur with most adhesives). The rest are physical in nature, and merely attract—but they are so numerous, their combined strength is more than adequate for wood bonding.

All these adhesion and cohesion forces have one characteristic in common: They function only over very short distances, measured in angstroms, the distances existing between atoms and molecules in a solid or liquid. Obtaining angstrom distances between adhesive molecules and substrate molecules is the crowning achievement in bond formation. Factors in both the adhesive and in the wood enter strongly into achieving this goal. Some notion of how these forces operate at surfaces is useful.

Atoms at surfaces behave differently than atoms in the bulk of a material. One reason for this is that atoms in the bulk of a material are entirely surrounded by other atoms, all exerting attractive forces on each other in every direction, whereas atoms at the surface are attracted only in an inward direction. Surface atoms therefore have unused attractive forces on one side. These unused forces lead to the existence of "surface free energy," which ultimately plays a role in bond formation.

In a liquid, the inward attraction is the force that causes drops to form into spheres. A sphere represents the minimum surface area for a given mass, and is therefore preferred because of the consequences of the universal law of conservation of energy—in this case, the less surface, the less surface energy. Because the action of reducing surface area is one of tensile forces, the force is called *surface tension*. Surface tension in liquids is measured as a force per unit of length under tension. This is equivalent numerically to energy per unit of area, thus making surface energy also a measureable term.

Inward forces also exist in solids, but these are not strong enough to deform the solid. Surface tension of solids cannot be measured directly, but it can be assessed indirectly.

In both liquids and solids, the outward-directed attractive forces at surfaces are in limbo, as it were, and are ready to attract a passing atom. It is because of this very attractiveness, a situation greatly to be desired, that surfaces become contaminated before they can freely participate in adhesive bond formation.

In order to make adhesive bonds, the attraction of solid molecules for liquid molecules must be greater than the attraction of liquid molecules for liquid molecules. Otherwise, the liquid molecules would associate with their own molecules rather than with solid molecules. Under conditions most favorable to adhesion, a liquid placed on a clean surface would be drawn to a mono-molecular (one-molecule-thick) film over the surface. In this case the liquid spreads spontaneously over the surface. Under conditions totally unfavorable to adhesion, a liquid placed on a solid surface draws itself into little spheres, remaining as isolated as possible. Most adhesives behave in a manner between these extremes but closer to the spontaneous spreading situation.

When a drop of liquid is placed on a solid, its surface energy finds complementary energy fields on the surface of the solid. To the extent that they complement each other they reduce the total amount of surface energy (again, conservation of energy). This then becomes the preferred state, and the two materials remain in some attached relationship, bonded to the degree that energy has been conserved. The energy that becomes liberated as the result of this new association can be evidenced as heat, a "heat of wetting." It is readily sensed in the case of water being adsorbed on wood. If a quantity of dry sawdust is placed in a plastic bag and water is added, the heat liberated as the water bonds (wets), the wood can be felt by hand. In order to remove the water from the surface of the wood, energy in the form of heat must be reintroduced. This is why heat is used to accelerate drying when seasoning wood.

All natural systems are motivated toward the lowest possible state of energy they can achieve. Many chemical reactions and crystallization processes are driven by this universal compulsion to achieve the lowest possible energy state. It has been said that adhesion therefore is as guaranteed as gravity. The task for adhesives is achieving adhesion to the right molecules, those attached directly to the substrate, and not to other mobile molecules that may have chanced to be attracted first.

Liquids express their attraction for a solid in a most definitive way. As indicated above, a drop of liquid placed on a surface will either be drawn to a thin film, remain in drop form, or assume an intermediate shape. The possibilities are shown in Figure 3–7, which is a schematic representation of a drop of a liquid on a solid surface. The respective surface tension forces are shown as vectors (arrows) emanating from a point at the edge of the drop where spreading motivation is expressed. Three vectors are shown, one for the solid, one for the liquid, and one for the interfacial region where the liquid and the solid have interacted. The vector for the liquid is drawn as a tangent to the curve of the drop at the point where it meets the solid. At equilibrium, that is, when spreading movement of the drop has stabilized, some equality of forces can be assumed. This equality is rationalized by Young's equation, which transforms the forces into one plane and puts them in opposition at the edge of the drop as if a tug of war were going on between molecules in the liquid and molecules in the solid surface:

$$F_S = F_{SL} + F_L \cos \theta$$

where F_S is the surface tension of the solid; F_{SL} is the surface tension of the interfacial region; F_L is the surface tension of the liquid; and θ is the angle between the tangent and the solid surface, measured through the liquid.

The idealized situation on which the equation is based calls for a pure liquid, a smooth solid, and no physical or chemical effects of the liquid on the solid. None of this is true for wood adhesives on wood substrates. The liquid is not pure, the surface is not smooth, and there is an effect of the liquid on the solid. Nevertheless, and although the equation implies that the situation can be reduced to quantitatively manageable numbers, useful information can be derived simply by observing the angle θ. Several quick deductions can be drawn from such an observation:

1. When the angle is zero, affinity is high, the liquid will spread spontaneously over the solid, and maximum adhesion and wetting will ensue.
2. When the angle is greater than 90°, there is relatively little affinity, the liquid will

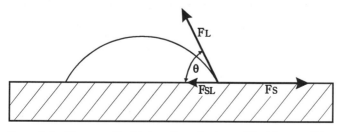

Figure 3–7   The generalized interaction of a drop of liquid on a solid surface.

not spread over the solid, and very little adhesion or wetting will result.

3. When the angle is between zero and 90°, affinity is roughly inversely proportional to the size of the angle, as is the amount of spreading, adhesion, and wetting that will occur.

The above interactions between a liquid and a solid are depicted as those of a drop on a surface. It should be emphasized that the same interactions occur if the liquid is deposited as a film, a foam, or a mist, though perhaps not as visible. It should also be noted, with even greater emphasis, that this interaction determines to what degree the adhesive is drawn to and into a porous surface such as wood. When the liquid carrier is preferentially drawn in, as often happens, it has a strong bearing on the permissible assembly time, and on the execution of other adhesive actions in bond formation.

The viscosity of the adhesive also influences the extent to which adhesive is drawn over or into the surface, and the time required for this to happen. The higher the viscosity, the less a liquid is inclined to move, and the longer it will take to move. Adhesives normally are formulated to have rather high viscosities in order to keep them on the glue line to form link 1. When pressure is applied to a mating joint, one of its purposes is to overcome the limiting effects of this high viscosity and help drive the adhesive to its task of achieving proximity with the wood. Viscosity also increases during the assembly time, making the wetting action even more dependent on pressure.

It was mentioned above that one of the reasons Young's equation does not hold strictly for wood is that the liquid affects the surface. Water, a component of many glues, swells wood. This is a beneficial action in bonding because it opens up the wood structure and allows adhesive molecules greater access to the minute interstices of wood. But water or other solvent component of adhesives also affect wood surfaces in another important way: They dissolve or otherwise remove some of the contaminants that may interpose themselves between the adhesive per se, and the solid wood substance. The ability of some adhesives to deal with contaminants in this way sometimes accounts for their better performance in bond formation. It is generally recognized that alkaline adhesives function better in both opening wood structure and removing contaminants than do low pH adhesives. Solvent-based adhesives may have a similar effect.

pH. Every adhesive has a pH that is generally specific to its type, or is varied to fit a particular need.[1] It governs not only its original developmental mechanisms, but also its liquid characteristics, its solubility, its hardening characteristics, and to a certain extent its bond performance characteristics. The regulation of pH during manufacture of the adhesive is one of the important functions of quality control, for it dictates not only speed and degree of reactions but also the configuration of molecules. This dictation extends into shelf life where a different pH may be necessary to maintain properties for a sufficient period of time, and on into bond formation where a different pH may be needed for hardening. Often the purpose of the catalyst is to change the pH and allow hardening to proceed. On the glue line itself, where the pH of the thin film may be swamped by the pH characteristics of the wood, it may be necessary to add a buffering agent to protect and maintain the adhesive pH.

The glue user has little discretion with regard to pH other than to add catalysts, and this is usually done by prescription of the adhesive manufacturer. However, it is well to understand the interactions, and to be aware of sources of variability that may affect results. For example, glues that cure on the alkaline side may not do well on woods of low pH such as oaks. Also, woods that are high

---

1. pH is a measure of the acidity or alkalinity of a substance and is measured on a scale of 1 to 14. A pH of 7 is neutral; pure distilled water has a pH of 7. Most substances, however, are either acidic (pH less than 7) or alkaline (pH greater than 7).

in extractives tend to bond better with highly alkaline adhesives. For this reason, and because alkaline liquids tend to swell and open up the cell wall structure, glues with high pH have a better chance of gaining a foothold in the wood substance.

After the bond has been formed, pH has further influences, especially at extremes, and mostly detrimental. At high pH, the attractiveness for moisture increases. Hence for products where moisture absorption is a criterion, such as comminuted materials, specifications are more difficult to meet with high-pH glues. At low pH, acid attack on the wood may occur, destroying the bond on the wood side of the glue line. Some glues, e.g., cold-setting ureas, can sometimes destroy themselves with their own residual low pH due to hydrolysis promoted by acid conditions.

Effects Due to the Mechanism of Solidification

There are, as mentioned earlier, four mechanisms by which adhesives convert from the liquid to the solid state. This action ostensibly merely confers strength to the bond. However, because the solidifying action also controls the fluid functions that actually create the bond, an important interaction exists. Although there are many variables, all the fluid functions can be reasoned by applying the mobility concept to the mechanisms of solidification. At this point, only the barest essentials are needed to sense the evolving concept.

Adhesives that solidify by heat loss are called *hot melts*. These adhesives become fluid when heated and resolidify when cooled. There are no solvents in them to affect mobility. Control of temperature and heat loss therefore are the main determinants of quality of bonding. The viscosity and hence the mobility is determined by the temperature to which they are heated: The higher the temperature, the greater will be the mobility. Maintaining the optimum temperature throughout the bond-forming stages is important. This means that during the interval between placing the adhesive on the wood surface and the application of pressure to close the joint, the adhesive must remain mobile. Any cooling will inhibit wetting, penetration, and transfer, but probably not flow.

The problem is that these adhesives cool very quickly. The resulting rapid solidification reduces the window of time during which adhesive motions can take place. For this reason, heat-loss adhesives are most likely to suffer from insufficient mobility during bond formation unless measures are taken to prolong the liquid state. The hydrocarbon nature of heat-loss systems produces materials that tend to be somewhat less attracted to the hydrophilic surfaces of wood. Accordingly, the bond depends more on mechanical interlocking into the pore structure of wood than on adhesion. Moreover, because of rapid cooling at the interface, penetration may be inhibited and a shallow bond may result (links 6 and 7 inadequately formed).

Since these motions occur after pressure is applied, the proper temperature must be maintained for as long as possible before cooling is allowed to consolidate the joint. This obvious requirement is almost universally compromised in gluing operations using hot melts, because quality of bonding depends on prolonging high temperature, and rate of production depends on how fast the glue line can be cooled. Hence there are conflicting imperatives in the use of these glues, and some adjustments between quality of bond and speed of operation is necessary.

Adhesives that solidify by loss of solvent or liquid carrier are solutions or dispersions of molecular or particulate material which produce a strong solid upon removal of the solvent. Removal of solvent is accomplished by evaporation, diffusion, or capillarity, sometimes accelerated with heat. Mobility is totally dependent on the solvent. Hence programming and control of solvent loss is the main determinant in bond formation. With some adhesives on impervious materials, most of the solvent has to be removed before closing the joint. In this case the adhesive is applied to both surfaces, so that transfer, penetration, and wetting can occur with solvent help. Flow, and the bonding of adhesive to adhesive, then occurs under pressure. When water is the solvent, wood is expected to absorb it and solidify the bond.

The use of solvent or liquid carrier rather than heat to liquefy provides two added advantages to bond formation: (1) Polar (but not nonpolar) solvents help open up the wood structure through swelling action, thus achieving better anchorage; and (2) the solvent helps carry the adhesive material deeper into the wood for building links 6 and 7—i.e., provides more mobility.

Adhesives that solidify by chemical reaction typically contain synthetic resins dissolved or dispersed in a solvent, usually water. Two solidifying processes exist: polymerization and cross-linking. In the former, the resin molecules are relatively small, so they are soluble, fluid, and mobile. They solidify by growing to large molecules which are insoluble and locked together chemically. In the latter, the molecules begin large and are made larger by joining them with chemical links. The action is "triggered" at the proper time by the addition of catalytic agents and/or heat. When heat is used to cause solidification, it may be considered as the opposite to loss-of-heat solidification. Therefore it is logical that a different set of effects applies. Additionally, when both catalytic agents and heat are used, the effects tend to compound and become more complex.

On the glue line, while these reaction adhesives are trying to form a bond by flowing, transfering, penetrating, and wetting, two other actions (associated with hardening) are also going on independently, both of which are inimical to the finer bond-forming actions. These two independent actions are loss of the liquid carrier and growth of the resin molecule. Both reduce mobility, and in some circumstances, the loss of liquid can also affect the polymerization process. Although ultimately it is necessary to lose all liquid elements during polymerization, the finest degree of molecular mobility is required. Too

rapid loss of liquid (the reaction medium) thwarts to some extent the ability of molecules to gain the proper relationship with each other and to execute their own cohesion-generating activity. Hence considerable interdependence exists between solvent loss and polymerization, which is often not sufficiently accounted for but which requires close systems control in order to optimize all actions.

When heat is used to accelerate hardening, the effect of lowered viscosity (and therefore increased mobility), experienced by all liquids, must be factored in, keeping in mind that the different ingredients in the adhesive may respond differently to temperature. For example, filler material will not have the same mobility as resin or solvent; it will always have less. (Its purpose is in fact to reduce mobility.)

Combination solidification systems incorporate two of the three hardening mechanisms. In one combination, loss of heat and loss of solvent operate together. One can see that loss of mobility on the glue line might be rather fast in this case, threatening adhesive actions and reducing opportunities to develop all links to the maximum degree. When loss of solvent and chemical reaction combine, the risk to bond formation is somewhat less because the two hardening actions occur partially in sequence. The other possible solidification combination, heat loss and chemical reaction, is difficult to achieve because the two hardening mechanisms are not as compatible. However, it is desirable to have such a system because it would result in a stronger and more durable bond than is now possible with hot melts, and it would retain the speed of the hot melt, its greatest virtue.

The basic consideration in these combined solidifiers is that since there are two solidifying actions at work, their combined actions must be taken into account in understanding the effects on mobility. In general, heat loss results in a gel being formed rapidly, terminating some of the bond-forming actions before solvent (water) loss occurs. This means that all the bond-forming motions must occur quickly, because the solvent-loss phase confers only final strength. In other words, bond formation depends on mobility while the glue is still hot. The same is true of the solvent loss/chemical reaction systems. Bond formation must occur before too much solvent escapes; the chemical reaction creates only the final strength.

Effects Due to the Speed of Solidification

The rate at which adhesives harden not only determines the rate of production, but also controls some mobility effects, particularly their magnitude. Most of the problems in the use of adhesives that harden by heat loss, for example, are due to hardening so fast that some adhesive actions cannot take place, or are curtailed.

Adhesives harden over a very wide range of speeds, from seconds to minutes, hours, days, weeks, and months, depending on their composition. Heat-loss adhesives harden in seconds to a gel state sufficient to hold parts together at low loads. Some chemical reaction adhesives also harden in seconds at room temperature when properly catalyzed. Others require hours to harden at room temperature but only seconds when heated. Still others do not harden at all at room temperature but harden in minutes at elevated temperature. Adhesives exist that form a holding bond on contact and increase in strength over time.

Speed of hardening, being primarily a matter of composition, is a built-in characteristic of the adhesive. However, it is modified at point of use, intentionally or unintentionally, by (1) the use of additives, such as catalysts and solvents; (2) the condition of the wood, its temperature, its moisture content, the roughness of its surface, and state of inactivation; (3) the species of wood, its age, and its thickness; (4) the age of the adhesive (from time of manufacture and from time of mixing); (5) the amount applied to the surface; and (6) the temperature of the glue line, and the rate of temperature increase.

The effects of these factors in changing the rate of hardening, of course depend on the mechanism of hardening. For example, using a heat-loss adhesive on cold lumber may accelerate hardening to the joint where the necessary motions are inhibited or extinguished altogether, placing links 6 and 7 at risk and resulting in a shallow bond at best. Contrarily, an adhesive that hardens by chemical reaction may have its hardening slowed on cold lumber, demanding, at best, longer time under pressure, and at worst, curtailing solidification to an unacceptable degree. With some glues, the inadequate solidification may never be recouped by subsequent heating, especially if solvent is necessary to carry out the chemical hardening reaction. The latter is particularly insidious because the bond will have deceptive strength, and fail at a later time.

It is important to distinguish between the *rate* of sorption of water and the *capacity* for sorption. Thin wood reduces primarily the capacity for sorption and to a lesser degree the rate (as it approaches saturation), whereas wood of high moisture content reduces both the rate and the capacity. Heartwood, summerwood, and high-density wood reduce the rate of sorption but not the capacity. When capacity is the limiting factor, the adhesive will not harden completely. On the other hand, if sorption is merely slow, total hardening will only be delayed. The amount and kind of wood behind the glue line has a bearing on the rate of hardening of solvent-loss adhesives.

Adhesives that harden by chemical reaction owe their speed of solidification primarily to catalysts or heat or both. The addition of catalysts may increase or decrease the speed, whereas increasing the temperature always increases the speed of solidification of a chemically hardening adhesive.

It was mentioned earlier that increased speed of hardening reduces mobility-dependent adhesive actions because of increasing viscosity and reduced time for actions to occur. In the case of heat-hardening adhesives, there is a countereffect due to the lowering of viscosity with increasing temperature. Thus there are competing fluidity effects, which sometimes produce beneficial results, sometimes not, depending on the nature of the adhesive and the precise circumstances. If, for example, hardening has proceeded so fast and so far that motions are being curtailed, the viscosity decrease that comes with temperature increase could save the situation. On the other hand, if hardening is slow in developing, the viscosity decrease presents double jeopardy, producing a temporary high-mobility phase that could result in overpenetration and loss of link 1. The temperature effects, it should be noted, occur not only on the glue line but also in storage, in the hopper, and on the spreading apparatus. This can result in some mobility loss before the adhesive is applied to the wood. This mobility loss, coupled with long assembly time and overly dry wood, can lead to poor bond formation.

A confounding effect also occurs with these types of adhesives. Since they also contain solvents, primarily water, a solvent loss effect must be considered as affecting not only mobility but also the degree of hardening, as mentioned earlier. Solvent is not only a carrier of the adhesive, but also a reaction medium in which the solidification reaction occurs. With premature loss of solvent, some adhesives experience insufficient polymerization, the process by which small soluble molecules become the large, insoluble ones necessary for strength and durability of a joint. An extreme example illustrates the situation. Powdered adhesives are sometimes prepared from their liquid state by flash-drying, a process of very fast heating to evaporate the solvent, water. Little, if any, polymerization occurs during this drastic heating procedure; the solvent can be reintroduced with virtually no change in original viscosity. The activities in a glue line are not as drastic, but since the situation deals with molecular motions, some inhibition, registering perhaps obscurely at links 2 and 3, seems within the realm of possibility. One notes that molding powder resins function in the absence of moisture, or else problems with steam pockets would occur; whereas adhesives used in powder or film form re-

quire moisture to be provided from the wood to produce the proper mobility for bond formation.

As a general rule, glues that set too fast risk insufficient penetration and wetting; and glues that set too slow will achieve good wetting but risk overpenetration.

Effects Due to the Degree of Solidification

Like rate of solidification, degree of solidification or hardness is also basically a characteristic of the composition of the adhesive, and since it reflects more on bond performance than on bond formation, it is discussed at greater length in Chapter 4. However, it should be noted again that while degree of solidification does not influence bond formation, the inverse is true: Factors that operate during bond formation influence the degree of solidification such as undercure. Some adhesives never really harden, such as mastics and rubber-based adhesives, which always remain deformable under stress. At the opposite extreme are adhesives that reach a glass-hard state when fully cured. In between there is a full range of hardnesses.

One of the great perplexities of wood adhesive joints is that performance is not related to degree of hardness. We instinctively accept that the harder an adhesive becomes, the stronger it is, and the better a joint will be. However, just the opposite may be true in some cases.

Adhesives that are less deformable than the wood tend under load to concentrate stresses at the boundary of the glue line. Such adhesives can be expected to yield joint strengths at or somewhat under the strength of solid wood. It is a strange fact that adhesives that are more deformable than wood (and, by intuition, less strong) can produce a joint that is or appears to be stronger than solid wood. What is happening in this case is that, under load, the more deformable glue line is distributing stress more uniformly over the joint areas (not concentrating it) and is thus able to carry more load. Unfortunately, the latter adhesive could fail under continuous loading at a value considerably lower than that shown by a relatively quick test in a laboratory. The ability to tailor an adhesive to match the properties of the wood exactly remains a challenge for polymer chemists.

In fact it can be shown that in some types of joints, the hardest adhesives reach maximum strengths before they are fully cured, after which strength tends to lessen. One reason for this is that wood itself is an elastoplastic material that has certain rheological (deformation) properties. Generally, when the rheological properties of the hardened adhesive exactly match those of the wood, maximum strength will be achieved. Paradoxically, many of the glassy-hard adhesives cure to maximum strength before they cure to maximum durability.

Effects Due to the Integrity of Solidification

The ability to maintain bond strength (durability) against environmental factors is the most distinct and most important difference among glues. While durability is essentially a quality derived primarily from the composition of the adhesive, it is an inexorable fact that it is also greatly dependent on factors that play during bond formation. A poorly made bond will certainly perform poorly no matter how much durability is built into the adhesive chemically. All the factors that influence the creation of each link in a bond have a bearing on the performance of the adhesive.

Many factors are involved in bond deterioration. They include all those itemized in the equation of performance, some of which can be decisive even though they are only poorly understood. Arising in the wood to which the adhesive is attached, such matters as grain direction on either side of the glue line, wood density, thickness, pH, and forced curvature all have a bearing on the durability of glue lines that are otherwise equal in quality.

The durability of an adhesive is therefore not a fixed quantity. Every specimen that might be used to measure resistance to deterioration is itself a specific construction with its own set of wood factors, geometry factors, and bond-forming factors. Information obtained with a test specimen therefore cannot be easily translated to other types of structures. However, by standardizing testing procedures, it is possible to obtain some sense of the inherent durability characteristics of different glues. This subject is discussed in Chapter 11.

In summary, adhesives exhibit motions on the glue line of (1) flow, (2) transfer, (3) penetration, (4) wetting, and (5) solidification. These motions have *sequence* because of differing sensitivities to changing levels of *mobility,* which occur as the adhesive hardens on the glue line. Since the adhesive hardens over a definite span of time, the actions that depend on different degrees of fluidity can be associated with specific time periods.

The motions also have specific locations where their actions take place:

- *Flow* occurs throughout the bond line and in the plane of the surface.
- *Transfer* occurs on the unspread surface of the joint.
- *Penetration* occurs in the wood pore and capillary structure adjacent to the bond line (links 6 and 7).
- *Wetting* occurs at the wood surface (links 4 and 5).
- *Solidification* occurs throughout the adhesive layer.

Motions have *magnitude*, which produces *consequences*. These are attributable to the composition of the adhesive and the particular conditions of use. The consequences can be serialized on the basis of magnitude of motions or degree of mobility:

- *Starved*—too much mobility
- *Bonded*—optimum mobility
- *Unanchored*—limited mobility
- *Prehardened*—no mobility

These consequences, since they are fairly observable to the unaided or slightly aided eye, form the basis for monitoring and controlling the performance of all glues in forming a bond. They are important to much of the discussion that follows.

Adhesives owe their bond-forming mobility and much of their bond performance qualities to properties that relate to or affect rheological behavior. From an adhesive perspective these include:

1. Liquid properties
2. Mechanism of solidification
3. Speed of solidification
4. Degree of solidification
5. Integrity of solidification

4

Characteristics and Composition of Adhesives

In this chapter, interest centers on what it is about adhesives that make them stick and form strong bonds. Attention is drawn primarily to link 1, the adhesive film, but also with its interfaces with wood, links 4 and 5. The other four links to be assured, 2 and 3, the boundary within the adhesive film, and 6 and 7, the subsurface region of the wood, also have dependencies in the properties of the adhesive, and need correlation.

Recalling that glues are basically liquids that harden, they promise only two things: (1) that they will adhere, i.e., "stick," to the wood, and (2) that they will solidify to form a strong material. Both of these promises are made by the glue manufacturer. However, since there are many operational factors interacting with the glue, glue users can often help themselves by understanding glue behavior that derives from its composition.

"Sticking" is an easily accepted phenomenon, but one that is rather difficult to understand because its basis is in the atomic structure of the material. However, it is possible to develop a sense of the process without delving too far into the arcana of physics and chemistry. Similarly, hardening also depends on complex physical and chemical processes but can be sufficiently understood without getting too deep in technicalities.

In the context of bond formation and bond performance, adhesives display both liquid and solid properties. The properties of liquids are relatively easy to observe and conceptualize, while those of solidified adhesives, on the other hand, may be easy to observe but difficult to understand. This is because a solidified adhesive is part of a composite structure, and a decidedly minor part of it volumetrically, and therefore its properties are blended inexorably with those of the substrate.

It is more important to know the proper use of glue formulations than to know how they are formulated. Thus, although the basic distinctions between glues are chemical, the most useful information is functional. The importance of chemistry in understanding glues is that it ordains the *maximum* bond performance that can be expected. The importance of knowing how a glue functions is that it determines the actual bond performance that is actually achieved under the various conditions existing in a particular gluing operation.

Glues are chosen primarily for the performance level expected of the bond—strength and durability. A second consideration is the so-called handling properties, that is, how the glue must be prepared for use, applied, and hardened to form a good bond. A third consideration is cost. Cost includes not only the price of the glue per pound, but also the amount needed per unit of glue line area, and the special conditions necessary to harden the glue. Under production conditions it is also prudent to include as costs the value of rejects produced as a result of varying use conditions.

As a concession to brevity and the interests of most adhesive users, only the more common wood adhesive systems will be described here, and mainly to the extent that the description pertains to the general utility of the glue. In other words, many technical details have been omitted in deference to the practical use objectives of this book.

This chapter will conclude, by way of summary, with an intuitive rating of the different glue types, based first on their bond performance promise, and then on their potential for dysfunction in bond formation. A selection chart based primarily on manufacturers recommendations provides the bottom line for decisions and at the same time summarizing the descriptions of the various glues.

The organization of the following discussion is based on division of adhesive systems into groups or families that have similar hardening characteristics, because this also organizes them on the basis of similar uses. The following organization emerges:

Partial hardening:
  Mastics
Heat loss:
  Hot melts
Solvent or water loss:
  Polyvinylacetate, animal proteins, polysaccharides, rubbers
Chemical reaction:
  Addition polymerization:
    Epoxies, isocyanates
  Condensation polymerization:
    Phenol-formaldehydes, resorcinol-for-

maldehydes, urea-formaldehydes, melamine-formaldehydes
Combined hardening systems
  Solvent loss + heat loss:
    Hot animal glue
  Solvent loss + chemical reaction:
    Cross-linked polyvinyl acetate, caseins, soybean adhesives
Combined adhesive systems

It is important to bear in mind that all adhesive molecules experience similar motivating forces, but of varying magnitude in interaction with wood and process factors. Many of the main differences in bond formation arise from molecular size and configuration (i.e., linear, branched, or cross-linked). The basic consideration here is that small molecules with limited branching have a better chance to diffuse into, orient, and gain wetting proximity to wood material than highly branched, large molecules. Moreover, linear molecules respond plastically and generally enjoy a slight edge in bond formation, though the edge is often lost in the bond performance stage, because the same plasticity makes it yield under relatively lower loads.

Considering the transition from a liquid to a solid, the many mechanisms that will be discussed reduce to only two on a basic functional basis: those in which molecular size does not change during the transition, and those in which molecular size does change. In polyvinyl acetates, for example, the molecular structure is fixed from beginning to end of the bonding process. During bond formation the molecular aggregates in these adhesives merely coalesce to form a solid material. In the other kind, phenolics, for example, the molecules are in a continuous state of change, increasing in size until the final cured condition is reached.

The latter kind offers the greatest range of compositional interactions with wood and process. For this reason, and because they form the most durable bonds, we will consider phenolics to be the classic adhesives, and accordingly cover them in greater detail as a means of illustrating the kinds of interactions that can take place. The effects are essentially the same for other resins of similar chemistry.

GENERAL CHARACTERISTICS OF ADHESIVES

Before delving into the chemical composition of various adhesives, it is helpful to sketch in advance the kinds of behavior to be expected from the different systems. Using as a framework the rheological properties discussed in Chapter 3, we will view from a different angle the effects that arise on the glue line due to ingredients in the adhesive.

Hardening Mechanisms

After adhesion has been achieved through liquid properties, solidification takes over to establish the bond between the two surfaces. Several solidifying mechanisms have been mentioned: heat loss, solvent loss, chemical reaction, and combinations of these. Special effects derive from each.

Heat-loss adhesive systems depend on solids that can be melted to a liquid by heat. These are normally thermoplastic resins, which, because of their molecular structure, are hard and strong at ambient temperatures, but become soft and eventually liquid when heated above a critical temperature. Heat-loss adhesives therefore must be applied hot. Their chief advantage is that they cool back to a solid quickly when in contact with wood, providing for relatively fast assembly operations.

No change in molecular size or structure occurs during melting and resolidifying. As a result, a compositional compromise must be struck between a meltable compound for bond formation, which requires rather small molecules, and a strong, heat-stable compound for bond performance, which requires large molecules.

Solvent-loss adhesives are liquefied by dissolving, emulsifying, or suspending rather than by heating. The term "solvent" is here used as a general term to refer to liquids in

which adhesive materials are carried. In industry, a distinction is made between those that actually dissolve the adhesive material and those in which the adhesive material is only suspended. In the latter, the liquid is usually water and they are referred to as water-based or water-borne adhesives. The former are referred to as solvent-based or solvent-borne adhesives, and are formulated with liquids other than water.

As in the heat-loss systems, in solvent-loss systems there is no change in molecular structure during hardening; molecules begin and end at the size and configuration needed for strength as a solid, compromised in this case for the molecular architecture needed for a liquid vehicle. Modern adhesive technology, however, has greatly broadened this narrow concept of solvent-loss adhesives through the use of plasticizers, emulsifiers, cross-linking agents, and more efficient solvents. This makes it possible to engineer adhesives with a range of molecular and particulate sizes to suit particular needs. In all cases, the strength is attained when the solvent leaves the system, and the molecules or particles come close enough together to coalesce.

A disadvantage of solvent-loss systems is that solubility persists; that is, the solid state in the completed bond is reversible (except in cross-linking systems). Virtually all solvent-loss adhesive bonds remain sensitive to heat and will lose strength when subjected to elevated temperatures. Generally, those adhesive bonds that remain sensitive to heat also are sensitive to continuous stress, allowing joints to creep under load.

Adhesives that solidify by heat loss or solvent loss do so by physical changes. Many adhesives solidify by *chemical change,* in which the molecular structure is caused to grow in the glue line from small, soluble, and fusible molecules to large, insoluble molecules. Some types also become infusible, others remain fusible, depending on the nature of the molecule. The process is one of *polymerization,* of which there are two kinds: condensation polymerization and addition polymerization, each producing different effects that will be described as each is introduced.

Chemical reaction adhesives seem to incorporate many of the advantages and exclude many of the disadvantages of the heat-loss and solvent-loss adhesives. They can begin with smaller molecules, thus allowing maximum mobility for bond formation with the solvent carrying the adhesive material onto and into the wood, facilitating bond formation, and then hardening to impart strength and durability. Some polymerizing adhesives, such as epoxies, have no solvent, and depend entirely on molecular mobility to provide adhesive actions.

Combinations of these three adhesive systems also exist. In the case of hot animal glues, cooling results in a semirigid gel with some holding strength. Drying then completes solidification, and produces the ultimate strength. Because these glues remain chemically unchanged in the hardened glue line, their bonds resoften in water and therefore lose strength at elevated moisture content, and fail altogether when immersed in water.

Protein glues of casein or soybean origin combine a chemical change with water loss. In this case, drying provides the first increment of solidification for strength. The chemical reaction then provides durability by reducing solubility with, for example, the formation of calcium salts.

With most adhesive systems, one of the more fractious consequences of solidification is the shrinkage that often accompanies hardening. Because the shrinkage of the adhesive is seldom equal to any movement that may occur in the wood as a result of moisture changes, it is always a source of stress. Add to this the fact that wood movements differ with grain direction under the same moisture change and it is possible that considerable stress can be generated on the glue line in addition to the external loads that might be placed on it. Glues shrink upon hardening for two reasons: loss of solvent, and contraction as a result of crystallization or intermolecular bond formation which draws molecules closer together.

Speed of Solidification

The speed with which glues convert from a liquid to a solid derives partly from the chemical or physical mechanisms by which they do so, and partly from the conditions present in the glue line. Glues that harden by chemical action are normally controlled by catalysts which alter the rate at which molecular interactions occur. The speed of hardening can therefore be controlled by the amount of catalyst added. Speed is further controlled by the temperature of the glue line. The cure time is usually related to pot life, since the catalyst begins its activity at the time of mixing. Moisture in the glue line also has a bearing on speed of cure. Its dilution effect reduces speed of curing, and its heat of vaporization affects temperature rise in hot pressing.

Glues that harden by loss of water depend primarily on the absorbency of the wood to function as a dehydrating medium. Since this varies with moisture content, slower hardening will occur on high-moisture wood, and vice versa on low-moisture wood of the same species. The action reduces to the amount of water that can be drawn from the glue line in a given time. This being so, the amount of water in the glue line also must be considered. Two situations exist; one involves the water content of the glue itself. A glue containing 35 percent water will harden faster than one containing 60 percent water, other factors being equal. The other situation involves the amount of glue applied to the wood surface. A glue applied at a rate of 40 lb per thousand square feet will harden faster than the same glue applied at 60 lb, simply because there is more water to be absorbed. Pressure on the glue line supplies a "squeegee" action with some glues that helps express water.

Glues that harden by cooling depend upon the wood to absorb heat. Just as the moisture content of the wood affects the rate of water absorption, so the heat content and diffusivity of the wood affects heat absorption. Hence cold wood will accelerate hardening, and warm or hot wood will delay hardening. Generally, placing a hot melt glue on a cold wood surface will result in shallow bonding because the rapid hardening deprives the glue of mobility too soon, and not enough penetration occurs.

Two special cases fall out of the above discussion. One involves rapid loss of water by glues that harden chemically at room temperature. On very dry wood, water, which is needed to facilitate molecular activity in the glue, may be removed by the wood before polymerization can effect a full cure. Some strength and durability development will occur, but considerably less than the formulation otherwise promised.

The other special case involves certain glues that harden by loss of water. Glues that coalesce with loss of water, e.g., polyvinylacetate, are inhibited from doing so when applied to cold wood, below that recommended by the manufacturer. The ingredients, suspended in particulate or globular form, tend to precipitate rather than coalesce, producing a low-strength, chalky film.

Stages of Solidification

During solidification many glues pass through two distinct end points as they undergo the phase change from liquid to solid. The first of these is the *gel point,* at which the adhesive loses liquid properties and becomes rubbery. The second point, farther along the course of curing, is the *glass transition point,* which marks the end of the rubbery state and the onset of a rigid solid.

Speed of hardening is easily observed in glues that harden chemically by noting the time it takes to reach the gel point. This allows studying the effects of catalysts, pH changes, diluents, extenders, and other additives on the hardening action. Some glues and catalysts slow down with age. By checking the gel time, it is possible to determine whether storage life has been exceeded.

By graphing time versus temperature, a curve can be developed which can be extrapolated to give the gel time at any temperature. Using this information it is possible to

accelerate the test and thus allow quick checks of mixes in current use. Interpretation of the time-temperature curve also allows one to estimate how long it might take to achieve cure in a hot press.

Solid Properties

As adhesives harden on the glue line, most (except elastomerics) achieve a glassy state, the *glass transition point* (T_g), which marks the onset of dependable strength properties. On one side of this point the material is rubbery; on the other it is hard and brittle. This point has two important implications, one involving time and the other, temperature. The time to reach the glassy state from the liquid state has a bearing on press or clamp time; the longer it takes to reach this transition point, the longer will be the necessary press time.

Temperature has two effects on T_g. Heat affects the rate of transition from liquid to solid and thus reduces the pressure period in hot pressing. Heat also affects the stability of the glassy state in some plastics. Once the glass transition point has been reached, it may or may not be reversible with heat, depending on the chemical and physical structure of the solid. This provides a major distinction between two classes of plastics, thermoplastics and thermosets. Once they have hardened, the *thermosets*, brittle and cross-linked, cannot be reversed to the molten state by the application of heat. Instead, they will decompose without melting.

The other group, the thermoplastics, can pass back and forth over their T_g, reversing freely according to their characteristic softening temperature. They are linear polymers without a great deal of stable intermolecular attraction; heat disrupts the attraction and allows free movement.

The T_g of thermoplastics varies with the materials. Some have a T_g near room temperature, some below, and some above. A room-temperature T_g means that the glue will be rubbery or liquid at all temperatures above ambient. It is generally desirable to use an adhesive with as high a transition temperature as possible, in order to ensure strength under extreme service conditions. In the case of hot melts for example, the higher the T_g the greater the strength of Link 1.

Many methods can be used to manipulate the glass transition and the state of hardness it represents. These include the addition of plasticizers which in essence act as lubricants between molecules; changing the length of molecular chains, changing the amount of branching on a molecule, changing the inherent "lubricity" of molecules through introduction of substituents of varying attractiveness, and perhaps most efficacious, varying the amount of cross-linking between molecular chains. The use of sulfur in compounding rubber is a familiar example of the latter. A relatively small number of cross-links has a dramatic effect on the properties of a thermoplastic.

After adhesives harden, they acquire different properties and take on a new role. While maintaining the *adhesion* achieved during bond formation, they must have developed *cohesion*. In doing so, they become stress transfer mechanisms between the two surfaces being joined. Two new properties are thus acquired: strength and durability. Although both ultimately require definition in terms of the glued product and the service it is expected to deliver (i.e., bond performance) at this point, interest centers on the adhesive as a solid material.

A number of factors contribute to the resulting bond performance, some of which derive from the chemistry of the adhesive. They include the basic molecule-to-molecule (or particle-to-particle) connections that have been made in hardening. These connections, consummated as they are in the environment of a glue line, may be assumed to be the most vulnerable, if not the weakest, links within link 1. They are therefore critical to the strength and durability of the solidified glue film. Properties of the hardened adhesive that derive from the conditions encountered during solidification include the glue line states described in Chapter 3—undercure, overcure, grainy, frothy, filtered, chalky, crazed, stippled, and solid—all of which also affect strength and durability.

With respect to strength, most properties of the solid state apply, but three are particularly pertinent: tensile strength, shear strength, and creep. Characterizing the solidified state of the glue line as a solid is an oversimplification, and may be misleading. Actually, glue bonds represent a continuum in solid properties ranging from gummy to glasslike. While every point in the continuum can be described in highly technical terms, from a bond standpoint, they reduce to the three properties mentioned.

Moreover, the strength of an adhesive is a relative term, not an absolute one. While it is possible to obtain strength values for the adhesive as a material, such values do not readily translate into the glue line, where the substrate contributes, positively or negatively, to the apparent strength. Nevertheless, it is desirable to know something about the inherent strength of the adhesive as a material.

The solid properties of adhesives are determined using films cast on glass or other substrates on which adhesion can be prevented. Thin films can be tested in tension and flexure using micro procedures scaled to the size and strength of the film. Hardness testing is particularly useful in observing the effects of heat and solvents on the hardened adhesive. Shear strength can also be observed in a film by proper gripping and application of the force. Shear through the thickness may be estimated with a punching action in which a circular area is removed in a manner similar to that of a paper punch. Alternatively, of course, the adhesive can be sheared as a glue line between two blocks which have a greater shear strength than the adhesive.

Creep, as a time function of strength, is also observable in thin films by the same methods as above if duration of load is programmed into the test. Varying the load at various levels below maximum, the time to fracture is noted. A plot of load versus time will then produce a profile of the creep characteristics of the adhesive film. Since creep increases with increasing moisture and temperature, these two factors are essential components of a given profile, and should be included in any characterization of an adhesive. With respect to creep, only one categoric statement can be made: The glassy, cross-linked, brittle adhesives do not creep; the rest do, the amount depending on composition and moisture and temperature.

The durability of the solidified adhesive is basically dependent largely on its sensitivity to various factors present in its use environment. These include primarily moisture (or solvents) and heat. Attack by organisms is important to glues of natural origin. Sensitivity to moisture manifests as softening, dissolving, and hydrolytic breakdown. Sensitivity to heat manifests by softening and breakdown.

Softening tendencies can be detected by including moisture, solvents, and/or heat in strength tests of films. Generally the cross-linked, glassy, brittle adhesives do not soften with heat; the rest do, the amount depending on composition and temperature. Dissolving properties can also be determined on cured films by grinding and subjecting to water or solvent immersion, with or without heat, and then testing the liquid for evidence of base ingredients in solution. Breakdown usually produces molecular fragments, which can be detected by various chemical means.

Although the properties of adhesives in film form read directly into the glue line, it is possible to predict bond performance only to assure that there are conditions where link 1 is certain to fail and there are conditions where link 1 will not deteriorate. In between there are many conditions where the integrity of link 1 may or may not be adequate, and this can only be determined by testing the glued structure where wood and its geometry interact with the bond. These considerations are discussed more fully in Chapter 11.

ADHESIVES

The general characteristics of adhesives discussed in the previous paragraphs become specific when they are associated with particular chemical compositions. They become even more specific when formulated to function in a particular product or process. The latter is in the realm of a special technology that bridges chemistry to the rest of the gluing art. Its overall objective is to adjust adhesive mobility requirements to the factors in the equation of performance as they exist in a given situation. While formulation is beyond the scope of this book, it is instructive to examine briefly the chemical factors with which such a technology might work.

Partial Hardening Systems

Mastics. The chief characteristic of mastics is their very high viscosity and tack. Their pastelike consistency is due to a high proportion of solid materials and low solvent or carrier content.

From the user's standpoint, mastic-type adhesives are the simplest in terms of bond formation. They generally come out of a can or tube, ready to be applied. Parts are assembled and pressed together, and a bond forms immediately. In some applications no change of state occurs; tack forms the bond and to a certain extent retains bond strength. In other applications some hardening occurs with time due to loss of volatile fractions and/or molecular activity, particularly in the newer formulations as manufacturers strive to broaden the range of use. In the main, however, mastics remain fairly plastic and deformable under load; i.e., the strength of link 1 is generally considerably less than that of links 8 and 9. Their use is therefore limited to situations where stress on the bond is relatively low.

Although they are easy to use, mastics are complex mixtures of materials. They are compounded with various kinds of rubber (natural, synthetic, or reclaimed), either as a latex in water or as a dispersion in a solvent. In addition, many other ingredients are needed to create specific properties. These include tackifiers such as rosins and resins, some of which, such as the phenolics and resorcinolics, may also be subsequently cross-linked to increase rigidity. Fillers such as carbon black, chalk, zinc oxide, and clay are used to provide "body" and reduce cost. Plasticizers, antioxidants, and curing agents are incorporated to enhance bond performance. Pitch and bitumen may also be added to alter physical properties.

The possibilities for varying the properties of mastics thus are infinite. Special formulations have been developed for such diverse applications as ceramic tile, plastic tile, metal tile, asphalt tile, acoustical tile, hardboard, wood paneling, gypsum wallboard, furring strips, flexible wall coverings, cementitious materials, linoleum, wood block flooring, countertop laminates, roof, deck, and interior insulation, and sandwich panels.

In applying the mastic, its high viscosity necessitates troweling or the use of a pressure gun. The exact procedure depends largely on what is being glued, and is usually stipulated in instructions accompanying the material. For some applications an area is covered by using a trowel appropriately serrated to apply the proper amount and give an open assembly time for some evaporation to occur. The tile or overlay is then positioned, and roller pressure is applied to assure contact. The serrated edge on the trowel allows the mastic to be spread in ribbons or bead form rather than as a smooth film. By placing the tile with a slight rubbing motion, any skinning effects that may have occurred are broken, and fresh mastic makes the contact.

The ease with which bond formation is achieved with mastics has prompted improvements to broaden their usage. Formulations that harden are used in construction applications, and are called construction adhesives; and because of their gap-filling properties, their use in bonding rough lumber is particularly attractive. The substitution of mastics for nails in assembling pallets is appealing as it would eliminate troublesome nail popping. The bonding of green lumber offers a further challenge. With this hope in mind, considerable research has been devoted to improving bond performance. In the meantime, an application that is semistructural in nature is fully operational. This is the bonding of subfloor panels to joists. Two desirable effects are produced: greater stiffness of the floor, less bounce; and elimination of squeaky floors and nail

popping. Application is fairly simple: A bead of mastic is extruded onto the edge of the joists, and the panel is nailed in place as is normally done. This ease of use that mastics offer, coupled with their forgiving nature with respect to bond formation, portends greater use in assembly gluing when link 1 is further enhanced.

Heat-Loss Systems

Hot Melts. Heat-loss adhesives or "hot melts" are molten at high temperature and achieve strength upon cooling. Their basic chemical constituent is therefore thermoplastic in nature. Thermoplastics, being composed of long linear or slightly branched molecular chains, derive strength from intertwining and adhering to each other through a large number of relatively weak intermolecular bonds. Heat disrupts these bonds and permits flow to take place. Cooling restores the intermolecular bonds.

Chemically, the molecular chains are linear hydrocarbons of high molecular weight. By replacing hydrogens with other chemical groups such as acetates, hydroxyls, or halides, strength, durability, and temperature-related properties can be modified. Additives such as rosin, rubber, fillers, and other thermoplastic materials contribute a variety of properties. Polyamides are also useful in hot melts.

$$\cdots\text{---}\underset{\overset{|}{H}}{\overset{\overset{|}{H}}{C}}\text{---}\underset{\overset{|}{R}}{\overset{\overset{|}{H}}{C}}\text{---}\underset{\overset{|}{H}}{\overset{\overset{|}{H}}{C}}\text{---}\left[\underset{\overset{|}{H}}{\overset{\overset{|}{H}}{C}}\text{---}\underset{\overset{|}{R}}{\overset{\overset{|}{H}}{C}}\right]_n\underset{\overset{|}{C}}{\overset{\overset{|}{H}}{C}}\text{---}\underset{\overset{|}{H}}{\overset{\overset{|}{H}}{C}}\text{---}\underset{\overset{|}{H}}{\overset{\overset{|}{H}}{C}}\text{---}\cdots$$

The length of the chain "n" and substituent groups "R" are selected according to the temperature at which it is intended to melt, or the temperature it is expected to resist when serving as a bond. In general, the higher the melt temperature, the stronger the solidified material will be at room temperature. It should be noted, however, that a stronger hot melt does not necessarily lead to a stronger bond. The reason is that when the hot melt is placed on wood at room temperature, the layer next to the wood quickly cools below its melting point. Penetration is inhibited, and the adhesive may attain only a shallow grip on the wood. A procedural change will circumvent this problem. Preheating the wood surface reduces the heat sink effect and delays the cooling of the melt. This approach can also be used to increase the assembly time if necessary. The downside is that the pressure period may also have to be extended. Trial and error may be necessary to optimize the system. In any case, the open assembly time should be kept as short as possible in order to avoid "skinning over" and interferring with the transfer motion. Since hot melts do not contain chemical elements that attract it to a wood surface, most of its penetration must be achieved by applying pressure.

The viscosity of a particular hot melt depends entirely on its temperature: the higher the temperature, the lower the viscosity. Since viscosity is controlled by the amount of heat applied, systems for delivering molten adhesive must contain means for controlling temperature. Hot melts can be applied as a spray, bead, or curtain through heated or insulated hoses from a pressurized heated tank. In the craft shop, the glue is fed as short cylindrical rods into a glue gun, which then delivers a molten bead of glue.

Hot melts are sold in a wide range of solid forms, including flakes, granules, chunks, or ropes for industrial operations, or in rod form for bench work. A film form is also available for paper bonding, for which it is specially suitable since it contains no solvent that might alter the properties of the paper.

Since these glues harden by losing heat, they are very fast setting. Sufficient strength for most applications is developed in a matter of seconds. The process brings very rapid gluing to the assembly of wood parts, analogous to soldering in bonding of metals. However, since hot melts are comparatively soft and waxy, their maximum strength in a bond is less than the tensile strength of wood perpendicular to the grain. For this reason, when joints made with these glues are forcibly broken, they usually break in the adhesive film (link 1) rather than in the wood. Because softening by heat is built in the ad-

hesive, it is to be expected that bonds exposed to heat will lose strength, the amount of loss depending on the temperature.

Hot melts are not soluble in water, and therefore bonds are unaffected directly by it. However, indirectly, through its effect on the dimensional stability of wood, moisture changes tend to loosen bonds, due to stresses that develop as a result of the shrinking and swelling of the wood. The larger the pieces of wood being held together by a hot-melt bond, the greater are the stresses that are likely to be imposed on the bond. It is partly for this reason that hot-melt bonding is usually restricted to rather thin wood pieces.

The effects of heat and solvents on bond strength poses a caution in finishing operations, where both these destructive agents may exist. This can be an advantage when the bond may for some reason need to be intentionally destroyed, perhaps to reposition the parts or to repair the article. The thermoplastic nature of the glue also imposes a limitation due to its tendency to creep. Products in which long time loading may be a factor are therefore unsuitable for hot melts—laminated lumber, for example, or curved parts in which the bond is responsible for maintaining the curve against the straightening forces of the wood.

Because of their very fast strength development, hot melts can be used only where rapid assembly and pressurizing is possible. The most appropriate situations are those that can be automated. These include edge banding of panels and factory assembly of overlays. The assembly of small objects that require only a drop or two of glue also find advantage in hot melts. A unique use is a combination of hot-melt spot welds to hold an assembly together while a more durable or stronger glue achieves strength, thus avoiding the long pressure periods otherwise needed for the stronger glue.

Solvent- and Water-Loss Systems

Adhesives that set by loss of the liquid that provides their fluidity come from a variety of natural and synthetic sources. Six are briefly discussed as representative: polyvinylacetate (PVAc), aliphatics, animal protein, polysaccharides (starch and cellulose), and rubber. Like the hot melts and mastics, they undergo no molecular (chemical) changes on the glue line. All depend on evaporation, diffusion, and capillarity to remove dispersants or solvents from the glue line and allow solidification. For this reason, the solid state is reversible. This means that glue bonds are weakened by water or solvents, and in many cases also by heat.

Differences among these glues reside primarily in the basic molecular structure of the solidifying material, and the chemical derivatives into which they can be formed or reformed.

Polyvinylacetate (PVAc) Adhesives. Many adhesives as well as plastics in general are based on the polymerization potential offered by the carbon-to-carbon double bond.

Ethylene, $\begin{matrix} H & H \\ C=C \\ H & H \end{matrix}$, is a basic starting molecule, a monomer, for building many plastics. Its chief property is its capability for self-addition to form a linear, thermoplastic polymer, polyethylene, familiar as a wrapping film.

$$-\underset{\underset{H}{|}}{\overset{\overset{H}{|}}{C}}-\underset{\underset{H}{|}}{\overset{\overset{H}{|}}{C}}-\underset{\underset{H}{|}}{\overset{\overset{H}{|}}{C}}-\left[\underset{\underset{H}{|}}{\overset{\overset{H}{|}}{C}}-\underset{\underset{H}{|}}{\overset{\overset{H}{|}}{C}}\right]_{n}\underset{\underset{H}{|}}{\overset{\overset{H}{|}}{C}}-\underset{\underset{H}{|}}{\overset{\overset{H}{|}}{C}}-\underset{\underset{H}{|}}{\overset{\overset{H}{|}}{C}}-\underset{\underset{H}{|}}{\overset{\overset{H}{|}}{C}}-$$

By replacing one of the hydrogens in ethylene with an acetate group, a monomer is produced which also self-adds to form the linear polymer, polyvinylacetate:

$$-\overset{H}{\underset{H}{C}}-\overset{H}{\underset{O}{C}}-\overset{H}{\underset{H}{C}}-\overset{H}{\underset{O}{C}}-\left[\overset{H}{\underset{H}{C}}-\overset{H}{\underset{O}{C}}\right]_{n}\overset{H}{\underset{H}{C}}-\overset{H}{\underset{O}{C}}-\overset{H}{\underset{H}{C}}-\overset{H}{\underset{O}{C}}-$$

The basic monomer is called a *vinyl*. After polymerization, the results are called *polyvinyls* with the further designation of whatever substituent group is included, e.g.,

polyvinylchloride (PVC), polyvinyl alcohol (PVA), etc. The acetate group imparts a high degree of polarity to an otherwise nonpolar molecule.

Polymerization is carried out under emulsion conditions that produce globules of resin suspended in water. Given various additives, the result is the familiar "white glue" used in industry and craft shops. In its basic form, it comes ready to use, a fairly fluid liquid with a pH of 4.5 to 5.0, easily applied, and forming bonds with most porous substrates. The emulsion is delicately stabilized so that only a small amount of water needs to be lost in order to break the emulsion and cause the globules to coalesce and form a rigid mass. This accounts for the fast bond-forming properties of these glues. Although they have little initial tack, they can form bonds in as short a time as 10 minutes on dry, low-density woods. Consequently, assembly times must also be short. Full bond strength is achieved in about 24 hours.

Bonds are as strong as (and sometimes apparently stronger than) maple by standard shear tests. However, being thermoplastic, PVAc deforms under sustained stress. Creep and failure can occur at relatively low loads. Also, in common with thermoplastics, glue bonds soften with elevated temperature, losing strength proportionally. Moisture also softens with proportional loss of strength. At elevated moisture and temperature, creep is therefore heightened, and failure can occur at even lower loads. High-density lumber glued with these adhesives cannot withstand wide swings in moisture content.

Otherwise, this type of glue is freely used in secondary and tertiary gluing to assemble components and structures. In assembly joints where grain directions often appear in perpendicular orientations across the glue line, PVAc bonds permit slight differential movement without breaking, and in this respect, the deformability of these glues is an advantage. Gap filling is not promoted with PVAc, but since they bond at low pressure, if enough solid glue is deposited on the glue line, it can form a rather thick bond. However, because of the large volume of water dissipated, shrinkage is inevitable, and the bond may form as islands in the glue line.

As with many emulsions in water, freezing destroys the emulsion and renders the glue unusable. Applying the glue to a cold surface, below about 60°F, also adversely affects the coalescence of the PVAc particles. The failure to fuse is evidenced by a failure of the glue to become translucent, resulting instead in a telltale chalky appearance and a poor bond.

Properties of PVAcs cannot be generalized because of the many modifications that are possible, both chemically and by the addition of emulsifiers, plasticizers, stabilizers, fortifiers and fillers. When conditions seem to preclude the use of these glues, some modification may be engineered to compensate.

One chemically engineered modification that has resulted in a major improvement involves the introduction of other substituents into the polymer. These glues are classified as aliphatics, a term handed down from earlier times when such compounds were derived from fats. They are the "yellow glues," also known as "carpenter's glues," and they function in a manner similar to PVAcs. However, they are faster setting, can tolerate lower temperatures, are more resistant to creep, and are less sensitive to heat and moisture. They are also more thixotropic, a property that reduces the tendency for squeeze-out to drop down edges and foul up those surfaces. Because of the harder nature of the set film, sanding is less troublesome.

Animal Protein Adhesives. Perhaps among the oldest of glues, those derived from animal hides, bones, and sinews, and from fish, seem to have inherent bond-forming properties. Proteinaceous in nature, they are derived by hydrolysis from the collagen present in these sources. (The term "collagen" in fact comes from the Greek word for glue, *kolla,* or the French word, *coller,* to glue.) Chemically, the constituent units of protein

are classed as amino acids. They are linked together into large molecules with amide groups (polypeptide chains) through the reaction of an amine with an acid, the same reaction that produces polyamides (nylon). The chains terminate with carboxyl groups or amine groups and are only slightly amphoteric with an isoelectric point near neutral pH. The different sources, of course, produce different properties, since they were originally intended to perform different functions in the animal.

Collagen has the general formula:

$$--R1--\overset{\overset{\displaystyle H}{|}}{\underset{\underset{\displaystyle X}{|}}{C}}\left[\overset{\overset{\displaystyle H}{|}}{\underset{\underset{\displaystyle H}{|}}{C}}--\overset{\overset{\displaystyle H}{|}}{\underset{\underset{\displaystyle H}{|}}{C}}--N--O--\overset{\overset{\displaystyle O}{\|}}{C}--\overset{\overset{\displaystyle H}{|}}{\underset{\underset{\displaystyle H}{|}}{C}}\right]_n\overset{\overset{\displaystyle X}{|}}{\underset{\underset{\displaystyle H}{|}}{C}}--R2--$$

where n is the repeating peptide connector between molecular segments R1 and R2 of similar configuration, and X either amino—$NH2$ or carboxylic $-O-CO-CH3$ substituents which are responsible for amphoteric properties. After extraction of the collagen, preparation involves chemically breaking the connectors by hydrolytic reactions, reducing the size of the original molecule and making it more soluble in water. In other words, preparation involves a "breakdown" operation rather than a "build up" operation as in synthetic resins. These derivatives easily form colloidal solutions which gel at room temperature. They can be inhibited from doing so by incorporation of gel depressants such as thiourea and thus be prepared in liquid form for ready use by the consumer.

No chemical changes occur in the glue line. Strong adhesion to the substrate develops due to the high polarity conferred by the amino and carboxyl groups, and strength develops as the substrate absorbs the water in which the glue is dispersed. The molecules then reachieve the natural cohesion they had originally. The solidified state however is sensitive to water, and can be resoftened with moisture and heat. Therefore, although the bond is very strong, its usage is restricted to situations where dryness can be assured.

No matter the source, the outstanding

characteristic of animal glues is that they gel. Two types of animal glue are marketed for wood gluing. Solid glues, in flake, powder, or cake form, are intended to be used hot and are discussed in the section on combined hardening glues. The liquid types (e.g., hide glues) are solubilized and kept in solution with gel inhibitors to prolong shelf life. In use they are applied directly from the container by brushing, dipping, or spraying. Because of their high mobility, some assembly time is needed to firm up the film and prepare it to receive pressure. The presence of gel inhibitors, however, means that the speed of setting is long, 2 to 8 hours normally, longer for ultimate strength.

When fully hardened and kept dry, the bonds are as strong as the wood, and they do not creep significantly under load. However, slight increases in moisture weaken them, and if accompanied with elevated temperature, severe strength loss occurs. While generally a disadvantage, this sensitivity to heat and moisture is useful when structures such as chairs, toys, and musical instruments need to be disassembled for repair.

Liquid glues harden to a near glassy state and can be sanded without loading the sandpaper. With a pH of 7, they do not discolor wood. The glue line is amber in color, and is not affected by solvents used in finishing.

Polysaccharides

Through a very minor change chemically, a major structural and physical change has been engineered by nature in polymerizing the sugar, glucose, into two of the most widely different, as well as widely distributed products of the plant kingdom: starch and cellulose. These two natural polymers outweigh in tonnage all other materials of commerce put together. In many respects, they

also contribute more to humankind than any other material. Although their properties as wood adhesives are not outstanding, they have been well researched and represent approaches to future developments as economic considerations compel greater attention to their potential use.

With exactly the same glucose repeating unit in the molecule, starch is organized to be relatively amorphous and fairly soluble, while cellulose is highly crystalline and insoluble except after drastic chemical treatment. The chemical difference consists of the left and right (alpha and beta) nature of the linkage between adjacent glucose units in the polymer chain. The configuration of the cellulose molecule, with its segments organized in alternating orientations along the chain, allows close nesting of adjacent chains in a regular manner. Regularity leads to the development of crystallinity, and the close nesting leads to the formation of strong hydrogen bonds between cellulose chains. The resulting tight packing of cellulose chains is responsible for the resistance of cellulose to solution. Even though the chains are composed of highly soluble glucose sugar, they are not free to behave individually but must act as a particulate material impenetrable by solvents.

The configuration of the starch molecule, on the other hand, in which all glucose units have the same orientation, does not allow such close packing, and individual chains can express their own solubility properties. Nevertheless, in the natural state, they are gathered into clusters that require some energy to dissociate. In both starch and cellulose, it is the presence of three free hydroxyl groups on each glucose molecule that is responsible for their high polarity and reactivity. In the accompanying molecular structures compare starch A with cellulose B.

B   OH H   CH2OH   OH H   CH2OH
         C—C   O—C   C—C   O—C
      H HO \H   H   H \   H HO \H   H   H \
  O—   C   C   C   C —O— C   C   C   C   —O—
    H   H   /   O—   C   \H HO /H   H   H   /   O— C   \H
      O—C   C—C   O—C   C—C
      CH2OH   OH H   CH2OH   OH H

Starch Adhesives. The primary source of useful starch is grains and roots (e.g., potatoes and cassava). It is extracted in the form of globules or granules that are too large to be dispersed into a usable mixture. When cooked in water, the globules swell and burst, releasing starch molecules. Their high viscosity in solution is due to their high molecular weight. Starch occurs in two configurations, linear and branched, which have different solubilities. In *amylose* the molecule is linear and readily soluble in hot water, whereas in *amylopectin* the molecule is branched and insoluble in hot water. Acid hydrolysis reduces chain length and produces more soluble fractions, necessary in adhesives. In *dextrins,* the smaller molecules produced by hydrolysis are allowed to repolymerize into highly branched but still relatively small molecules. This configuration facilitates adhesion while retaining strength.

Since strength increases with molecular size, starch adhesives are hydrolyzed to a lesser extent for wood bonding than for paper bonding. The wood glues therefore have higher viscosities. High molecular weights with spreadable viscosity is achieved by use of heated mixes, and by addition of alkali. The latter tends to discolor woody substrates, but provides strong bonding with a ready-to-use adhesive. The use of heat to lower viscosity adds the heat-loss mechanism to the water-loss mechanism of hardening. This provides instant tack, and allows high-speed operations such as the making of corrugated board and paper tubing.

Starch bonds are not durable against moisture, and have the extra vulnerability of being susceptible to fungi, bacteria, and insects. The addition of preservatives is therefore necessary to protect against these agents of destruction. Sensitivity to moisture is reduced by the addition of synthetic adhesives such as urea and resorcinol. Although the use of starch for wood gluing has declined since the introduction of synthetics, it is still used in edge-gluing lumber for core stock, and to glue box shooks and paper overlays.

Cellulose Adhesives. The crystalline structure of cellulose requires harsher methods to place it in solution. After isolation from its source in wood or other plants by pulp chemistry for paper making, the cellulose fiber must be treated to reduce the strong hydrogen bonding between molecular chains so that solution can take place. Treatments to solubilize cellulose include esterification, etherification, xanthation, and cupperammoniation. Of these, esterification represents a general approach. Taking advantage of the reaction of acids with alcohols to form esters, cellulose with its three hydroxyl groups is reacted with nitric acid to form cellulose nitrate. If all three hydroxyls are nitrated, the resultant material is highly explosive (smokeless gun powder). However if only one or two are nitrated, the compound becomes soluble in organic solvents. When the solvents evaporate, a very strong, tough film is formed. The presence of polar hydroxyls, and even more polar nitrate groups result in a material having high affinity for polar surfaces such as wood and paper.

By appropriately formulating with volatile solvents and other additives, solutions are made that contain sufficient solids to form a glue line. Bond strength development depends on evaporation of the solvents, and hence porous substrates are easiest to glue.

Because these mixtures are extremely flammable both from the solvent standpoint and the solids standpoint and because of toxicity of the solvents their use as adhesives is limited. A less flammable adhesive is made by esterifying with an organic acid such as acetic or butyric acid. The resulting cellulose acetate or butyrate is also soluble in organic solvents but is nonflammable.

These materials are not widely used for bonding wood because of cost, marginal bonding, and their thermoplastic nature with its susceptibility to solvents and heat. However, being good film formers, they appear in paints and finishes for wood, in pastes for repair work, and in additives for other adhesives.

Rubber Adhesives. Many polymeric materials have rubberlike properties, one of their main characteristics being the mobility of individual segments of the molecule, particularly the deformation of covalent bonds by bending. This kind of action is provided by long linear molecules that are twisted and coiled or have a zigzag structure. The molecular bonds then can act as springs, with easy extension and immediate retraction.

The first observation of this behavior was made on natural rubber. Natural rubber, derived from the sap of the rubber tree, Hevea braziliensis, is a polymer of 2-methyl-1,3-butadiene, (isoprene), $CH_2 = CCH_3 —CH = CH_2$, in which the conjugated double bonds participate in forming a long chain, but leave one in the molecule (see bottom of page).

It is generally accepted that the adhesion of the rubber molecule is due primarily to its rheological properties, particularly its flexibility. This allows it to deform and conform intimately with the atomic irregularities of a surface, and thus make maximum use of what little adhesion forces it has. Natural rubber, while having many qualities that make it useful for a wide variety of applications, is characteristically low in adhesion and even lower in cohesion. In bond strength tests, failure is usually within link 1, cohesive, if only a few molecules from the interface.

Spurred by World War II shortages and the needs of the auto and shoe industries, research efforts led to a large number of advances, which include chemical changes, additives, synthetics, and copolymerizations. The first of these, achieved accidentally by Charles Goodyear more than 150 years ago, resulted in a major improvement in the properties of natural rubber. The process of "vulcanization" was born. It made possible the variation at will of hardness and stiffness of the rubber mass. This was accomplished by utilizing the double bonds prevalent in the rubber molecule to provide a means of cross-linking. Sulfur was the first substance used in vulcanizing rubber, and is still the major cross-linking agent, though other means have been developed. Properties vary in direct proportion with the amount of sulfur reacted, up to the maximum of 35 percent, which produces the so-called hard rubbers or ebonite. Additions as low as .5 percent produce marked changes, true also in cross-linking synthetic elastomers.

Once the nature of the rubber molecule was determined, chemists were able to develop monomers that could be polymerized to produce similar properties. These include polychloroprene (neoprene rubber), in which chlorine is substituted for the methyl group in isoprene. This change imparted greater resistance to oils, chemicals, air, light, heat, and flame, representing a major improvement over nature. Other synthetics followed: polysulfide rubber, butyl rubber, isocyanate rubber, butadiene-styrene rubber (SBR), and butadiene acrylonitrile (Buna rubber). All the synthetics offer improvement in one or more properties while retaining rubberlike characteristics. They are used in mastics and other elastomeric applications.

The most important additives to rubber

are tackifiers, which, as the name implies, improve tack, one of the main properties of rubber-base adhesives. The "instantaneous adhesion" or "quick grab" associated with tack is a phenomenon that is not well understood but is easily demonstrated. A drop of a tacky adhesive pressed between the fingers produces instant resistance to separation, and leaves a stringy mass clinging to the fingers. There are many kinds of tackifiers, each providing some advantage in method of incorporation or specificity to some application. They include wood rosins, polyterpene resins, and phenolic resins, among others.

Tack allows two surfaces, previously coated with the rubber adhesive, to adhere instantly even though dry to the touch. This is the basis for the gluing procedures common to rubber adhesives: application of adhesive to both surfaces, driving off the solvent, mating the surfaces, and applying rolling pressure.

Rubber adhesives are of two types, solvent and emulsion. Solvents are hydrocarbon in nature, and include benzene, toluene, and naphtha for the natural rubbers, and blends of ketones or ethylene chloride with hydrocarbons for the more polar synthetics. In devising the solvent system, a major consideration is the rate of vaporization of each ingredient. This needs to be well coordinated in order to avoid *blushing,* a condition in which the more volatile fraction of a solvent evaporates too fast, cooling the surface, and causing moisture from the air to condense on the glue film. Premature skinning over can also occur, trapping solvent in the film. Tack is destroyed, and strength development is delayed.

The emulsion systems are dispersions in water (latexes) of small rubber globules. In order to achieve and maintain dispersion of a nonpolar material in a polar liquid, an emulsifier is needed. An emulsifier is a material the molecules of which have both polar and nonpolar properties. Its function in this case is to surround the rubber globules and make them act physically as if they were polar and thus miscible in water. Because emul-

sifiers in a sense encapsulate the rubber, they influence both adhesion and cohesion, decreasing them slightly. The use of rosin-based tackifiers improves both properties.

Water-base and solvent-base rubber systems are essentially similar in use. Some differences may be noted. Solvent systems have a lower concentration of rubber solids, 10 to 25 percent, compared to 40 percent or more for latexes. However, despite the higher solids content of latexes, they have a much lower viscosity. The higher solids content and the larger particulate sizes in latexes leads to less penetration and more "holdout" when applied to porous surfaces. Consequently, the surface area covered per gallon is much greater for latexes than for solvent-based systems. On the other hand, the greater penetrating ability of solvent-based systems allows a deeper grip on the substrate. The lower solids content and the greater penetration of solvent systems may require that more than one coat be applied in order to build a film of sufficient thickness to reach the two surfaces.

The solvent-base systems may have a much shorter open assembly time, 5 to 10 minutes, because of the high volatility of the solvents. However, since they are also toxic and flammable, ventilation is necessary, and precautions must be taken against fire. For the price of a longer assembly time, about 60 minutes at ambient, water-based systems avoid this danger. Latexes also can withstand a longer period before assembly must be made, since they retain tack for weeks in some cases.

Rubber-base adhesives in general exhibit useful adhesion to almost all substrates. In their most appropriate applications, they make use of the instant bond they achieve, and the relatively low pressure needed to achieve it. Hence on-site gluing is a popular use, particularly the gluing of high-pressure laminate to panels in making countertops. Automated operations, where the solvent can be flashed off with hot air or radiant heat and pieces quickly pressed under rolls, find production advantages in rubber adhesives. The bonding of paper with solvent types eliminates wrinkling. The flexible bond also lends itself to the bonding of cloth, backing materials, and films.

Bonds are permanently deformable, and hence are unable to resist stresses of long duration. However, because they are elastic, small deformations are recoverable, permitting glued parts some movement. Edges of assemblies are most vulnerable, and whenever possible, some form of protection or reinforcement should be provided to prolong service. Bonds made with solvent-base cement are not affected by water at link 1, but water-base cements are vulnerable because of their emulsifiers. Water may destroy adhesion links 4 and 5 under long exposure. On wood substrates, the effect of moisture will register more strongly through the dimensional changes of the wood and the stresses they generate across the glue line. Heat reduces strength in all cases.

The compatibility of rubber with other adhesives has led to many combinations where its flexibility is used to modify the stiffness of stronger and more durable adhesives. Blending with phenolics or caseins, for example, produces an adhesive suitable for bonding metal to wood or even metal to metal. The rubber allows the adhesive bond to conform to the dimensional changes of the metal, which otherwise would break the more rigid bonds of casein or phenol alone.

Chemical Reaction Adhesives

Adhesives that harden by chemical reaction are the inventions of molecular scientists applying knowledge of the organic chemistry to that learned from adhesives of natural origin. The resulting man-made polymers then are manipulated chemically to yield properties that are in some cases superior to the natural ones. One of the earliest synthetic polymers was phenol-formaldehyde, known as a compound for decades before it was first introduced as a plywood adhesive in the mid-1930s. Time and experience has shown it to be one of the most durable adhesives ever developed. Its outstanding properties

are responsible for the rise of the softwood plywood industry and the establishment of plywood as a reliable construction material. Its usefulness in the panelization of wood continues in the consolidation of comminuted wood intended for severe exposures. Phenol-formaldehyde resins with molecular modifications also find extensive use as molding powders and as impregnants for imparting dimensional stability, stiffness, and resistance to weather. Total consumption of these resins is the highest of all resinous materials. For this reason, it is logical to treat phenolic resins as the flagship of the adhesive industry, and accord them greater attention in this chapter. Much of what applies to these resins also applies to other resins of similar chemistry and use characteristics.

Phenol-Formaldehyde Adhesives

Chemical Composition and Properties. Phenolic resin is the reaction product of phenol with formaldehyde. The reaction proceeds in stages, the first being one of addition of formaldehyde to phenol to form methylol phenol. The methylol phenol then reacts with itself (condenses) in a second stage, ultimately forming very large molecules under the proper conditions of temperature and pH. Several interesting chemical processes are illustrated by these reactions.

In the first stage of the reaction, formaldehyde merely "adds" itself onto the phenol molecule without losing any atoms (Hence the term "addition" reaction).

In the second stage of the reaction the formaldehyde add-on splits off its $-OH$ group (hydroxyl group), which then takes a $-H$ atom from a passing phenol molecule and forms a water molecule, HOH. Meanwhile the points or sites on the two phenol molecules which originally held the $-OH$ and the $-H$ are extremely reactive and strongly attracted to each other. They form a bond as soon as they can get close enough together and in the proper orientation with respect to each other. The union is facilitated by catalysts which help the formation of the water molecule and maintain the receptivity of the bonding sites. The resulting connection produces what is called a "methylene bridge" between two phenol molecules. This type of connection, being covalent and carbon to carbon, is considered to be the strongest and most durable that can be formed between two organic molecules (see bottom of page).

It is important to note that the reaction of the methylol group on one phenol with the hydrogen on another is the only means by which phenolic adhesives gain strength and durability. All use technology and most formulation technology are aimed at controlling or accelerating this process.

Although the formation of methylene bridges is the only means by which phenolic adhesives polymerize and gain strength, there are thousands of different phenolic resins and hundreds of patents pertaining to their formation. These represent a wide range of molecular sizes and configurations produced by varying the molar ratios of phenol to formaldehyde, the pH, the manner of adding the ingredients, the programming of the reaction conditions, and the state of the final product.

Two distinctly different types of resins are made depending primarily on the molar ratio of phenol to formaldehyde. If the ratio is less than 1, that is, there is less phenol than formaldehyde, the resin will be rich in

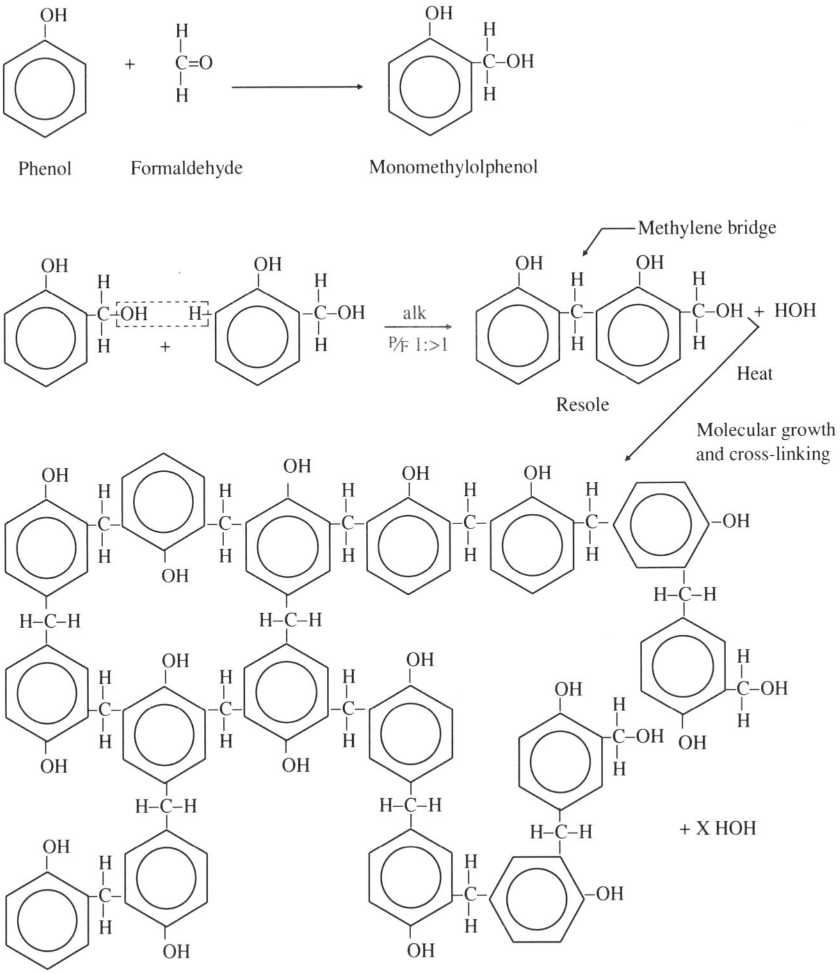

Note: Molecular structures shown as two dimensional are actually three dimensional.

methylol groups and capable of polymerizing to a cross-linked, insoluble state without addition of other ingredients. Such a resin is called a *resole*. Its chief characteristic is that it is always ready for hardening, needing only the proper environment (e.g., heat). A disadvantage is that it continues to harden in storage and therefore has a limited life, the liquid resins more so than the powders or films.

When the ratio of phenol to formaldehyde is greater than 1, that is, there is more phenol than formaldehyde, the resulting resin will be deficient or lacking in methylol groups and its molecules will be linear in nature under acid conditions. Such a resin is called a *novolac*. Its chief characteristic is that since it has little or no methylol, it is unable to polymerize on its own. This gives it an indefinite shelf life. A major disadvantage is that in order to convert this resin to an insoluble state, a source of formaldehyde must be added as a hardener at the point of use (see bottom of page).

Both the resole resins and the novolac resins go through distinct stages from the initial reaction to final set. The A stage represents the first reaction products of phenol and formaldehyde, the formation of methylols. The B stage is the early condensation products of the A stage in which polymerization has begun, but the process is arrested to preserve a relatively low molecular weight and to retain solubility and fusibility. The C stage is the final cured state, achieved on the glue line—the state that must be attained in

order to assure maximum strength and durability.

Each stage can be visualized as containing a rather specific range of molecular sizes. The A stage is essentially a mixture of monomers and dimers having a waterlike viscosity. The B stage may contain all sizes of molecules that still retain solubility in water or solvents, or may be dispersed in a liquid; they still retain fusibility, that is, the ability to flow under heat and pressure. The C stage may be considered as one giant molecule, the result of the chemical linking and cross-linking of the B-stage molecules.

The molecular state of polymerization associated with each stage confers various important characteristics with respect to adhesion and adhesive action. The A-stage resole resin molecules are very small and highly attracted to the cell wall material of wood. Being small, they also have high mobility and penetrating power, so great in fact that they are able to penetrate cell walls like water, perhaps better than water. (When strips of wood are placed in a water solution of A-stage resin, the concentration of the resin in the solution gradually decreases, indicating that the wood preferentially adsorbs the resin over water.) Adsorption of this kind, in which molecules or atoms in solution exhibit a preference for the molecules of a surface, is the very basis of adhesion. The action in this case results in one of the best means of reducing the dimensional instability of wood. Since the resin penetrates deeply into the cell wall structure of wood, subse-

Novolac

Heat and Formaldehyde donor

Molecular growth

+ X HOH

quent curing "freezes" it there and the wood achieves a permanently swollen state, unable to shrink and therefore unable to swell. "Impreg," a highly stabilized wood, and "papreg," a highly stabilized paper (seen in high-density laminates on countertops), are made by impregnating with A-stage resole resins and curing to the C stage. "Compreg" is impreg that has also been compressed to a high density. It is popularly seen as knife handles.

The ability of such small resin molecules to move into and penetrate cell wall material under their own power has strong implications in bond formation. It suggests, as mentioned in the previous paragraph, that they can automatically form links 4 and 5, the adhesion function. In addition, the action that carries them into wood material means that they are in a position to repair subsurface damage and form links 6 and 7. Ironically, it is precisely because of this ability that A-stage resins do not by themselves make good bonds. Their penetrating power robs the region between joint surfaces of adhesive, starving the joint and preventing the formation of link 1.

B-stage resins contain larger molecules. Resoles of this stage are used for bonding purposes. Many options and compromises are involved in B-stage preparation, and often their constitution is specifically "tailored" to a particular application. Among the important variables that can be manipulated are molecular weight, or degree of polymerization. Normally, the higher the molecular weight, the faster the resin will cure on the glue line, and the less it will move or penetrate the wood. Also, solubility decreases with increasing molecular weight. When the advantages of large molecules are important to a particular situation, more powerful solvents are used to make a spreadable adhesive. With wood adhesives these would usually be alcohols or alkalies or both. Both have advantages and disadvantages. Alkalies are relatively inexpensive, but their inherent hygroscopicity invites water into the cured glue line. Alcohols evaporate quickly, thereby reducing the time during

which critical viscosities can be maintained, e.g., on the spreader and on the glue line during assembly times. One compromise that is often made occurs in flakeboard manufacture, where, in order to reduce hygroscopicity of the board, alkali must be reduced without reducing solubility. In this case, reducing molecular size solves the problem, but with a penalty of slower cure, longer press times, and decreased productivity.

The effect of molecular weight on the mobility of the adhesive has consequences in bond formation that spur other compromises. We have seen that small molecules migrate, penetrate, and wet the wood substance readily, but do not therefore perform the space-filling and bridging function of link 1. Large molecules tend to be less mobile; they do not penetrate wood substance but remain on the glue line, helping to form the linkage between the two wood surfaces. What, then, is the ideal molecular size? Logic says that a distribution of molecular sizes is needed. This can and is being done by adhesive chemists. But an overriding question remains: What proportion of each size should be included? This is where the adhesive technologist must take over and harmonize the composition of the adhesive with the particular substrate and conditions of use. Different wood species, qualities of surface, and moisture histories demand different compositions to assure that all necessary adhesive motions optimize and all links are formed.

As is common with most wood adhesives, certain additives are incorporated into the basic resin matrix to modify properties or to lower cost. These include other polymer species such as rubber and resorcinol discussed in a following section. Of increasing importance is the substitution of natural phenolic materials, particularly tannins and bark extracts, for some or all of the higher-priced petroleum-based phenol. Tannins, being condensed phenols, react vigorously with formaldehyde, and produce resins that have properties similar to synthetic phenol resins.

Lignin, with its basic phenolic composition, is also an important additive. Long used in its natural state as the binder in the Masonite process for making fiberboard, lignin has also been used in its derived state of paper mill waste as a binder in particle board manufacture. Recent attempts to activate derived lignin by oxidizing and/or reacting it with formaldehyde has resulted in resins that can be substituted for a portion of the phenol resin in adhesive formulations without serious loss of durability. The longer curing times needed for lignin-based resins, however, seem to outweigh their lower cost.

Solvents and diluents are used to facilitate the application of the adhesive and deliver it to the molecular interface with wood. After the resin reaches this point, their presence can cause as much harm as good.

Phenolic resins for wood are carried primarily in water solution or dispersion, the cheapest and perhaps the best solvent. The water has a great affinity for wood and thus can carry the resin all the way to primary adhesion sites. It can later be diffused out of the wood to reequilibrate the moisture content of the glued product. The water also plays a strong role in bond formation, first as a reaction medium in which polymerization takes place, and second to promote flow of the glue mixture. Both these actions thus depend on the amount of water in the glue line at the time the bond is forming.

Too little water results in inadequate flow and penetration. Too little water also tends to inhibit molecular movement and orientation with resultant reduction of durability and perhaps strength. Too much water, on the other hand, will have the opposite effect on flow and penetration, leading to excess mobility and starved glue line conditions. Too much water also has a dilution effect, slowing the condensation reaction and retarding the cure. Since the polymerization of phenolic resin involves the splitting off of water molecules, the less water in the system as the resin approaches ultimate cure, the better the cure will be.

Thus a crucial interaction involving water molecules occurs in the final stages of curing when the condensation reaction is taking place. The formation of water molecules by reaction of methylol groups with ring hydrogens to form the necessary methylene bridges tends to be suppressed by the presence of free water. Hence a drying action is essential to reach maximum cure. A conflict may exist in which a reaction medium is needed to facilitate molecular movement, primarily orientation at this stage, but at the same time, or almost immediately after, a newly formed water molecule must leave the scene.

Fillers and Extenders. Common fillers and extenders used with phenolic resins include cereal flour, nut shell flour, wood flour, bark flour, glass flour, cereal hulls, clay, lignin, and dried blood. Old, expired powdered resin has also been used when it is available. The latter is particularly appropriate because it is both compatible and at least partially reactive. Each additive imparts a definite characteristic to the adhesive mixture in controlling flow and penetration properties, lowering cost, or affecting durability, usually adversely at high additions. Compatibility with the resin and ability to remain in suspension are necessary attributes. Bonding to the resin, and even reacting with it, are desirable but not essential. Fillers and extenders function by adding "body" to the adhesive mixture, or by blocking the pore structure of wood to control penetration. By contributing body, these additives increase the bulk of the adhesive and permit a more uniform spread of a lesser amount of resin. They also inevitably control the flow properties of the glue mixture by the way they interact physically with the liquid system in which everything is suspended.

Some of these materials require preparation or treatment prior to incorporation into the adhesive mix. Cereal flours having a high gluten content tend to produce gummy lumps if they are not treated to make the gluten more soluble. The addition of sodium bisulfite is one means used to accomplish this. Soft, clear winter wheat flour, however, normally requires no treatment. More intensive treatments are given to the cereal hull flours and the bark flours to improve performance. Cooking with or without added chemicals reduces the otherwise coarse nature of these fillers, producing a more free-flowing mixture.

The amount to add depends on many factors; the topography and porosity of the surfaces being glued, and the balance between cost and durability are two initial considerations. Generally, 10 to 15 percent of the weight of the resin is added in any case to control flow and penetration, the primary function of a filler. Beyond this amount the addition is considered an extender to lower cost. In a well-formulated phenolic glue mix, it is feasible to add fillers and extenders equal to the weight of resin solids without serious loss in durability. Specifications, however, usually limit the inclusion of the so-called amylaceous (cereal) flours to a low percentage because they reduce durability the most.

The use of extenders in high percentages, besides reducing durability, also reduces the tolerance of the adhesive to certain operational factors. Since they tend to reduce mobility, their presence in the mix exacerbates the mobility reductions due to long assembly time and dry wood. Moreover, since they usually require additional water in mixing, they may create problems at short assembly times and high moisture content wood in hot pressing, one cause of the filtered state (see Chapter 3).

As with most materials of natural origin, variability in extender properties is a universal problem that requires constant attention. The properties of the mix therefore also vary accordingly, particularly water content and viscosity. These in turn affect the spreading operation, the assembly time actions, the pressing actions, and the final performance of the bond.

pH. Most phenolic resins used for wood bonding are alkaline in nature, with a pH range between 9 and 12. The sodium hydroxide used to create the pH environment is both a catalyst and a solubilizer of the resin. The latter action permits the development of large molecules that nevertheless remain in solution. Thus, at lower pH the molecules

are smaller, with the concomitant property of needing more time or temperature to cure on the glue line than resins at higher pH. In the production of phenolic resins, chemical engineers program the addition of sodium hydroxide as well as formaldehyde into the reaction kettle to control some of the molecular size and configuration distributions in the final resin.

On the glue line, the high pH of phenolic resins serves two other important functions. One of these is cleaning the wood surface by dissolving off certain contaminants, making them part of the cured film. While this has a weakening effect on links 2 and 3, the boundaries of the glue line at the wood-adhesive interface, the beneficial effect of cleaning outweighs the weakening effect on the cohesion of the film. The second effect of high pH is a swelling of the wood substance, and an opening up of the cell wall structure for better penetration and anchoring of the resin.

At the opposite end of the pH scale, where phenolic resins are even more reactive, the cure rate is so fast that they usually have unworkably short mix lives. It is possible to circumvent this obstacle by using small mixes or by mechanics which mix, apply, and pressurize the assembly within the setting-time constraint. Research to avoid the wood-degradation effect of low-pH phenolic resins would open new ways of producing glued objects with this durable adhesive. Currently this fast curing system is used in patching or repair compounds and other applications where speed is important.

Catalysts. In normal use, phenolic adhesives do not require addition of catalyst at the point of application. Under certain circumstances, however, such as when it is necessary to counter an adversely low pH in the wood, the alkalinity of the resin may be increased by adding sodium hydroxide to the adhesive mixture. Otherwise, only the addition of heat is needed to solidify the resin.

The novolac phenolic resins, lacking methylol adducts, cannot be cured with heat alone. They need in addition a source of formaldehyde before they can be driven to full cure. The source commonly used is either formalin, a 37 percent solution of formaldehyde in water, or paraformaldehyde, a condensed version of formaldehyde in powder form. A less noxious source of formaldehyde is hexamethylenetetramine, a reaction product of ammonia and formaldehyde in powder form. Added to the resin mixture, it breaks down into its two constituents when heated in the press, releasing the formaldehyde precisely as needed.

Toxicity. When it is fully cured, phenolic resin is not considered toxic. However, it is composed of two very toxic chemicals, phenol and formaldehyde, as well as caustic soda. To the extent that there are free or unreacted constituents in the resin mixture, and it should be assumed that there are, the manufacturer's precautions should be observed in applying them. Gloves and eye protection should be worn, and prolonged contact with the skin should be avoided. Fortunately, resins intended for wood gluing are water-soluble, and therefore are easily washed off. Eyewashes and emergency showers are prudent where there is possibility of overspray or spillage.

Vaporized formaldehyde, and to a less extent phenol, may be released in small amounts near hot presses. People who are sensitive to such vapors should remove themselves from the area, or wear protective gear. The 50-year history of phenolic resin use in plywood and particleboard manufacture has led to the feeling that the resin used in wood gluing is rather benign, and requires only careful handling.

Physical Properties. Many of the choices and use characteristics of phenolics are dictated by their physical properties—their form, viscosity, shelf life, solids content, color, odor, durability, and strength.

Form. Phenolic resins are marketed in three basic forms: liquid, powder, and film. Each has advantages and disadvantages in cost and application, though their strength, durability, and color are essentially the same.

Liquid resins, generally of 40 percent resin solids content, are preferred primarily by large users, and in tank truck quantities. They are handled by pump, measured by sight-gauge or meter, and are ready to use without mixing for some applications.

One of the disadvantages of liquid phenolic resins is that when it is desired to increase resin content on the glue line, there will be a concomitant and unavoidable increase in moisture content, increasing the risk of blows and steam blisters during high-temperature pressing. The recourse in this case is to dry the wood to a lower moisture content, but this increases wood preparation costs, reduces the output of dryers, changes mechanical properties, and increases surface inactivation of the wood. The higher strength that comes with lower moisture content has detrimental effects in this case because it means poorer conformation of the joint surfaces during pressing. The lower wood moisture content also results in greater breakage of such elements as veneers and flakes through the various transfer operations, and most important, during pressing.

A disadvantage of liquid phenolic resins is their rather short shelf life, 1 to 3 months. During this time, the resin continues to polymerize on its own, though very slowly. An end point is reached when the increased molecular size raises the viscosity and lowers the mobility beyond useful limits.

Powdered phenolic resins are essentially liquid resins with the water removed. The water is removed by spraying the liquid resin into a column of hot air, flashing off the water, and collecting the remaining solids below. The process is so efficient that the resin sustains very little additional polymerization; it can be reconstituted to the same consistency by adding back the water. Powdered resins are more expensive per pound of solids, but are cheaper to ship because there is no water contributing weight. Because of the absence of water, powdered resins have longer shelf lives—as much as a year or more in cool, dry conditions.

Powdered resins present two options in their application. First, mixing them with water makes them the same as liquid resins with all their properties. Second, they can be applied to the wood surface in powder form, avoiding mixing operations entirely and introducing no additional water to the assembly. However, in order for the resin to achieve mobility for bond forming, it must have moisture as well as heat. The moisture is provided by the wood, which therefore can be of higher moisture content than with liquid resin. The higher-moisture-content wood can be a blessing to the entire process of producing glued assemblies: more through-put in drying, less breakage of fragile wood elements, better surface contact in pressing, and greater dimensional stability of the product due to lower built-in stresses. (See Chapter 5 with regard to the mechanical properties of wood.)

A major disadvantage of powdered resins lies in their application to the wood surface. In one of their important uses, flakeboard manufacture, the resin is applied by tumbling flakes and resin in a blender. The resin particles adhere loosely to the surfaces. The amount that adheres is dependent on, and therefore limited, by the nature of the surface. Although this makes it more difficult to apply the higher levels of resin, they perform better at the lower levels than liquid resins because they tend to penetrate less and make a more substantial link 1.

Films of phenolic resin are made by dipping tissue-thin paper in liquid resin and driving off the water without advancing the resin. The film is then cut into standard widths and rolled into bolts for shipment. Phenolic films have the same operational properties as powdered resins, except that they are not reconstituted to liquid form. They too require no mixing, but unlike powdered resins, film resins can be applied to the surface precisely and uniformly, and without waste. In its most appropriate use, the film is cut to the exact size of the glue line and merely interleaved with veneer, especially thin decorative veneer. As with powdered resins, film resins derive their mobility from heat and moisture

in the wood. These resins also are expensive to buy by the pound, but the additional cost is often redeemed by greater simplification of the assembly and bonding process. Typically, film resins do not penetrate as much as liquid resins, a disadvantage as far as repairing subsurface damage, links 6 and 7, is concerned. This same property, however, is an advantage on thin decorative veneers, where bleed-through and blotching of the surface may be a problem.

Odor. Phenolic resins, particularly the liquid forms, have a characteristic odor reminiscent of liniment. However, this is sometimes masked by formaldehyde if any is being released, or by alcohol if it is present in the mixture. Extenders also tend to obscure the odor of the resin, and lend their own.

Color. Phenolic resins may range in color from a straw or beer color to various shades of red. In the cured glue line they usually appear dark red. The dark red color is a reasonably good indicator of a phenolic bond.

Characteristics After Hardening. Fully cured, neat phenolic resin is glassy hard and brittle. It also shrinks as it cures, a result not only of the loss of mix water but also of the elimination of the water formed in condensation polymerization. The formation of methylene bridges resulting from this condensation also produces a tighter molecular structure. The ensuing shrinkage, sometimes accompanied by crazing, is the main reason that phenolic glues cannot serve as gap fillers.

The outstanding characteristic of a fully cured phenolic resin is its resistance to degradation, which is attributable to the resonant stability of the benzene ring, and the carbon-to-carbon bonds formed in polymerization. Its resistance to heat, to fusion, and to solvation are due to its cross-linked structure. Because of this resistance, tests for bond quality must employ extreme conditions such as immersion in boiling water in order to distinguish good from better.

Resorcinol-Formaldehyde Adhesives

With only a very small change on the benzene molecule, the addition of one oxygen atom, this cousin of the phenolic resins achieves profound changes in solidifying properties. The resin becomes curable at room temperatures. The addition of an oxygen atom to form a second acidic hydroxyl at the meta carbon to make the resorcinol chemical, is not, however, a simple operation. The many chemical reactions and physical treatments result in a costly product. However, its room temperature curing combines with its high durability to make it indispensable for gluing thick products that are difficult to heat but require durable bonds.

Chemically the presence of the second hydroxyl on the benzene ring renders it extremely reactive to formaldehyde in forming the methylol compound, and subsequently to the formation of methylene bridges in solidifying. The reactions are in fact exothermic (heat-producing) and virtually spontaneous, precluding use as a wood adhesive until special procedures for their control were devised. One of these involved the making of the base resin. By maintaining a formaldehyde to resorcinol molar ratio of less than 1, a novolac resin is produced in which molecules are deficient in methylol groups, terminating instead with resorcinol groups. With

this structure, no further polymerization can occur, and the resin is relatively stable. Alcohol may be added to further improve stability during long storage, years in some formulations (see bottom of page).

In preparing for use, the additional formaldehyde needed to complete the cure is introduced as a catalyst (actually a hardener in this case, since it participates in the curing reaction). The formaldehyde may be added in solution as a liquid, formalin; or as a powder, paraformaldehyde, a condensed form of formaldehyde. Formalin provides faster and somewhat uncontrolled curing, since it is free and ready to react. Paraformaldehyde, on the other hand, provides a controlled release, and can be modified through degree of condensation to release at different rates as the use demands.

As with phenolic resins, the speed of cure of resorcinolic resins is related directly to pH and temperature, the higher the faster. Normally, the pH ranges from 7 to 9, giving pot lives of 2 to 4 hours, and pressure periods of 4 hours to overnight at room temperature. A caution is always in order with respect to pot life: The exothermic reaction begins with the addition of hardener. In large batches, the heat generated cannot be dissipated fast enough without a cooling procedure. The additional heat causes faster curing, which causes more heat to be generated, and an upward spiral of heat and curing combine

Resorcinol   Formaldehyde   Monomethylol Resorcinol

Novolac

Fomaldehyde
Alkali

Molecular growth and cross-linking

to drastically shorten the anticipated pot life.

The cured state is essentially the same as the highly cross-linked structure of phenolics, since resorcinol has the same three reactive hydrogens, which allow molecular growth in three directions. It is hard and brittle, brown-red in color, and as durable as phenolics if it is fully cured. The "if" represents a further caution. Resorcinolics do not cure fully below 70°F. The resulting undercure leads to reduced durability. Some of the reduced durability can be recouped by subsequent heating. However, if the glue line has dried out, there may not be enough mobility left for molecular movements to take place.

The "if" also extends to the gluing of very dry wood, e.g., below 6 to 8 percent moisture content. In this case the glue line may dry out before it has a chance to cure fully. This can happen even if the assembly time is reduced accordingly. Some evidence of this is seen in the greater difficulty of passing specification tests in the winter, when heated shop conditions tend to produce lower moisture contents in wood. The dry-wood effect also operates to curtail the dissolution of the hardener which is providing the curing impetus. One can easily demonstrate this effect by spreading a film of adhesive on a dry wood surface and covering with a thin glass plate such as a coverglass for a microscope slide. After the glue cures, viewing through the glass with the aid of some magnification will reveal the still undissolved hardener.

A further "if" ensues in gluing woods that are low in pH, such as oak. The thin glue line cannot maintain its own necessary pH against that of the wood. In such cases, heat must be introduced to restore or boost the curing potential. The gluing of fire-retardant or preservative-treated wood also calls for special attention, and requires advice from the adhesive manufacturer.

Otherwise, resorcinolic resin adhesives are considered to be among the best available for bonding wood. An indication of their bonding power is their use to bond compreg planks (phenolic impregnated and compressed veneer laminates) into blanks from which aircraft propellers were carved during World War II. It seems safe to say that were their price lower, they would be the adhesives of choice for most applications.

The similarity of chemical reactions invites the possibility of combining the properties of phenolic resins and resorcinolic resins to produce a lower-cost resorcinol or a faster-curing phenol. Both have been accomplished either by copolymerization or by physical blending. In copolymerization, a base molecule is formed with phenol. Resorcinol is then added to form the terminal novolac end groups to react with the paraformaldehyde hardener as in the normal manner. The resulting phenol resorcinol formaldehyde resin (PRF) has a lower cost because of its phenol base, and a room-temperature cure because of the resorcinol end groups. In the physical blending approach, the phenolic resin is made as a normal hot-press resole resin with methylol end groupings. At the point of use, resorcinol is added as a hardener. Its cross-linking ease with methylol groups then speeds the cure of the phenolic resin. Instead of adding raw resorcinol in powder form, it is also possible to add resorcinol resin or phenol-resorcinol resin as a liquid hardener with a similar effect.

The resorcinolic adhesives are manufactured as aqueous solutions of 60 to 65 percent resin solids as one part, usually part A of a two-part system. Part B contains the powder paraformaldehyde hardener along with fillers, commonly walnut shell flour and/or wood flour. In preparing for use, a weighed amount of hardener, for example, 20 percent by weight, is added to the liquid resin, and the mixture is stirred briefly. A waiting period of about 15 minutes is recommended to allow the paraformaldehyde to dissolve and begin the hardening action. The mixed adhesive is applied by brush, roller, or extruder, preferrably to both sides of the joint surface. Both minimum and maximum open and closed assembly times are specified, the minimum to assure some viscosity buildup before pressure is applied, thus reducing mobility and avoiding excessive squeeze-out. The maximum assembly time, sometimes as long as 90 minutes, assures that sufficient mobility remain for all glue motions to occur. Pressures of 100 psi for soft woods and 200 psi for hard woods are generally recommended.

The most exemplary use of resorcinolic adhesives is in gluing lumber into beams and arches for heavy construction. In this size product it is impractical to introduce heat for curing other adhesives with sufficient durability to withstand the exposure conditions and the stress conditions that may be imposed on the structure. They are also use-

ful in boat building and in gluing any solid wood pieces for extreme exposures. The red glue line signals its presence, and for this reason it is not widely used for veneering due to bleed-through tendencies which represent blemishes on the face of panels. This same bleed-through tendency, however, suggests that resorcinols have good penetrating ability and will produce good links 6 and 7.

Urea-Formaldehyde Resins

The reaction of formaldehyde with available hydrogens as exemplified in the case of phenol and resorcinol has several counterparts in other adhesive compositions. One of these is related to the reaction of formaldehyde with ammonia to form hexamethylenetetramine, the "hexa" hardener used in curing phenolic novolacs. If the ammonia is part of an organic molecule, the reaction occurs in a similar fashion. Urea contains two ammonia groups (technically "amino" groups, and hence the resulting compounds are classed as amino resins). The four reactive hydrogens allow that many opportunities for forming methylol groups, and cross-linking to produce a hard, brittle, insoluble resin when fully cured.

The adhesive is made by reacting formalin with urea under acid conditions at varying molar ratios, pH, and heat until a prescribed viscosity is reached. Cooling and pH adjustment assures useful stability. The adhesive is rich in methylol groups, needing only to be triggered into condensing, splitting off water, and forming methylene bridges. Triggering is accomplished by lowering the pH. This can be done by direct addition of acid, but would result in too rapid cure and too short a pot life. Hence a method of controlled release of acidic elements is designed into the formulation.

The timed release of acidic elements is accomplished by indirect means. A catalyst is selected that contains salts of strong acids and weak bases such as ammonium chloride, ammonium sulfate, or ammonium phosphate. When such salts dissolve, they produce the corresponding acid. Dissolution is encouraged and controlled by the removal of the ammonia fragment. For this purpose, an extra amount of free formaldehyde is included in this resin to carry out its spontaneous reaction with ammonia as in producing "hexa." For this reason, all urea resins have varying amounts of free formaldehyde as part of the formulation. Because part of this formaldehyde escapes to the atmosphere during processing and during service in products, manufacturers are finding ways to eliminate it or reduce it to tolerable limits.

Formaldehyde can usually be identified during use of the adhesive, and sometimes in the glued product. The pungent and acri-monious odor of formaldehyde has raised concerns about the safety of urea adhesives. Although formaldehyde is a toxic chemical, its actual effect on humans is still being debated; dermatologic or allergic effects are acknowledged, however. The author notes that he was in constant daily contact with formaldehyde fumes, sometimes of suffocating intensity, for 5 years during the mid-1940s, and although rashes and watery eyes appeared, no other ill effects have manifested themselves during the following 45 years.

Urea resins are remarkably versatile. They can be formulated to cure at room temperature, or at elevated temperatures; they can be highly diluted with extenders or fortified with other resins; they are used to bond all the wood elements, from lumber to fibers. These resins are considerably lower in cost than phenolics or resorcinolics, but they are more sensitive to heat and moisture; they cannot resist boiling and therefore are in a lower durability category. Hydrolysis by moisture accelerated by heat appear to be the major mechanisms reducing performance.

The cured resin is hard, brittle, and creep-resistant. It is relatively colorless in the pure state, but becomes tan when fillers are added. Sometimes a color is added to the catalyst as a marker or to confirm that catalyst has been incorporated in mixing. Substantial shrinkage occurs during cure, due to loss of both mix water and the water given off in the condensation reaction. The tightening three-dimensional structure of the cured resin produces additional shrinkage. The shrinkage characteristic results in a tendency to craze, and therefore precludes use as a gap filler. Gaps greater than 0.015 in. are beyond the reach of urea adhesives; less than 0.010 in. is the preferred limit. Additives are available, however, that improve crazing and gap-filling properties. These include fibrous fillers and reactive liquids, such as furfuryl alcohol. The former act to reinforce the cured structure, and the latter reduce the amount of water in the system while contributing to the solids content.

Urea resins are marketed in two forms, powder and liquid, the latter in various resin solids concentrations. In preparing the liquid for use, the user adds catalyst, fillers, extenders, and fortifiers as a particular application demands. Thus, with a single resin in inventory, the user (with the advice of the manufacturer) becomes the formulator, custom designing mixtures to suit the need. The liquid form is also more attractive to the large user, where pumping and metering can be used to advantage and stock turnover is rapid. The powder form has two variations. The "neat" form, without any additives, is used like the liquid form in that the user becomes the formulator. It has the advantage of a longer storage life, it is cheaper to ship, and it is somewhat more convenient for the small user. In the other powder variation, catalyst, fillers, and fortifiers are already incorporated as the intended use demands. The user merely adds the necessary water. This form is preferred by craftspeople and small job operators. Because all forms of the resin contain highly reactive methylol groups which are ready to polymerize, storage time and conditions need to be closely monitored. Cool, dry conditions are recommended, and FIFO (first-in, first-out) scheduling are urged to prevent waste. Generally, the neat powdered resins have the longest shelf lives, followed by the catalyst-incorporating powder resins. Liquid resins have the shortest lives, 2 to 3 months.

Fillers for urea resins include nut shell flours and wood flours added to a limit of about 15 percent of the resin solids. Fillers are intended to control flow and penetration of the adhesive on the glue line. However, above 15 percent filler the mixture begins to lose fluidity, restricting flow too much and making it more difficult to apply. Extenders such as the amylaceous (e.g., starchy) flours have a greater degree of compatibility, and therefore do not interfere as much with the physical actions of the adhesive. Consequently, they can be added in greater amounts, over 200 percent in some applications. A notable formulation is known as a 100-100-100 mix, being 100 parts liquid resin, 100 parts wheat flour extender, and 100 parts water. Used in hardwood interior paneling, this mix costs much less per pound and has adequate strength, but some durability against water is sacrificed.

Highly extended resins such as the 100-100-100 mix require extra cautions in gluing procedures. Amylaceous extenders, though more assimilated into the glue mixture, retain a high degree of filler action. Since fillers tend to reduce flow and penetration of the adhesive, other factors that also reduce these adhesive actions—for example, long assembly times, low-moisture-content wood, warm wood, and low pressure—can combine to induce bond formation toward the unanchored or even prehardened condition. Thus links 6 and 7, the subsurface damaged region, may not be adequately formed. Links 4 and 5 may also be jeopardized to the degree that flour particles occupy, block, or straddle adhesion sites on the wood.

Another source of caution in the use of highly extended mixes arises from the extra parts of water added to accommodate the flour. In hot-press operations, this extra glue water adds to the extra water that may already have been added to the glue line to achieve the higher spread rates recommended for rough veneer or to prolong assembly times. Total water content of the assembly entering the hot press thus may become more than can be dissipated during the press cycle. Two consequences can result; one is blisters as trapped water forms steam pockets, an outcome that is intensified on veneer with a high moisture content. A more insidious consequence, especially on inactivated surfaces where insufficient water is drawn from the glue line during the assembly time, is the inclination to filtering of the resin into the wood, leaving the bond line with an even higher ratio of flour to resin, and predisposing it to delayed failure.

In pressing at room temperature, when the adhesive is formulated to cure without heat, the main cautions center around wood moisture content, wood temperature, and assembly time. As with any adhesive that depends on chemical reactions to cure, urea resins require a reaction medium in which to carry out their molecular activities. If water is drawn out of the glue line too swiftly, which may happen on very dry wood, cure proceeds with difficulty and may be incomplete. Wood with less than 6 percent moisture content—and for some formulations, less than 8 percent—can produce this dry-out problem. It occurs regardless of whatever else the bond line may encounter. Assembly time usually has a maximum of about 20 minutes, and generally no minimum. However, in gluing lumber a minimum of 5 to 10 minutes of closed assembly time is beneficial in reducing squeeze-out. Warm or hot wood speeds the cure, but also shortens the permissible assembly time. Cold wood, below 70°F, reduces the cure rate and inhibits full cure.

The cautions with respect to highly extended mixes point to the general conclusion that they are more sensitive to some operational factors, i.e., more difficult to use, and more inclined to dysfunction. Therefore the reduced adhesive cost must be balanced against the cost of increased process control or defective product. When problems arise, one of the solutions is to use a less extended mix that will be more tolerant of use conditions.

As mentioned above, urea resins are very versatile, and find usage throughout the wood gluing field. Their typical use is in producing hardwood plywood for interior applications, and for reconstituted wood paneling, particularly particleboard for furniture corestock and floor underlayment. They are also widely used as a binder for medium-density fiberboard, a premium material for overlays in producing furniture or paneling. The acidic nature of urea adhesives makes them less suitable for gluing oily woods or woods with high extractive contents. They have been found to be inconsistent in bonding Douglas fir, and therefore are seldom used to produce plywood with this species.

Melamine-Formaldehyde Adhesives

With the same general chemistry as urea but operating from a different molecular structure, melamine resins provide vastly improved bond performance properties, rivaling those of phenolics. The chemistry involves the same reaction of formaldehyde with amine groups to form methylol groups, which then condense to methylene bridges as with ureas. In this case, however, the amine groups are part of a cyclic structure of alternating carbons and nitrogens called melamine. It is postulated that the cyclic structure imparts greater stability to the resulting linkages. The three amine groups assure a three-dimensional, cross-linked molecular structure when fully cured. Because of the instability of melamine resin in liquid form, it is generally spray-dried to powder form (see below).

Fully cured melamine resin is very hard, glassy, and durable against heat and moisture. It is also colorless, being clear as glass. Its hardness, clarity, and durability have made it a common kitchen accessory, as in Melmac tableware and Formica countertops. With regard to bond performance, melamine resins undergo the same specification testing as phenol and resorcinol resins, and are generally considered to be equally durable for

Melamine   Formaldehyde   Monomethylol melamine
up to hexamethylol if
M/F ratio is 1 : 6

most practical purposes. Differences in durability appear only after extended periods of severe exposure.

While its chemistry is similar to that of ureas, its operational characteristics are different; it can only be hot-pressed, and in this respect it is more related to phenolics in terms of bond formation. However, because it is more expensive, its use is justified mostly by a need for a colorless glue line. Thus it is used in gluing decorative veneers for exterior exposures such as doors and boats. Because decorative veneers are made mostly from hardwoods, melamine resins have come to be associated with the gluing of exterior-quality hardwood plywood.

Because of their high cost, melamine resins are usually blended with urea resins to produce a bond durability somewhat less, but still of exterior quality for some uses. The melamine is sold separately as a fortifier and blended by the user, or blended by the manufacturer along with fillers and catalysts, so that only the addition of water is needed to produce a usable mix. It has been noted that powder blends achieve different levels of durability with assembly time; longer times produce better results. This is attributed to differential mobility of the two resin fractions during the heating cycle in the press. At longer assembly times the reduced mobility tends to keep the fractions together. Since durability is decreased at shorter as-

sembly times, it is assumed that the melamine is the more mobile fraction.

In contrast to phenolic glues, which normally are pressed at temperatures of 300°F or more, melamine glues are more like hot-press ureas, curing at 240 to 260°F. The lower temperature makes them less liable to produce steam blisters than phenolics, although combinations of high-moisture-content veneer and heavy spreads are still to be avoided. Because of their acidity, melamine glues may perform less well on woods of high extractive content.

Epoxy Adhesives

The outstanding property of epoxy resins is that they are liquids with 100 percent solids content. They therefore contain no solvents to be dissipated, and they experience very low shrinkage upon curing. With their excellent adhesion, they thus perform in situations where gaps exist or where substrates are nonporous and impervious.

Chemically, the resin is based on the reactions of the epoxide ring or ethoxylene group. This three-member ring is highly reactive, being easily opened by either basic or acidic agents. A typical reactant is methyl alcohol, which opens the ring and adds on to produce methoxyethanol. The reaction produces a solid when reactants are bi- or multifunctional. Most commercial epoxies depend

Epoxide ring   Methanol   Methoxyethanol

Epichlorohydrin   Bisphenol-A   Catalyst   Epoxy resin

on epichlorohydrin to provide the initial bifunctionalism for the epoxide element, and bisphenol-A to provide the bifunctionalism for the hydroxyl-bearing element. With a molar excess of epichlorohydrin, the resulting molecule becomes terminated at both ends with epoxide rings. This product reacts with another bisphenol-A to continue polymer growth until reactants are exhausted, or until the desired molecular size or viscosity is reached. The hydroxyls that are produced along the chain provide further points for reactions, cross-linking to produce an insoluble, relatively infusible, thermoset resin.

The resins are cured with catalysts or reactive hardeners which form the ultimate cross-links for a strong, durable molecular structure. In order to simplify the addition of curing agent, a two-part system is used; one part contains the resin, and the other part the hardener and any additives to be incorporated. The system is contrived for easy use; equal portions of each part are mixed to produce the curing adhesive, which then sets at room temperature. Curing agents for epoxies include polyamines, polyamides, and polysulfides. Urea and phenolic resins, as well as acid anhydrides, may also be used.

Toxicity may be a problem, as both the resin and the curing agents are highly reactive chemicals. Amines in particular can cause dermatitis and asthma attacks in some people. Ventilation and cleanliness can minimize ill effects.

As with all synthetic polymers, opportunities abound for modifying properties by changing substituent groups; by varying functionality to control cross-linking and rigidity; by molecular size and viscosity control; by blending or copolymerizing with other polymer species; or by using various hardeners.

Epoxies have a high degree of adhesion for many materials, including wood. Forming a bond between two wood pieces, however, is not as simple as it would seem. The formulations that are generally available lack sufficient "body" to keep them from penetrating into the wood and starving the glue line. Consequently, fillers such as titanium dioxide and ferric oxide are added up to 50 percent by weight without loss of shear strength at room temperature. Filler additions up to 150 percent of the weight of resin are feasible in some instances. The addition of fillers also reduces costs and increases heat resistance.

The cost of epoxies precludes their widespread use in wood gluing except for specialty work as in making novelties or repairing broken joints. Pot life can be varied by the amount of hardener added. Users generally desire fast bond formation, and this can be as quick as 5 minutes, varied by the amount and kind of hardener added. Pot life varies accordingly. Short pot life goes with fast setting. Those that have a short pot life are also more exothermic, requiring that mixes be made in small batches.

After hardening, the epoxy resins resist breakdown by heat and solvents, and are also relatively insensitive to moisture (the room-temperature-cured systems less so than the heat-cured ones). However, they are not as resistant as phenolics, the accepted standard. Improved resistance is developed by combining epoxies with phenolics.

Isocyanate Adhesives

Developed during World War II in a search for better adhesives to bond tire cords and rubber, isocyanates quickly became known as adhesives that can bond "anything to anything." They have been slow in being applied to wood because of their high cost and toxicity. Nevertheless, they represent a leading edge to technologies for bonding wood, promising less demanding processes and improved products.

The adhesive is based on the high reactivity of the isocyanate radical, $-N=C=O$. Coupled with strong polarity, compounds that carry this radical have not only good adhesion potential, but also the potential for forming covalent bonds with substrates that have reactive hydrogens. When a molecule contains two isocyanate radicals, as in diisocyanate, it combines adhesion with the ability to develop cohesion by polymerizing with itself. Reaction of a bifunctional isocyanate with a bifunctional alcohol produces linear molecules, while tri- and tetrafunctional molecules allow cross-linking. Thus the properties of these materials can be varied through a wide range from elastomeric to rigid, making possible such diverse products as powder puffs, synthetic shoe leather, and structural car parts.

The basic reaction that typifies the isocyanate functionality is with an alcohol:

$$R_1{-}N{=}C{=}O \;+\; R_2{-}OH \longrightarrow R_1{-}\overset{\displaystyle H}{N}{-}\overset{\displaystyle O}{C}{-}O{-}R_2$$

With water:

$$R_1{-}N{=}C{=}O \;+\; HOH \longrightarrow R_1{-}NH_2 \;+\; CO_2$$

With a dihydric alcohol and diisocyanate:

$$HO{-}R_1{-}OH \;+\; O{=}C{=}N{-}R_2{-}N{=}C{=}O \longrightarrow HO{-}R_1{-}O{-}\overset{\displaystyle O}{C}{-}\overset{\displaystyle H}{N}{-}R_2{-}N{=}C{=}O \longrightarrow polymer$$

Since both ends of the product are still reactive, the molecule can continue to grow in a linear manner. Many different polyfunctional molecules bearing reactive hydrogens are available for use in this reaction, providing a range of properties. Three or more such hydrogens, as in glycerol or sugar, would produce a cross-linked, rigid material.

Two diisocyanates dominate this field: Toluene diisocyanate (TDI)

and diphenylmethane-4, 4diisocyanate (*MDI*)

Prepolymers are made by reacting polyfunctional alcohol (polyol) with a diisocyanate. When the polyol is in molar excess, the prepolymer is hydroxyl-terminated and can-

not grow further until more diisocyanate is added. This then would become a two-part system as might be used in producing foamed products. With diisocyanate in excess, the prepolymer is isocyanate terminated. As such it can react with moisture, forming amines, which then can react with themselves to grow and cross-link, becoming thus a one-part adhesive system, with moisture provided by the substrate or the atmosphere.

The development of these almost spontaneous reactions into usable adhesives involves as a first consideration controlling reactivity during storage and application, and releasing it at the appropriate instant in the glue line. A second consideration is reducing the toxic effects. The latter is approached by reducing volatility, primarily through making larger molecules which have a lower vapor pressure. Proper ventilation, protection, and housekeeping are also necessary, and cautions are usually conveyed as part of use instructions.

Control of reactivity reflects the genius of molecular engineering. A number of systems have evolved that solve many of the problems. The isocyanate that is most commonly used because of its lower volatility is diphenylmethane diisocyanate (MDI). Another common isocyanate is toluene diisocyanate (TDI), which is more volatile and therefore not used as a wood adhesive.

The simplest adhesive system is the direct application of MDI in some polymeric form by spraying onto wood elements, chiefly flakes or strands. The process is otherwise the same as for producing flakeboard, a hot-press operation. Although the cost per pound is higher than for the phenolics that are normally used, several advantages accrue that offset the initial cost:

1. Less MDI is needed to produce the same board properties.
2. Lower press temperatures can be used.

3. Faster press cycles are possible.
4. More moisture can be tolerated in the flakes.
5. Therefore less drying energy is needed.
6. There is somewhat greater dimensional stability in the boards.
7. No formaldehyde is emitted.

There are also some disadvantages:

1. Panels stick to cauls or press platens, necessitating the application of release agents (sometimes incorporated in the MDI prepolymer).
2. Greater care in use is necessary to avoid health hazards.
3. There is too little tack to maintain the integrity of the mat during transport.
4. Special storage conditions are required to avoid premature moisture reactions.

To overcome some of these disadvantages, several approaches are used:

1. Diisocyanate can be prepolymerized (PMDI) to form a larger molecule with lower volatility.
2. The diisocyanate can also be "blocked" chemically to prevent premature polymerization. The blocks break down under the heat of the press and the isocyanate is free to perform its bonding actions.
3. Emulsifying MDI reduces volatilization, and permits more uniform distribution of the binder on the flakes, since in being less concentrated, a greater volume of material passes through the spray nozzles for the same amount of solids deposited.
4. Two component systems widen applicability by allowing room-temperature curing.

Fast-setting, two-component systems that cure at room temperature offer a breakthrough that could alter some of the traditional processes for bonding wood. Patterned after the use of the two-component polyester systems for bonding fiberglass where resin and catalyst are mixed at the spray nozzle and immediately blended with the fiber, a similar process seems possible for wood.

The reaction of diisocyanate with multifunctional molecules produces urethane resins of varying properties, from foams of low density for insulation and packaging to foams of intermediate density that simulate wood, to solids of great strength. In the two-part system, one part contains a polyester, polyether, or polyhydroxy compound, generally referred to as the resin side, the polyol side, or simply the B side. The other part is the A side: It contains the cross-linker, MDI or TDI, either in "neat" form or as isocyanate terminated polymers. Catalysts are usually incorporated in the B part along with any other additives necessary to the properties desired. Mixing is done immediately before application. Machines have been developed that meter out the two liquids in correct proportions to a high-speed mixer which mixes and delivers to the surface in less than a second. There is never a reservoir of mixed glue; it goes to the surface as fast as it is mixed.

Mixers are of three types: static, dynamic, and impingement. The static mixer consists of a short, slender tube containing an arrangement of baffles that split and interflow the two incoming streams of A and B many times before emerging as one liquid. In the dynamic mixer, the short tube is fitted with baffles on a spindle which turns at high speed, thoroughly blending the liquids as they pass through. The output from these mixers is fed through a nozzle for spraying or simply poured or extruded onto a surface or into a form. Mixers of this type require purging with solvents after each use to prevent polymerization in the tube and plugging of the system. An alternative is impingement mixing, in which the two streams, A and B, come together from opposite directions and impinge against each other at high velocity, causing instant mixing. The impingement occurs just behind the orifice, and the mixed liquid immediately emerges. Plugging does not occur, and the need for purging solvents is eliminated.

These systems of simultaneous mixing and delivering to the joint surface bypass the pot or tank which holds a batch of mixed glue until it is needed. In order to maintain spreadability for long periods of time, the adhesive in a pot has to be slow in curing, which then also means longer time under pressure to cure in the glue line. Since pot life and press time are directly related, simultaneous mixing and application offers a unique means of greatly shortening press cycles by permitting the use of adhesives that are too fast-curing for batch mixing. With specially engineered materials handling systems for the wood-element part of the product, gluing processes could reach the ultimate goal of fast, continuous pressing without the need to introduce heat through layers of wood. In the case of veneer gluing, this also means higher permissible moisture content, less breakage, and higher yields.

The reaction of isocyanates with hydroxyl groups produces covalent primary bonds. This raises the possibility of the formation of such bonds through the hydroxyls on wood substance. If so, it would explain the excellent adhesion of isocyanates and also the improved dimensional stability of panels made with them. While isocyanate bonds are considered to have less resistance to heat and water than phenolics, the performance of flakeboards bonded with these binders seems to be slightly better in accelerated exposure tests. It seems plausible that the improved dimensional stability reduces the development of the interparticle stress that destroys bonds.

As a general conclusion, isocyanate-based adhesives seem to bear out the proposition that the route to the highest degree of bond formation begins with the application of small, polar molecules, which are then polymerized in situ to an insoluble, infusible state with as little shrinkage as possible.

Adhesives with Combination Hardening Systems

A major outcome of combined hardening mechanisms is not so much the improvement of bond formation, but the improvement of bond performance. Generally, liquid properties remain the same in terms of handling and application. However, the hardened glue line is more durable, though not necessarily stronger. This is true of three of the four glues discussed in this section: cross-linked PVAc, casein, and soybean. The fourth, hot animal glue, is different. Bond performance is not improved, but bond formation is much faster due to the rapid heat-loss component of the glue.

Hot Animal Glues. Unique in the speed with which these glues achieve bond strength and raise it to a high level, hot animal glues also employ a unique two-stage mechanism for hardening in the glue line. In the first stage, a strong gel forms as the glue cools in contact with the surface. In the second stage, water diffuses out of the glue line, raising the strength proportionately. The first stage occurs quickly, as in a hot melt, conferring handling strength in seconds or minutes. The second stage is governed by the laws of diffusion, and can vary depending on the wood and its ability to absorb water. Other variables include the water content of the glue formula and the amount of glue in the joint.

In producing hot animal glues, collagen is extracted from bones, hides, and tendons (as for liquid hide glue), hydrolyzed to a soluble state, and dehydrated to a solid. It is then ground to flakes or granular powder, and sold in that form. The user prepares it for application by soaking in cold water until swollen, and then heating to 150°F to reduce viscosity. The amount of water added depends on the grade of the glue, which in turn depends on the jelly strength at a standard concentration and temperature. The strength is measured in grams needed to force a plunger to a certain depth in the jelly. Hence, grades are rated in grams, and assigned names, numbers, and letters, such as A Extra, 1XM, etc., to indicate each of the many grades available. Jelly strength is primarily a function of the extraction procedure. Fractions that come off early in the extraction have lower grades than those that come later. The source and the amount of hydrolysis also have a bearing on the jelly strength.

Higher grades require more water to reach the proper consistency for the intended use. In general, although the higher grades give slightly greater bond strength, choice of grade is made on the basis of rate of gelation for the assembly time required. The higher grades provide the fastest gelation, the shortest assembly time, and the maximum speed of operation, while the lower grades provide longer assembly times and longer high-tack periods. Thus choice of grade relates more to handling properties than to bond performance.

These animal glues are modified in many ways to obtain a wide range of properties suitable not only for wood but also for bookbinding, gummed tape, and paper gluing. Additives include defoamers, preservatives, wetting agents, plasticizers, and insolubilizers. A gel depressant, thiourea, is sometimes added to lengthen the time to gelation and thus increase assembly time.

While hot animal glues are among the easiest to use because of their inherent high adhesion to many materials (including glass), they have their own cautions against failure. The most serious of these is a consequence of one of their most desirable attributes: fast gelling. Maximum permissible open assembly time is relatively short. Once placed on the surface, these glues, like hot melts, begin immediately to cool, gel, and skin over. Fast mating of the parts is recommended, after which some delay before pressure is applied allows a little thickening and helps reduce excessive flow and squeeze-out. Also, as with the hot melts, placing these glues on cold surfaces accelerates gelling and further shortens assembly time, perhaps curtailing all bond-forming actions, including flow, transfer, penetration, and wetting.

Another, often violated caution is overcooking, in terms of both temperature and time. As in food preparation, cooking is a breakdown process. A batch of glue that remains in the pot at 150°F all day suffers some breakdown, and will not be as strong

if it is reheated and used the next day. Higher temperatures accelerate the breakdown. Hot animal glues tend to thicken toward the end of the day as a result of the evaporation of water. Adding water restores viscosity, but does not restore broken molecules. The general admonition is to discard old glue and start fresh every day. Old glue can be used for other purposes, such as sizing.

Bonds made with hot animal glues tend to be somewhat stronger than those made with liquid animal glues. Both types are brittle, unaffected by solvents, and do not load sandpaper. Both soften with moisture and more so with heat, though they are fairly resistent to heat alone. Hot animal glues find their most appropriate use in assembling furniture, where the quick bonding permits fast assembly with short dwell times in pressing equipment. Because animal glue bonds are not durable against moisture, this characteristic is an advantage when bonds need to be broken to make repairs.

Cross-Linked Polyvinylacetate. By adding cross-linking potential to water-loss hardening, polyvinylacetates take a step toward greater durability and creep-resistance. Cross-linking is accomplished by incorporating small quantities of molecular fragments that carry extra functionalism into the main molecule. These can then be activated to form bonds with adjacent molecular chains. Activation is initiated with appropriate catalysts or cross-linkers added at the time of use.

These adhesives are therefore two-part systems that require mixing before use, but they are otherwise used like conventional PVAc. Cross-linking, like most chemical reactions, can be accelerated by the addition of heat, making it both a hot-press and a cold-press glue. The cross-linking imparts greater rigidity to the cured glue film, reducing the mobility of individual molecules and making them less susceptible to softening by heat and moisture. Bond performance is thus improved, allowing use in structures that produce a higher level of stress or that are subjected to more extreme exposure conditions. However, they are not as resistant to stress and severe exposure as resorcinolics.

Casein Adhesives. A protein derived from skim milk, casein, like animal glues, is composed of amino acids bound together with amide or peptide bonds. Though of similar chemical constitution, the protein derived from animals differs markedly from that derived from milk. Animal protein is gelatinous, made up of linear polypeptide chains with few free amine or carboxyl groups. As a result, it is nearly neutral in solution and is more difficult to insolubilize. As a gelatin, it is easily liquefied by addition of heat and a small amount of water. Casein, on the other hand, does not liquefy or soften when it is heated. It has many free amine and carboxyl groups, more of the latter, making it amphoteric and reactive with both acids and alkali.

These glues have similar proteinaceous chemical compositions as animal glues, with peptide bonds

$$HO-\overset{\overset{O}{\|}}{C}-R\left[\overset{\overset{H}{|}}{\underset{\underset{H}{|}}{C}}-\overset{\overset{H}{|}}{N}-O-\overset{\overset{O}{\|}}{C}\right]_{n}RNH_2$$

connecting molecular segments which contain varying numbers of amino and carboxylic side groups, but in greater number than animal protein. After the casein curd has been isolated, it is solubilized with sodium hydroxide, hydrolyzing the peptide and reacting with the carboxyls. Dried, it is formulated into powdered adhesives with various additives to modify properties, chief among which is lime. The user adds only water.

Casein is separated from milk by reducing the pH to about 4.5, its isoelectric point. The curd that precipitates is washed, dried, and ground to a fine powder. One hundred pounds of milk yields about 3 lb of casein. Casein is freely soluble in an alkaline medium. The particular alkali used has a bearing on the properties of the hardened film. For example, if the alkali is ammonia or a volatile derivative, it will evaporate during cure and leave an almost pure, insoluble casein. Sodium hydroxide produces the salt, sodium caseinate, after reacting with the carboxyls. This salt is soluble, and the glue bond thus remains sensitive to moisture. Calcium hydroxide, on the other hand, forms the insoluble salt calcium caseinate. Being bivalent, calcium also acts as a cross-linking agent to further insolubilize the casein and increase durability.

In an adhesive for maximum durability, the casein curd is brought into solution with sodium hydroxide, which is then converted to sodium caseinate and finally to calcium caseinate in the glue line. However, the powder that is sold as an adhesive contains not sodium hydroxide, for that would make it too hygroscopic and unstable, but a weak salt of sodium such as sodium carbonate or trisodium phosphate. An excess of lime is also incorporated. The casein, the carbonate, and the lime, being all powders, are premixed dry and sold ready for the addition of water by the user. The water activates the reaction of lime and carbonate, releasing the sodium hydroxide, which then solubilizes the casein, forming sodium caseinate. This action takes place during mixing, and accounts for the waiting period after mixing specified in handling instructions. In the glue line, the calcium gradually replaces the sodium and develops cross-links, forming first a gel and then hardening as water diffuses out and into the wood.

In addition to these chemicals, casein glues may contain a small percentage of oil to control dust and to retard the reaction between casein and the incorporated chemicals. Fillers such as wood flour and nut shell flours are included to increase "body" and restrain overpenetration. Viscosity is adjusted with such chemicals as calcium chloride or formaldehyde to thicken the glue, and sodium sulfite or sugar to thin it down. Since casein is of natural origin, it is subject to biologic attack. To protect against molds, fungi, and bacterial decomposition, preservatives are also incorporated when maximum durability is desired.

Casein wood glues usually require two parts of water to one part of casein to prepare for use. They are applied at the rate of 60 to 100 lb per 1000 ft^2 of glue line. The wood therefore is required to absorb considerable water to harden the glue. In bonding veneer of medium density, $\frac{1}{8}$ in. thick, for example, the wood moisture content could be raised by as much as 15 percent. If the wood began at 10 percent, it would be 25 percent coming out of the press.

Like other glues, casein glues have advantages and disadvantages. They are reasonably tolerant of some of the more vexing factors operating against bond formation and performance. Their high alkalinity, around pH 12, helps overcome some surface inactivation. The high alkalinity also makes casein one of the better glues for bonding oily or high-extractive-content woods. Casein glues are tolerant of wide temperature variations, accepting the range between just above freezing to the temperatures used in hot pressing. However, bonds made at lower temperatures take longer to form and may be less durable than those formed at higher temperatures. Because of their relatively high viscosity and apparent "body," they are considered to be good gap fillers. While they do form reasonable bonds on rough surfaces, their 66 percent water content ensures that some shrinkage will occur as hardening progresses, and therefore they may not sustain a bond across a wide gap. The resistance of casein to heat is good; it does not soften with heat due to the molecular makeup of the glue. As mentioned above, strength and creep resistance is excellent, although on higher-density woods, wood failure averages only about 50 percent in shear tests.

One of the major disadvantages of casein glues is that they develop a grayish stain on woods such as maple and oak due to the high alkalinity. Formulations are available that produce less stain, but they are also less water-resistant. Casein glues are considered water-resistant rather than waterproof, since upon soaking in water for 48 hours in a standard test, specimens lose approximately 40 percent of their original dry strength. Assembly time is relatively short, 20 to 30 min-

utes, enough for most operations in which they are used.

Casein glues can be modified in many ways by adding, subtracting, or substituting a number of chemicals to alter viscosity, setting time, staining, abrasiveness, water resistance, and resistance to biologic attack. Blending with other adhesives such as blood albumin improves water resistance, but only if the glue is hot-pressed. Mixtures with elastomeric latexes reduce brittleness, and allow better bonding to dissimilar materials.

One of the more taxing uses of the higher-quality casein formulations that illustrates their reliability in bond performance is in lumber laminating for beams and arches to be used in heavy construction. While they need to be protected from severe environments by painting or used in interior locations where moisture content can be controlled to less than 16 percent, they are able to sustain the continuous loads of structural members. Other structural uses are fabricating box beams, and attaching gusset plates in making trusses. Since they are tolerant to varying temperature and rough surfaces, casein glues are particularly desired for on-site fabricating where nail-gluing is the only form of pressure. Some formulations are sufficiently fast setting to allow use in secondary gluing such as attaching door skins to door frames.

Recent events have curtailed the availability of casein glues for structural purposes in the United States. Since they no longer can be adequately protected against microorganisms, most such formulations have been withdrawn from the market until new methods of preservation have been developed.

Soybean Adhesives. A close chemical cousin of casein, the protein in soybean meal is made up of similar amino acid peptide chains. This protein, however, has two distinct features. Its molecules are tightly coiled, and its polar groups are more or less occupied internally and not available to express their polar nature in the natural state. Second, a large amount of carbohydrate is included with the soybean meal, of the order of 30 percent. Protein itself comprises about 50 percent of the total mass. The rest is cellulosic fiber, ash, and water. Soybean meal is flour that has been made from soya beans by grinding and extraction of oil.

The chemistry of adhesive production with soybean meal follows closely that for casein: solubilizing with sodium hydroxide, and then insolubilizing with calcium hydroxide. The solubilizing accomplishes two important changes. It permanently uncoils the protein molecules, and it renders the carbohydrate fraction tacky. The latter action contributes useful properties to the liquid adhesive, but not to the hardened adhesive. In the glue line, carbohydrates have the same effect as extenders, reducing strength and water resistance. It is partly for this reason that soybean glues are less durable than casein glues. There is evidence that if the carbohydrate is removed from the system, the remaining protein would have improved water resistance, but not enough to equal casein, and not enough to warrant the extra expense of removing it.

Many other chemicals are added to improve or modify properties. These include silicates, which help develop a flowable viscosity and prolong pot life as well as assembly time. Carbon disulfide, thiourea, and formaldehyde donors denaturize the protein and increase the cross-linking for improved durability. Pine oil reduces foaming during power mixing and roller spreading. Preservatives that protect against molds and bacteria are sometimes added when necessary. In uses that do not require high strength and durability, extenders such as clay and flours of various kinds may be added to reduce cost even further. Soluble blood can be incorporated to boost water resistance in hot-press formulations.

The ability of soybean glue bonds to resist occasional soaking for short periods of time led to their widespread use in producing interior grades of softwood plywood in the early days of the industry. Both hot-press and cold-press systems were used, the latter requiring the clamping of plywood assemblies in bales overnight. This meant the handling of heavy head boards, I-beams, and clamps for each bundle, a rather large burden of labor. A minor change in formulation led to the elimination of the clamping period and revolutionized the cold-press process almost overnight. By modifying the glue mix so that a firm gel developed with the loss of only a small amount of water, the pressure period was reduced dramatically, making possible a "no clamp" process. In the "no clamp" process a bale of plywood assemblies is made, given a few minutes of assembly time, and then placed in a hydraulic press at the prescribed pressure for 15 minutes to allow development of holding power, after which the bale is set aside for the glue to harden the rest of the way. The reduction in labor and clamps produced an immediate economy in the manufacture of interior plywood.

With the advent of synthetic resins and the desire on the part of industry to produce only exterior-quality plywood and thus avoid failures due to inappropriate applications, the use of soybean adhesives for structural plywood has declined sharply. However, uses for other, less demanding purposes such as box shooks, laminating low-density fiberboards, and paper overlaying continue due to the low cost, acceptable glue bond strength and durability, and working properties of these glues.

Combined Adhesive Systems

The few adhesive systems described above, and the hundreds of possible modifications, suggest rather convincingly that no universal adhesive exists. Each modification inevitably strives to maximize some particular quality, but it also inevitably engenders *exclusiveness,* in being less applicable to other uses. One approach to becoming more *inclusive* is to *blend* different kinds of adhesives, and thus *combine* properties of each. A few examples follow.

Phenolic resins and rubber combine the strength, durability, and adhesiveness of phenolics with the elastomeric properties of rubber to produce a tough, durable bond that can accommodate the differential movements of dissimilar materials and make possible the bonding of such materials as wood to metal and metal to metal. Resorcinolic resins and rubber provide the same properties but with room-temperature setting and a higher price.

Casein and rubber provide the same ability to bond dissimilar materials, without the need for hot pressing, but at a lower level of durability.

Phenolics and resorcinolics convert phenolics to room-temperature-setting materials while at the same time reducing the cost of resorcinolics and maintaining the durability of both.

Melamine and urea resins reduce the cost of melamines while increasing the durability of ureas; i.e., melamine resins become a fortifier for urea resins. The combination, however, loses the cold-press versatility of ureas.

Polyvinyl acetate and urea resins combine the fast bond formation of PVAc with the more durable urea. The brittle urea stiffens the PVAc bond, reducing creep. Room-temperature hardening is retained, as well as the fast bond formation of the PVAc.

Starch and urea resins retain the fast tack and bond formation of starch with additional durability contributed by the urea.

Starch and resorcinolic resins do the same as ureas but contribute a higher level of durability.

Considerations in Adhesive Selection

With thousands of glues on the market, each with promotional literature suggesting the many uses to which they are applicable, the user often feels left to the flip of a coin in selecting one for a particular use. This may be one way to choose between brands, but some logic is available to help choose between types of glues. Leaving economics aside, two technical considerations need to be applied: One, the most important, deals with bond performance, the strength and durability requirements of the glued product. The other deals with bond formation, handling properties that make the glue fit the operating circumstances with a minimum risk of misuse and failure, i.e., tolerance to the specifics of the process.

Given the endless modifications that multiply the available choices, practicality demands reducing attention to a few generic glues, and broadly categorizing them on the basis of their reaction to several factors from the equation of performance that are particularly discriminating.

With respect to bond performance, resistance to four degradative factors provides a means of rating glues on the basis of durability: stress, heat, moisture, and organisms, identified by the acronym SHMO. Lacking a universal test that might impose quantum increments of these factors, such that each glue might be given a single number to compare with others, the author has intuitively averaged personal observations to yield a comparative assessment. Using letter grades for resistance to each factor, averages emerge which produce a useful ranking according to SHMO. These ratings are shown in Table 4-1 for the various generic types of woodworking adhesives. In confirmation of these ratings, Figure 4-1 shows the results of long-term aging tests on the first four types of adhesives.

Table 4-1 is useful only to the extent that it represents the best bonds that can be made with each generic type of glue—unmodified, unfortified, and unextended. Modifications engender changes that would move the ranking up or down depending on their impact on durability. The "best bond" reservation implies that these glues may not produce their best bond consistently in practice due to their level of tolerance for factors that normally vary throughout a range in gluing processes. In other words, some glues are easier to use than others, either because they are more forgiving or because they eliminate some procedure that carries a source of variability. Although carelessness has no place in gluing operations, some glues are more foolproof or more tolerant than others and are not as likely to dysfunction if misused.

The factors that are most likely to subvert bond formation include gaps, pressure, moisture content, temperature, assembly time, catalyst addition, mixing, and pot age. Using the same intuitive ranking procedure,

Table 4–1.   Bond Performance of the Generic Woodworking Adhesives Ranked According to SHMO[a]

ADHESIVE	STRESS[b]	HEAT	MOISTURE	ORGANISMS	INTUITIVE RANKING[c]	TYPICAL USE
		RESISTANCE TO				
Phenol	E	E	E	E	1	Exterior softwood plywood
Resorcinol	E	E	E	E	2	Exterior laminated beams
Melamine	E	E −	E −	E	3	Exterior hardwood plywood
Urea	E	G −	G −	E	4	Interior hardwood plywood
Casein	E	G	F	P	5	Interior laminated beams
Soybean	E −	G	F −	P	6	Interior softwood plywood
Starch	F	G	P	P	7	Lumber cores, paper
Animal	G	F	P	P	8	Furniture assembly
PVAc	F	F −	P	E	9	Furniture assembly
Rubber	P	P	F	E	10	Overlays, paper

[a]E = excellent, G = good, F = fair, P = poor.
[b]Stress = continuous stress.
[c]Assuming that each represents the best bond achievable (see Figure 4–2).

glues can be rated on the basis of resistance to varying operational factors in bond formation, i.e., tolerance (see also Chapter 10). Figure 4–2 shows the result of such a ranking. It is distressingly inverse, mostly, of the order for bond performance. Although rubber is lowest in terms of bond performance, it is highest in terms of ease and certainty of bond formation. Contrarily, phenolic adhesives are highest in bond performance, but the bond is most difficult to achieve.

Let us examine these two extremes. Rub-

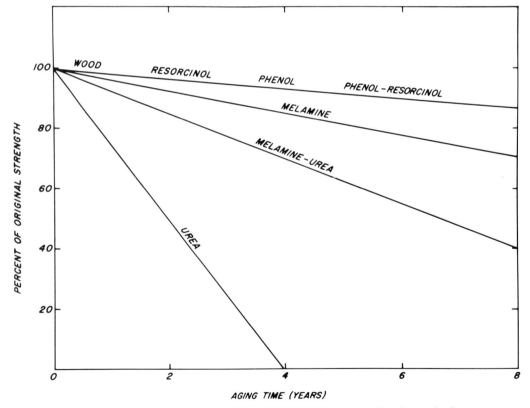

Figure 4–1   Long-term aging of adhesive bonds exposed unprotected to the weather in plywood specimens. (Courtesy of USFPL.)

ber cement is applied to the surface to be bonded directly from the container as received; no catalyst to be accurately measured and added, no solvent to be measured and added, no fillers, no extenders, no mixing (i.e., no opportunity for error in preparing it for use). The glue is applied by any convenient means. A long open assembly time is permissible, in fact necessary, to allow solvents to escape; the surfaces are mated and a bond immediately forms. Care must in fact be taken to place the surfaces together in exact position the first time, because the instant bond does not allow a second chance to reposition them. Hand or roller pressure overcomes bridging, ensures more contact, and further improves the bond. No heat needs to be driven through the wood, and no time needs to be spent in press or clamps.

Overpenetration is not a problem since rubber cements do not migrate far beyond the glue line except on coarse woods (on which more than one coat may be needed due to the low solids content of solvent-based systems). *Starved* joints therefore do not occur unless insufficient glue was applied in the first place. Excessive *squeeze-out* also does not occur since the glue is allowed to set to a firm gel before pressure is applied; link 1 is always there. The motion of *transfer* is never aborted because glue is applied to both surfaces. *Wetting* is not dependent upon pressure; it occurs spontaneously with the aid of solvent, which also helps overcome surface contamination with certain extractives. Thus links 4 and 5 are automatic, albeit comparatively weak. *Solidification* of link 1 requires only loss of solvent; no molecular changes need to be executed in the environment of the glue line. Because of this factor, sources of many negative effects are out of play with rubber cements. Only two main cautions apply: sufficient spread of glue, and sufficient open assembly time.

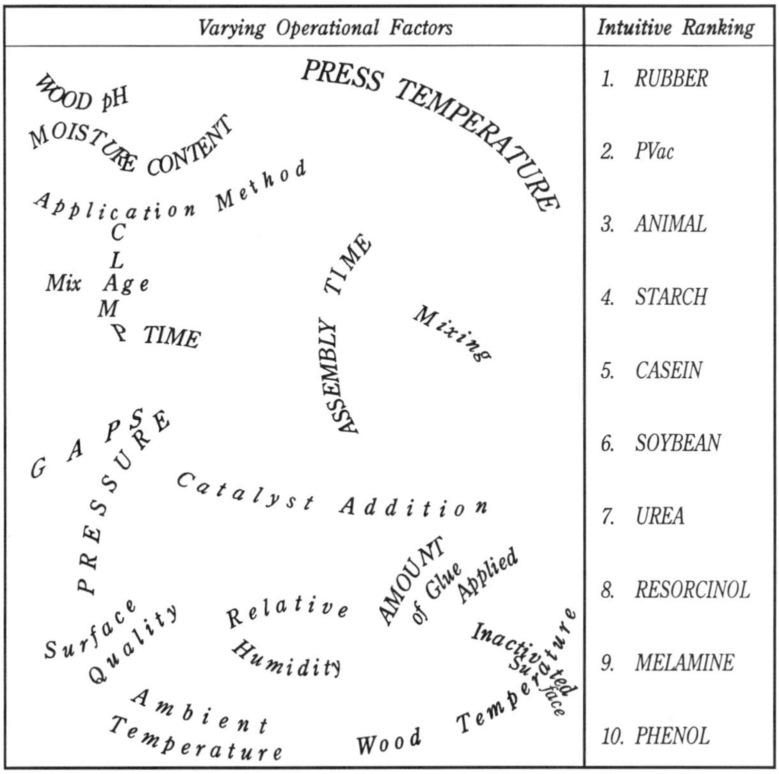

Figure 4–2 Types of woodworking adhesives ranked according to tolerance to variability, ease of use, or tolerance to misuse. In general, the more chemistry that needs to occur on the glue line, the more likely it will be that some adhesive action will be affected by operational variables.

Table 4−2. Selection of Adhesives: Glued Products and the Glues that Bond Them[a] Special Formulations Adjust the Same Base Glue for Different Applications.

PRODUCT	GLUE	REMARKS
Lumber, face-to-face:		
Beams and arches	Resorcinol	Exterior use
	Phenol-resorcinol	Exterior
	Casein	Interior
Posts and turnings	Urea	Interior
	Aliphatic	Interior
Lumber, edge-to-edge:		
Furniture paneling	Urea	Clamp or RF
	X-PVAc	Clamp or RF
Core stock	Urea	Clamp or RF
	Aliphatic	Clamp
	PVAc	Clamp
	Casein	Clamp
Lumber, end-to-end:		
Scarf joints	Phenol-resorcinol	Exterior
	Melamine-urea	Exterior and RF
Finger joints	Phenol-resorcinol	Exterior
	Melamine-urea	Exterior and RF
	X-PVAc	Interior and RF
	Urea	Interior and RF
	Aliphatic	Interior
	PVAc	Interior
Veneer:		
Plywood	Phenol	Exterior softwoods
	Melamine	Exterior hardwoods
	Urea	Interior hardwoods
	Extended phenol	Interior softwoods
	Soybean	Interior softwoods
Laminated veneer lumber	Phenol-resorcinol	
Splicing (edge-to-edge)	Urea	
	Hot melts	
	Tape	
Flakes, strands, wafers:		
Panels	Phenol	Exterior
	Isocyanate	Exterior
Particles, fibers:	Urea	Interior
	Phenol	Exterior
Assembly joints:	X-PVAc	Interior
	Aliphatic	Interior
	PVAc	Interior
	Animal glue	Interior

Phenolic adhesives, on the other hand, have the opportunity to encounter almost all of the factors that can produce negative effects, beginning with storage time and conditions; and in the case of plywood glues, mixing to incorporate additives, sometimes cooking extenders, and measuring caustic, with timing and order of addition stipulated. The adhesive is applied to only one surface, leaving transfer, penetration, and wetting of the opposite surface, and formation of links 4, 5, 6, and 7 dependent upon glue line conditions. All of the chemical action of polymerization from small molecules to large insoluble ones of link 1, occurs in the varying environment of the glue line. This environment is strongly influenced by moisture content and moisture gradients, temperature and temperature gradients, assembly time, pressure, and press time. All are factors that vary (except press time when it is on automatic timer). The glue, however, has tightly fixed chemical and physical requirements that do not tolerate wide swings in mobility-changing factors. Thus both the opportunity for unfavorable conditions, and the sensitivity to them is inherent in bond formation with phenolics. This mandates very close control of the entire process, and justifies the ranking that phenolics are the most difficult to use.

Glues between these two extremes have various combinations of operational risks and tolerance to them. Unfortunately, it is still a scientific dream to develop an adhesive that combines the bond formation of one, rubber, with the bond performance of the other, phenolic, in a single glue. Beyond this dream is one that would bring costs of such a universal glue down to the level of starch or soybean.

It should be noted in passing that while the most difficult-to-use glues are those which require molecular reaction on the glue line, these same glues also provide the most durable bonds. This poses the question of the role of molecular or particulate size on the establishment of the crucial links that depend upon penetration and proximity to achieve bond integrity. It would appear that adhesive components, so small and so mobile as to require efficient filler action to ensure link 1, are also most able to create links 4, 5, 6, and 7.

Thus the choice of a glue for a given application inevitably involves trade-offs. Until the super, all-purpose glue appears, Table 4–2 may be helpful in making a first approximation choice based on the type of product being glued. From that point selection of a specific adhesive formulation may require trial and error and/or consultation with a manufacturer.

5

Composition and Characteristics of Wood

The most basic description of wood is that it is a cellular, ligno-cellulosic material. To an adhesive chemist, accustomed to bonding various kinds of materials, this description provides enough information to begin devising a compatible liquid-to-solid system for adhering to wood. Note that we *did not* say ". . . for producing a bond between pieces of wood." The latter is the work of an adhesive technologist, who can harmonize specific characteristics of wood and gluing operations with the rheological properties of a potential adhesive to assure that a strong bond ensues from the hardening actions promised. In this chapter, the properties of wood that affect bond formation and bond performance are discussed. The subject matter divides naturally into four major categories: anatomical properties, physical properties, chemical properties, and mechanical properties.

The properties of wood are well documented in the literature. How these properties affect bond formation and bond performance, however, requires elaboration in the context of the gluing principles developed in Chapter 3. Many studies have sought to relate species (and, by inference, their properties) to gluability. Certain species, such as those used in the manufacture of plywood, laminated beams, and particleboard,

have received considerable attention. Often the study involves an adhesive variable, a process variable, and a species variable, but seldom a wood property variable other than moisture content. Generalizations and extrapolations are rarely attempted or justified.

Because of the highly variable nature of wood, both within and between trees and within and between individual pieces of wood, and because of the evolutionary nature of adhesives (i.e., always changing, always improving), statements about adhesives and species may have a limited utility or life span.

In order to provide information that is more widely useful, and more lasting against adhesive changes, we have taken a more specific, and at the same time, more generalizable, approach. The description of the different species of wood, with their infinity and complexity of detail can be found in textbooks written specifically for that purpose. Here we shall focus instead on the effects of the properties of wood on the actions of adhesives in bond formation and bond performance. Which species have which properties then becomes a separate question in determining the actual performance in a given instance.

Similarly, rather than associating properties of wood with all possible glues, we shall associate properties with their effects only on the mobility parameter, where effects are discrete, limited, and relatively precise. The simplifying presumption here is that an open pore, for example, is an open pore, no matter what species it is in; and mobility is mobility, no matter what glue is providing it. The effects of density, pH, strength, and other properties follow from the same reasoning.

ANATOMICAL PROPERTIES OF WOOD

In considering the anatomical properties of wood, we are interested primarily in their effects on the movement an adhesive makes into the wood structure, i.e., *penetration.* Hence the approach aims to sense the degree of openness that exists in the wood being glued.

Since all wood properties begin with the standing tree, it is useful to keep in mind an image of a tree that defines its parts and locates the sources of characteristics. The traditionally useful part of a tree, the trunk or bole, appears to be cylindrical in nature. Actually it is a truncated cone; that is, it tapers, being widest at the ground line and becoming gradually smaller in diameter in the upper portion of the tree. Often the tree is neither round nor straight. Taper, roundness, and straightness all have a bearing on the quality of the surface presented to the glue,

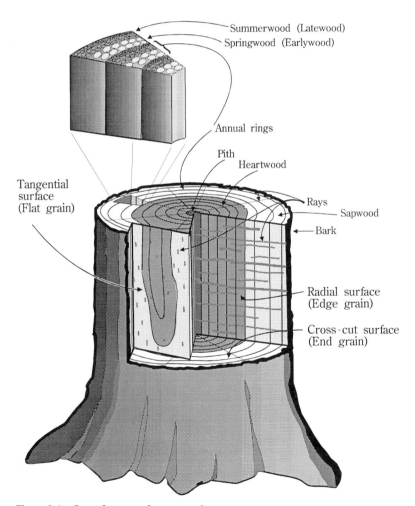

Figure 5-1   Gross features of a tree trunk.

quite apart from the basic anatomy of the wood. We will discuss this aspect further in Chapter 6. For now, the parts shown in Figure 5–1 represent the gross features of a tree trunk with which we will be concerned. These include all the wood from the bark inward, divided into two zones, sapwood and heartwood; and the annual rings, also divided into two zones, springwood and summerwood. The two major species categories, softwoods and hardwoods, have their own anatomical differences. All the parts have significance to the gluing process. The significance lies in the finer detail of these parts, and deeper study is therefore needed.

The simplest representation of the anatomy of wood is as a bundle of soda straws, aligned mostly in one direction, connected endwise in various ways, and glued together sidewise. The passageways in the straws, called *lumens,* vary in size, and the walls vary in thickness. Some passageways are clear, some are partly obstructed, some are plugged, and most are coated with the residues of a once-living cell. All the straws also have openings or *pits* connecting them along their length. These vary in size, number, and location, as well as permeability, both within and between species, and within the same piece of wood.

Two basic kinds of straws—*cells,* technically—exist. One, called a *fiber,* is primarily a strength-giving element; the other, called a *vessel* or simply a *pore,* is a cell for conducting liquids (Figure 5–2). These two kinds of cells are the distinguishing characteristics between the two major kinds of wood, *softwoods* and *hardwoods.*[1] Softwoods, coniferous or needle-bearing trees, are composed almost entirely of fibers, more precisely, *fiber trachieds,* which serve the dual role of strength and conductance. Softwoods have no vessels, although some, such as the pines,

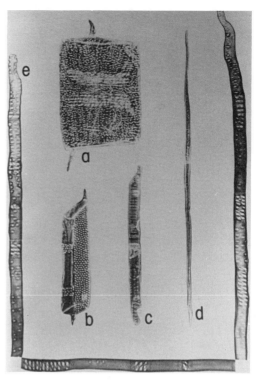

Figure 5–2   Wood cells: a, b, and c are hardwood vessels (pores); d is a hardwood fiber; and e is a softwood fiber. (Courtesy of USFPL.)

do have resin canals for the transport of resinous materials. (The presence of excessive resin in wood can cause some problems in gluing through its effects on sorption and adhesion.)

Hardwoods, deciduous or broad-leaved trees, contain vessels—tubelike structures stacked one upon another like barrels—that form a continuous vertical conduit. However they also contain fibers for strength. The fibers in hardwoods, however, are somewhat different than those in softwoods. Softwood fibers are rectangular in cross section, two to three times longer than those in hardwoods, and occur in fairly straight radial rows. Hardwood fibers, on the other hand, tend to be circular in cross section, and are dispersed randomly among the vessels. The fibers in hardwoods are mostly thick-walled, with few interconnecting pits. Those in softwoods are generally more permeable to adhesives than the fibers of hardwoods. In hardwoods, the vessels are the main avenues for penetration of adhesive.

Both hardwoods and softwoods contain other types of cells with specialized functions, such as ray tissue for horizontal conduction, and parenchyma for food storage. These cells or tissues play a lesser, though sometimes a devastating role in bond formation because of the inhibiting materials they may contain. The tubular nature, and the basic cellular differences between hardwoods and softwoods, are shown in all three dimensions in Figure 5-3.

From a bond-forming and performance standpoint, the major anatomical differences between species of wood reside in the sizes of fibers and vessels, their amounts, their distribution, and the porosity they provide. These differences are discussed below.

Springwood and Summerwood

One of the differences between species results from the growth pattern of a tree. During the growing season, different kinds and sizes of cells are formed in some trees, depending on genetically driven demands that change over time. This produces a ring effect wherein large conductive cells are formed early in the season, *springwood,* and strength cells are formed later in the season, *summerwood.* The resulting rings can be wide or narrow, and can have different proportions of early and late wood depending on growth conditions (water, temperature, nutrients, age, disease, insects, sun, wind, competition) at the time (see Figure 5-4). Tree rings develop cellular configurations that are characteristic of their species, and thus are of diagnostic interest also.

Some trees appear to grow uniformly throughout the growing season and therefore produce a less distinct ring. This is the case for spruce among the softwoods (Figure 5-5e) and maple among the hardwoods (Figure 5-5f). The term *diffuse porous* categorizes the hardwoods like maple that have a uniform growth, whereas the hardwoods that have a sharply distinct ring such as oak and ash are defined as *ring porous*. An intermediate ring structure occurs in some hardwood species, where the change between

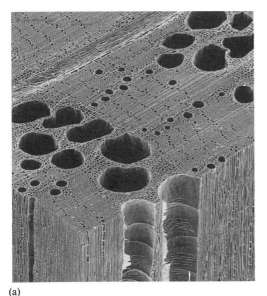

(a)

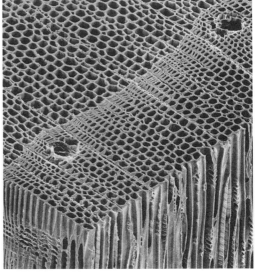

(b)

Figure 5-3 Three-dimensional micrograph of (a) hardwood red oak, × 100; and (b) softwood, Eastern white pine, × 150. (Courtesy of Wilfred Côté, N. C. Brown Center for Ultrastructure Studies, State University of New York, College of Environmental Science and Forestry, Syracase, N.Y.)

Figure 5-4   Growth rings reflect growing conditions in the forest. (Reprinted with permission of Champion International Corporation.)

early growth and late growth occurs gradually. This results in a ring that is termed *semidiffuse* (Figure 5-5k). Thus there are three distinct types of ring architecture in hardwoods: ring porous, diffuse porous, and semidiffuse porous. Among softwoods, the ring structure is referred to as having *abrupt transition* (Southern pine, Figure 5-5m) or *gradual transition* (balsam fir, Figure 5-5d).

More important than architecture is the *proportion* of springwood and summerwood present in the ring, because this is a strong indicator of wood strength. In hardwoods, wood formed late in the season usually has more fibers than early wood, so the more summerwood is present, the stronger the wood will be (and vice versa). In softwoods, which are all fiber, the summerwood fibers have thicker walls, making that part of the ring stronger also. The rate of growth of a tree, because of its effect on the proportions of different cells in the ring, is therefore often used, in terms of rings per inch, as a criterion for strength in wood (see Figure

5-6). In softwoods neither very fast growth nor very slow growth produces the strongest wood. Rather, an optimum rate of growth exists, and this is usually indicated in specifications when strength is of paramount importance.

The apparent drastic difference in porosity between springwood and summerwood, and the close proximity of these two zones on the surface of wood, poses one of the greatest problems in the formulation of adhesives. Optimizing all adhesive motions in the face of such major surface variability calls for a great deal of tolerance in the mobility characteristics of an adhesive. In many situations, the adhesive cannot do it all, but must be aided by some operational factor under the control of the user, such as assembly time.

Sapwood and Heartwood

Another major source of variability within species of wood, and a cause of many gluing problems, is related to the age and growing

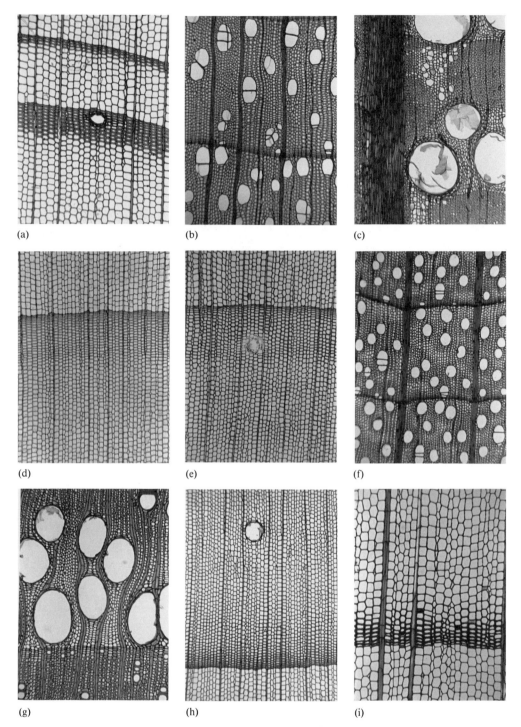

Figure 5-5 Cross sections of various woods commonly glued, × 75. (a) douglas fir (b) yellow birch (c) white oak (d) balsam fir (e) eastern white spruce (f) sugar maple (g) white ash (h) white pine (i) redwood (j) yellow poplar (k) black walnut (l) eastern red cedar (m) southern pine (n) aspen (o) American beech (p) black cherry (q) red oak (r) pecan hickory. (Bruch Hoadley, University of Massachusetts.)

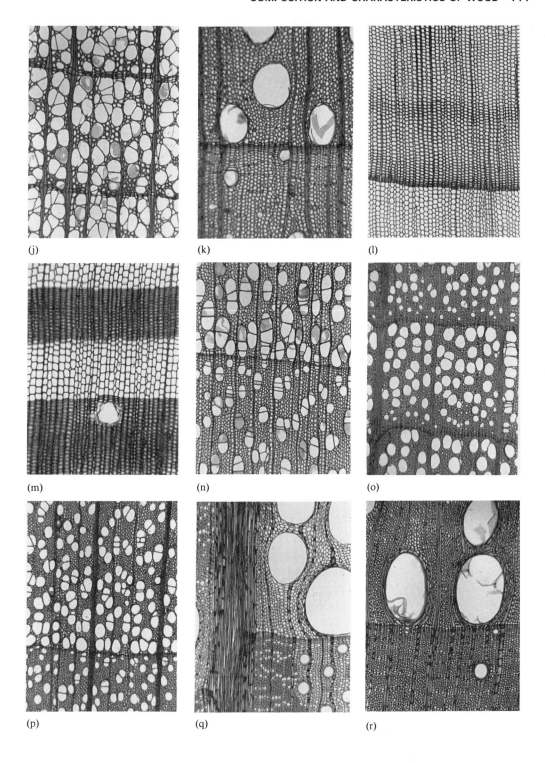

(j)

(k)

(l)

(m)

(n)

(o)

(p)

(q)

(r)

(a)   (b)   (c)   (d)

Figure 5-6 Variations in ring composition due to growth conditions: (a) fast-growth oak; (b) slow-growth oak; (c) fast-growth white pine; (d) slow-growth white pine.

conditions of the tree (see below). With age comes chemical changes in the wood already formed. While these appear primarily as physical changes, they also alter the apparent porosity of the wood, affecting the mobility of adhesives.

The changes produce two distinct regions in the stemwood of a tree, *sapwood* and *heartwood*. Sapwood is the part of the tree where the cells are still physiologically functioning, a band of wood from the inner bark to a point several inches in that varies with species. Inside this sapwood band and extending to the pith is heartwood, composed of cells that have ceased to transport but may store food. Heartwood cells are dead, but they continue to serve the tree as part of the support system holding it up. Both sapwood and heartwood contain the same cell structure and organization; they differ primarily in the rather remarkable way the tree disposes of metabolic waste or unused food while at the same time providing some resistance against decay.

After they die, the cells that were once sapwood slowly become infused with extraneous materials such as oils, waxes, and phenolic compounds, derived from products of life processes. These materials change the color of wood, its permeability, its hygroscopicity, its equilibrium moisture content, its shrinking and swelling, its durability, its density in some species, its rate of sorption, but fortunately, not its strength.

Another transformation that occurs in heartwood formation, but only in the case of certain hardwoods, is the occlusion of springwood vessels or pores with a frothy material called *tyloses*. Tyloses have the effect of clogging the pores and drastically reducing the permeability of the wood to liquids. One of the important differences between white oak and red or black oak is the presence of tyloses in the former. The preference for using white oak to make beer barrels is attributable to the tyloses in its pores, which help make for a watertight container. This same property, however, by affecting the penetrability of the wood, also affects gluing properties, often negatively.

Age of the Tree

Trees, like people, have more or less distinct stages in their life cycles somewhat equivalent to childhood, adulthood, and senility. There are deficiencies at both ends. Young, vigorous trees grow fast under proper conditions. The rings are wide, but the wood produced tends to be inferior in some respects to that grown later in life: It is weaker, and it shrinks and swells more along the grain. This wood is called *juvenile wood*. It is relatively easy to glue because of its low density and open structure, but the low strength and instability may result in unsatisfactory performance of the product. Old trees, on the other hand, tend to grow slowly and produce narrow rings. They also have a higher percentage of heartwood, and concomitantly, a lower amount of sapwood. The wood from old trees of some species—Douglas fir, for example—is considered more difficult to glue than wood of younger trees, suggesting that other changes have also occurred.

Abnormal Wood

When trees grow under conditions normal for their needs, we would expect the wood to also be "normal." However, conditions vary from location to location, and from time to time. Under certain adverse conditions, wood is produced which is regarded as abnormal. In addition to water, sun, and soil ingredients, which affect the quality of wood, one of the more insidious variables is the upright stance of the tree. When, for one reason or another, trees are out of plumb with the direction of gravity, the genetic imperatives of the tree compel corrective actions. The response seems to be a reaction to stress which develops on one side of the tree when its weight is off center as would be the case in a leaning tree. Limbs of trees and consequently their wood are especially subject to the effects of gravity. Because of their origin in highly stressed areas of a tree the wood of both softwoods and hardwoods is called *reaction wood*.

Hardwoods and softwoods react differently to this situation. Softwoods tend to form extra wood on the underside of the lean, as if to make a fillet or brace to support the tree. Since this region is in compression as a result of bending stresses in the leaning tree, the wood formed there is called *compression wood*. Hardwoods, on the other hand, tend to produce extra wood on the opposite or tension side of a leaning tree; the wood formed there is called *tension wood*. In both cases one would think the wood formed in a high-stress region would be especially strong. Perhaps it is for the purposes of the living tree, but for many of our purposes the wood is abnormal. It is abnormally high in shrinking and swelling along the grain, abnormally high in density, and abnormally low in strength. When stressed to fracture, such wood breaks in a brash manner, i.e., like a carrot. The abnormally high instability along the grain causes excessive warping which, if restrained, results in fractures across the grain. While such wood is not particularly difficult to glue, the instability is a source of stress in glued products, and therefore it has an adverse effect on performance.

Compression wood is characterized visually by wide annual rings compared to the width of the same ring on the opposite side of the tree (Figure 5–7). The ring seems to be composed mostly of summerwood, with a gradual rather than an abrupt transition to springwood. To experienced woodworkers the wood looks lifeless. Tension wood also has wide rings. Its presence can often be seen on the side grain, where machining usually produces a fuzzy grain surface. An unusually shiny side grain surface that glistens in the light is also indicative of tension wood. It is virtually impossible to eliminate the fuzzy grain by sanding or other surfacing methods; the fibers tend to behave in a rubbery manner.

Grain

The word *grain* has many meanings in the wood field. When a wood is said to have "grain," the implication is that it has texture and patterns of decorative value. In this con-

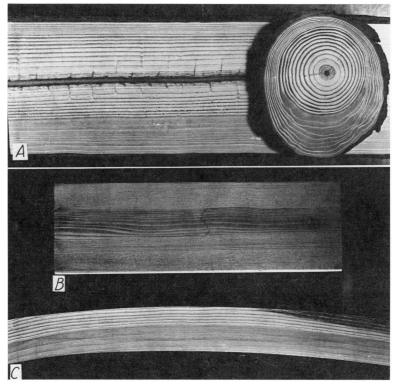

Figure 5-7 (a) Compression wood in softwoods as viewed in cross section and on the face of a board cut from the center of the log. (b) Compression wood in the center of a board, which has ruptured upon drying. (c) Compression wood along on edge of a board, which has caused distortion. (Courtesy of USFPL.)

text a wood surface with "grain," such as oak, has distinctive bands of different cells or tissues, which produce a pleasing appearance. On the other hand, wood with "no grain," such as maple, yellow poplar, and some gums and aspen, produces no or only slight distinctive patterns. These woods are also diffuse porous and therefore have indistinct ring structure. In terms of gluing properties, the woods that have good grain in this context pose a greater challenge to the adhesive because of the differing mobility potential encountered at different points on the surface.

Terms such as *hard grain, soft grain, open grain,* and *closed grain* have meanings approximately as they indicate. In gluing terms they associate with degrees of penetrability, hard and closed being less permeable, and soft and open implying more permeable.

The terms *end grain* and *side grain,* as used in the previous section, indicate the plane of cut with respect to the direction of the fibers in the wood. Thus end grain is the surface in which the cut ends of the fibers occur; it is sometimes also called the *cross section* or *cross grain,* since the cutting tool traverses across the log or board in producing the surface. End grain is the surface one sees on a stump. Side grain, on the other hand, is a surface that exposes the long dimension of fibers; it is also called the longitudinal surface. Two specific side-grain surfaces are recognized: *radial,* a surface cut along the radius of a tree; and *tangential,* a surface cut tangent to the cylindrical trunk of the tree (or perpendicular to the radius) (see Figure 5-1).

The radial surface is also referred to as *quartered,* because it is the surface that would appear if the log were cut into quarters. The term *edge grain* also refers to this cut, because the edges of the annual rings appear on the surface. Radial surfaces feature the broad sides of rays in perfectly quartered cuts, or flecks and splashes of ray tissue when the cut is made slightly off the radius. The tangential surface has an alternate as well: *flat cut,* in which the rays are not as prominent because they are cut in their smallest cross section. They appear as faint, short lines along the grain; their length as well as their width is characteristic for a species. Most wood surfaces that will be glued are neither truly radial nor truly tangential, but are cut at some intermediate angle, producing some of both in the same piece of wood.

Differences in gluing properties due to these longitudinal surface cuts are minimal except as they expose different expanses of summerwood. The surface containing the most summerwood would be one cut on a true tangent. Veneer produced by lathes has the potential for producing the maximum in summerwood surface, because the knife can skim along the ring in the summerwood zone and cleave it onto both surfaces in wide bands. Veneering of Southern pine is especially prone to do this.

Besides angles with respect to either the true radius or the true tangent there is another angle which is of even greater importance. This is the angle of the surface with respect to the true direction of the fibers in the wood. Its importance derives from the strong influence of fiber angle on the mechanical and physical properties of wood, and therefore on the gluing properties and the performance of the glued product. Moisture movement, dimensional stability, strength, and surfacing properties are all related directly to grain angle. Again, several terms reflect this meaning of the word "grain." *Straight grain* implies that the fiber direction coincides with both of the two longitudinal surface planes of the wood piece (Figure 5–8a). *Diagonal grain* refers to the situation where the fiber angle is inclined with respect to the edges of the piece (Figure

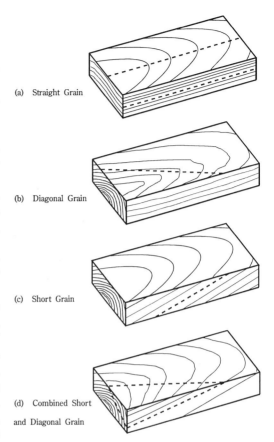

(a) Straight Grain

(b) Diagonal Grain

(c) Short Grain

(d) Combined Short and Diagonal Grain

Figure 5–8 Grain direction in wood elements.

5–8b). In *short grain* the fiber angle is inclined with respect to the faces of the piece (Figure 5–8c). Diagonal grain and short grain can occur alone, or they may occur in combination to produce a *combined grain angle* (Figures 5–8d). These deviations in grain direction are collectively termed *deviant grain*.

Deviant grain has three major sources in the tree. One is that for some reason the fibers in the tree do not form parallel to the long axis of the trunk, but tend to spiral around it. In the tree or log this is called *spiral grain* Fig. 5–9(b). When the log is sawn or veneered, it produces diagonal grain. The second source is crooked logs. Here the fibers may or may not be parallel to the bark and the long axis of the tree, but because the saw or veneer knife must follow a straight path, the cut surface always contains differ-

(a)

(b)

(c)

(d)

Figure 5-9 Sources of defects originating in the tree: (a) deviant grain around knots; (b) spiral grain in the tree; (c) compression failure incurred when the tree hit the ground over a hump of some kind; (d) warping due to abnormal wood or deviant grain. (Courtesy of USFPL.)

ing grain angles. The milder forms of short grain come from logs with pronounced taper. Spiral grain, crook, and taper together produce the combination grain angles.

A frequent source of deviant grain is knots. Knots produce grain that is totally perpendicular to the faces, or all angles in between. The deviation can extend for some distance from the knot itself, being steepest at the knot and leveling off as it merges with the normal trunk wood (Figure 5–9a).

The effects of grain in bond formation involve primarily the porosity that occurs in the different planes of the cut. Cross- or end-grain surfaces are never glued directly for two reasons. First, such surfaces are too porous. This allows excessive penetration, starving the joint, and making it difficult for the glue to form link 1. Figure 5–10 shows the penetration effects of an end-grain surface on Douglas fir with a resorcinolic adhesive. Most of the resin has migrated up the fiber cavities, more so in springwood than summerwood where both porosity and miniscus effects are more favorable.

The second reason for avoiding end-grain bonding is that links 8 and 9 are too strong in this aspect. They carry high along-the-grain loads of a structure, loads that are normally higher than an adhesive bond can support. Moreover, part of the load is in tension, the weakest direction for glue bonds. When wood must be glued endwise, special joint configurations are contrived which mitigate the porosity effect, increase the side-grain bond area, and convert tension loads to more resistant shear loads. This kind of gluing is discussed in Chapter 9.

In side-grain gluing, some deviation from straight that produces a mild form of short grain is desirable. Contrary to expectation, it may be more difficult to obtain a strong bond if the grain is absolutely straight than if the grain deviates a little. One reason for this is that because wood glues are compounded to deal with porosity, they do not have the mobility to penetrate through cell walls. In order to form a strong bond, glues must reach the undamaged wood below the surface. They do this mostly by penetrating through cell lumens or through pits or cracks in the walls. Only when the fibers are at a slight angle with the joint surface can the glue be assured of access to undamaged wood.

Diagonal grain has no bearing on bond formation, but it does cause much unwanted behavior in the glued product. Dimensional changes are greater across the grain than along the grain. Wood that contains diagonal grain always projects a component of instability in the direction that is supposed to be stable. Similarily, wood is weakest across the grain. Wood that contains diagonal grain always projects a component of weakness in the direction that is supposed to be strong.

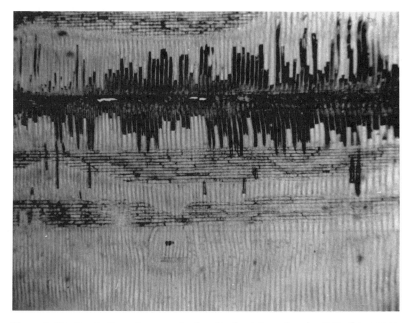

Figure 5–10 Penetration of a resorcinolic adhesive into end grain. Note the greater penetration of adhesive into springwood. Note also the miniscule differences in springwood and summerwood. (Photo by author of microsection by D. A. Stumbo, California Forest Products Laboratory.)

The two components of weakness and instability in the long direction can combine to cause warpage with changes in atmospheric conditions (Figure 5-9d).

Porosity

Porosity may be considered a converse of strength or density from a bond formation standpoint, since it deals with the openness of the wood to the passage of liquids (or gasses). Porosity derives from the term *pore,* which is used by anatomists as a synonym for vessel elements. As used here the term does not refer to the presence or absence of vessels, but only to the degree to which the wood structure permits the flow of liquids through any element. The converse nature of porosity is based upon the fact that wood tissue which is produced primarily for strength has a higher density, and therefore has cells with thicker walls, smaller lumens, and fewer connective openings to other cells. These characteristics tend to restrict the mass mobility of adhesive into the wood structure. (The term *wood structure,* as used here, refers to the cellular nature of wood; *wood substance* would refer to the material of which the cells are made.) While modern adhesive theory discounts the need for penetration in producing strong bonds, as demonstrated by the successful bonding of metal, glass, and plastics, in the case of wood the outermost surface is usually weakened in preparation, or contaminated during processing, or both. This means that stronger wood is less permeable to adhesive and, further, that the stronger the wood, the less likely it is to allow penetration of the adhesive. By implication, then, the stronger the wood, the shallower, and perhaps weaker, will be the bond. Thus an adhesive has to be compounded differently for strong (dense) woods than for porous woods in order to achieve optimum penetration into each.

One can also see that an adhesive would seem to need mutually exclusive properties in order to function in a wood such as red oak or Southern pine, where maximum porosity and maximum density occur within millimeters of each other on the surface of the same piece. Such differences are responsible for much of the variability in bond quality in such species. Since a glue user has very little control over the anatomical variability of the wood being glued, it is up to the adhesive technologist to compound a compromise formulation and teach its proper use for optimizing results over both extremes of porosity. This may involve, at the first approximation, compositions that incorporate fractions of lesser mobility together with fractions with maximum mobility.

There are also other factors of a physical nature, and some of a chemical nature, associated with density and porosity differences in wood that influence bond formation. These affect the finer mobility actions of an adhesive, which is discussed in later sections.

PHYSICAL PROPERTIES OF WOOD

The physical properties of wood play a less direct and more subtle role in bond formation and bond performance. Two physical properties are the primary protagonists in glue function: density, already mentioned, and moisture content. These two factors create many different effects that derive from the ways they affect the mobility of the glue and the subsequent stress on the bond. Moisture content, which is a consequence of the hygroscopicity of wood, affects both bond formation and bond performance through its effect on sorption characteristics and on the dimensional stability of wood. These and other factors are discussed here and integrated with the mobility parameter in terms of their effects on gluing. The changes in dimension that wood undergoes with changes in moisture content produces stress across glue lines. The stress that is produced represents one of the most severe loads that can be placed on a bond. For this reason (among others), moisture content changes are often included in testing procedures for bond quality. The magnitude of this force is illustrated by the practice in early times of splitting stones by driving dry wooden pegs into cracks or holes and pouring water on them.

Table 5–1. Functions Relating Mechanical Properties to Specific Gravity of Clear, Straight-Grained Wood

	SPECIFIC GRAVITY-STRENGTH RELATION*	
PROPERTY	GREEN WOOD	AIR-DRY WOOD (12% MOISTURE CONTENT)
Static bending		
Fiber stress at proportional limit (psi)	$10,200G^{1.25}$	$16,700G^{1.25}$
Modulus of elasticity (million psi)	$2.36G$	$2.80G$
Modulus of rupture (psi)	$17,600G^{1.25}$	$25,700G^{1.25}$
Work to maximum load (in.-lb/in.^{3})	$35.6G^{1.75}$	$32.4G^{1.75}$
Total work (in.-lb/in.^{3})	$103G^2$	$72.7G^2$
Impact bending, height of drop causing complete failure (in.)	$114G^{1.75}$	$94.6G^{1.75}$
Compression parallel to grain		
Fiber stress at proportional limit (psi)	$5,250G$	$8,750G$
Modulus of elasticity (million psi)	$2.91G$	$3.38G$
Maximum crushing strength (psi)	$6,730G$	$12,200G$
Compression perpendicular to grain, fiber stress at proportional limit (psi)	$3,000G^{2.25}$	$4,630G^{2.25}$
Hardness		
End (lb)	$3,740G^{2.25}$	$4,800G^{2.25}$
Side (lb)	$3,420G^{2.25}$	$3,770G^{2.25}$

*The properties and values should be read as equations; for example: modulus of rupture for green wood = $17,600G^{1.25}$, where G represents the specific gravity of oven-dry wood, based on the volume at the moisture condition indicated.
Source: Wood Handbook, USDA.

The resulting swelling force produced the desired split.

Density

Some aspects of density[2] were alluded to in the previous section. The relationship of strength to density has two levels of significance. One is general. It includes all species and shows an upward trend in strength with increasing density. However considerable variation exists suggesting that something besides density, perhaps cellular architecture, affects strength also. A much higher level of significance applies *within* species where the relationship of strength to density is so strong it can be predicted mathematically. See Table 5–1, above. Sources of vari-

ability within species include reaction wood, extractive content, and moisture content. Reaction wood has a higher density, but lower strength than normal wood. Extractives increase density but have little or no effect on strength. High moisture content mostly confounds the density measurement. Its effect on strength is drastic but reversible.

While strength and density are directly related, they have separate effects with respect to adhesive. Strength has a bearing on links 8 and 9 as a bond performance factor in establishing the maximum load the bond has to carry. Density, on the other hand, has its primary effect on bond formation, involving links 6 and 7, and the penetration action of the adhesives. Since cell wall material has a relatively constant density of about 93 lbs/ft^3, variations in the density of different pieces of wood can be presumed, other factors being equal, to be due to variation in the *volume* of cell wall material present. Under the same presumption, there must exist a complementary volume of voids associated with any density below 93 lbs/ft^3. Hence, void volume var-

---

[2]*Density* is expressed as weight per unit of volume, and is measured in pounds per cubic foot or kilograms per cubic meter, or grams per cubic centimeter. *Specific gravity* is an expression comparing the density of a material to the density of water. In a ratio, the units cancel, producing a dimensionless, and therefore a more universal term.

ies inversely with cell wall volume (and density). To the extent that voids are one of the means by which adhesives penetrate wood, one may conclude that penetration will be greater in low-density wood than in high-density wood, a conclusion reached from another direction in the previous section.

Density, however, also plays a powerful role in bond performance because of its strong relationship with dimensional stability. High-density woods not only tend to shrink and swell more with changes in moisture content, they do so with greater force than low-density woods. Bonds between high-density woods therefore are subject to greater stresses with moisture content changes.

This latter fact has strong implications for the apparent performance of otherwise similar-quality bonds. For example, bonds made with the same adhesive in yellow poplar, a low-density, diffuse porous species, and in sugar maple, a high-density diffuse porous species, could conceivably have the same bond quality. If this were a cross-grain construction and subjected to a change in moisture content, bonds in the sugar maple might fail while those in the yellow poplar would remain intact because the higher density maple has a greater potential for creating stress than the yellow poplar. Densities of a number of common woods can be found in Tables 5–2, 5–3, and 5–4, together with other properties.

Hygroscopicity

Wood is a *hygroscopic* ("water-loving") material by virtue of its chemical composition. Wood attracts water from its surroundings in much the same manner as salt or brown sugar draws moisture from the air during the humid days of summer. The attraction is reversible; wood will give up water under dry conditions. Wood therefore takes on and gives up moisture according to the moisture content of its surroundings. The action is predictable and quantifiable: There is a definite relationship between the amount of moisture in the atmosphere, expressed as rel-

ative humidity, and the amount of moisture a wood can attract. *Relative humidity* is the amount of moisture in a volume of air, compared to the maximum amount it could hold at a given temperature, reduced to a percent. Thus for every relative humidity there is a definite wood moisture content in equilibrium with it. The relationship produces the *equilibrium moisture content* or *EMC,* a fairly predictable quantity for most woods. Figure 5–11 shows this relationship as an average for all species.

Several important points need to be made with respect to the relationship between relative humidity and moisture content. One is that the equilibrium moisture content is different depending on whether moisture is being lost or gained. When moisture is being lost in response to a lowered humidity, the equilibrium is higher than when moisture is being gained. This produces the characteristic hysterisis loop shown in Figure 5–11. The reasons for the differing response are hidden in the physiochemical and organizational nature of the cell wall, which deals with sorption energy and accessibility of hydrogen bonding sites. The practical aspects of the situation, however, suggest that when the actual moisture content is crucial, within about 3 percent, conditioning procedures need to account for the variation that might occur.

Another important point is that temperature affects the equilibrium value. Increasing temperature lowers the EMC, the lowering being greater at high relative humidities than at low. Representative values of EMC at various temperatures and relative humidities are listed in Table 10–9.

A third point is that some species have different EMC values. This is attributable mostly to the presence of extraneous materials in the wood occupying sites where water molecules might otherwise be attracted. Such materials have the effect of reducing the hygroscopicity and hence lower EMC values.

Some processing regimes also reduce hygroscopicity, particularly if heat is involved; drying wood at high temperature is one of the more common of these. Such processing

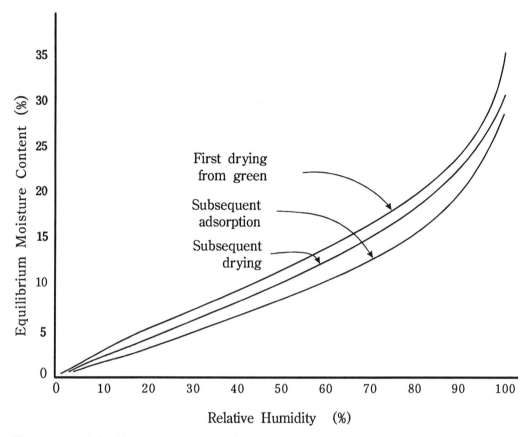

Figure 5-11 Relationship between relative humidity and the moisture content that wood may attain at equilibrium (EMC). Hysterisis is due to restrictions in adsorption sites created during drying.

is common for many wood elements that are destined for gluing, such as veneer, and more particularly flakes, which normally are subjected to temperatures exceeding 900°F. Products that have been consolidated under high heat, often in excess of 350°F, lower the EMC even further. The surfaces of these hot-pressed products may be so severely modified by the heat that they become difficult to glue in secondary gluing processes. A number of disruptive events may occur, inimical to bond formation: changes in molecular configuration of wood substance, such as loss of water of constitution (see next section), translocation of extractive materials to the surface, and closing the porosity of the surface by compression. The latter produces a glossy smoothness that looks good but it inhibits adhesive penetration. Deliberate

treatment of wood elements to improve fire and decay resistance or to reduce shrinking and swelling behavior also have an effect on EMC. In these cases adjustments in adhesive or gluing procedure may be necessary.[3]

The losses in hygroscopicity or absorbency have both good and bad effects on bond formation and bond performance. Lower hygroscopicity means less water absorption under high-humidity conditions, therefore less dimensional change, therefore less stress on the glue line, and therefore better performance of the bond. On the other hand, the same forces that attract water also attract glue, and so a loss of hygroscopicity also

---

[3]Wax is added to many reconstituted products to reduce the rate of water absorption and thus improve bond performance in wet environments. The wax, however, produces little change in the hygroscopicity of the wood.

Table 5–2.   Mechanical Properties[1] of Some Commercially Important Hardwoods Grown in the United States

COMMON NAMES OF SPECIES	SPECIFIC GRAVITY	STATIC BENDING			IMPACT BENDING—HEIGHT OF DROP CAUSING COMPLETE FAILURE (IN.)	COMPRESSION PARALLEL TO GRAIN—MAXIMUM CRUSHING STRENGTH (PSI)	COMPRESSION PERPENDICULAR TO GRAIN—FIBER STRESS AT PROPORTIONAL LIMIT (PSI)	SHEAR PARALLEL TO GRAIN—MAXIMUM SHEARING STRENGTH (PSI)	TENSION PERPENDICULAR TO GRAIN—MAXIMUM TENSILE STRENGTH (PSI)	SIDE HARDNESS—LOAD PERPENDICULAR TO GRAIN (LB)
		MODULUS OF RUPTURE (PSI)	MODULUS OF ELASTICITY[2] (MILLION PSI)	WORK TO MAXIMUM LOAD (IN.-LB/IN.³)						
Alder, red	0.37	6,500	1.17	8.0	22	2,960	250	770	390	440
	.41	9,800	1.38	8.4	20	5,820	440	1,080	420	590
Ash:										
Black	.45	6,000	1.04	12.1	33	2,300	350	860	490	520
	.49	12,600	1.60	14.9	35	5,970	750	1,570	700	850
White	.55	9,600	1.44	16.6	38	3,990	670	1,380	590	960
	.60	15,400	1.74	17.6	43	7,410	1,160	1,950	940	1,320
Aspen:										
Bigtooth	.36	5,400	1.12	5.7	—	2,500	210	730	—	—
	.39	9,100	1.43	7.7	—	5,300	450	1,080	—	—
Quaking	.35	5,100	.86	6.4	22	2,140	180	660	230	300
	.38	8,400	1.18	7.6	21	4,250	370	850	260	350
Basswood, American	.32	5,000	1.04	5.3	16	2,220	170	600	280	250
	.37	8,700	1.46	7.2	16	4,730	370	990	350	410
Beech, American	.56	8,600	1.38	11.9	43	3,550	540	1,290	720	850
	.64	14,900	1.72	15.1	41	7,300	1,010	2,010	1,010	1,300
Birch:										
Paper	.48	6,400	1.17	16.2	49	2,360	270	840	380	560
	.55	12,300	1.59	16.0	34	5,690	600	1,210	—	910
Yellow	.55	8,300	1.50	16.1	48	3,380	430	1,110	430	720
	.62	16,600	2.01	20.8	55	8,170	970	1,880	920	1,260
Cherry, black	.47	8,000	1.31	12.8	33	3,540	360	1,130	570	660
	.50	12,300	1.49	11.4	29	7,110	690	1,700	560	950
Cottonwood:										
Balsam poplar	.31	3,900	.75	4.2	—	1,690	140	500	—	—
	.34	6,800	1.10	5.0	—	4,020	300	790	—	—
Eastern	.37	5,300	1.01	7.3	21	2,280	200	680	410	340
	.40	8,500	1.37	7.4	20	4,910	380	930	580	430
Elm, American	.46	7,200	1.11	11.8	38	2,910	360	1,000	590	620
	.50	11,800	1.34	13.0	39	5,520	690	1,510	660	830
Hackberry	.49	6,500	.95	14.5	48	2,650	400	1,070	630	700
	.53	11,000	1.19	12.8	43	5,440	890	1,590	580	880

Hickory

Species										
Bitternut	.60	10,300	1.40	20.0	66	4,570	800	1,240	—	—
	.66	17,100	1.79	18.2	66	9,040	1,680	—	—	—
Pecan	.60	9,800	1.37	14.6	53	3,990	780	1,480	680	1,310
	.66	13,700	1.73	13.8	44	7,850	1,720	2,080	—	1,820
Shagbark	.64	11,000	1.57	23.7	74	4,580	840	1,520	—	—
	.72	20,200	2.16	25.8	67	9,210	1,760	2,430	—	—
Locust, black	.66	13,800	1.85	15.4	44	6,800	1,160	1,760	770	1,570
	.69	19,400	2.05	18.4	57	10,180	1,830	2,480	640	1,700
Magnolia:										
Cucumbertree	.44	7,400	1.56	10.0	30	3,140	330	990	440	520
	.48	12,300	1.82	12.2	35	6,310	570	1,340	660	700
Maple:										
Red	.49	7,700	1.39	11.4	32	3,280	400	1,150	—	700
	.54	13,400	1.64	12.5	32	6,540	1,000	1,850	—	950
Silver	.44	5,800	.94	11.0	29	2,490	370	1,050	560	590
	.47	8,900	1.14	8.3	25	5,220	740	1,480	500	700
Sugar	.56	9,400	1.55	13.3	40	4,020	640	1,460	—	970
	.63	15,800	1.83	16.5	39	7,830	1,470	2,330	—	1,450
Oak:										
Northern red	.56	8,300	1.35	13.2	44	3,440	610	1,210	750	1,000
	.63	14,300	1.82	14.5	43	6,760	1,010	1,780	800	1,290
Southern red	.52	6,900	1.14	8.0	29	3,030	550	930	480	860
	.59	10,900	1.49	9.4	26	6,090	870	1,390	510	1,060
White	.60	8,300	1.25	11.6	42	3,560	670	1,250	770	1,060
	.68	15,200	1.78	14.8	37	7,440	1,070	2,000	800	1,360
Sweetgum	.46	7,100	1.20	10.1	36	3,040	370	990	540	600
	.52	12,500	1.64	11.9	32	6,320	620	1,600	760	850
Sycamore, American	.46	6,500	1.06	7.5	26	2,920	360	1,000	630	610
	.49	10,000	1.42	8.5	26	5,380	700	1,470	720	770
Tupelo, Black	.46	7,000	1.03	8.0	30	3,040	480	1,100	570	640
	.50	9,600	1.20	6.2	22	5,520	930	1,340	500	810
Walnut, black	.51	9,500	1.42	14.6	37	4,300	490	1,220	570	900
	.55	14,600	1.68	10.7	34	7,580	1,010	1,370	690	1,010
Willow, black	.36	4,800	.79	11.0	—	2,040	180	680	—	—
	.39	7,800	1.01	8.8	—	4,100	430	1,250	—	—
Yellow-poplar	.40	6,000	1.22	7.5	26	2,660	270	790	510	440
	.42	10,100	1.58	8.8	24	5,540	500	1,190	540	540

[1]Results of tests on small, clear straight-grained specimens. [Values in the first line for each species are from tests of green material; those in the second line are adjusted to 12 pct. moisture content.] Specific gravity is based on weight when ovendry and volume when green or at 12 pct. moisture content.

[2]Modulus of elasticity measured from a simply supported, center-loaded beam, on a span-depth ratio of 14/1. The modulus can be corrected for the effect of shear deflection by increasing it 10 pct.

[3]Coast Douglas-fir is defined as Douglas-fir growing in the States of Oregon and Washington west of the summit of the Cascade Mountains. Interior West includes the State of California and all counties in Oregon and Washington east of but adjacent to the Cascade summit. Interior North includes the remainder of Oregon and Washington and the States of Idaho, Montana, and Wyoming. Interior South is made up of Utah, Colorado, Arizona, and New Mexico.

Table 5-3.   Mechanical Properties[1] of Some Commercially Important Softwoods Grown in the United States

COMMON NAMES OF SPECIES	SPECIFIC GRAVITY	STATIC BENDING			IMPACT BENDING—HEIGHT OF DROP CAUSING COMPLETE FAILURE (IN.)	COMPRESSION PARALLEL TO GRAIN—MAXIMUM CRUSHING STRENGTH (PSI)	COMPRESSION PERPENDICULAR TO GRAIN—FIBER STRESS AT PROPORTIONAL LIMIT (PSI)	SHEAR PARALLEL TO GRAIN—MAXIMUM SHEARING STRENGTH (PSI)	TENSION PERPENDICULAR TO GRAIN—MAXIMUM TENSILE STRENGTH (PSI)	SIDE HARDNESS—LOAD PERPENDICULAR TO GRAIN (LB)
		MODULUS OF RUPTURE (PSI)	MODULUS OF ELASTICITY[2] (MILLION PSI)	WORK TO MAXIMUM LOAD (IN-LB/IN.[3])						
Baldcypress	.42	6,600	1.18	6.6	25	3,580	400	810	300	390
	.46	10,600	1.44	8.2	24	6,360	730	1,000	270	510
Cedar:										
Alaska	.42	6,400	1.14	9.2	27	3,050	350	840	330	440
	.44	11,100	1.42	10.4	29	6,310	620	1,130	360	580
Eastern red	.44	7,000	.65	15.0	35	3,570	700	1,010	330	650
	.47	8,800	.88	8.3	22	6,020	920	—	—	900
Northern white	.29	4,200	.64	5.7	15	1,990	230	620	240	230
	.31	6,500	.80	4.8	12	3,960	310	850	240	320
Western red	.31	5,200	.94	5.0	17	2,770	240	770	230	260
	.32	7,500	1.11	5.8	17	4,560	460	990	220	350
Douglas-fir[3]:										
Coast	.45	7,700	1.56	7.6	26	3,780	380	900	300	500
	.48	12,400	1.95	9.9	31	7,240	800	1,130	340	710
Interior North	.45	7,400	1.41	8.1	22	3,470	360	950	340	420
	.48	13,100	1.79	10.5	26	6,900	770	1,400	390	600
Interior South	.43	6,800	1.16	8.0	15	3,110	340	950	250	360
	.46	11,900	1.49	9.0	20	6,220	740	1,510	330	510
Fir:										
Balsam	.34	4,900	.96	4.7	16	2,400	170	610	180	290
	.36	7,600	1.23	5.1	20	4,530	300	710	180	400
Grand	.35	5,800	1.25	5.6	22	2,940	270	740	240	360
	.37	8,800	1.57	7.5	28	5,290	500	910	240	490
Pacific silver	.40	6,400	1.42	6.0	21	3,140	220	750	240	310
	.43	10,600	1.72	9.3	24	6,530	450	1,180	—	430
Hemlock:										
Eastern	.38	6,400	1.07	6.7	21	3,080	360	850	230	400
	.40	8,900	1.20	6.8	21	5,410	650	1,060	—	500
Western	.42	6,660	1.31	6.9	22	3,360	280	860	290	410
	.45	11,300	1.64	8.3	23	7,110	550	1,250	340	540
Larch, western	.48	4,900	.96	10.3	29	3,760	400	870	330	510
	.52	13,100	1.87	12.6	35	7,640	930	1,360	430	830

Species										
Pine:										
Eastern white	.34	4,900	.99	5.2	17	2,440	220	680	250	290
	.35	8,600	1.24	6.8	18	4,800	440	900	310	380
Jack	.40	6,000	1.07	7.2	26	2,950	300	750	360	400
	.43	9,900	1.35	8.3	27	5,660	580	1,170	420	570
Loblolly	.47	7,300	1.40	8.2	30	3,510	390	860	260	450
	.51	12,800	1.79	10.4	30	7,130	790	1,390	470	690
Lodgepole	.38	5,500	1.08	5.6	20	2,610	250	680	220	330
	.41	9,400	1.34	6.8	20	5,370	610	880	290	480
Longleaf	.54	8,500	1.59	8.9	35	4,320	480	1,040	330	590
	.59	14,500	1.98	11.8	34	8,470	960	1,510	470	870
Pitch	.47	6,800	1.20	9.2	—	2,950	—	860	—	—
	.52	10,800	1.43	9.2	—	5,940	—	1,360	—	—
Ponderosa	.38	5,100	1.00	5.2	21	2,450	280	700	310	320
	.40	9,400	1.29	7.1	19	5,320	580	1,130	420	460
Red	.41	5,800	1.28	6.1	26	2,730	260	690	300	340
	.46	11,000	1.63	9.9	26	6,070	600	1,210	460	560
Shortleaf	.47	7,400	1.39	8.2	30	3,530	350	910	320	440
	.51	13,100	1.75	11.0	33	7,270	820	1,390	470	690
Sugar	.34	4,900	1.03	5.4	17	2,460	210	720	270	270
	.36	8,200	1.19	5.5	18	4,460	500	1,130	350	380
Western white	.35	4,700	1.19	5.0	19	2,430	190	680	260	260
	.38	9,700	1.46	8.8	23	5,040	470	1,040	—	420
Redwood:										
Old-growth	.38	7,500	1.18	7.4	21	4,200	420	800	260	410
	.40	10,000	1.34	6.9	19	6,150	700	940	240	480
Young-growth	.34	5,900	.96	5.7	16	3,110	270	890	300	350
	.35	7,900	1.10	5.2	15	5,220	520	1,110	250	420
Spruce:										
Engelmann	.33	4,700	1.03	5.1	16	2,180	200	640	240	260
	.35	9,300	1.30	6.4	18	4,480	410	1,200	350	390
Sitka	.37	5,700	1.23	6.3	24	2,670	280	760	250	350
	.40	10,200	1.57	9.4	25	5,610	580	1,150	370	510
Tamarack	.49	7,200	1.24	7.2	28	3,480	390	860	260	380
	.53	11,600	1.64	7.1	23	7,160	800	1,280	400	590

[1]Results of tests on small, clear straight-grained specimens. [Values in the first line for each species are from tests of green material; those in the second line are adjusted to 12 pct. moisture content.] Specific gravity is based on weight when ovendry and volume when green or at 12 pct. moisture content.

[2]Modulus of elasticity measured from a simply supported, center-loaded beam, on a span-depth ratio of 14/1. The modulus can be corrected for the effect of shear deflection by increasing it 10 pct.

[3]Coast Douglas-fir is defined as Douglas-fir growing in the States of Oregon and Washington west of the summit of the Cascade Mountains. Interior West includes the State of California and all counties in Oregon and Washington east of but adjacent to the Cascade summit. Interior North includes the remainder of Oregon and Washington and the States of Idaho, Montana, and Wyoming. Interior South is made up of Utah, Colorado, Arizona, and New Mexico.

Table 5–4. Mechanical Properties*† of Some Woods Imported into the United States

COMMON AND BOTANICAL NAMES OF SPECIES	MOISTURE CONTENT (%)	SPECIFIC GRAVITY	STATIC BENDING MODULUS OF RUPTURE (PSI)	MODULUS OF ELASTICITY§ (MILLION PSI)	WORK TO MAXIMUM LOAD (IN.-LB/IN.³)	COMPRESSION PARALLEL TO GRAIN—MAXIMUM CRUSHING STRENGTH (PSI)	SHEAR PARALLEL TO GRAIN—MAXIMUM SHEARING STRENGTH (PSI)	SIDE HARDNESS—LOAD PERPENDICULAR TO GRAIN (LB)	SAMPLE NUMBER OF TREES	ORIGIN
Andiroba (Carapa guianensis)	12	—	15,600	1.85	13.4	7,900	1,680	1,220	2	BR
Angelique (Dicorynia guianensis)	—	—	17,400	2.19	15.2	8,770	1,660	1,290	2	SU
Apamate (Tabebuia rosea)	12	—	13,800	1.60	12.5	7,340	1,450	960	9	CS
Apitong (Dipterocarpus spp.)	12	—	16,200	2.35	—	8,540	1,690	1,200	53	PH
Avodire (Turraeanthus africanus)	12	0.51	12,700	1.48	9.4	7,180	2,040	1,080	3	AF
Balsa (Ochroma pyramidale)	12	0.17	2,800	0.55	—	1,700	300	100	**	EC
Banak (Virola koschnyi)	12	—	10,800	1.72	8.1	5,720	1,300	640	8	CA
Capirona (Calycophyllum candidissimum)	12	—	22,300	2.27	27.0	9,670	2,120	1,940	2	VE
Cativo (Prioria copaifera)	12	—	8,700	1.15	7.2	4,490	1,040	610	4	PA
Courbaril (Hymenaea courbaril)	12	—	19,400	2.17	17.6	9,680	2,470	2,440	9	CS
Gola (Tetraberlinia tubmaniana)	14	0.66	16,700	2.21	—	9,010	—	—	11	AF
Goncalo alves (Astronium graveolens)	12	—	17,100	2.17	10.4	10,560	2,060	2,230	4	CS
Greenheart (Ocotea rodiaei)	14	0.93	25,500	3.70	22.0	13,040	1,830	2,630	1	GY
Ilomba (Pycnanthus angolensis)	12	0.44	8,900	1.75	—	5,510	—	750	††	AF
Jarrah (Eucalyptus marginata)	12	—	16,200	1.88	—	8,870	2,185	1,915	28	AU
Jelutong (Dyera costulata)	16	0.38	7,300	1.18	6.4	3,920	840	390	3	AS
Kapur (Dryobalanops lanceolata)	12	—	17,400	2.02	15.5	9,700	1,710	1,230	5	AS
Karri (Eucalyptus diversicolor)	12	—	19,200	2.76	—	10,400	2,140	2,030	21	AU
Kerving (Dipterocarpus spp.)	16	0.69	14,500	2.63	13.3	8,000	1,360	1,160	11	MI
Khaya (Khaya anthotheca)	12	—	11,500	1.41	9.8	6,300	1,700	900	9	AF
Kokrodua (Perciopsis elata)	12	—	18,400	1.94	18.5	9,940	2,090	1,560	6	AF
Lapacho (Tabebuia heterotricha)	12	—	22,600	2.32	26.0	10,930	2,280	3,010	3	PA
Lapacho (T. serratifolia)	12	—	26,300	3.31	23.0	13,420	2,070	3,670	3	SM
Lauan: Dark red Red lauan (Shorea negrosensis)	12	—	11,300	1.63	—	5,890	1,220	680	15	AS

Species										
Light red										
White lauan (Pentacme contorta)	12	—	11,700	1.69	—	6,070	1,200	700	18	AS
Laurel (Cordia alliodora)	12	—	12,100	1.49	—	6,280	1,220	790	13	CA
Lignumvitae (Guaiacum sanctum)	12	1.09	—	—	—	11,400	—	4,500	—	—
Mahogany (Swietenia macrophylla)	12	—	11,600	1.51	7.9	6,630	1,290	810	77	CS
Meranti, red (Shorea dasphylla)	12	—	12,100	1.63	11.7	6.970	—	630	2	AS
Oak (Quercus costaricensis)	12	0.68	17,600	1.64	16.8	—	—	1,570	2	CR
Obeche (Triplochiton sclerozylon)	12	—	7,500	0.86	6.9	3,930	990	430	2	AF
Okoume (Aucoumea klaineana)	12	0.37	7,300	1.14	—	3,900	—	380	§§	AF
Palosapis (Anisoptera spp.)	12	—	12,800	1.82	—	6,630	1,410	920	16	AS
"Parana pine" (Araucaria angustifolia)	12	—	13,500	1.62	12.2	7,650	1,730	780	***	SM
Pau marfim (Balfourodendron riedelianum)	15	—	18,900	—	—	8,200	—	—	5	BR
Peroba de campos (Paratecoma peroba)	12	0.75	15,400	1.76	10.2	8,920	2,140	1,600	(11)	BR
Pine, Caribbean (Pinus caribaea)	12	—	15,200	2.03	15.3	8,000	1,870	1,150	14	CA
Primavera (Cybistax donnell-smithii)	12	—	10,900	1.22	10.3	6,140	1,710	700	4	HO
Ramin (Gonystylus bancanus)	12	—	18,400	2.17	17.0	10,080	1,514	1,300	9	AS
Rosewood, Indian (Dalbergia latifolia)	12	—	16,900	1.78	13.1	9,220	2,090	2,630	5	AS
Sande (Brosimum utile)	12	0.44	—	—	—	6,310	—	500	3	EC
Santa Maria (Calophyllum brasiliense)	12	—	14,800	1.82	13.2	8,060	1,910	1,210	18	CA
Spanish-cedar (Cedrela angustifolia)	12	—	11,300	1.42	12.5	6,010	1,200	570	2	BR
Teak (Tectona grandis)	12	0.63	12,800	1.59	10.1	7,110	1,480	1,030	56	IN
Walnut, European (Juglans regia)	8	—	13,090	1.54	9.8	7,320	1,320	860	10	AS

*Results of tests on small, clear, straight-grained specimens. Property values were taken from world literature (not obtained from experiments conducted at the U.S. Forest Products Laboratory). Other species may be reported in the world literature, as well as additional data on many of these species.

†Some property values have been adjusted to 12% moisture content; others are based on moisture content at time of test.

‡Specific gravity based on weight when oven-dry and volume at moisture content indicated.

§Modulus of elasticity measured from a simply supported, center loaded beam, on a span-depth ratio of 14/1. The modulus can be corrected for the effect of shear deflection by increasing it 10%.

¶Key to code letters: AF, Africa; AS, Southeast Asia; AU Australia; BR, Brazil; CA, Central America; CH, Chile; CR, Costa Rica, CS, Central and South America; EC, Ecuador; GU, Guatemala; GY, Guyana (British Guiana); HO, Honduras; IN, India; MI, Malaysia—Indonesia; NI, Nicaragua; PA, Panama; PE, Peru; PH, Philippine Islands; SM, South America; SU, Surinam; and VE, Venezuela.

**1,500 board feet.
††1 bolt.
‡‡195 tests.
§§21 tests.
¶¶26 planks.
***11 planks.
Source: Wood Handbook 72, USDA.

means a loss of gluability. The loss goes beyond reduction in wetting (links 4 and 5); it also reduces water absorption from the glue. This not only reduces the rate of solidification in those glues that harden by loss of water, but it may also upset a crucial water balance in those glues that harden with heat. Some of these glues need to lose a portion of their mix water in order to compensate for the extra mobility that will be induced by heat. When the wood fails, or is too slow, in this absorption during assembly time, the increased mobility leads to starved or filtered bond lines.

A test that quickly shows the absorbency of a wood surface can be performed with a drop of water. A drop of water placed gently on the surface will disappear into the wood at a rate that is suggestive of the degree of absorbency of the surface. On a very absorptive surface, the drop will disappear almost as fast as it is placed in contact with the wood. On a wood surface that is nonabsorptive, the drop will remain indefinitely, sometimes evaporating rather than being absorbed. The drop in this case may be responding to a combination of factors: minimum porosity, low grain angle, diminished hygroscopicity, or surface inactivation. In the latter case, the drop of water will retain a spherical shape with a high angle of contact between it and the wood surface. Whatever the situation, the effect on the glue line is registered as an influence on mobility, and of course, the effect will differ with the type of glue and other factors. The resulting interactions will also need to be included in assessment of the consequences.

Moisture Content

The moisture content of wood, and the distribution of moisture within and between individual places, influences both bond formation and bond performance in ways that demand close control. Several ways have already been mentioned: the effect of moisture on the dimensions of wood, and the concomitant shrinking and swelling with changes in moisture content. This means that dimensions are always subject to change. If a

change in moisture occurs just prior to gluing, joint surfaces may not mate, especially in lumber gluing. If a change occurs after gluing, joint dimensions and surfaces may *try* to change, as they would if not glued, and in so doing, create stress across the bond line. This aspect of gluing is discussed further in Chapter 7. As mentioned previously, the moisture in wood also interacts with the water balances in the glue during bond formation.

The water in wood is considered to be of three different types depending on how or where it is held. All three types have a bearing on the gluing process. *Free water* is that which is contained in the cell lumens. It is not "held" there, but merely "contained," like water in a drinking glass or a garden hose. *Bound water* is water that is present in the cell walls. It is held there by molecular attractive forces, the same forces that play in adhesion actions, and with which adhesive molecules must compete to achieve anchorage to the cell wall. This is also the water that controls the dimensional changes of wood and affects strength properties. *Water of constitution* is, strictly speaking, a misnomer, since it is water that exists as part of the molecular structure of wood, in the form of hydrogen and hydroxyl groups (see the section on chemical properties) which split off and form water under prolonged high heating. Because each of these kinds of waters plays a role in wood gluing, it is well to establish some benchmarks that can be observed with respect to wood moisture content.

The moisture content of wood, besides being defined in different types, is divided into four categories on the basis of amounts, which, though they have loose boundaries, are sufficiently descriptive for practical purposes. They are:

- *Green.* The moisture content of wood in standing trees or when first cut. The cell lumens may be totally filled or only partially filled.
- *Fiber saturation point (FSP).* The moisture content when lumens are just empty but the cell walls are fully saturated (the

equilibrium moisture content when wood is exposed to 100 percent relative humidity, taken to be 30 percent as a first approximation).

- *Air dry.* The equilibrium moisture content that ensues from exposure to ambient atmospheric conditions, assumed to be 12 percent on average.
- *Oven dry (OD).* The moisture content of wood subjected to a dry atmosphere at 212°F until stabilized, considered to be zero moisture.

When free water is present, the cell walls are always saturated with bound water and the wood is fully swollen, the green condition. At the fiber saturation point (FSP), all the water is bound water, and the wood is still fully swollen. Shrinkage begins as the moisture content falls below FSP, and continues proportionally to the oven-dry condition.

The amount of free water in green wood varies over a very wide range depending not only on species, but also on time of year (springtime being higher than wintertime). Table 5–5 lists the green moisture content of a number of species for both sapwood and heartwood. Variability in the amount of water in green wood is the source of some problems in wood gluing, mainly because it is difficult to season wood to a uniform moisture content when different pieces begin the drying process with different moisture levels.

It is often confusing to note that the moisture in wood can exceed 100%. This is because moisture content is calculated as the ratio of the weight of water to the weight of *oven-dry wood,* reduced to a percent:

$$\text{Percent moisture content} = \frac{\text{weight of water}}{\text{weight of OD wood}} \times 100$$

On this basis, a wood such as red cedar, which has thin-walled fibers and therefore large lumen volume, can hold more water than the wood weighs.

The effects of wood moisture on bond formation and bond performance are many and varied. They can be brought within reasonably understandable limits for bond formation by focusing primarily on the effects of moisture on the mobility of the adhesive. The key factor for reasoning the consequences of moisture is that *increased moisture in the glue always causes increased mobility, and vice versa.* In terms of bond performance, the key factor is that *moisture content changes always cause dimensional changes,* and dimensional changes always, with minor exceptions, cause stress on the glue line.

Bond formation effects of moisture content begin with the amount and rate of water absorption. It seems patent that the more water there is in the wood, the less it can absorb and the slower it will absorb. This means that a film of water-dispersed glue placed on a dry wood surface will dry out faster than a similar film on wood of high moisture content, other factors being equal. In practice, the effect will play throughout the assembly time period, causing reduced mobility in accordance to the moisture content. The length of the assembly time then determines how prepared the glue is to receive pressure or heat. The maximum permissible assembly time may have to be shorter for dry wood than for wood of high moisture content in order to avoid dry-out or prehardening. With high moisture wood, the minimum assembly time is generally more important, and it may need to be lengthened in order to ensure optimum mobility and avoid overpenetration and a starved joint.

After pressure is applied, the same effects of moisture on mobility continue, and they now control the rate of solidification in the case of moisture-loss adhesives; the rate will be faster on dry wood, slower on wetter wood. Adhesives that cure by chemical processes are also affected but in different ways depending on the mode of hardening. In cold press systems low moisture content wood may rob the adhesive of its reaction medium before it has a chance to cure fully.

When heat is applied to hasten the rate of hardening, some very strong interactions with moisture occur that often destroy or upset the normal functions of the glue in bond

Table 5-5. Average Moisture Content of Green Wood, by Species

SPECIES	MOISTURE CONTENT*		SPECIES	MOISTURE CONTENT*	
	HEARTWOOD (%)	SAPWOOD (%)		HEARTWOOD (%)	SAPWOOD (%)
Hardwoods			Softwoods		
Alder, red	—	97	Baldcypress	121	171
Apple	81	74	Cedar:		
Ash:			Alaska	32	166
Black	95	—	Eastern red cedar	33	
Green	—	58	Incense	40	213
White	46	44	Port Orford	50	98
Aspen	95	113	Western red cedar	58	249
Basswood, American	81	133	Douglas fir:		
Beech, American	55	72	Coast type	37	115
Birch:			Fir:		
Paper	89	72	Grand	91	136
Sweet	75	70	Noble	34	115
Yellow	74	72	Pacific silver	55	164
Cherry, black	58	—	White	98	160
Chestnut, American	120	—	Hemlock:		
Cottonwood, black	162	146	Eastern	97	119
Elm:			Western	85	170
American	95	92	Larch, western	54	110
Cedar	66	61	Pine:		
Rock	44	57	Loblolly	33	110
Hackberry	61	65	Lodgepole	41	120
Hickory, pecan:			Longleaf	31	106
Bitternut	80	54	Ponderosa	40	148
Water	97	62	Red	32	134
Hickory, true:			Shortleaf	32	122
Mockernut	70	52	Sugar	98	219
Pignut	71	49	Western white	62	148
Red	69	52	Redwood, old-growth	86	210
Sand	68	50	Spruce:		
Magnolia	80	104	Eastern	34	128
Maple:			Engelmann	51	173
Silver	58	97	Sitka	41	142
Sugar	65	72	Tamarack	49	
Oak:					
California black	76	75			
Northern red	80	69			
Southern red	83	75			
Water	81	81			
White	64	78			
Willow	82	74			
Sweetgum	79	137			
Sycamore, American	114	130			
Tupelo:					
Black	87	115			
Swamp	101	108			
Water	150	116			
Walnut, black	90	73			
Yellow poplar	83	106			

*Based on weight when oven dry.
Source: Wood Handbook, USFPL.

formation. The introduction of heat sets in motion several events, which have both positive and negative effects:

1. *Reduced viscosity,* the normal effect of heat on most liquids, with a consequent increase in mobility of the adhesive on the glue line, and a tendency to overpenetrate and starve the joint. When heat arrives at a glue line that may have lost too much mobility, due, for example, to long assembly time, the reduced viscosity it causes may restore sufficient mobility to allow bond formation to occur. This action is widely used with resin adhesives that contain little or no fillers or extenders. The resin can be applied and allowed to dry completely, as in the case of phenolic film adhesives, which are revived on the glue line by heat and the moisture contributed by the wood.

2. *Increased viscosity,* due to increased rate of drying or hardening of the adhesive, particularly those that harden by chemical action. This action counteracts that in (1), but not equally, depending on the curing rate of the adhesive. The faster rate of hardening shortens the time the adhesive is vulnerable to excessive mobility and therefore is a beneficial action.

3. *Migration of moisture* in a direction away from the heat source. This action changes the moisture distribution, producing moisture gradients throughout the assembly, and disrupting again the water balances in the glue lines. Glue lines in the interior will receive an additional amount of moisture, and therefore will be disposed to the effects of higher mobility, which can lead to starved or filtered conditions.

4. *Temperature gradients* develop in accordance with the thermal properties of wood, being hottest nearest the heat source and diminishing inward until, with time, the gradient flattens out and the total assembly arrives at platen temperature.

5. *Temperature gradients* also develop as a result of moisture movement. The moisture is both a heat sink and a carrier of heat. The heat-sink effect occurs because of the extra calories needed to raise the temperature of water, and especially because of the extra calories needed at the transition point to change water into vapor. The latter is the cause of many of the problems in using glues, such as phenolics, that require high temperatures to cure. During the time that water is in transition to vapor, the temperature cannot rise, and will not rise until all the moisture has been vaporized. During this time, the adhesive is in a high state of mobility, perhaps its highest, and is most likely to be drawn excessively into the wood, starving the glue line.

6. Temperature and moisture gradient effects are not as pronounced in heating with high frequency energy because of the more uniform heat distribution.

In-service EMC

After a wood product has been placed in service, it experiences all the moisture content changes induced by its environment. Some of this may involve soaking by precipitation or immersion, but the majority involves only exposure to varying relative humidities of a seasonal or temporary nature. It is surprising to note the relatively short time it takes for a piece of wood to come to an equilibrium moisture content when exposed to a change in relative humidity. Figure 5–12 plots the changes in moisture content over time of small pieces of wood. They are carried through both an adsorption and a desorption atmosphere. This shows the sorption through all surfaces of the piece. It should be noted that sorption through end grain can be many times faster than through side grain. Hence, the size and configuration of the piece of wood has a bearing on the rate of change.

Since end-grain edges of wood are quickest to change, they represent areas where effects due to moisture gradients, such as checking and splitting, are most likely to occur. The repeated shrinking and swelling that

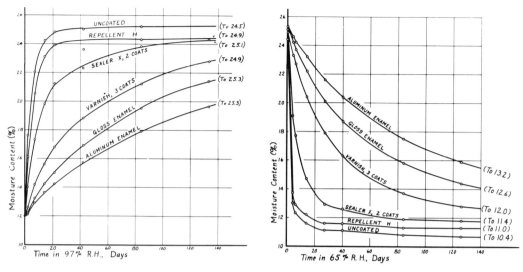

Figure 5-12 Rate of change of moisture content of small wood blocks, variously coated, exposed to changes in relative humidity of 65 percent to 97 percent and back to 65 percent. (Courtesy of USFPL.)

accompany swings in moisture content produce cyclic tension and compression forces at the surface that ultimately result in end checking. Moisture gradients also develop in side-grain surfaces, especially under accelerated drying conditions, resulting in checking, though less obvious, on these surfaces.

As an aid to establishing end-use targets for moisture contents, the U.S. Forest Products Laboratory has charted the continental United States in terms of isobars which show the equilibrium moisture content that wood might attain in a building during summer and winter seasons (Figures 5-13 and 5-14). Summer and winter readings were then averaged to show regions where a given moisture content would be likely to fluctuate the least, this being the target moisture content for products destined for those regions (Figure 5-15).

The information represented by Figures 5-13 and 5-14 plays a significant role in the manufacture and performance of all wood products but especially those that involve gluing. They not only indicate the swings in moisture content that wood might experience in service, but more importantly they indicate the potential moisture content status of wood awaiting processing. Many prob-

lems in gluing can be associated with change of seasons when wood is processed at one extreme or the other. If wood in process is carried into the next season it is particularly prone to create problems because each piece may be riddled with moisture gradients and no dimension can be considered established.

Shrinking and Swelling

The dimensions of any piece of wood change with changes in moisture content, one of the more important characteristics of wood. (Of equal importance is the change in strength properties that accompanies changes in moisture content, which will be discussed later.) The relationship between change in moisture content and change in dimension has been well studied. These characteristics of the shrinking and swelling of wood bear on bond formation and performance: (1) the amount, (2) the differential, and (3) the force.

The amount of dimension change differs among species from about 2 percent to 12 percent across the grain. Within species it differs in each of the three principal directions. The force with which shrinking and swelling occurs depends on the density of the

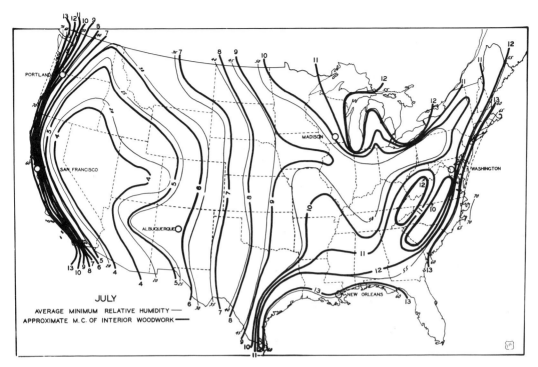

Figure 5–13   Equilibrium moisture content that wood can attain indoors during summer in the United States. (Courtesy of USFPL.)

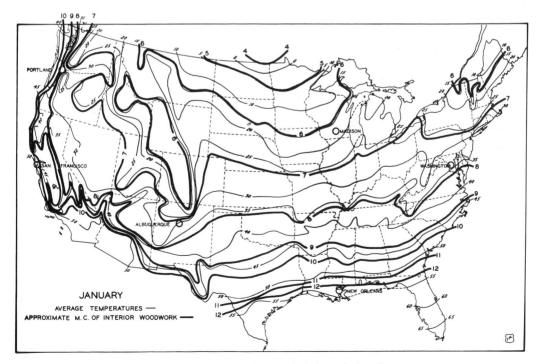

Figure 5–14   Equilibrium moisture content that wood can attain indoors during winter in the United States. (Courtesy of USFPL.)

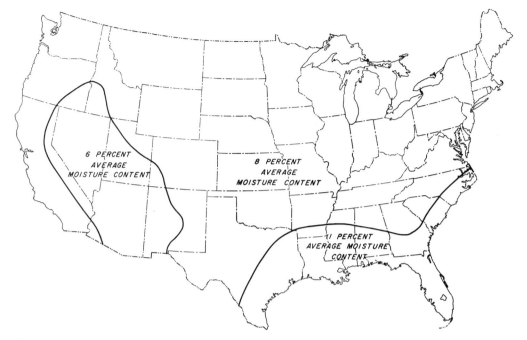

Figure 5–15 Average equilibrium moisture content for wood indoors in the United States over a yearly time span. (Courtesy of USFPL.)

wood, its moisture content, the temperature, and the amount of restraint imposed.

Amount of Shrinking and Swelling. Tables 5–6 and 5–7 list the shrinkage values for woods used in the United States. These values represent the total shrinkage from green to oven dry. Values to intermediate moisture contents can be approximated on the assumptions that (1) a straight-line relationship exists, and (2) the fiber saturation point, the moisture content at which shrinkage begins as drying proceeds, is 30 percent. On this basis, one-thirtieth of the total shrinkage will occur for each percent moisture content below 30.

For example, to determine the shrinkage that might occur in drying beech to 12 percent moisture content, a decrease of 18 percent from 30, multiply 18 × 1/30 × 5.5 (the maximum for radial shrinkage or 11.9 for tangential shrinkage, Table 5–6), giving 3.3 percent and 7.14, respectively. (Of course, if the actual fiber saturation point is known,

that value should be used in place of the 30 that was assumed.) Reduced to a formula,

$$Sp = St \frac{(30 - MCp)}{30}$$

where Sp is the partial shrinkage desired, St is the total shrinkage for the species and direction, and MCp is the moisture content for which the shrinkage is desired.

It often is more important to learn how much dimension change will occur with a given change in moisture content. The same relationship can be used. For example, a change in moisture content from 12 percent to 4 percent (or from 4 percent to 12 percent), a change of 8 percent, will incur 1/30 times 8, or 8/30th of the total change. Multiplying 8/30 by 5.5 or 11.9 (again from Table 5–6 for beech) produces a dimension change of 1.47 percent radially, or 3.17 percent tangentially. In actual dimension terms, a 6-in.-wide beech board will shrink (or swell) 0.015 times 6 or 0.09 in. in width if it is a quarter-cut board,

Table 5-6. Shrinkage Values of Domestic Woods

SPECIES	SHRINKAGE FROM GREEN TO OVEN-DRY MOISTURE CONTENT*			SPECIES	SHRINKAGE FROM GREEN TO OVEN-DRY MOISTURE CONTENT*		
	RADIAL (%)	TAN-GENTIAL (%)	VOLU-METRIC (%)		RADIAL (%)	TAN-GENTIAL (%)	VOLU-METRIC (%)
Hardwoods							
Alder, red	4.4	7.3	12.6	Honeylocust	4.2	6.6	10.8
Ash:				Locust, black	4.6	7.2	10.2
Black	5.0	7.8	15.2	Madrone, Pacific	5.6	12.4	18.1
Blue	3.9	6.5	11.7	Magnolia:			
Green	4.6	7.1	12.5	Cucumbertree	5.2	8.8	13.6
Oregon	4.1	8.1	13.2	Southern	5.4	6.6	12.3
Pumpkin	3.7	6.3	12.0	Sweetbay	4.7	8.3	12.9
White	4.9	7.8	13.3	Maple:			
Aspen:				Bigleaf	3.7	7.1	11.6
Bigtooth	3.3	7.9	11.8	Black	4.8	9.3	14.0
Quaking	3.5	6.7	11.5	Red	4.0	8.2	12.6
Basswood,				Silver	3.0	7.2	12.0
American	6.6	9.3	15.8	Striped	3.2	8.6	12.3
Beech, American	5.5	11.9	17.2	Sugar	4.8	9.9	14.7
Birch:				Oak, red:			
Alaska paper	6.5	9.9	16.7	Black	4.4	11.1	15.1
Gray	5.2		14.7	Laurel	4.0	9.9	19.0
Paper	6.3	8.6	16.2	Northern red	4.0	8.6	13.7
River	4.7	9.2	13.5	Pin	4.3	9.5	14.5
Sweet	6.5	9.0	15.6	Scarlet	4.4	10.8	14.7
Yellow	7.3	9.5	16.8	Southern red	4.7	11.3	16.1
Buckeye, yellow	3.6	8.1	12.5	Water	4.4	9.8	16.1
Butternut	3.4	6.4	10.6	Willow	5.0	9.6	18.9
Cherry, black	3.7	7.1	11.5	Oak, white:			
Chestnut, American	3.4	6.7	11.6	Bur	4.4	8.8	12.7
Cottonwood:				Chestnut	5.3	10.8	16.4
Balsam poplar	3.0	7.1	10.5	Live	6.6	9.5	14.7
Black	3.6	8.6	12.4	Overcup	5.3	12.7	16.0
Eastern	3.9	9.2	13.9	Post	5.4	9.8	16.2
Elm:				Swamp chestnut	5.2	10.8	16.4
American	4.2	7.2	14.6	White	5.6	10.5	16.3
Cedar	4.7	10.2	15.4	Persimmon, common	7.9	11.2	19.1
Rock	4.8	8.1	14.9	Sassafras	4.0	6.2	10.3
Slippery	4.9	8.9	13.8	Sweetgum	5.3	10.2	15.8
Winged	5.3	11.6	17.7	Sycamore, American	5.0	8.4	14.1
Hackberry	4.8	8.9	13.8	Tanoak	4.9	11.7	17.3
Hickory, pecan	4.9	8.8	13.6	Tupelo:			
Hickory, true:				Black	5.1	8.7	14.4
Mockernut	7.7	11.0	17.8	Water	4.2	7.6	12.5
Pignut	7.2	11.5	17.9	Walnut, black	5.5	7.8	12.8
Shagbark	7.0	10.5	16.7	Willow, black	3.3	8.7	13.9
Shellbark	7.6	12.6	19.2	Yellow poplar	4.6	8.2	12.7
Holly, American	4.8	9.9	16.9				
Softwoods							
Baldcypress	3.8	6.2	10.5	Cedar (cont.)			
Cedar:				Incense	3.3	5.2	7.7
Alaska	2.8	6.0	9.2	Northern white	2.2	4.9	7.2
Atlantic white	2.9	5.4	8.8	Port Orford	4.6	6.9	10.1
Eastern red cedar	3.1	4.7	7.8	Western red cedar	2.4	5.0	6.8

(*continued*)

Table 5-6. Shrinkage Values of Domestic Woods (continued)

SPECIES	SHRINKAGE FROM GREEN TO OVEN-DRY MOISTURE CONTENT*			SPECIES	SHRINKAGE FROM GREEN TO OVEN-DRY MOISTURE CONTENT*		
	RADIAL (%)	TAN-GENTIAL (%)	VOLU-METRIC (%)		RADIAL (%)	TAN-GENTIAL (%)	VOLU-METRIC (%)
Douglas fir:[†]				Pine (cont.)			
Coast	4.8	7.6	12.4	Lodgepole	4.3	6.7	11.1
Interior north	3.8	6.9	10.7	Longleaf	5.1	7.5	12.2
Interior west	4.8	7.5	11.8	Pitch	4.0	7.1	10.9
Fir:				Pond	5.1	7.1	11.2
Balsam	2.9	6.9	11.2	Ponderosa	3.9	6.2	9.7
California red	4.5	7.9	11.4	Red	3.8	7.2	11.3
Grand	3.4	7.5	11.0	Shortleaf	4.6	7.7	12.3
Noble	4.3	8.3	12.4	Slash	5.4	7.6	12.1
Pacific silver	4.4	9.2	13.0	Sugar	2.9	5.6	7.9
Subalpine	2.6	7.4	9.4	Virginia	4.2	7.2	11.9
White	3.3	7.0	9.8	Western white	4.1	7.4	11.8
Hemlock:				Redwood:			
Eastern	3.0	6.8	9.7	Old-growth	2.6	4.4	6.8
Mountain	4.4	7.1	11.1	Young-growth	2.2	4.9	7.0
Western	4.2	7.8	12.4	Spruce:			
Larch, western	4.5	9.1	14.0	Black	4.1	6.8	11.3
Pine:				Engelmann	3.8	7.1	11.0
Eastern white	2.1	6.1	8.2	Red	3.8	7.8	11.8
Jack	3.7	6.6	10.3	Sitka	4.3	7.5	11.5
Loblolly	4.8	7.4	12.3	Tamarack	3.7	7.4	13.6

*Expressed as a percentage of the green dimension.
[†]Coast Douglas fir is defined as Douglas fir growing in the states of Oregon and Washington west of the summit of the Cascade Mountains. Interior west includes the state of California and all counties in Oregon and Washington east of but adjacent to the Cascade summit. Interior north includes the remainder of Oregon and Washington and the states of Idaho, Montana, and Wyoming.
Source: Wood Handbook, USFPL.

and 0.032 times 6 or 0.19 in. if it is a flat-cut board.

One can see from these calculations the importance of placing wood products in service at a moisture content at the midrange of the variability expected for a particular location. By beginning at midrange, the product will undergo only half the shrinkage or swelling it would sustain if it began at one extreme or the other.

Differential Shrinking and Swelling. Wood shrinks and swells different amounts for the same moisture change in the three principal directions. Radial change is about half of tangential change (Tables 5-6 and 5-7). Longitudinally, the dimension change with a change of moisture content is negligible, except in tension wood and compression wood.

These three dimension factors are responsible for much of the warpage that occurs in wood, both in the original state and in the reconstituted state. They affect bond formation by distorting surfaces before bonding. This happens because most pieces of wood contain one grain plane or another that is not oriented truly with the cut planes of the piece—i.e., they are to a certain degree bias cut. As dimensions change in accordance with nature's true planes, the cut planes must change. Figure 5-16 shows dramatically how shrinkage distorts wood cut with different orientations to the radial and tangential directions. The distortions can be removed only by restoring to the original moisture content, or by resurfacing.

When wood pieces are glued together, the same propensity for distortion with moisture change remains in each piece. One can imag-

Table 5–7. Shrinkage for Some Woods Imported into the United States*

SPECIES	SHRINKAGE FROM GREEN TO OVEN-DRY MOISTURE CONTENT† RADIAL (%)	TANGENTIAL (%)	SPECIES	SHRINKAGE FROM GREEN TO OVEN-DRY MOISTURE CONTENT† RADIAL (%)	TANGENTIAL (%)
Andiroba (*Carapa guianensis*)	4.0	7.8	Mahogany (*Swietenia macrophylla*)	3.7	5.1
Angelique (*Dicorynia guianensis*)	5.2	8.8	Nogal (*Juglans* spp.)	2.8	5.5
Apitong (*Dipterocarpus* spp.)	5.2	10.9	Obeche (*Triplochiton scleroxylon*)	3.1	5.3
Avodire (*Turraeanthus africanus*)	3.7	6.5	Okoume (*Aucoumea klaineanv*)	5.6	6.1
Balsa (*Ochroma pyramidale*)	3.0	7.6	Parana pine (*Araucaria angustifolia*)	4.0	7.9
Banak (*Virola surinamensis*)	4.6	8.8	Primavera (*Cybistax donnell smithii*)	3.1	5.2
Cativo (*Prioria copaifera*)	2.3	5.3	Ramin (*Gonystylus* spp.)	3.9	8.7
Greenheart (*Ocotea rodiaei*)	8.2	9.0	Santa Maria (*Calophyllum brasiliense*)	5.4	7.9
Ishpingo (*Amburana acreana*)	2.7	4.4	Spanish cedar (*Cedrela* spp.)	4.1	6.3
Khaya (*Khaya* spp.)	4.1	5.8	Teak (*Tectona grandis*)	2.2	4.0
Kokrodua (*Pericopsis elata*)	3.2	6.3	Virola (*Dialyanthera* spp.)	5.3	9.6
Lauan (*Shorea* spp.)	3.8	8.0	Walnut, European (*Juglans regia*)	4.3	6.4
Limba (*Terminalia superba*)	4.4	5.4			
Lupuna (*Ceiba samauma*)	3.5	6.3			

*Shrinkage values in this table were obtained from world literature and may not represent a true species average.
†Expressed as a percentage of the green dimension.
Source: Wood Handbook, USFPL.

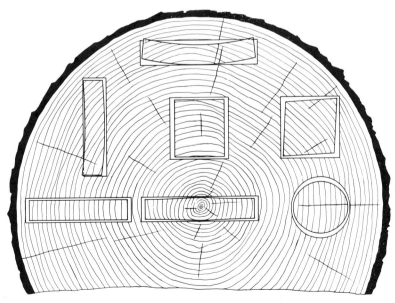

Figure 5–16 Distortions of cut wood pieces due to differential shrinkage between radial and tangential surfaces. (Courtesy of USFPL.)

ine the stress across the glue line as each piece seeks to distort in its own direction. This is the principle source of the stress that destroys bonds in accelerated test procedures or in service when they are below quality for the intended exposures. This source of stress can be generated, for example, when a radial surface is glued to a tangential surface. It increases as the thickness of the wood increases, reaching a maximum when a tangential surface meets another tangential surface with the grain crossed at 90°.

Force of Shrinking and Swelling. A piece of wood lying on a table creates no external force as it shrinks and swells. (Internally, some forces are generated as moisture gradients create differences in dimension change, a cause of surface checking.) Force is generated when the dimension change is resisted, in our case by being glued to something that changes differently or not at all. Stress due to shrinkage is inherently limited by the tensile strength of the wood perpendicular to the grain. For American beech at 12 percent moisture content, the tensile strength averages about 1000 psi; whereas for spruce, it averages less than 350 psi. Bonds therefore do not need to be stronger than this to withstand shrinkage stresses.

Swelling force is a different matter. While the *structure* of the wood limits its shrinkage force, the *substance* of the wood powers swelling forces. The fundamental attraction of wood for water energizes the entry of water molecules into the molecular structure of the wood, adding its volume to the volume of wood substance and swelling the mass.

The magnitude of the forces engaged in the attraction of wood for water is beyond easy comprehension. In order to compress bound water to the degree that has been estimated, a compressive force of 200,000 psi would be required. These are intermolecular forces, not directly transmittable externally. However, volumetric considerations have shown that wood of low original moisture content is capable of exerting pressures of about 10,000 psi. This is the force that in ancient times was used to split rocks as mentioned earlier. The amount of this pressure that might be transmitted outwardly depends on the structure of wood. Thick cell walls that do not buckle easily would transmit more of this pressure. Thin cell walls that permit internal collapse would transmit less. Thus the swelling pressure of wood is related to the density of the wood; high-density woods create greater swelling as well as shrinking stresses than low-density woods. The greater difficulty of achieving good bond performance with high-density woods reflects the operation of this factor.

Because swelling forces that originate in the cell wall are transmitted by the cell wall itself to the exterior of the piece, factors that affect the strength of the cell wall also affect the stress transmitted. Two factors in addition to cell wall thickness (density) are moisture content and heat. The higher either of these two factors is, the lower will be the stress that can be transmitted. Thus a change in moisture content at low moisture content may not only create more stress, it may also transmit more of it to the exterior. This fact demands greater consideration when developing accelerated tests for bond performance which include moisture.

Finally, the *direction* of swelling forces with respect to the glue line has interesting effects. During cold-press bond formation, swelling induced by adsorbed water from the glue line tends to increase the pressure provided by clamps, at least near the beginning of the pressure period. After the bond has been made, swelling forces perpendicular to the glue line, though seemingly of no consequence, can cause problems especially if they are unequal from point to point, as they are in reconstituted products. One source of swelling forces on the glueline is the uneven compression of wood during pressing, which causes the wood to swell unevenly. Those areas that swell the most produce tensile forces on neighboring areas which have swollen less, rupturing bonds in the latter. Swelling forces in the *plane* of the glue line cause shear stresses when they are unequal across the glue line, as they are in cross-laminated constructions.

Color

Color per se has no direct bearing on bond formation or performance. However, it is a strong indicator of conditions in the wood that may influence the results. The most obvious of these is the color difference between heartwood and sapwood: The former is usually darker in color than the latter. Sapwood is generally light colored, white to yellow except when sapstained, in which case it will be various shades of grey, black, or blue. Sapstain itself does not cause problems, but it may indicate a high-moisture condition. The presence of mildew increases further the suspicion of high moisture content. It also should raise the suspicion of decay, a strength-reducing factor.

The color of decayed wood depends on the type of decaying organism and the extent of decay. Two distinct types occur, known by the color they produce in wood. *White rot* is produced by fungi that consume the lignin fraction in wood. *Brown rot* is produced by fungi that destroy the cellulose or carbohydrate fraction of wood. In either case the loss in strength of links 8 and 9 will act against the performance of the glued product, perhaps to a greater extent than performance of the bond.

Thermal Properties

The interaction of heat with wood produces a number of effects which have a bearing on adhesive functions. The most critical are the containment and transmission of heat in bond formation, and dimension changes due to temperature change in bond performance. These effects are discussed in this section.

Thermal Conductivity. The importance of thermal conductivity in bond formation is due to its effect on glue lines that have to be heated to solidify them. The time it takes to heat a glue line through wood some distance from the heat source has an obvious bearing on rate of curing and consequently on rate of production. It also has a bearing on the mobility functions of the glue. For example, for an adhesive that needs to be heated to 250°F, the crucial factors are not only the attainment of 250°F, but also the time it takes to get there. During this upheat time, a heat-curing adhesive undergoes opposing viscosity changes which vary with time. Viscosity decreases as a physical response to heat; at the same time, viscosity increases as a result of loss of water and the chemical reactions of hardening. Eventually, the chemical reactions dominate, and the hardening reaction proceeds to completion. Before then, however, a critical point occurs when the lowered viscosity has not yet been compensated by the hardening action. At this point, the adhesive has its greatest potential for flowing, transfering, penetrating, and wetting. The length of time during which the adhesive has greatest mobility thus determines the magnitude of these actions. Recalling that these actions need to be *optimized,* not *maximized,* there is a danger that they could be over or under optimum depending on the length of this time interval. Consequences of too much mobility or too little mobility are thus cast at this precise point in the press cycle. It therefore represents an important point of attack in troubleshooting.

The temperature range between about 210 and 220°F is of particular interest. In this range the temperature rises at a reduced rate until all the water has vaporized (due to the extra heat needed to vaporize water, 540 calories per gram). Some adhesives, phenolics especially, tend to be vulnerable at this temperature; the prolonged high-mobility phase promotes overpenetration and starved glue lines. In most cases, the length of the maximum mobility phase is determined by (1) the rate of temperature rise, and (2) the reactivity of the adhesive.

The rate of temperature rise is controlled by many factors, including the temperature of the heat source, the capacity of the heat source, the distance of the glue line from the heat source, and of course the nature of the heat source itself, such as heated platen, hot air, infrared, dielectric, microwave. The first three of these depend on the conductivity of wood to deliver the heat inward.

Wood, particularly dry wood, is known

for its good insulating properties, part of its appeal in house construction. Good insulation value implies that the material does not transmit or conduct heat readily. Thus one would expect wood to act as a heat shield, so that the side away from a heat source would be cooler than the side facing the heat source. The difference in temperature between the two sides under steady input of heat energy is a measure of the heat conductivity of the material. This is expressed as the coefficient of thermal conductivity k, being the amount of heat energy in BTUs that would flow through 1 ft² of material in 1 hour if the temperature difference were 1°F per inch of thickness. These factors are related by the equation:

$$k = \frac{Qd}{At\,(T_1 - T_2)}$$

where k is the coefficient of thermal conductivity, Q is the quantity of heat, d is the thickness of the material, A is the area, t is the time interval, T_1 is the temperature of the heated side, and T_2 is the temperature of the cool side. The English units for k therefore are (BTU) (in.)/(ft²) (°F) (hr). The k values

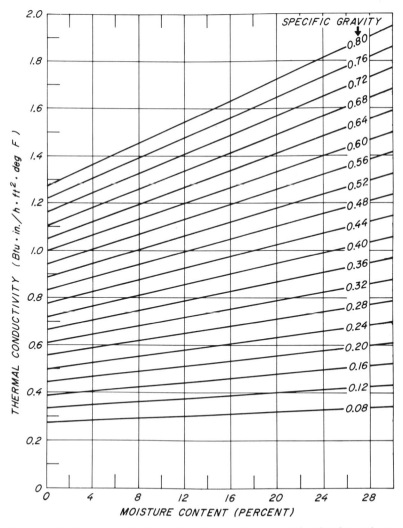

Figure 5-17   Thermal conductivity of wood perpendicular to grain related to moisture content and specific gravity. (Courtesy of USFPL.)

averaged for a number of wood species are presented in Figure 5–17. The values are for side-grain conductivity of wood covering a range of moisture contents and specific gravities. Conductivity is two to three times greater along the grain than across it. This means that areas in a board with knots or steep grain angles will transmit heat faster than areas with straight grain. Speed of transmission also increases with increasing density, moisture content, and extractive content.

In summary, a maple board transmits heat faster than an aspen board; a maple board at high moisture content transmits heat better than one at low moisture content; and heartwood with its higher load of extractives transmits heat better than sapwood at the same moisture content. High heat conductivity results in faster temperature rise in heating glue lines, faster cure times, and shorter pressure periods.

Heat Capacity. If subjected to the same heat source for the same length of time, different materials will arrive at different temperatures. This is a result of their differing heat capacities. The material at the lower temperature will have a higher heat capacity because it can absorb or store more heat (BTUs or calories) before it experiences a rise in temperature. *Heat capacity* (equal numerically to specific heat, (c) is defined as the amount of heat required to raise a unit weight of the material by one degree, in units of calories per gram per degree Celsius or BTUs per pound per degree Fahrenheit.

The heat capacity is virtually the same for all woods; 1 lb of maple has the same heat capacity as 1 lb of aspen. However, a board or veneer of maple requires more heat to bring it to the same temperature as a piece of aspen of the same dimensions, because of its greater weight, i.e., more pounds in the maple piece than the aspen piece.

Heat capacity varies with moisture content, being greater at high than at low, due not only to the greater heat capacity of water, but also to the energy with which water is bound to wood. Thus a board or veneer of maple or aspen requires more heat to attain a certain temperature if at 30 percent moisture content than at 8 percent. Dry wood has a heat capacity of about 0.26 BTU/lb/°F.

Thermal Diffusivity. A piece of wood and a piece of steel at equilibrium with the temperature in a room will seem to be at different temperatures; the wood will feel much warmer than the steel. The wood is said to have a lower *thermal diffusivity* than steel. In other words, the steel can absorb heat from its surroundings, the hand in this case, faster than the wood. Conversely, wood is slower to give up heat. This fact is used in heat-cured glued products to store heat after a minimum time in the press. It provides residual heat to complete the cure. This is done by stacking panels on top of each other as they come out of the press, in a closed stack without stickers. The process is called "hot stacking" and is widely used in plywood and flakeboard manufacture as a means of shortening the time required in the press. Processes that use phenolic resins are particularly benefited by the slow diffusivity of wood.

In an ingenious inversion of the heating operation, called the stored-heat process, the wood is heated to a resin-curing temperature *before* the glue is applied. The glue is then applied swiftly, and the assembly is quickly put under pressure without additional heat. The heat stored in the wood then diffuses to the glue line, curing the glue. This process requires close, split-second timing of each step. Adjusting the viscosity and reactivity of the glue to the instantly hot environment of the joint surfaces is a cut-and-try procedure involving the total mechanics of the particular system. Although it is not widely used, this process of introducing heat as a first step in the gluing operation seems most appropriate for assembling thick constructions with the more durable and cheaper adhesives, such as phenols, which require heat to cure.

The same property of diffusivity operates in an adverse manner in hot-press processes which use urea resins. After urea resins are fully cured, further heating has a destructive effect, causing the adhesive molecules to break down and lose strength. In this situation, it is desirable to get the heat out as fast as possible after pressing. Rather than using closed stacking as with phenolics, hot-pressed urea bonded panels are open stacked, usually in a vertical aspect, to allow the heat to diffuse out and be wafted away as fast as the wood diffusivity permits.

With respect to other wood properties, diffusivity decreases with increases in both density and moisture content. Thus a hot-pressed maple panel will cool at a slower rate than an aspen panel starting at the same temperature. The question of moisture content versus diffusivity generally does not apply to hot-press processes, because the wood comes out of the press fairly dry. Grain direction affects diffusivity, being greater along the grain than across the grain. Hence, areas in a panel having short grain, such as around knots, will cool faster than areas of straight grain.

Diffusivity is not usually measured directly, but is derived from other thermal properties using the equation

$$h = \frac{k}{cD}$$

where h is the diffusivity in square inches per second, k is the coefficient of conductivity, c is the specific heat (heat capacity), and D is the density. By substituting h into a differential equation of diffusion parameters (Fick's second law) and integrating, the temperature levels in a material can be calculated for any cooling or heating condition.

For purposes of comparison, the diffusivities of several materials are given in Table 5–8. While the units of diffusivity, square inches per second, are difficult to associate with reality, the numbers in the table are suggestive of how different materials heat up or cool down in response to a change in ambient temperature.

Thermal Expansion. Like other materials, wood expands upon heating and contracts upon cooling. This property has a greater ef-

Table 5–8. Thermal Diffusivities of Several Wood-Based Materials

MATERIAL	DIFFUSIVITY (IN.^{2}/SEC)
Southern pine	245×10^{-6}
Insulating fiberboard	320×10^{-6}
High-density fiberboard	177×10^{-6}

Source: Koch, "Utilization of the Southern Pines," *Agricultural Handbook* 420, United States Department of Agriculture (USDA).

fect on bond (or product) performance than on bond formation. The most common effect usually evidences itself as warping of the product. Warping in this context is caused by components of a glued construction undergoing different dimensional changes in response to temperature changes. Although along the grain wood has a very low coefficient of thermal expansion (inches of dimension change per inch of length per degree of temperature change), Table 5–9, it varies in the three principal grain directions, being 5 to 10 times greater across the grain than along the grain. Moreover, along the grain the expansion is independent of density and species, but across the grain there is a direct relationship with density and therefore with species.

The effect of moisture content on thermal expansion is a complex, one since it involves both the effect on mechanical properties and the effect on dimensional change. Because moisture would be leaving cell walls during temperature increases causing shrinkage, it tends to offset the specific thermal expansion of the wood, sometimes producing an apparent negative effect of temperature.

Table 5–9. Coefficients of Thermal Expansion of Several Materials

MATERIAL	COEFFICIENT OF THERMAL EXPANSION, $\times 10^{-6}$ (°C)
Aluminum	29
Copper	17
Cast steel	13
Wood parallel to grain	4
Wood perpendicular to grain	35
Plywood*	7
Particleboard*	11.0

*Computed.

One can sense that, in a plywood-type construction, which contains perpendicular grain directions by design and varying grain direction by nature, different species, and different thickness, some warping tendencies due to differential thermal expansions are bound to arise. These are small in most cases, and are far overshadowed by the effects of moisture changes on the same factors.

Greater thermal effects occur as a result of materials glued to wood which have a higher coefficient of expansion, such as high-density overlays. This can result in severe dimensional imbalance between parts, creating stress across the glue lines. When this happens, either the glue line fails, the component materials undergo some accommodating deformation, or the product warps to relieve the stress.

Adhesives play a role in minimizing stress and warping. When different thermally induced dimension changes occur in the substrates or between the substrates and the glue film, some rupturing of bonds may occur unless one or the other can yield or warpage relieves the stresses. This is one of the main difficulties in bonding dissimilar materials. If the glue line can deform without failing, it can absorb or otherwise relieve the stresses, and warpage does not occur. Two methods of accomplishing this are available, though much more needs to be done in tailoring the rheological properties of solidified adhesives to the exact change potential of the adherends. One method is to coat the surface of the material having the higher coefficient with an elastomeric adhesive compounded to accord with the wood. The assembly is then glued with the appropriate wood adhesive. In the other method, the entire adhesive is compounded so that it deforms in response to dimension changes on both sides, but still has the strength to hold surfaces together.

Electrical Properties

The response of wood to electrical energy takes three primary forms, two of which are of interest to wood gluing: (1) conducting a current, or *conductivity,* (2) polarizability to sustain a charge, or *dielectric constant,* and (3) converting a percentage of the charge to heat, or *power factor.*

Wood is regarded as a nonconductor of electricity, an insulator when dry (a property useful in utility poles and towers). Conductivity increases a million times as wood increases in moisture content from zero to fiber saturation point. From there to fully saturated, conductivity increases another 50 times. As far as gluing is concerned, wood is not a good enough conductor of electricity to permit heating it by direct current or ordinary alternating current to cure a hot-setting glue line.

However, used as a dielectric material in a high-frequency field, useful energy can be developed in wood to heat-cure glue lines at a very rapid rate. This use invokes the dielectric constant and the power factor of wood. In practice, the assembly to be glued becomes part of the condenser in a high-frequency circuit, occupying the space between the condenser plates as shown in Figure 5–18(a). In an ideal condenser, this space would be filled with air (or a vacuum). With current flowing in the circuit, no energy would be lost in such a condenser. But with wood or any other dielectric material in the system, it becomes less than a perfect condenser to the extent of the polarizability of the intervening material. Under the influence of an alternating current, each plate of the condenser becomes oppositely charged—polarized—during one-half of the cycle, and reversed during the other half. When a material having polar constituents is between the plates, they also become polarized in accord with the polarity of the plates, reversing as they do. The result is that the plates can receive a greater charge than with air in between. A condenser with wood in it is thus said to have a greater *capacitance* than the one with air. The ratio between the two capacitances is the *dielectric constant* of the material, and since the dielectric constant of air is 1, the dielectric constant of materials is always greater than 1.

Some of the electrical energy that was used to polarize constituents in the wood is re-

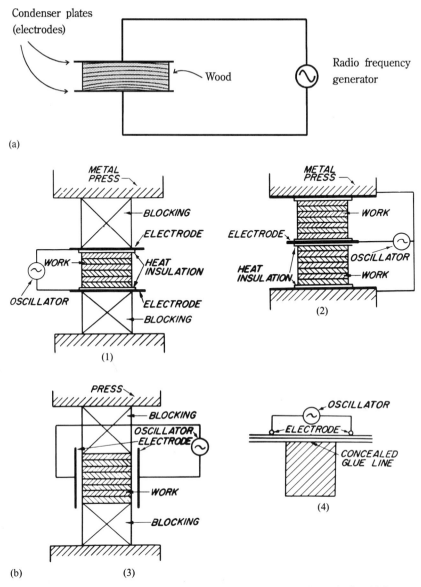

Figure 5-18   (a) Schematic of wood as dielectric material between electrodes in a high-frequency field. (b) Electrode arrangement for heating glue lines: 1 and 2, perpendicular heating (current heats both wood and glue); 3, parallel heating (current heats primarily the glue line); 4, stray field heating combining some of both. ((b) Courtesy USFPL.)

turned to the circuit, some stays in the material. It is this latter energy that creates the desired heat. The amount of energy that is not given back but is converted to heat varies with the dielectric constant of different materials. This value is called the *power factor,* and is defined as the ratio of the energy retained to the total energy supplied.

The conversion of electrical energy to heat in the condenser material is due to molecular friction. This happens because as the polarity of the plates oscillates with the imposed high-frequency current, the polarized molecules in the wood also oscillate as they try to adjust to the rapidly changing electrical field. Friction develops between molecules, and this generates the heat. It is important to note that the heat is generated in situ, it does not need to be conducted, and therefore it is not subject to the limitations of the thermal conductivity of wood.

Several wood factors affect the rate of heating that can be generated in this manner. The most important of these is moisture content. Water is a highly polar material. Consequently, as the moisture content of the wood increases, so does the dielectric constant, and hence the power factor. The same relationship occurs with wood density. Figure 5–19 shows these relationships. They indicate that wood at high moisture content and/or high density will generate more heat than wood at low, under the same electrical input. The actual temperature increase then depends on the heat capacity of the wood.

The efficiency of converting high-frequency electrical energy to heat energy is of the order of 50 percent, and while equipment and operational costs may not always be justified, two major advantages accrue: Heating is very rapid, and temperature gradients with their resulting moisture gradients are greatly reduced. An example of speed is the edge-gluing of lumber, where press times are customarily reduced to seconds. The reduction in temperature gradients is particularly advantageous in the manufacture of reconstituted products, where moisture gradients cause variations in properties through the thickness.

There are also some disadvantages. One is greater sensitivity to variables in the gluing operation. Because of the strong influence of water on the dielectric constant, heating can be only as uniform as the moisture content of the wood, or as uniform as the density. Mixed species in the same assembly may need testing to determine compatibility in the electrical field. The high speed of dielectric curing demands efficient in-feeding and off-bearing to capture the greater rate of production. The fast rate of heating and the electrical manner in which it is generated also demand special considerations in the composition of the adhesives to avoid frothing and arcing in the glueline.

These and other application factors are discussed further in Chapter 9.

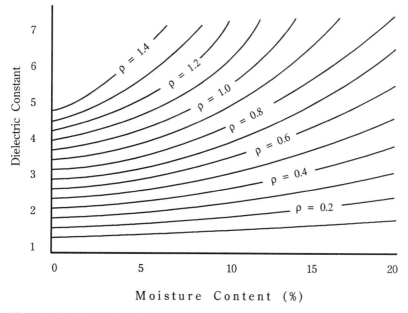

Figure 5-19 Relationship of moisture content and specific gravity to the dielectric constant of wood. (From C. Skaar, New York State College of Forestry, Technical publication 69, 1948.)

CHEMICAL PROPERTIES OF WOOD

Chemically, wood is made up of two major constituents: cellulose and lignin. It also has two minor constituents, collectively termed extractives, and ash. These constituents are present in wood in the following approximate amounts:

- Cellulose, 60–75 percent
- Lignin, 20–30 percent
- Extractives, 1–10 percent
- Ash, 0.1–0.5 percent

The first three of these figure strongly in wood gluing, primarily in bond formation, but also to some extent in bond performance. The effects register most critically on links 4 and 5, the adhesion links, and to lesser extent on links 2 and 3, the cohesive strength of the adhesive surface boundary. Links 6 and 7 also become involved through the effect of extractives on penetration. Links 8 and 9 are affected by these constituents primarily as they relate to strength and dimensional stability.

Lignocellulosics (the combined term for lignin and cellulose) comprise the most abundant, most widely available, and most intensely utilized materials on earth. It is one of the great wonders of nature that this material is produced by plants from water and carbon dioxide, using solar energy for power. The basic elements of its composition are therefore carbon, oxygen, and hydrogen. These elements are ingeniously organized into molecules of the two major constituents, each however with drastically different properties. The molecules are formed in different sizes and configurations, and deposited in the cell wall with amounts and orientations to best suit the functions of the cell.

In the cellulose fraction, the three elements, carbon, hydrogen, and oxygen, are first organized into sugar molecules for easy transport in the vascular system of the tree, and these are then connected end to end to form long chains, becoming the strength-producing component of the cell wall. Their strength is primarily in their length direction, and therefore they are oriented predominantly in the long direction of the fiber (with some notable exceptions).

In the lignin fraction, the same three elements are organized first into benzene molecules. These molecules are not, however, connected end to end to form long chains as in cellulose, but rather are connected in a more random manner, such that the molecule has no predominant direction. This produces a highly cross-linked, three-dimensional, amorphous material which contributes not strength, but stiffness to the cell wall.

Extractives comprise a miscellaneous group of complex polymeric materials that infiltrate cell walls and cell lumens. They play little if any role in the gross strength functions of the cell, but they do have a profound effect on the physical properties. They can be removed—extracted—from wood by appropriate solvents, hence the name *extractives*. Because of their solubility, they are sometimes classified on the basis of the solvents in which they are soluble.

The ash content of wood represents the minerals taken up from the soil. Less than 1 percent in most cases, its role centers on metabolic processes, and seems to have only minor effects on wood properties.

These chemical constituents of wood are not distributed uniformly throughout the structure. Though they are interdiffused to a certain extent, they are more or less concentrated in certain cells or certain parts of the cell. Because of this, they sometimes produce effects far above, as well as far below, what their relative proportions would suggest. The commonly accepted composition and organization of a typical woody cell is depicted in Figure 5–20. One may deduce that, depending on where the knife passes through the cell wall in creating a surface, widely differing surface properties may be presented to the glue. This is especially true in creating fiber elements, where prior conditioning of the wood predisposes the surfaces to be either lignin-rich or cellulose-rich. One may also deduce the tenuous nature of the lumen walls as a bonding surface, given their coating of residue materials.

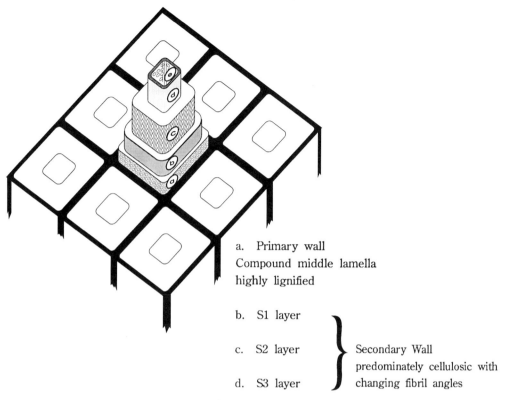

a. Primary wall
Compound middle lamella
highly lignified

b. S1 layer

c. S2 layer } Secondary Wall
predominately cellulosic with
d. S3 layer } changing fibril angles

Figure 5-20   The organization and distribution of cell wall constituents.

e. Warty layer   Residues of living cell

With this distribution in mind, the individual constituents will be described in greater detail, while at the same time discussing their implications for the gluing process.

Cellulose

The properties of cellulose derive from its molecular composition, structure, and organization. Recalling its genesis as a sugar molecule, one would expect cellulose to be soluble in water. The fact that it is not is due to its structure and organization. A sugar molecule has the equivalent of one water molecule for each carbon atom, hence its designation as a "carbohydrate," a hydrate of carbon. In being transformed into cellulose, however, each sugar molecule loses parts of two water molecules to provide connection points. Another is lost in forming a ring structure from the original straight carbon chain. Thus each original sugar molecule (glucose), in the ultimate cellulose chain, contains only the elements for three water molecules (Figure 5-21a). Nevertheless, they are able to confer on cellulose its attractiveness for water (hygroscopicity) and also the sites for adhesion, polarity, hydrogen bonding, and reaction with chemicals. The polar nature of cellulose results in organizations which help defeat its chemical inclination to dissolve in water. This they do by attracting each other in a rather precise manner, forming areas of tightly packed microcrystals of pure cellulose. Thus a high proportion of cellulose is inaccessibly crystalline in nature. The remainder is in disoriented and amorphous chains.

Two characteristics of this arrangement are of direct interest. One results from the long chain structure of cellulose. Each chain may run through many crystallites, produc-

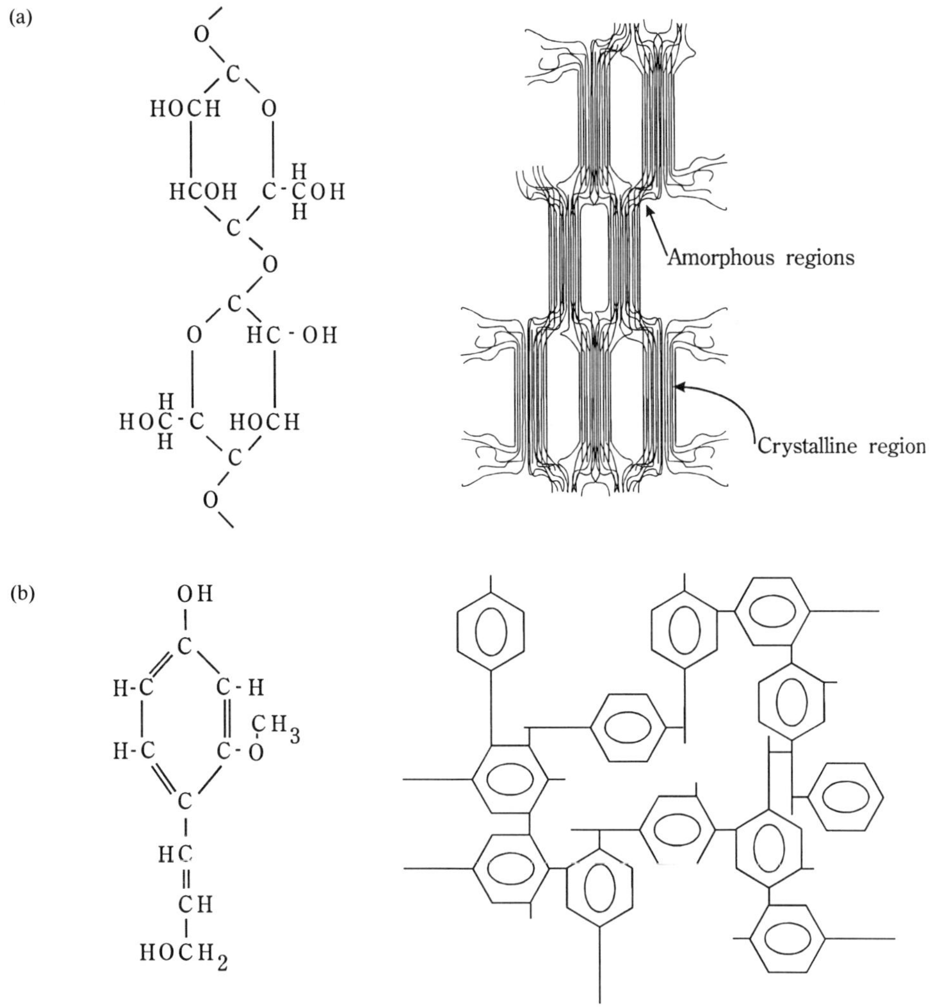

(a)

(b)

Figure 5–21 (a) Cellulose; (b) Lignin.

ing regions of crystalline cellulose interrupted by regions of total disorientation. The mass may be likened to a dish of spaghetti, where a group of strands may line up side by side and stick together for a short distance, then wander around on their own for a short distance, become parts of different crystallites, and then wander off again (Figure 5–21a, right). The net effect is a solid material with a limited degree of chemical and physical activity, since water and most chemicals cannot enter the crystallites. All reactions must occur either on the surface of the crystallites or in the disoriented, amorphous regions, since only there are the water elements on cellulose accessible and available for reaction.

The second characteristic that ensues from the amorphous-crystalline-amorphous organization of cellulose chains is that free movement is possible but limited. As water, for example, seeks to enter the system, swelling occurs because the water molecule adds its volume to the space between the amorphous chains and crystallites. But the chains can move only so far before they run out of slack laterally, thus limiting the amount of space and therefore the amount of water that can enter, and of course limiting the amount of swelling that can occur.

Hemicellulose

Although generally considered part of the cellulose fraction of wood (holocellulose), hemicelluloses, because of their carbohydrate nature, act more like extractives in being soluble or easily degraded into solubility. Hemicelluloses are composed of five- and six-carbon sugars in relatively short chains, like leftover parts of the main production. Distributed throughout the cell wall and lumens, they influence to a large extent the hygroscopic activity of wood, some more than others. Solubility ranges from cold-water soluble to soluble in dilute sodium hydroxide. In quantity, it ranges from a few percent in some species to about 25 percent in species such as larch.

There is little a glue user needs to do about hemicelluloses, since they are part of the wood structure. They are particularly contentious in the production of hydraulic cement boards, since they, and the sugars they break down to, interfere with the hydration of the cement. In this case the offending elements can either be washed out in advance or neutralized with additives in the adhesive. Although they are known to foul up some reconstituted wood processes, notably wet-process fiberboard production, they possess adhesive properties in themselves and represent a potential source of chemicals for modification into adhesives.

Most of our conjectures about the interactions of glue with wood in achieving adhesion involve the polar and chemically reactive hydroxyl (OH) groups on cellulose and their presumed availability. We will dwell on this point further after the other constituents have been discussed.

Lignin

The molecular structure of lignin resembles that of cellulose except that the six-member ring is made up only of carbon atoms and the characteristic alternating single and double bonds of benzene. Though a complement of hydrogen atoms are present, hydroxyls and oxygen are scarce. One must conclude that lignin cannot be as reactive as cellulose.

Another constant characteristic of lignin is a side chain of propane. The basic building block of lignin, equivalent to the glucose molecule for cellulose, is phenylpropane, (Figure 5-21b). Occasionally a methoxyl group appears as a side chain, and a phenolic hydroxyl group also is evident, sometimes free and sometimes as part of a connecting link to another molecule.

These building blocks are attached to each other in many ways, producing many different kinds of lignin, but all have essentially the same chemical composition. The right side of Figure 5-21b shows a schematic representation of a lignin. Although the structure seems to lie in the plane of the paper, it actually extends in all directions in an apparently random manner, forming the highly cross-linked structure mentioned earlier.

The chemical makeup and structure of lignin suggest that, unlike cellulose, it is a rather brittle, inert material, a primary function of which is both as an adhesive, bonding fibers together, and as a stiffening agent. Like all good adhesives (and a superb example for adhesive technologists), the lignin has formed the nine links of the bond chain to such an optimum degree that in fact they support each other. Links 2, 3, 6, and 7, the weak boundary layers of man-made bonds, are engineered out of existence by an interdiffusion of molecules from both components. The compound middle lamella (Figure 5-20) comprises links 1, 2, and 3, which are composed predominantly of the lignin adhesive, but, especially toward its outer boundaries, is increasingly suffused and reinforced with randomly oriented cellulose chains. In the next layer of the fiber, called S1, there is a higher proportion of cellulose and a lower proportion of lignin. In this layer also the cellulose chains are oriented horizontally, providing a sort of casing for the rest of the fiber. The S2 layer, farther toward the lumen, contains the maximum amount of cellulose (with its chains oriented vertically, the direction of maximum stress in a tree), and the minimum amount of lignin—equivalent to links 8 and 9. In the S3 layer, the cellulose chains are again oriented horizontally, as in the S1 layer, and the lignin content is again high, forming an inner casing for the fiber.

From a bond standpoint, the diffusion of the lignin into the fiber wall not only increases the stiffness of the cell, it also provides for a gradual transfer of stress from one fiber to the next. The mechanism minimizes the concentration of stress that would occur if there were an abrupt line of demarcation between the two phases. In man-made bonds using the more rigid adhesives, an abrupt transition between wood and glue sometimes allows stresses to accumulate at this very sensitive point in the bond, precipitating failure earlier than would otherwise be the case.

The actual nature of the bond between cellulose and lignin is still in dispute, though the difficulty of removing the last traces of lignin in papermaking suggests that the bond is a very good one. Inferring from the genetic imperatives of a tree, it is reasonable to assert that it would avail itself of all means possible to produce the most effective bond— in which case one can assume that chemical, physical, and mechanical adhesion are all present, and to varying degrees depending on the exigencies being experienced at the time of formation. Knowledge about lignin as it exists in wood is largely intuitive. Because of its cross-linked nature, it cannot be extracted for study without destroying some aspect of its makeup. The lignin studied in the laboratory is composed of breakdown fragments modified by the necessarily severe extraction process. When these are analyzed, they present endless possibilities as to how they might have been connected originally. Despite this lack of specific knowledge, the general chemical composition of lignin suggests that it does not have nearly the attraction potential of cellulose. It has a lower polarity and hygroscopicity, and essentially is more like a hydrocarbon than a carbohydrate. Its resistance to solvents reflects its highly cross-linked structure, which creates the difficulty in studying it.

Though lignin is cross-linked, brittle, and insoluble, characteristics of thermoset polymers, it displays some anomalous behavior under heat and pressure, deforming thermoplastically. This is the basis of the Masonite process, in which wood fibers are consolidated by making the lignin flow and rebond. While this type of bond is sufficiently strong for that type of product, it does not have the durability of the original lignin fiber bond.

Judging by the chemical nature of cellulose and lignin, the one rich in reactive groups and the other somewhat inert, it would seem logical to conclude that a ligneous surface would be more difficult to glue with conventional adhesives than a cellulosic surface under conditions that do not promote the flow of lignin. Some confirmation of this conjecture would be useful.

Extractives

The term *extractives,* as mentioned earlier, is derived from the method used to remove them from the wood. In being dissolved out, they are differentiated on the basis of solubility. An initial distinction is made on the basis of water solubility versus solvent solubility. Water solubility is further divided into cold-water solubility and hot-water or steam solubility. Solvent solubility includes alcohol, benzene, and ether, singly or in combination. Extractives comprise a wide range of organic materials, including waxes, oils, fats, tannins, carbohydrates, acids, gums, and resins.

Extractives are considered extraneous as far as the cell is concerned. They are deposits—impregnants—in the cell lumen or cell wall. Made primarily during heartwood formation, their greatest effects are on the hygroscopicity, permeability, and durability of the wood, decreasing the first two and increasing the third. Despite their presence as minor ingredients, extractives have a major effect on wood gluing, challenging in one way or another the formation of every link in the bond system. They do this through their pH effects (discussed subsequently), contamination effects, and penetration effects.

Because extractives are deposits, not strongly bound anywhere, they are relatively free to move. They move by diffusion, either as a volatile material or as a solute carried along when its solvent moves. Thermal and moisture gradients prompt this movement. They also move with capillary and surface tension forces.

Extrapolating into the gluing process, one can perceive a source of gluing problems stemming from the drying or conditioning of wood in preparation for applying the glue. As moisture leaves the wood, it carries small amounts of certain extractives with it. When heat is used to accelerate drying, more extractive is solubilized, and some is volatilized, promoting even more of it to move. Heat, however, can also have a beneficial effect in "setting" certain resinous materials, restricting their mobility. Rapid heat input can also break or otherwise interrupt the moisture track in the capillaries, and thus cut off the further migration of extractives by this route.

Practitioners of the gluing art have long known that time alone has a deleterious effect on gluability. The more time passes between surfacing and gluing, the more time extractives have to move. In some species with mobile extractives, overnight standing alone can decrease the ability of the surface to absorb water. (A second factor during such waiting periods is the possibility of moisture change and concomitant dimension changes, which cause their own set of problems.) For these reasons, specifications governing gluing processes (especially lumber) often prescribe gluing the same day as surfacing.

However they move, their destination is usually a surface, a freshly cut or broken surface being particularly attractive. Here they preempt the strongest attractive sites just created, if airborne contaminants have not already beaten them to it. The newly formed surfaces, which carry the hopes for good wetting (links 4 and 5) now are covered by a nonadhesive material. In order for glue to attach itself directly to the wood molecules, it must have the power somehow to displace the usurping material. If the offending contaminant is soluble in the adhesive or

the solvent of the adhesive, this displacement has a chance of occurring, and a degree of wetting will take place. If not, links 4 and 5 will have incurred a weakness. Unfortunately, what is true of contaminated freshly cut surfaces is also true of uncut surfaces, the lumen walls, which not only receive newly transported materials but also harbor residues from previous cell activity. In any case, the lumen walls do not present a favorable surface for bonding, because of their chemical nature and their physical form as a crustaceous and loosely bound layer. (Lumens, of course, can still function as a receptacle into which glues can penetrate and mechanically lock themselves.)

Gluing processes that are particularly vulnerable to extractive contamination are those in which creation of the gluing surface *precedes* drying, e.g., manufacture of plywood, flakeboard, and fiberboard. In these processes all the factors which contribute to inactivation of surfaces are in operation. It should be noted in passing that the amount of material needed to inactivate a surface is very small and probably could not be measured by even the most sensitive gravimetric methods.

As a general rule, alkaline adhesives tolerate contaminated surfaces better than acidic adhesives, and solvent-based adhesives better than water-based ones. This is predicated on the solvent power of the adhesive for the contaminant. Alkaline adhesives also have the power to swell or otherwise open up the surface for better penetration to uncontaminated areas. When the adhesive is expected to overcome the presence of contaminants on a surface, some side effects on the hardening of the adhesive should be anticipated, either on the rate or on the degree of hardening. Particularly susceptible are links 2 and 3, where dissolved extractive molecules can interfere with cohesion development in the adhesive and produce a weak boundary layer at this point in the system. Any pH changes that occur as a result of extractive takeup will also affect cohesion development.

A more subtle effect of extractives results

from their lowering the hygroscopicity and water absorption properties of wood. During assembly-time periods, some adhesives are designed to lose a portion of their mix water in order to develop the right mobility characteristics for the pressure period. High extractive or otherwise inactivated surfaces reduce the rate of glue water loss. The result is that, for a given assembly time, the adhesive will have more mobility than may be good for it when pressure and/or heat are applied. Starved or filtered glue lines then occur. For those glues that harden by loss of water, an additional interaction occurs, in which the slower water loss means a longer pressure period may be necessary.

A particularly vexing problem arises when the distribution of extractives varies from piece to piece of the same species, or from point to point of the same piece. Differences in extractive content between heartwood and sapwood are the chief causes of such variability. Within annual rings, there are also opportunities for differing extractives depending on the location of the tissues that harbor them, such as resin canals and storage cells. It is possible and in fact common for single glue lines to contain both high and low extractive contamination. Dramatic demonstrations of this are sometimes observable in plywood, where part of a glue line contains a delaminated section coincident with a heartwood area.

pH

On a scale of 1 to 14, with 7 being neutral, 1 to 7 being acid, and 7 to 14 being alkaline, wood has a pH[4] range of about 3 to 6. (For comparison, common carbonated beverages have a pH of 3, black coffee a pH of 5, and household ammonia a pH of about 12.) In addition to pH, materials also exhibit a buffering potential, essentially a resistance to pH change when acid or alkaline ions are introduced.

[4]pH is calculated as the negative logarithm of the hydrogen ion concentration. Each unit therefore represents a 10-fold change. That is, a pH of 5 is 10 times more acid than a pH of 6; a pH of 9 is 10 times more alkaline than a pH of 8.

Adhesives operate through their own highly controlled pH's, designed to produce a solubility, a rate of hardening, and a degree of hardening that is specific for each formulation. The pH of wood can be and often is in conflict with that of the adhesive. Though link 1 may be directly affected through coagulation or hardening effects, a more likely point of interference is at links 2 and 3, where the strongest interaction of wood and glue occurs. Only the first layer of glue molecules or particles needs to be disturbed in their secondary mission of developing cohesion for the entire bond system to be weakened. Moreover, anything that affects the solubility of the glue is also bound to affect links 4 and 5, where we expect molecule-to-molecule conformity to take place across surface boundaries pitted with irregularities. Failure in these four links, of course, also guarantees failure in links 6 and 7.

The potential conflict between a thin glue line, designed to harden at a pH of 8 on a wood with a pH of 4, is easily imagined. It is well illustrated in the gluing of oak with resorcinolic adhesives, where the low pH of the wood retards and actually inhibits the curing of the alkaline adhesive. Some heating of the glue line, to aid the curing, is usually urged by the adhesive manufacturer in this situation.

Glues that cure with acid catalysts, such as ureas, run into the opposite problems of curing too fast on high-acid woods. This is particularly true in particleboard manufacture when strongly acidic woods are used. Precure of the resin can occur before the mat reaches the press. Since not all resin-coated wood particles reach the press in the same elapsed time span, this effect produces a sporadic downgrading of the product. To counteract this effect, resins are specifically tailored to the wood being used, reducing the acidity of the resin accordingly.

To further protect the pH of adhesives from change by the wood, they too are sometimes provided with a buffering capacity of their own. This widens their utility and avoids the necessity of precise tailoring to fit particular applications.

Highly acidic adhesives can also affect the

wood. When the adhesive has a pH less than 3, there exists the possibility of acid attack on the wood. This is observed in long-term durability tests in which decreasing strength and increasing wood failure occurs over time. Although there is some evidence that the attack on the wood is also related to the specific ions present as well as the pH, low-pH adhesives are usually avoided for wood gluing. Acid-catalyzed phenolics, for example, attractive because they cure very rapidly without heat, have pH's between 1 and 3, and are generally proscribed in specifications for critical products.

For similar reasons, the practice of applying an acid catalyst to one side of a joint and an uncatalyzed resin to the other side (called separate-application gluing) is not recommended despite some commercial successes. In addition to the danger of acid attack, there is also reduced penetration on the catalyzed side, and reduced catalysis on the opposite side, especially in the tendrils of resin that penetrate the wood. The process is attractive for noncritical products, however, because it has the advantage of fast curing with a resin of long pot life.

The pH of the adhesive has yet another effect on wood: discoloration. Highly alkaline adhesives such as casein or soybean tend to discolor some woods. Oak and maple are particularly subject to this problem. The wood in the vicinity of the glue line turns a grayish color, which is unsightly in some products. Polyvinyl acetate adhesives occasionally produce black or grayish glue lines. In this case, however, the problem is not the pH of the adhesive but more likely iron contamination of the resin, either from applicators or from containers. Adhesives that have color, such as the reddish phenolics and resorcinols, add their color to the wood they contact.

Ash

The ash content of wood, generally less than 0.5 percent, has no direct bearing on the performance of glues, except as they might affect the pH or machining characteristics. Woods that are high in silica are perhaps the most fractious, by dulling tools which then produce poor gluing surfaces. Certain woods, particularly maple, sometimes contain mineral streaks: short, dark areas elongated along the grain. These are generally regarded as blemishes and pose no gluing problem, though occasionally a poor bond occurs due to hygroscopicity differences.

MECHANICAL PROPERTIES OF WOOD

The mechanical (strength) properties of wood, like the mechanical properties of all materials, define in numerical terms the behavior of the material when it is subjected to stress and associated strain. Stresses and strains are also inherent in bonded systems, though the behavior is much more complex and unpredictable, due to the composite structure, each element of which may behave in a different manner.

In discussing the mechanical properties of wood, we are interested in how its strength affects both bond formation and bond performance. It is obvious that there is a direct effect on the loads a bond must carry or transmit as it performs, but it is less obvious how wood strength affects the formation of a bond. The properties act directly through links 8 and 9, and indirectly through their effect on machining during preparation of the gluing surfaces. The direct and indirect effects register on all links of the bonded system.

Although wood has been part of the human culture for as long as there have been humans, its properties are still not fully understood. New, highly sophisticated testing procedures are adding daily to our knowledge of wood, improving its performance and enlarging its scope of utility. Because wood is both a material and a structure, it has material properties as well as properties due to its structure. Its material is both elastic and plastic, both crystalline and amorphous, both hydrophyllic and hydrophobic; its properties are therefore time dependent and treatment dependent. Its structure is a mass of cells varying in configuration, size, orientation, function, and distribution; its properties therefore depend on species, growing conditions, and processing methods. Hence, the mechanical properties of wood cannot be defined by single number, but only by ranges of numbers, and by numbers qualified by time, history, and treatment.

Strength of Clear Wood

In order to establish some sense of order in the properties of wood, it is necessary to have a base from which to calculate deviations due to special circumstances. The base is presented in Tables 5-2 and 5-3, a compilation of the characteristic mechanical properties of the commercial woods of the United States; and in Table 5-4, some foreign-grown woods. These tables were abstracted from the *USDA Forest Service Wood Handbook*, which may be consulted for a more complete listing. The values given are averages of many tests on many trees, and may be considered the defect-free, clear wood norm for the species, unmodified by strength-reducing factors.

Variability

Contrary to the impression of precise properties implied by Tables 5-2, 5-3, 5-4 and 5-10, wood is exasperatingly variable and complex due to its anisotropic nature and its botanical origin, modified by environmental conditions and preparation for use. Being anisotropic means that the properties are different in each of the three perpendicular directions, in some instances 10 to 20 times greater in one direction than the others, resulting in a strong direction along the grain and a weak direction across the grain. In the weak direction, some properties can vary depending on whether the stress is directed radially or tangentially.

The botanical genesis of wood ordains basic differences between species. Each species has its own set of mechanical properties. Within a species, strength properties vary markedly depending on growing conditions. Conditions which produce excessively fast growth or excessively slow growth, high density or low density, result in wood that is

either at or near the extremes of the strength range for the species. Conditions that produce compression wood or tension wood result in wood that is outside the lower extreme and therefore considered abnormal. A representative display of the distribution of strength values taken from a random sample of clear wood specimens is shown in Figure 5-22. Given the coefficients of variation for different properties of wood (Table 5-10) similar distributions can be drawn for other properties, assuming the values are normally distributed. Such a display allows the prediction of the percentage of pieces in a given sample that might fall within certain strength classes. By implication, it also provides an indication of the variability to be expected in bond strength tests that produce 100 percent wood failure, i.e., links 8 and 9.

Variation Due to Density. Density is perhaps the major factor arising in the tree that

Table 5-10. Average Coefficient of Variation for Some Mechanical Properties of Clear Wood

PROPERTY	COEFFICIENT OF VARIATION* (%)
Static bending	
Fiber stress at proportional limit	22
Modulus of rupture	16
Modulus of elasticity	22
Work to maximum load	34
Impact bending, height of drop causing complete failure	25
Compression parallel to grain	
Fiber stress at proportional limit	24
Maximum crushing strength	18
Compression perpendicular to grain, fiber stress at proportional limit	28
Shear parallel to grain, maximum shearing strength	14
Tension perpendicular to grain, maximum tensile strength	25
Hardness	
Perpendicular to grain	20
Toughness	34
Specific gravity	10

*Values given are based on results of tests of green wood from approximately 50 species. Values for wood adjusted to 12% moisture content may be assumed to be approximately of the same magnitude.
Source: Wood Handbook, USDA.

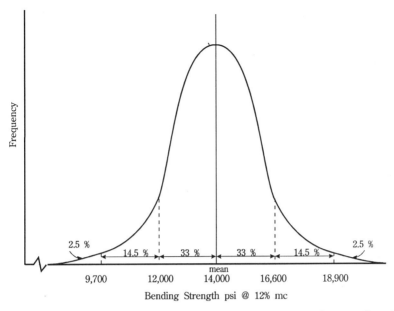

Figure 5-22 Theoretical distribution of strength values from a random sampling of clear wood boards showing the variability to be expected in normal wood having no apparent defects.

is a fairly reliable indicator of strength, even between species, but especially within species. This is because, as mentioned earlier, density is strongly related to the amount of cell wall substance present, and since cell walls are the chief elements producing strength, one would expect a direct correlation. Figure 5-23 shows this direct relation with respect to modulus of elasticity. Different species would show the same relationship with only slight shifts vertically. Relationships of density with other properties can be drawn with the help of the constants and exponents given in Table 5-1. The strength at any specific gravity can then be determined from the plot. Alternatively, strength at a different specific gravity can be calculated using the formula.

$$\frac{SpG1}{SpG2} = \frac{strength\ 1}{strength\ 2}$$

It should be borne in mind that there are also variables that alter the strength/density relationship. These include moisture content, extractive content, and abnormal wood (compression wood and tension wood). De-

cay, grain direction, and other defects also destroy the validity of this calculated relationship.

Grain Direction. Wood is strongest along the grain. When stresses are imposed at any other grain angle, wood has less ability to resist, and it will appear to be weaker than table values predict. Measured as the angular deviation from the edges or face of a wood piece, in inches of deviation per inch of length, reduced to whole numbers, the drastic effect on strength is displayed in Table 5-11. Calculations of strength loss with angle of the grain using Hankinson's formula are shown in Figure 5-24. The grain in a high-stress area of a glued product therefore must be factored in when predicting the performance of that product.

Apart from the effect of grain angle on bond formation mentioned earlier with respect to porosity at the glue line, the lowered strength accompanying deviations in grain drastically reduces the amount of bending that may be borne by a lamination being glued into a curved shape. Otherwise the decreased strength plays in the performance of

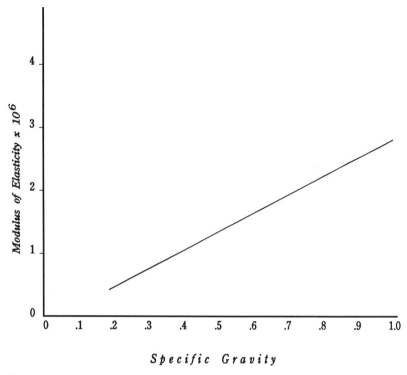

Figure 5–23 A general relationship of specific gravity to modulus of elasticity for clear wood. Relationships to other properties are curvilinear in nature (see Table 5–11).

the glued product both in terms of strength to be expected, which is understandable, and also in terms of warp, which is less obvious initially. Grain angle effects operate in two different ways in affecting warp, in both cases by destroying the "balance" about the neutral plane in some products. (1) By lowering strength, deviant grain reduces the ability to restrain deformation in the part of the assembly in which it is located, e.g., in one of the crossbands of a lumbercore panel. (2) Stability in the long direction is no longer negligible, since a component of across-the-grain shrinking and swelling now is directed lengthwise. Thus there is shrinking and swelling in a direction where there should be

Table 5–11. Strength of Wood Members with Various Grain Slopes Compared to Strength of a Straight-Grained Member, Expressed as Percentages

MAXIMUM SLOPE OF GRAIN IN MEMBER	MODULUS OF RUPTURE	IMPACT BENDING— HEIGHT OF DROP CAUSING COMPLETE FAILURE (50-LB HAMMER) (%)	COMPRESSION PARALLEL TO GRAIN—MAXIMUM CRUSHING STRENGTH (%)
Straight-grained	100	100	100
1 in 25	96	95	100
1 in 20	93	90	100
1 in 15	89	81	100
1 in 10	81	62	99
1 in 5	55	36	93

Source: Wood Handbook, USDA.

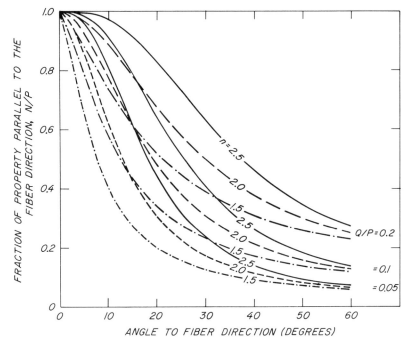

Figure 5-24 Effect of grain angle on mechanical properties of clear wood according to a Hankinson-type formula. Q/P is the ratio of the mechanical property across the grain (Q) to that parallel to the grain (P); n is an empirically determined constant. (Courtesy of USFPL.)

stability, and there is reduced mechanical restraint to control the forces that may be transmitted across the glue line.

Springwood versus Summerwood. The vastly differing density, and therefore strength, of springwood and summerwood causes variations within an individual piece that sometimes exceeds that found between species. Summerwood, when present as a distinct band, is at least twice as strong as springwood. Hence, the behavior of a piece of wood to an applied load depends on the direction of the rings with respect to the direction of the resulting forces. For example, if the forces are in compression across the grain, resistance to compression will be greater in the tangential direction, where the summerwood is in position to receive and carry the load, than in the radial direction, where the entire load also bears on the springwood (Figure 5-25). Similar considerations apply in bending loads with the addi-

tional factor of location of each part of the ring in the highest stress areas of the beam.

Springwood and summerwood exert their strength differences also in processes for producing the gluing surfaces. Rotating knives create stresses perpendicular to the surface, i.e., in compression across the grain. Hence, in Figure 5-26, the planer knife passing over summerwood exposed at the surface deforms the layer of springwood underneath, which later springs back and raises the summerwood above the general plane of the board. A similar situation occurs in veneering, although in this case the knife may also deflect as it encounters a band of summerwood, maximizing the effect.

Processing. Finally, the process of converting a tree to useful objects modifies strength properties, some advantageously, some disadvantageously. For example, seasoning the wood, i.e., removing moisture the

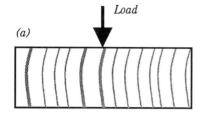

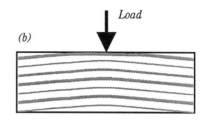

Annual rings are parallel to direction of load. Springwood and Summerwood share the load, although summerwood takes most due to its higher modulus.

Annual rings are perpendicular to direction of load. Both springwood and summerwood sustain the entire load equally. Springwood, being weaker, compresses more.

Figure 5-25 Compression perpendicular to the grain as affected by ring direction.

tree needed in growing, increases some strength properties two- to threefold. Storing the logs or cut pieces improperly can result in, at best, warping, and, at worst, decay, a serious loss of strength. The seasoning process itself creates its own variations in final moisture content, which produce variations in strength as well as in dimensional change. Processing factors are discussed more fully in Chapter 6.

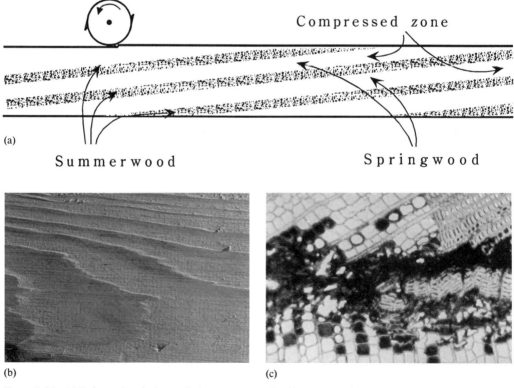

Figure 5-26 (a) Deformation during surfacing as a result of knife pressure. (b) Compressed zone springs back, raising summerwood above surface plane. (c) Adhesive attempts to repair the damage.

Moisture Content

As noted in the discussion of physical properties, the moisture content of wood varies according to the environment to which it has been subjected, having a range from green, or totally water-soaked, to oven dry. The fiber saturation point is a pivotal moisture content in that, above this point, strength properties are independent of the actual moisture content, i.e., they are constant. Below this point, strength is strongly related to the moisture content, increasing in strength with decreasing moisture content (Figure 5–27). While not all properties are affected the same amount with moisture content, the deviation from the values for wood at 12 percent can be calculated using the following equation:

$$P = P_{12} \left(\frac{P_{12}}{P_g} \right) - \left(\frac{M - 12}{M_p - 12} \right)$$

where:

P is the property to be calculated

M is the percent moisture content for which the strength is desired

P_{12} and P_g are the property values at 12 percent and green, respectively, from the clear wood strength tables

M_p is the moisture content at which strength properties begin to change, generally taken to be 25 percent if it is unknown (Table 5–12 lists a few known M_p)

This equation can be used for most properties except toughness or impact, work to maximum load, and tension perpendicular to the grain, which tend to respond interactingly with other factors. The last, unfortunately, is important in bond performance, both in testing and in service, and should be accounted for by direct experimentation.

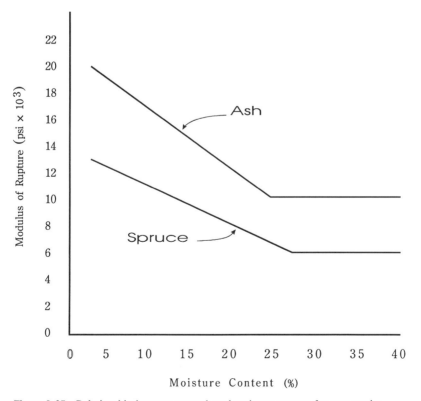

Figure 5–27   Relationship between strength and moisture content for two species.

Table 5-12. Moisture Content at Which Properties Change Due to Drying for Selected Species

SPECIES	M_p (%)
Ash, white	24
Birch, yellow	27
Chestnut, American	24
Douglas-fir	24
Hemlock, western	28
Larch, western	28
Pine, loblolly	21
Pine, longleaf	21
Pine, red	24
Redwood	21
Spruce, red	27
Spruce, Sitka	27
Tamarack	24

Source: Wood Handbook, USDA.

Heat

Many processes for converting timber utilize heat in one form or another to accelerate actions, particularly drying wood and curing the adhesive. Heat has two distinct effects on strength, one immediate, temporary, and recoverable, the other producing irreversible damage. The immediate effect of temperature on strength is illustrated graphically in Figures 5-28, 5-29, and 5-30. The permanent effects of heat on strength of wood are shown in graphic form in Figure 5-31 for dry heat alone, and in Figure 5-32, for heat and moisture combined. These effects operate in preparing the wood, steaming logs for veneering and kiln-drying lumber, or treating in autoclaves, in consolidating by hot pressing, and in testing with procedures that include steaming, boiling, freezing, and force-drying. Despite the strong effects of heating, this factor generally receives very little accounting except in critical structural applications.

The beneficial effects of heating wood outweigh the weakening effects for most applications. Since the effects are primarily on links 8 and 9, strength reductions may simply make wood appear easier to glue. (The effect of heat in creating surface inactivation is a separate matter.) In the main, the effects of heat and moisture represent some of the most useful properties of wood in making it more amenable to gluing by rendering it

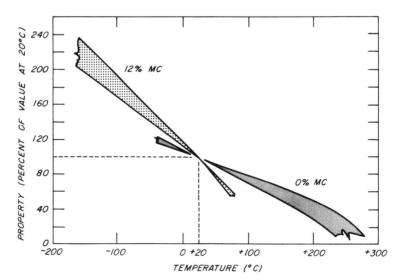

Figure 5-28 The immediate effect of temperature on strength properties, expressed as percent of value at 68°F. Trends illustrated are composites from studies of three strength properties—modules of rupture in bending, tensile strength perpendicular to grain, and compressive strength parallel to grain—as examined by several investigators. Variability in reported results is illustrated by the width of the bands. (Courtesy of USFPL.)

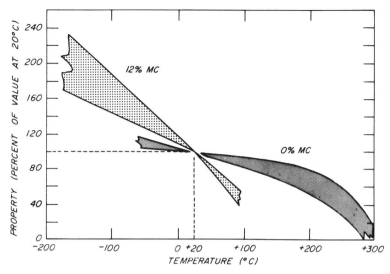

Figure 5-29   The immediate effect of temperature on the modulus of elasticity, relative to the value at 68°F. The plot is a composite of studies on the modulus as measured in bending, in tension parallel to the grain, and in compression parallel to grain by several investigators. Variability in reported results is illustrated by the width of the bands. (Courtesy of USFPL.)

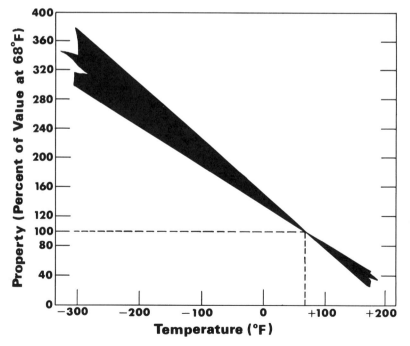

Figure 5-30   The immediate effect of temperature on the proportional limit in compression perpendicular to grain at approximately 12 percent moisture content relative to the value at 68°F. Variability in the reported results of several investigators is illustrated by the width of the band. (Courtesy of USFPL.)

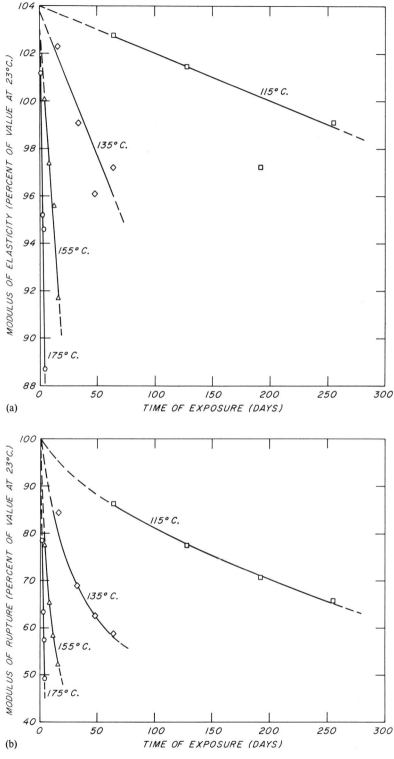

Figure 5-31   Permanent effect of oven heating at four temperatures on (a) modulus of elasticity and (b) modulus of rupture, based on four softwood and two hardwood species. (Courtesy of USFPL.)

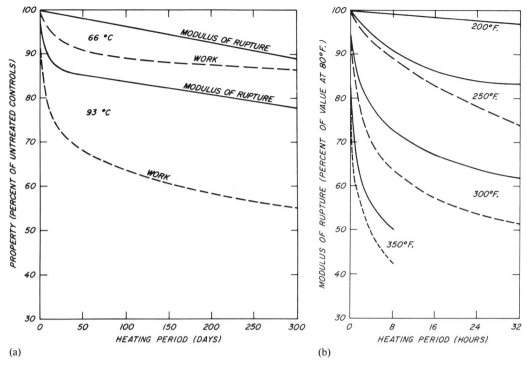

Figure 5-32   (a) Permanent effect of heating in water on work to maximum load and on modulus of rupture; and (b) permanent effect of heating in water (solid line) and in steam (dashed line) on the modulus of rupture. All data based on tests of Douglas fir and Sitka spruce. (Courtesy of USFPL.)

more compliant and easier to achieve closely mating surfaces for bonding. The performance of glue bonds, however, is always reduced by heat and moisture, particularly when combined with stress.

Stress, Strain, and Time

A board left leaning against a barn will some time later be observed to have a slight curvature. What has happened is a phenomenon called *creep,* which is progressive deformation under loads far below those needed to cause breakage in a short time. If the same curvature had been produced by forcibly bending the board and fastening it with a wire to hold the bent position, the wire would experience less and less tension as time went on. This phenomenon is called *relaxation,* a progressive lowering of resistance to deformation.

Because of creep, situations which involve long time loading must be engineered to a higher strength in order to avoid sagging. Ceiling and floor joists are particularly subject to creep deformations, due to the constant loads they carry. Creep is a function of load and time, and is usually calculated as the maximum load a beam can carry—for example, for 50 years—without deflecting more than some allowable amount. Figure 5-33 shows the effect of loading time on the apparent strength of wood. Note that short-time loading, such as impact, increases the apparent strength of wood. This means that wood can sustain a greater load for short periods than for long periods. Creep is accelerated by moisture, particularly moisture change, by heat, and by heavier loads.

Relaxation is involved in steam bending of wood to create curved parts. The process uses steam to soften the wood (i.e., lowering its strength). The wood is then forced into the desired shape and held there until relax-

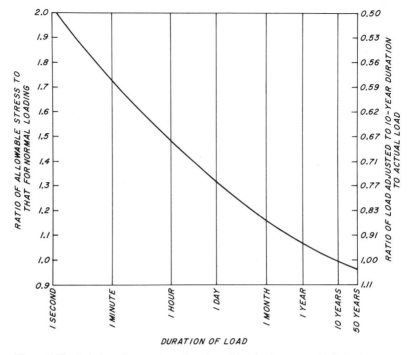

Figure 5–33   Relation of strength to duration of load. (Courtesy of USFPL.)

ation takes place and the wood no longer has the ability to straighten out. Relaxation is accelerated by heat and by moisture. The amount of deformation also has a bearing on the rate and amount of relaxation.

Both creep and relaxation have a substantial bearing on bond formation, even though the time frame of action is usually short. Conformability of mating surfaces may be beneficially affected. In product performance, both are active. Creep affects any application that is under constant load, such as floor joists, and the same cautions apply as for solid wood. It is understood that only thermoset adhesives should be used in such a situation, since they do not creep. Relaxation effects are generally beneficial in some glued products which contain residual stresses built in, for example, by moisture gradients that have since equalized, or in laminating curved shapes where the wood has been forced into the curvature. Compressive stresses created by excessive pressure may also partially dissipate through relaxation.

Fatigue

Wood, like a person, tires under repeated flexing. The concept of fatigue in wood therefore seems simple. If wood is held as a cantilevered beam and flexed up and down repeatedly, it can sustain millions of stress reversals at fairly high load levels. It has good fatigue resistance, a property useful in railroad crossties, which must flex every time a wheel passes over. It must also flex in glued constructions where stresses and strains pass through every time a moisture gradient passes through. Wood, and especially plywood, exposed to the weather with its moisture and temperature changes, undergoes severe fatigue effects that lead to surface checking and glue-line stress.

Logging Defects

Cutting the tree down, an event too early in the conversion process for close supervision, can produce irreparable defects. When the tree comes crashing down, it hits the ground with considerable force. If the ground is uneven or strewn with other logs, this force can produce very high bending stresses, resulting in fractures of varying degree. One rather insidious fracture, called *compression failure,* is difficult to see in the finished product with an unaided and untrained eye. It is not a total break, but an incipient fracture which is extremely weak in transmitting tension stress along the grain that may later be encountered (see Figure 5–9c).

Milling

More variabilities in strength are created by the manner in which logs are reduced to usable sizes and shapes. Two very influential wood strength effects are produced at this stage, which affect bond formation as well as bond performance: grain direction and subsurface damage. The processing factors which produce these forms of strength downgrade are discussed in Chapter 6.

WOOD STRENGTH AND BOND FORMATION

The strength properties that affect bond formation are compression perpendicular to the grain and bending strength. The former produces effects through its ultimate crushing strength, its proportional limit in compression, its recovery from compression, and its creep or relaxation. Some of these act primarily during the press cycle, though some linger to affect bond performance. Bending strength also acts during the pressing cycle, especially in curved products or with warped lumber. Bending forces used in achieving conformation of bonding surfaces remain in the glued product and interact with other bond performance factors.

A deeper insight into what happens on the glue line must be deduced from the behavior of wood to stress. The response of wood to imposed stress is shown graphically in Figure 5–34. Under increasing stress, wood experiences three stages or kinds of deformation. In the first stage, wood deforms elastically; that is, molecular bonds are merely stretched or bent slightly as a spring would be, and therefore the deformation will totally re-

cover upon removal of the stress. In the second stage, the deformation has a plastic component, as in a dashpot, a result of intermolecular displacements, but without breaking primary chemical bonds. Plastic deformations have no restoring force, and therefore they retain their new positions. Deformations of the second stage are partly recoverable, because of the remaining elasticity. When displacement exceeds that which can be accommodated within molecular bond limits, fracture occurs, and the deformation is not recoverable.

However, there is a situation in which fracture deformations recover dimensionally. This is exemplified by the manner in which furniture repair people remove dents (wood fractured in compression) in the surface of finished pieces. The application of moisture, either by steaming or by direct contact with water in the dent, raises the depressed wood back to its original position. In particleboards a similar but less pronounced action occurs, though it is unwanted in this case. The compression of the wood during consolidation produces deformation which partially recover over time as moisture reenters the wood. The result is "springback," a permanent increase in panel thickness, with accompanying loss of some interparticle bonding. A less obvious recovery of compression occurs in other glued products; an explanation of one such follows.

To understand more thoroughly how strength properties act on the glue line, visualize two wood surfaces about to be joined as edge-glued boards. They do not mate uniformly due to machining irregularities, warping, or moisture-induced dimension change after machining. Pressure is therefore needed to bring the surfaces together, in order for the glue to operate. This pressure always works *against the strength properties of the wood,* bending and compressing the wood until the surfaces meet. In curved structures, additional bending force is needed to make the laminations conform to the mold. It is important to realize that all forces applied to bring joint surfaces into bond-forming proximity conspire to make

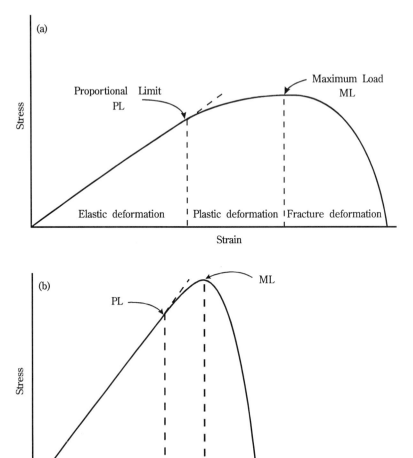

Figure 5-34   Generalized stress and strain behavior of (a) green wood and (b) dry wood.

glue-line pressure a most rampant variable. Despite the most sophisticated gauging of the press, what the glue actually senses varies dramatically from point to point on the glue line. It will be high on elevated areas, which then may also sustain crushing stresses, and there may be none in low areas, which then represent gaps in the bonded system.

Examining this idea further, in Figure 5-35(a) the area at (2) represents a high spot on the edge of a board about to be glued, and (1) and (3) represent low spots. After glue and pressure have been applied (b), the surfaces at (2) are in very close contact, and the wood has been compressed, while at (1) and (3) the surfaces are still separated, per-

haps by 0.015 in. or more. The rest of the glue line is in contact or at an acceptable 0.003 to 0.010 in. apart. As far as glue motions are concerned, the glue at (2), meager to begin with because of being on a high spot, receives a preponderant amount of pressure, and flows or penetrates out of the area; i.e., starved. At (3) the glue remains as applied when it hardens, perhaps undergoing some capillary penetration before it does so, if viscosity permits. At (1) the resistance of the board to bending in the edgewise direction must be overcome before the glue line senses any pressure. Meantime, the force delivered at point (1) is cantilevered to parts of the glue line farther in.

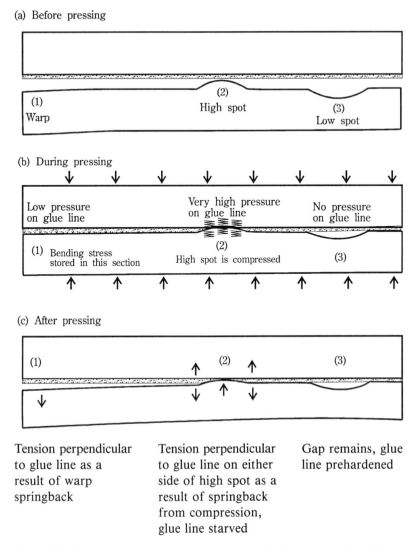

(a) Before pressing

(b) During pressing

(c) After pressing

Tension perpendicular
to glue line as a
result of warp
springback

Tension perpendicular
to glue line on either
side of high spot as a
result of springback
from compression,
glue line starved

Gap remains, glue
line prehardened

Figure 5-35   Wood behavior in response to surface irregularities in edge-gluing lumber.

Other events that affect bond performance occur after pressing but are discussed here to preserve the continuity of the discussion. At (2) the wood that has been compressed during pressing wants to recover from compression—immediately if it is below the elastic limit, or in varying amounts over time, depending on:

1. The amount of plastic deformation that has occurred
2. The amount of cell wall fracture that has occurred

3. The amount of deformation locked in by the glue

The degree of compression above the elastic limit depends on where the elastic limit is at the time, and the amount of deformation that has occurred. The elastic limit is lowered by moisture and heat. Compression strength (resistance to compression) is also lowered by moisture and heat. On the other hand, plasticity is increased by moisture and heat. Deformations that occur plastically do not recover with time but become permanent and are termed *irrecoverable strains*. (This fact contains the clue to controlling the springback problem in highly consolidated products.) Deformations that involve cell wall fracture are also considered irrecoverable. However, as mentioned earlier, they do undergo a degree of recovery that is induced by subsequent infusion of moisture, a form of delayed recovery.

Compression of the wood during bond formation produces consequences in bond performance because recovery from compression produces forces perpendicular to the glue line, its weakest direction. Point (2) therefore is not only starved and weak, it is creating forces that are inimical to the rest of the glue line. At point (3) there is no bond; the links on one side are totally missing. The glue lies as it was deposited, dried or hardened but otherwise unattached or only weakly bonded to the surface. Certain glues may shrink or craze in this situation if they have no gap-filling capability, or become frothy if they are cured dielectrically. At point (1) the board is trying to return to its original position, and in doing so creates a tensile stress perpendicular to the glue line, like a diving board recovering after the jump. It may open the joint up immediately, if it is going to, due to elastic recovery. In trying to judge the force that the board is exerting as it tries to recover its shape, it is necessary to consider all the factors that affect the bending properties of wood: stiffness, moisture content, density, grain, etc.; at most, the recovery force is equal to the force applied by the press in bending it toward the other surface.

The pressure variability that attends the edge gluing of lumber as described above is nevertheless considerably less than that in gluing comminuted wood products, particularly flakeboards. Here the source of high and low spots is multiplied by the operation of the felter. The operation of this machine determines how uniformly the flakes are deposited from point to point, and therefore how much wood there is at each point. Points that receive the most wood become high spots, and points that receive the least wood become low spots. As in edge-gluing

lumber, the high spots receive the most pressure and sustain the most compression. They are therefore the sources of later recovery forces which, as before, create tension stresses perpendicular to the glue lines of neighboring flakes. Again, the magnitude of recovery forces depends on the original wood properties, modified in this case by conditions in the press, the geometry of the wood, amounts of resin, moisture, and temperature while compression is taking place.

WOOD STRENGTH AND BOND PERFORMANCE

The main stresses a glue line sees and responds to are shear in the plane of the bond and tension perpendicular to the bond. The ability to deliver these stresses involve properties of wood that already have been discussed. Stresses that engage the completed bond arise from two distinct sources: internal and external. Both depend on the strength of the wood for the magnitude of the stresses delivered to the glue line. The strength of the wood also determines the locus of failure when stresses exceed the limits of the system. A decisive observation then must be made as to whether the failure occurred entirely in the wood, links 8 and 9; entirely in the glue line, link 1; or partially in both regions. Failures at links 4 and 5 indicate that neither region was totally engaged strengthwise.

Internal Stresses

Stresses that arise internally are due to differential inclination of the wood on either side of the glue line to strain under the influence of a moisture content change, or a thermal change. The inclination becomes differential because of grain differences, density differences, strength modulus differences, moisture change differences, coefficients of swelling or shrinking, or temperature differences in the wood bordering the glue line. The intensity of the stresses depends on the magnitude of these differences.

It is important to realize that all the sources of variability we have mentioned guarantee that no two pieces of wood will be alike. This being so, it is essential to realize that after they are bonded together they are still different, but under restraint of the glue line are now forced to act as if they were identical. The resulting conflict induces the stress that arises from within each piece. This internal stress then combines with any stresses imposed from without, putting to test the integrity of the glue line.

The magnitude of the stresses the glue line sees depends ultimately on the magnitude of the strength properties on both sides of the glue line, and on the amount of wood engaged. It follows, then, that minimum internal stress will occur when the wood on either side of the glue line is equal in every respect. This occurs most easily in parallel bonding of veneer or flakes, and most severely in constructions having perpendicular grain in adjacent pieces, such as plywood, and trusses.

External Stresses

Stresses that are delivered to the glue line from outside the structure, such as by load-ing in some way, accumulate or dissipate depending on the properties of the wood and the location in the structure. In any case, these stresses can never be greater than the wood, links 8 and 9, can carry. The question of whether the bond is stronger or weaker than the wood is resolved by observing where failure occurs when the bond is stressed to destruction. However, the deduction is often not clear-cut because of the stress concentrations and natural weaknesses that occur in the wood at critical points in the system, or because the rheological properties of the hardened glue diffuse stress concentrations.

In measuring the performance potential of a bond, wood properties are used both for creating internal stress and for delivering external stress. Besides choice of species, the careful selection of each individual piece of wood or at least characterizing each piece is as necessary as the gluing procedure used. Not only are the results more reliable but by reducing variability within samples, the precision of the conclusions is improved.

6

Characteristics Conferred in Preparing Wood

In preparing wood for gluing, two objectives command attention: the creation of the wood element to be glued, its size and shape; and producing a surface that will *mate* (achieve close proximity) with another with the least amount of pressure. A corollary objective is that the surfaces will represent conditions not seriously different from the original interior of the wood; i.e., that links 6 and 7 be as similar as possible to links 8 and 9. In meeting these objectives, attention is drawn to the following factors:

1. Smoothness
2. Trueness
3. Moisture content and distribution
4. Subsurface damage
5. Surface inactivation
    Chemical change
    Contamination
6. Grain angle
7. Plane of cut

These factors arise in the methods of creating surfaces, of seasoning, and of storing. Because each method has opportunities for imparting characteristics that strongly influence both bond formation and bond performance, all links in the bond system are at stake, not only the interfacial links. This chapter covers only the preparation of elements for primary gluing. Secondary and tertiary gluing either involve no preparation or undergo special procedures peculiar to the parts being glued. These procedures and the consequences in terms of performance are discussed as part of the more specific gluing details in Chapter 9.

The following discussion is aimed at unraveling the man-made nature of wood surfaces, and the effects generated in terms of adhesive actions. We begin with a description of the various surface conditions, methods of observation, and what the effects are on glue function, followed by an exploration of how these effects are produced in theory and practice.

Surface Conditions

Smoothness. While our perception of a surface evokes a sense of smoothness, it is better defined and measured as "roughness." Roughness is defined as deviations in elevation from a true plane. We would say a surface is perfectly smooth if there were no deviations. The magnitude of deviations from a true plane takes on different significance depending upon the area of interest: a playing field, a driveway, a piece of wood, a steel plate, the surface of water at rest in a bowl. In the latter case, there can never be a deviation greater than a water molecule, almost by definition of the nature of a liquid. In the case of steel, the surface is composed of crystallographic deviations, smooth to the eye but rough to a molecule. In the case of wood, the cell wall material may be in the same order as steel, but the cell lumens represent caverns by comparison. The cell lumens therefore determine a base smoothness for each species beyond which it is impossible to improve. The basic roughness presented by cedar and oak can be considered at a between-species level. However, the oak contains both extremes in every growth ring.

To this inherent roughness due to the cellular nature of wood must be added that due to the action of the knife or other means of creating a wood surface. Cutting devices introduce their own surface variabilities by the path their cutting edges follow. The deviations from plane, as in the case of revolving knives in a planer, are in the nature of peaks and valleys perpendicular to the surface, representing point to point differences in height and presenting a scalloped appearance (Figure 6–1(f)).

Trueness. In addition to quality of the surface per se, another characteristic relates to how well the surface is machined to the correct plane so it properly mates, or fits, to another surface. If one surface is straight and the other departs from a straight line because the knife or fence shifted during the cut, departures in fit occur that neither glue nor pressure can ovecome. In edge-glued lumber, for example, if edges are out of square or the ends are shy due to dubbing in the jointer, open joints are guaranteed even before gluing starts. Similar and more evident problems occur in edge gluing veneer to make up large sheets when edges are neither straight nor square. Less evident but equally serious problems occur in face gluing lumber or veneer when thick and thin areas occur in the same or neighboring pieces. This type of deviation superimposes itself onto all others,

producing gaps and misfitting joints that most glues are not designed to fill.

Integrity of Surface. The purely topographic characteristics of surfaces are invariably undermined by subsurface breakages resulting from the imposition of forces needed to drive the cutting tool through the wood. It is important to realize that these forces are imposed both on the piece of wood being removed and on the parent piece. Deformations, fractures, and partial fractures therefore occur on both surfaces being created. The resulting weakened wood becomes part of the bond system, links 6 and 7. Consequently the surfacing process includes considerations not only of smoothness, but also of soundness; i.e., its integrity in representing the strength of the wood back of the surface, links 8 and 9.

When the damaged wood is at the surface it is readily seen as torn or loose fibers (Figure 6–1(b)). Damage below the surface, the subsurface, is not as apparent, and is often unnoticed until the joint fails at loads less than anticipated, but showing a high degree of wood failure (shallow in most cases) in the fractured zone. (See also Figure 5–26.)

An untold amount of subsurface damage in surfacing lumber is not even incurred by the cutting edge of the knife, but by the heel of the knife back of the edge. Rotating knives in which the clearance angle is negative or has received too much jointing have a heel that protrudes beyond the circle of the cutting edge. This heel engages the wood after the edge has passed and imposes a crushing or polishing action, sometimes burnishing the surface (Figure 6–1(c)).

As with most machine-made abnormalities, the depth and severity of the damage occur in various degrees depending upon the nature of the cutting tool, its sharpness, its motion with respect to the grain of the wood, the properties of the wood, and associated factors. They range from torn surface fibers to deep checks in the wood that sometimes merge to form a continuous weak subsurface layer.

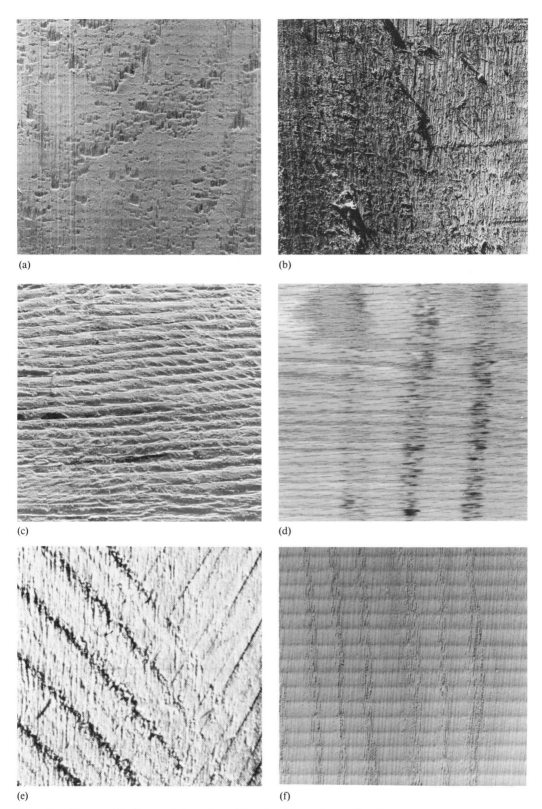

(a)

(b)

(c)

(d)

(e)

(f)

Figure 6-1 Topographic characteristics unfavorable to bond formation and bond performance produced in surfacing lumber and veneer: (a) chipped grain, (b) fuzzy grain (USFPL), (c) burnishing, (d) washboard veneer, (e) saw tooth scratches improve (Franklin Chemical brochure), (f) planer knife scallops.

Surface Inactivation. A freshly produced surface of any material contains all the molecular attractive forces that previously held the material together and gave it its strength properties. If the surface were created and maintained in a vacuum, most of the forces would remain active for a long period but some would dissipate or become internally directed over time. Exposed to the atmosphere, the molecular forces on the surface would very soon attract gaseous molecules or fine particulates, depending upon what is available. Thus the pristine surface can become contaminated to some degree from air alone. Consequently the forces become preempted and satisfied, leaving less to attract an adhesive.

Contaminants can also arise from inside the wood in the form of extraneous materials such as oils, waxes, fats, and sugars, which are not chemically or physically anchored and are relatively free to move. They move outward by several mechanisms: diffusion, migrating with the exiting water, or vaporization. Heat further aids the movement outward. Heat also stimulates the reaction of these forces with each other and, if severe enough, induces the loss of water of constitution, stripping hydroxyl groups from the backbone of the cellulose chain and making it less attractive.

There are two main sources of heat in the preparation of wood: that used to accelerate drying or seasoning, and that used to accelerate hardening of the glue. The former affects mostly veneer, flake, and fiber gluing, the latter secondary and tertiary gluing when surfaces presented to the glue have been previously heated and preparation has been omitted. A third source comes from surface preparation. Even sharp knives produce high temperatures at the cutting edge, but dull or clogged cutting tools, feed rolls, and improperly jointed knives aggravate the situation. The resulting "burnished" surface condition contains both structural damage and chemical damage.

Treatment of wood with chemicals to enhance performance against fire, fungi, and weather must be considered contaminants as far as gluing is concerned. Depending upon their chemical nature, concentration, and application factors, such treatments interfere to varying degrees with virtually all the actions an adhesive must perform. Although wood treated with creosote or oil-borne preservatives (the worst case) has been successfully glued by using specially formulated glues and closely followed procedures, the outcome is not freely certified. Wood that has increased hygroscopicity imparted by fire retardants or decreased hygroscopicity imparted by water repellents can also be glued but demands special attention.

These deliberate treatments can be accounted for in the gluing procedure, and appropriate measures can be taken. However, there are a number of contaminating additives to a wood surface that are incidental and enter the gluing operation unnoticed and unaccounted for. They include grade markings with wax crayons, grease and oil drippings from machinery, dirty hands, and even clean hands drawn over a surface to appreciate its smoothness. Localized reductions in adhesion potential are produced, and the adhesive may or may not cope.

Thus surface inactivation has a number of sources: air, wood, heat, chemical treatments, machines, and hands. While most of these impinge on links 4 and 5, the adhesion action zone, contaminants that originate in the wood or in chemical treatments also can affect links 2 and 3 by being absorbed into the adhesive and influencing its hardened properties. Links 6 and 7, on the other hand, could actually be improved, though at the expense of link 1, because glues remain fluid longer on inactivated surfaces, and are therefore more prone to penetrate when pressure is applied.

Observing the behavior of a drop of water on the surface of the wood to be glued gives a good early indicator of how the glue might later behave. If the drop disappears quickly, some glues may dry out before bonding actions are complete. If it remains in drop form for ten minutes or more, chances are that the glue will remain too fluid to respond properly to the application of pressure.

When the drop disperses slowly up the grain and a bit across the grain, flattening out, and finally disappearing in a minute or so, the wood can be judged appropriate for most ordinary gluing.

Moisture Content and Moisture Distribution. The role of wood moisture in bond formation and bond performance, as well as in glued product performance, has been emphasized many times. Here it is examined further in the context of variables that are associated with it.

Wood comes from the tree varying in moisture content, from a low in the heartwood to high in the sapwood, and from low in the winter to high in the spring. Most of this water must be removed for reasons that include

1. Increased strength
2. Decreased weight
3. Reduced shrinkage
4. Increased decay and insect resistance
5. Improved machining
6. Improved adhesive performance
7. Greater dimensional stability
8. Improved finish performance

However, the process of *wood seasoning* (removing water from wood) can also create some undesirable effects that must be minimized to ensure the above benefits. These include

1. Warping
2. Checking
3. Honeycombing
4. Collapse
5. Casehardening
6. Reverse casehardening
7. Differences in moisture content between pieces
8. Differences in moisture content within pieces
9. Overdried
10. Underdried
11. Thermal degradation
12. Surface inactivation

Removing water from wood is a technology in itself, and requires more complete study than can be covered in this book. Here we will briefly outline the methods used in drying wood. In succeeding sections greater specificity is introduced in connection with the production of the different wood elements, and some of the cautions that have implications in wood gluing are discussed.

There are two distinct methods for removing moisture from wood: air drying and forced drying. In air drying the wood is exposd to outdoor ambient conditions with some protection from sun and rain, and the water is allowed to diffuse out under the influence of the greater absorptive capacity of air. This process requires no additional energy input, but it is slow (months for lumber) and cannot bring the wood below about 12 percent moisture content on average. The slowness has advantages in that it avoids many of the defects that are caused by too-rapid drying. It also has been said, but without strong experimental proof, that the slow cyclic nature of air drying produces a more compact cell wall structure and higher strength. On the other hand, slowness encourages stain, fungi, and insects to gain a foothold, and it ties up the inventory for a long time. Air drying is only used for lumber; the other wood elements require—and can withstand—the more drastic accelerated drying using heat.

The drying process is greatly accelerated by introducing heat to the wood. For lumber this is done in kilns with the lumber properly stacked and stickered to allow uniform air circulation, and weighted down to reduce warping. The heat is introduced primarily as hot air or steam, with a carefully controlled temperature and relative humidity atmosphere to coordinate the rate of moisture movement with the concomitant shrinking that occurs. Dielectric and microwave heating have also been applied in an effort to produce a more uniform temperature in the wood and thus reduce the steep moisture gradients that damage wood. Once the moisture is out of the wood and into the surrounding air, it is purged by venting the kiln.

Application of freeze drying and vacuum drying techniques are also in use to control further the rate of moisture removal and produce a more uniform and stress-free product.

In veneer drying, the wood is not stationary as in lumber drying, but is moved forward in a long tunnel between steel rolls that keep it flat while it is surrounded by fast-moving, very hot air. The fragility of veneer, its sheetlike nature, and its many qualities that require sorting necessitate a great deal of manual handling. Not so in the case of flakes, particles, and fibers. These wood elements are simply blown through very hot air in cylinders or tubes. Because they are small pieces of wood, they are not subject to moisture gradient effects, and so the water can be virtually blasted out. However, they still need to be dried not to zero, as they can easily do, but to a precise moisture content. A main consideration is to keep them from burning while they are being dried and in the case of flakes to avoid breakage as much as possible.

Grain Angle. The grain angle (along with the plane of cut) is a recurring characteristic that usually becomes integrated or fixed into the wood element at the time of the first meeting of log and cutter at the mill. As the cutting tool travels in a straight line, it intercepts all the grain deviations that were formed in the growing tree. Crook, bow, sweep, twist, and taper all translate into some grain angle in wood pieces, sometimes constant, sometimes varying in the same piece. The different kinds of grain angle and methods of observation were discussed previously.

Because of its effect on apparent porosity of a wood surface, grain angle has a strong influence on the actions of a glue, particularly those that affect penetration. And because of its effect on strength and dimensional stability, grain angle has a strong influence on performance of the glued product.

While the occurrence of different grain angles in a wood piece seems to be controlled by tree growth, there are means of minimizing its more troublesome manifestations. One method commonly practiced in lumbering to eliminate the effect of taper is to run the saw parallel to the bark instead of parallel to the axis of the log. In rotary veneering, the same effect would be produced by advancing the knife parallel to the bark. The effects on grain angle of crook and bow of the tree are reduced by cutting logs at the point where they change direction. This results in shorter logs but straighter grain, and is sometimes a good option when short pieces are being demanded, as for furniture parts. There is no way to eliminate the effects of spiral grain in the tree from appearing as diagonal grain in a solid wood element unless the piece can be recut to square the edges with the grain. Spiral grain has no effect on particles or fibers because they form along grain anyway. The effect on flakes is minimal, perhaps more affecting the shape by cleavage in a diagonal direction.

Plane of Cut. The first meeting of log with cutter at the mill also determines which of the two vertical planes in a tree will become the gluing surface. In rotary veneering, the surface is always tangential without choice. In veneer slicing, the choice is unlimited from tangential to radial, with all planes in between depending upon how the cant is mounted in the slicer. In lumbering, choices between true tangential and true radial can be made but are severely limited by the cylindrical geometry of the log and the rectangular geometry of the lumber. A percentage of lumber pieces from a log will contain some of both in true orientation, but a majority will be somewhere in between, "bias cut" or "bastard cut," unless a deliberate attempt is made to maximize one or the other. In all cases the choice is made less from a gluing standpoint than from an appearance or performance standpoint, or from efficiency-of-cutting standpoint.

Given the same quality surface, the plane of cut does not affect bond formation to any significant degree. However, expanses of summerwood do present greater difficulty in gluing, and these are likely to be more expansive on tangential surfaces. Bond performance and product performance, on the other hand, have outcomes that are at least partially dependent on radial and tangential differences. Some of this may be due to ray tissue on radial surfaces that is weak and easily broken. Primarily the differences in performance arise from differences in dimensional stability. The effects are most pronounced in lumber gluing face to face, as in beams, and edge to edge, as in panels. Two facts are involved: that the tangential dimension changes approximately twice as much as the radial dimension for a given moisture change, and that moisture changes about ten times faster through end grain than through side grain. These two differences produce a source of stress across a bond line under changing moisture conditions.

Consider the situation of two boards glued together edge to edge (Figure 6–2). The boards in (a) are tangentially (flat) cut, and will shrink and swell more in width than the radial boards in (b) under the same moisture change. During a decreasing moisture

change, the ends of the boards will change faster than the middle, creating forces at the ends that tend to open the joint; i.e., tension perpendicular to the glue line. With (a) undergoing greater dimension change than (b), it will create the greater stress and be the most likely to fail, either in the wood or in the glue line depending upon which is weaker, and depending, of course, upon the magnitude of the change.

In the case of face-glued lumber, Figure 6–2(c) and (d), the same mechanism produces the opposite result, with the flat-cut boards performing better than the quarter-cut boards. When flat-cut and quarter-cut boards are mixed in the same assembly, as they frequently are, the changing moisture content introduces a shear stress to add to the tension stresses.

The interaction of moisture-induced dimension changes is explored further in Chapter 7 in connection with thickness of wood.

MECHANICS OF WOOD ELEMENT PRODUCTION

There are five basic mechanics in producing or surfacing the various wood elements for gluing:

1. Knife cutting
2. Sawing
3. Sanding
4. Grinding
5. Exploding

Each mechanism uses a different type of tool and produces a different kind of surface. Each mechanism also has many embodiments and is often used in combination with others, but is usually associated with the creation of a specific size and shape of wood element. Moreover, the wood element of interest may be either the parent piece of wood upon which a surface is being prepared, or the piece of wood being removed from the parent piece. For example, in planing lumber, the board is the object of interest, not the chips being removed. In cutting

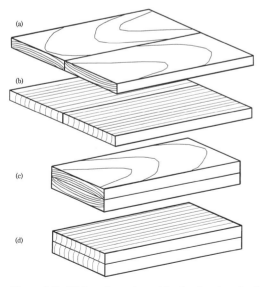

Figure 6–2   Flat- and quarter-cut lumber in edge-glued aspect, (a) and (b), and face glued aspect, (c) and (d), produce stresses of different magnitudes with moisture-content changes.

veneer or flakes, the chip is the chief object of interest. The block from which they are being cut of course carries one surface of the next piece to be cut. In either case, however, the condition of the wood entering this operation has a bearing on the quality of the product, as will be brought out during the course of this discussion.

This operation has no descriptive generic term covering all the methods used. It seems to fall into the general topic of wood machining as a technology since surfaces are created, but more than that, pieces of wood are created, and each with rather specific geometries. Consequently the subject is approached through a discussion of the different methods with an eye on the surfaces being created, but with the format of wood elements to be glued.

It should also be noted that only in lumber is the operation of surfacing for gluing separate from the operation of creating the element itself. In all other cases, the operation of creating the element also creates the surface. This means that only with lumber is a truely fresh, clean surface possible for the glue to see. The others, with the exception of wet-process fiber, face the inactivation influences of severe drying conditions.

A brief discussion of the different means by which wood is reduced to elements, and the surfaces they present to adhesives, will point up the essential mechanics and how they interact with the wood.

Knife Cutting

A knife is a wedge, ground to a very fine (sharp) edge. Its action therefore is predominantly that of a wedge with a low angle of inclination. The sharp edge is a stress concentrator. Since stress is numerically equal to the load divided by the area, an infinitely small area represented by a very sharp edge can magnify almost any force behind that edge to an infinitely high stress. The knife edge therefore has the ability to deliver a high stress, which then severs or breaks the material in front of it by cleavage action. Two factors are at work: the sharpness of the knife and the resistance of the wood to being broken (its strength). The inherent strength of the wood and all the factors that affect strength—such as moisture content, density, temperature, grain angle, and duration of load—all come into play at the knife edge, and have a bearing on how well or cleanly it cuts.

As they perform their action of severing, knives impart two kinds of surface characteristics: one due to the path of the knife edge as in Figure 6–1(f), and one due to the point where the wood chooses to break. Chipped grain, seen in Figure 6–1(a), is an example of the wood choosing to fail below the level of the knife edge, generally caused by a downward slope of grain ahead of the knife.

The action of the knife as it engages wood has been classified on two counts: the alignment of the knife edge with respect to the grain of the wood, and the path of the knife edge with respect to the grain of the wood. Four situations emerge.

Knife Edge	Knife Path	Example
Perpendicular	Parallel	Planer
Parallel	Perpendicular	Lathe, flaker
Perpendicular	Perpendicular	Cross-cutting saw, chipper
Parallel	Parallel	Rip-cutting saw

The simplest knife action is represented by a hand plane or draw shave, in which the edge of the knife is perpendicular to the grain and its path is parallel to the grain. This action is capable of producing the smoothest surface for gluing since the knife theoretically contributes no irregularities. In a lathe, the edge of the knife is parallel to the grain and its path is perpendicular to the grain. In both of these instances, the knife follows a straight path even though the wood may or may not be moving.

A circular path is followed by knives mounted on the surface of a rotating drum, as in a planer. Consequently the knife engages the wood in brief circular passes, leaving a surface with elevated blips. Depending upon the speed of rotation, the depth of cut, and the forward feed speed of the wood, the cutting action produces a series of scallops of varying elevation and distance between, as in Figure 6–1(f).

Sawing

A saw is a unique embodiment of cutting actions wherein small knife edges or points (teeth) are fashioned on the edge of a thin blade. The teeth function as chisels, cutting, gouging, and slashing small bits of wood (sawdust) out of the parent piece as they create a path (kerf) bounded by two new surfaces. This arrangement allows the cutting tool to move *through* the material rather than *over* it, and produce a large wood element, lumber.

While there are many versions of this mechanism, they reduce to three basic types: hand saws, band saws, and circular saws. Saws are called on to sever wood in either of two directions, along the grain (rip sawing), or across the grain (cross cutting). Because the wood offers a different resistance to cutting in the two directions, each has teeth specially designed to sever the wood according to the grain it engages.

In cross cutting, the teeth are predominantly pointed to better slash their way across grain. The teeth are inclined sideways slightly in order to cut a path wide enough for the body of the saw to slip through without producing friction on the edge of the cut. Cuts across the grain are not relevant to wood gluing, since end grain surfaces are not generally glued directly. However, it is important to realize that a saw designed for cross cutting may not produce a good gluing surface if used to cut along the grain.

Saws used for ripping, one approach to a surface for edge gluing lumber, have teeth that attack the end grain of the wood but also dress the side grain during the cut. Varying surface qualities are produced depending upon configuration of the teeth, sharpness, rim speed of the saw, feed speed of the wood, and stability of the saw against vibration and wobble. Stray tooth scratches and fuzzy grain (Figure 6–1(b) and (e)) are the result of improper operation or design of the saw. The latter is one of the more serious sawn surface conditions, produced at the side of a cut by tearing or shearing action. It is visible evidence of a weak surface that can later become a weak link in the bond. The condition is invariably associated with tension wood, but can also occur when cutting wood of high moisture content or using dull saws. Dull saws or saws with insufficient set, and dull planer knives also produce *burnished* (polished with high heat) surfaces. Burnished surfaces, though very smooth, must be regarded as hydrophobic and unreceptive to glue.

Sanding

Sanding is basically a process of cutting and tearing parts of fibers, whole fibers, or clumps of fibers out of the surface of wood by abrasive action of sharp stonelike grits drawn along the grain. Sanding can produce a surface smoothness that appears to exceed that of knife planing in being relatively free of machine-added deviations (except for a tendency to round off edges), and in being less affected by grain direction. For this reason it is usually the final operation in preparing wood surfaces for applying a finish, and for this reason it also has an appeal for preparing a surface for gluing.

A well-sanded wood surface, however, contains two conditions inimical to strong bond formation. First, because the severing action is somewhat indiscriminant and imprecise at the point where the surface is being created, the surface itself is not a sharply distinct plane, but grades through several layers of disturbed, crushed, or partly cut fibers. The adhesive is expected to rebond these ruptured fibers and restore their strength. The second condition unfortunately interferes with this action because the crushed cells either have collapsed lumens or the lumens are plugged with sander dust, and the

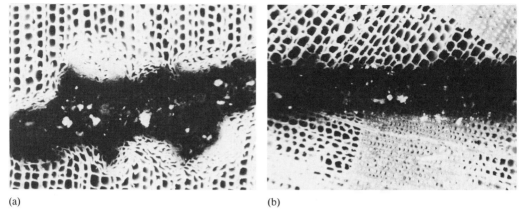

(a)   (b)

Figure 6–3 Photomicrographs of glue lines: (a) sanded, 60 grit, and (b) planed. Note crushed fibers in (a), weakening links 6 and 7. (USFPL)

adhesive is unable to penetrate to do the re-bonding. A layer of sander dust over the surface further thwarts the adhesive by robbing it of water and by creating a layer of what amounts to a concentration of filler. Figure 6–3(a) is a photomicrograph of a glue line made with sanded surfaces (60 grit), in which the damaged fibers have not been reinforced by the glue. Figure 6–3(b) shows a similar glue line made with planer cut surfaces in which the subsurface damage is less, and the repair is more complete.

The quality of a sanded surface depends upon grit size and sharpness, grit speed, pressure, and wood moisture content. Obviously, the larger the grit size, the more cutting it can do if given the pressure to imbed it in the wood. Along with more cutting comes deeper scratches and deeper subsurface damage. Sanding operations therefore go through a series of diminishing grit sizes in producing a smooth surface, ending with the finest.

Like any cutting tool, the sharpness of the grit declines with usage, becoming progressively duller. As the grit gets duller, the more frictional heat develops, accelerating the dulling action, loading the spaces between grits, and the cutting action reduces to polishing and burnishing. Grit speed accentuates frictional heating and therefore also accelerates the declining process.

Moisture content, through its effect on wood properties, also affects how the grit performs. At high moisture content, where

wood is weakest, the wood tends to yield and be displaced or deformed rather than cut. Such surfaces become fuzzy or woolly with partially cut fibers. Sanding quality improves with decreases in moisture content below the fiber saturation point, all the way down to zero moisture content.

Grinding

The production of wood elements with fibrous or granular dimensions has imperatives that differ from the production of larger elements. Because of the very large surface area to be created per pound of wood (see Chapter 7 on geometry factors), each unit of surface requiring a quantum of energy to produce, there is less concern for the topographic quality of the surface and more concern for the efficiency of the process. Grinding offers this efficiency.

Two distinctly different grinding processes typify the mechanics involved in reducing wood to small dimensions. In one, wood that has previously been reduced to chip size[1] is centrally introduced between two facing discs, one or both of which revolve. The discs are faced with teeth, flutes, or ridges that shear and tear the wood apart as it is forced radially outward. These attrition mills or disc refiners have many embodiments: single or double disc, vertical or hori-

---

1. A chip is a thumb-size piece of wood commonly used as the starting element in papermaking, cut by driving a knife diagonally across the end grain of a log with a depth of cut of approximately $\frac{3}{4}$ inch.

zontal disc, atmospheric or steam pressurized. Their job is to reduce wood to fiber or clumps of fibers (fiber bundles).

The other method of grinding utilizes a hammering action to disintegrate the chips. Two essential features characterize the mechanics: the hammers revolving like flails inside a drum; and a screen or grate on the bottom of the drum through which the pieces pass when they have been reduced to the size of the screen openings. The maximum size of the particles is controlled by the size of the holes in the screen; minimum size can be anything smaller than the openings that chance cleavages produce. Size distribution therefore can range from hole size to dust. Shape is controlled partly by the shape of the holes in the screen and partly by properties of the wood. Blunt, cubicle, or elongate rectangular particles are the usual output, although shredded particles can also be obtained under special conditions.

A major difference between the two grinding processes besides the mechanics is the moisture content of the entering wood. Attrition mills demand green wood (or dry wood that has been steamed), while hammermills require dry wood; green wood tends to clog the screens.

One can sense that the geometries and the surfaces are created by random forces operating along natural cleavage lines through the wood. Neither the geometries nor the surfaces have any regularity or uniformity. The question of subsurface damage is somewhat moot in the case of fiber size elements since fiber-to-fiber bonding will be necessary anyway. In the case of particles, subsurface damage is not only extensive, it is largely irreparable because of the very small amount of binder that can be allocated for this penetration action. In both wood elements, conformity to neighboring elements and proximity for bonding must be forced by high pressure.

Exploding

The explosion method of preparing fiberal wood elements was pioneered by George Mason in the 1920s as the basis for wet-process hardboard. Knowing that heat and moisture softens wood, and that steam pressure can be created inside chips by the same moisture, he heated green chips in a closed retort to a pressure of about 1000 psi. Releasing the pressure suddenly caused the chips to explode into fiber. Some further refining to break up large clumps produced a uniform fiber. Washing to remove sugars and other soluble materials that foul up the screens during hot pressing completed the preparation.

While still in use today for the unique flexibility imparted to the board, the industry has preferred the grinding process to create the kind of fiber currently popular.

PROCESSES FOR WOOD ELEMENT PRODUCTION

The five basic wood elements—lumber, veneer, flakes, particles, and fibers—invoke different mechanics or different embodiments of the same mechanics in their production. In the following discussion, we are not concerned with all aspects of each production process such as might be needed by anyone interested in operating such facilities; rather the emphasis is narrowed to the consequences that arise during the process of production that have a bearing on bond formation and bond performance. The degree to which gluing consequences are allowed to influence a given operation is a judgment call in the total context of a particular situation. Most of the details of how to change the mechanics so as to improve the consequences fall into the specific technologies of each process, and are best pursued through manufacturers of the equipment. However, some indication of the direction from which improvement might come is given as a first order of approach.

Preparation of Lumber

The preparation of lumber proceeds in three distinct stages: saw milling, in which the lumber is cut from the log; seasoning, in which the moisture content is reduced to that required for gluing (itself often a two-stage process of air drying followed by kiln drying); and finally, cutting to size, surfacing, and inspection for unacceptable defects. Often, these processes are carried out by separate companies. When carried out within the same company, the processes are separated as "green end" and "dry end."

Sawmilling. Logs, after having been previously graded and selected for size, straightness, and freedom from metal and unacceptable defects, are reduced to lumber using two operationally different means of cutting: circular sawing and band sawing (Figure 6–4). Neither of these two methods of producing lumber introduce characteristics peculiar to itself as far as subsequent gluing is concerned; the rather rough surfaces generated at this point are later replaced by fresh, new surfaces for the glue to see. Both, however, transfer to the board the grain features residing in the plane through which the saw is directed to pass. Two such features relate to the percentage of radial and tangential surface exposed on the board, and the angle of the grain with respect to the edges and surfaces of the board. While the saw operator has some control over the transfer of these features to the board, the control is seldom done with gluing in mind; rather strength, appearance, and wear or weathering performance dictate the procedure.

A third characteristic of the log that transfers to the board as a variable arises from the heartwood/sapwood regions. Because these two regions have a different appearance, contain different moisture, dry at different rates, and contain different defects, the sawyer often attempts to place his cuts so as to concentrate each region in separate boards.

Quality of cutting at this stage, as mentioned above, does not impact the gluing operation because a final surfacing is in store. However, cutting with dull or improperly fitted saws, coupled with saws that do not run true, can create depressions in boards from which the subsurface damage may not be removed by subsequent dressing.

Visual inspection for grade and tally is usually done on the "green chain" in conjunction with a trimming operation to remove major defects near the ends and to cut to a prescribed length. Sorting for quality segregation is then followed by stacking for storage or seasoning (Figure 6–5).

Seasoning. In seasoning the five major wood elements involved in gluing, lumber is by far the most difficult because almost all of the physical, and some of the mechanical, properties of wood come into play, many critically. The main property affecting the seasoning of wood is the rate of diffusion of water through wood along and across the grain. Methods of accelerating diffusion by increasing temperature invoke the thermal conductivity of wood, and the temporary or permanent decrease in strength of wood as-

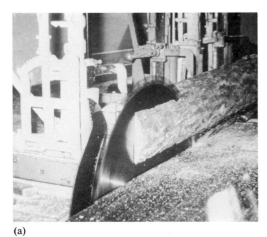

(a)   (b)

Figure 6–4 Cutting lumber: (a) circular sawing, (b) band sawing. (USFPL)

sociated with heat. Shrinkage effects begin to intrude when the moisture content reaches the fiber saturation point. These shrinkage effects are the culprits in many of the problems that arise in seasoning wood. Since during drying the surface of a board will reach the moisture content where shrinkage begins long before the center of the board will reach it, the two regions are put in conflict. The surface is restrained from shrinking by the still swollen center. This produces tension across the grain in the surface fibers and compression across the grain at interior fibers. These two stresses play with waxing and waning intensity throughout the remainder of the drying process. The stresses that arise during the diffusion of moisture out of a board can reach a magnitude where the wood is actually damaged. When the differences between the two stresses are at peak, four types of damage may occur: surface checking, casehardening, collapse, and honeycombing.

Checking occurs when the tension forces on the surface fibers exceed their tensile strength across the grain, most likely to occur when drying is too fast, and the surface moisture content passes below fiber satura-

Figure 6-5 Lumber properly stacked for seasoning. Note stickers (spacers) in vertical alignment. (USFPL)

tion too far ahead of the interior regions. Some woods, such as oak, are particularly susceptible to checking. End grain surfaces are the most prone to checking because of the steep moisture gradient that naturally occurs there. A moisture excluding coating is desirable on end-grain surfaces to retard drying, and to force the moisture to leave by side grain. Checking on side grain is more of a blemish, and of little significance to gluing, unless severe.

When the tension in the surface fibers is not relieved by checking, the fibers "set" in an expanded condition, and retain some of the stress. With further drying, the center enters the shrinking phase, setting the stage for reversing the stresses between surface and core. The surface has finished shrinking, therefore it resists the shrinking action of the core, placing it now in compression while the center experiences tension stresses. This situation continues to the end of the drying process, leaving the board encased in stresses. If left in that condition—known as "casehardening"—warping is certain to result if the stresses are later unbalanced by unequal planing or by resawing.

Because of this propensity for warping, accelerated drying processes include a procedure for relieving the stresses with a steaming period as a final conditioning treatment. When properly done the lumber is stress-free. If the procedure is overdone, however, the stresses reverse again, and the board remains permanently "reverse-casehardened."

Honeycombing results in internal checks, elliptical openings, usually along ray tissue, formed as a result of tension stress in the interior of the board during the later stages of drying. Sometimes they are extensions of surface checks. When severe, they destroy the utility of the board for many purposes, but do not interfere with gluing except when planing exposes them, in which case they provide openings into which the glue might be lost during the pressure period.

Collapse, as the term suggests, results from a flattening of the cell lumens as if by pressure. In this case, the source of stress is not only shrinkage, but also capillary forces created by the departing water. When the cell is completely filled with water, and the water is leaving as a liquid rather than as a vapor, tension forces develop across the receding meniscus that have the power to collapse cell walls. Both honeycombing and collapse are accentuated by high temperature during drying, as a result of the temporary weakening of the wood.[2] Collapse occurs mostly in species that have cell structures and pitting favorable to the creation of capillary forces. They include redwood, cedar, eucalyptus, red gum, and some oaks.

Warped lumber—though not, strictly speaking, damaged—is the bane of lumber gluing operations. Not only are true surfaces difficult to produce from warped lumber, they may not stay true. The state of warp is a function of different amounts of shrinking and swelling in different parts of the board. This in turn is due to differing grain directions, localized abnormal wood, or differing moisture contents, or all three.

If allowed to dry by itself, most lumber as it comes from the log will warp badly in drying; some species do so much more than others. In order to get dry lumber that is flat and straight, it must be weighted down during drying. And in order for it to dry uniformly from all sides, air must be free to circulate over all surfaces. When lumber is piled for drying, boards are separated by stickers (Figure 6–5). In any one pile, stickers are always placed in a vertical row approximately 16 inches aprart in order to avoid creating additional warp.

Lumber drying technology has vested the process in schedules that stipulate how time, temperature, and relative humidity must be programmed for each species to control the rate of drying and minimize defects. Scheduling decisions are based not only upon the

diffusion and mechanical properties of the wood, but also upon the level of moisture content that has been reached. Consequently, means of monitoring the process to track the moisture in boards is crucial because changes in temperature or relative humidity in the kiln at the wrong moisture content can produce major damage to the wood.

Monitoring is done in the most accurate way by tucking sample boards throughout the load and weighing them periodically. The accuracy is dependent upon how well the sample boards (usually about two feet long and end-coated) represent the whole load, and how well their actual moisture content can be calculated from weight-loss measurements. Procedures for doing this are part of the technology. More sophisiticated methods of following the progress of drying are also in use that depend upon the changes in electrical properties of wood with changes in moisture content.

A major source of difficulty in lumber drying arises when a load contains boards of widely different starting moisture contents, or have widely different drying rates. Sorting for moisture content, for incompatible species, for heartwood/sapwood, or for thickness is often necessary to insure defect-free drying. Alternatively, a mixed load can be dried at the schedule for the slowest drying boards, since slow drying does not injure the other boards.

Proper Moisture Content for Lumber Gluing

Two considerations control the optimun moisture content for lumber gluing: the nature of the glue and the environment in which the glued product will be used. As a general rule, wood should be glued at the average moisture content it will experience in service. Since this varies with location in the country, and with location inside or outside a building, and whether heated or unheated, specific recommendations require specific information. Charts of the summer and winter moisture conditions throughout the

---

2. Prolonged high temperature, as noted in a previous section, can result in thermal degradation, a permanent loss of strength. Consequently, the temperature is usually controlled to 180°F or less except in certain high-temperature processes where higher temperatures are used for shorter times.

United States were presented in Chapter 5. Abiding by the conditions identified in the charts will reduce the development of moisture-induced stresses in glued products and will improve bond performance from that standpoint. This subject is explored further in connection with gluing processes in Chapter 9.

Bond formation, however, has its own set of moisture-content optimums dictated by the manner in which the adhesives harden. Water-borne glues that harden by chemical reaction without heat cannot tolerate rapid water loss as would happen if they were applied to very dry wood. They prefer wood at the upper moisture content range, 8–12 percent. On the other hand, glues that harden by loss of water can better tolerate the lower end of the moisture content range, 6 to 8 percent.

A dilemma therefore arises when choices must be made. Does bond formation or bond performance control the moisture content? In the case of glued lumber for exterior uses, there is no problem since both the glue and the product demand the higher moisture content. For interior uses such as beams, columns, and furniture parts, where wood moisture content might drop to 4 percent or less in northern winter climes, the wood must start at the lower end of the moisture content range in order to avoid splitting and warping due to shrinking. Here the water-loss adhesives stand the best chance of forming a better bond. The chemically hardened water-borne adhesives would be at some disadvantage. However, accelerating their rate of hardening by heat or by catalyst will counteract to some extent the too-rapid loss of reaction medium, water.

The balancing of rate of water loss with rate of hardening is a delicate procedure, requiring a high level of cause-and-effect understanding. The quality of the hardened glue is affected by the speed with which it is attained, as is true with many chemical and physical reactions. In general, high speed leads to haphazard activity, and haphazard chemical and physical structures. Chance contacts are made, rather than preferred connections that might ensue at a more leisurely pace. Speed favors the production of small molecules, the formation of a looser structure. The growth of small molecules to large molecules in a tight structure is the ultimate objective in bond formation for maximum strength and durability. Although this is best done slowly, the need for speed in production situations demands compromises.

Surfacing Lumber for Gluing

After lumber has been seasoned to the specified moisture content, equalized to uniform moisture distribution, and treated to remove residual stresses, it is ready for the dry end operations. This usually begins with a "rough dressing" to provide a clear surface for closer inspection. The major amount of wood to be removed before final dressing is taken in equal amounts from both sides leaving $\frac{1}{32}$ to $\frac{1}{16}$ inch for the last cut. (It is deemed good practice to dress off equal amounts of wood from both sides to avoid warping in case some boards contain surface stresses that would otherwise be unbalanced if more wood were removed from one side.)

In face gluing of lumber, the final surface is preferably obtained by planer action, removing less than $\frac{1}{16}$ inch but not less than $\frac{1}{32}$ inch of wood, and leaving between 15 and 30 knife marks per inch. This amount of removal keeps the knives cutting rather than polishing, and eliminates feed-roll indentations.

Sanding, despite being widely advocated and practiced, is a less desirable method of surfacing for gluing, because of the weakening effect of crushed and partly torn fibers at or below the surface. While it is easier for the glue to establish a grip on these loose fibers and appear to form a strong bond, links 6 and 7 are in jeopardy (Figure 6–3). They will evidence their weakness sometime later as shrinking and swelling produces further weakening of the partially ruptured fibers. Sanding does, however, freshen up a surface that has been inactivated in some manner, and in this sense becomes a useful operation in secondary gluing, where such surfaces are most likely to occur.

In edge gluing, lumber is surfaced either by planing action (jointing) or by rip sawing. While planing produces the tighter (stronger) surface, either can produce a satisfactory surface, and either can produce a poor surface depending upon how the operation is conducted. All of the cautions applicable to face gluing apply to edge gluing, with the added concern for squareness and trueness.

Squareness depends upon maintaining a cutter at right angles to the surface. Boards that are warped cannot be cut square because the face against which they are referenced varies. Using force to hold a board flat while it is being edged will produce a square edge, but will not be able to present the squared edge to a mating edge unless the same lateral force is applied during pressing. Assuming a flattening force were applied during pressing and a suitable bond were formed, that bond would be called upon to resist that flattening force from then on, adding to whatever loads it might otherwise be called upon to resist. Warp is best removed by planing before edge jointing.

Trueness depends upon moving either the board or the cutter in a straight line. While suitable mechanics for doing this are well established, accidental deflections sometimes occur. One of the more common occurs at the ends, where the feeding mechanism may not yet have total control of the board as it begins to engage the knives. Too much or too little wood is removed, creating gaps that cannot be closed by pressure.

Cutting a tongue and groove along mating edges, as is sometimes advocated, serves primarily to ensure alignment of the edges. Although the tongue and groove also theoretically increase the gluing area, the poor quality of gluing on the extra surface may easily cancel the advantage.

The end joining of lumber demands the greatest precision in cutting. Not only is the quality of the surface important, but also the angles and depth of the cuts, which must be exact or all hope of surface mating is jeopardized. In scarf joining, programming a cutting tool at an angle through the thickness of a board demands firm hold-down action to ensure a stright cut and to prevent chattering as the knives come down to the feather edge. Fortunately, grain direction is almost always favorable in scarf cutting due to the angle being cut. However, the deviant grain around knots may extend into the scarf zone and create an unfavorable local grain direction. Ring orientation may also present a problem in species with abrupt transition from springwood to summerwood. On flat-cut boards, the tongues of summerwood extending over expanses of springwood cannot provide enough compressive resistance against the cutting tool because they depress into the soft springood and avoid being cut. They later rise and produce a "washboard" effect. This also happens in face planing, but is accentuated in scarf cutting.

Slope of the scarf is perhaps the most important design judgment. Some compromises must be made. The steeper the slope, the weaker the joint, but the less wood will be wasted in making the cut; in other words, more strength versus less waste. A slope of 1 in 12 to 1 in 15 is adequate for most high strength purposes. For a one-inch-thick board the scarf with a 1 in 12 slope would run 12 inches along the board. This will provide a little more than 12 square inches of glue line for every inch of board width, enough to carry up to about 90 percent of the axial load on the board.

The configuration of the tip of the scarf also demands consideration. In can either be drawn to a feather edge or left with varying degrees of bluntness. A blunt edge is less fragile than a feather edge, but more importantly, it will ensure delivery of pressure to the glue line should any end-wise slippage occur during pressing. A tongue and groove is sometimes cut across the face of a scarf to prevent this end-wise slippage. Alternatively, two grooves are cut that receive a spline to ensure positive location during pressing.

In cutting finger joints, the accent is on the cutters. This is essentially a moulding operation where the cutters must be precisely patterned to create both the fingers and the notches into which they fit. Thus wear on the cutters not only produces a poorer surface but also ruins the fit of the fingers. The fingers get thicker and longer, and the notches get narrower and shorter.

Design of finger joints includes consideration of slope, and length, number, and thickness of the fingers and corresponding notches. The bottom of the notch has special considerations. It contains end grain and therefore represents a discontinuity and zero strength. It also represents a stress riser, concentrating stress in adjacent wood. To reduce the ill effects of these factors, the notches should preferrably have rounded bottoms and be as narrow as possible. But this means thinner and more fragile fingers and, worse, cutters that are also thin and fragile. Consequently, compromises are again in order to optimize these factors. Joint strength is increased by long slender fingers of low slope, but this is conflicted by decreased knife performance. Moreover, since the working cross section of the board is being effectively reduced by the area of the notch bottoms, the overall strength of the board is unavoidably reduced. Taking this loss into account, the design of the fingers should be such as to generate enough glue line area to produce joint strength equal to the remaining wood strength.

The glue line is given pressure by an endwise thrust that forces the finger slopes together. If the fingers strike bottom before the slopes have come together, no bonding pressure will be delivered. Consequently, it is good practice to make the notches slightly deeper than the length of the fingers. Since no stress is transferred across the end grain at the bottom of the notches, this does not reduce the strength of the joint.

In judging the fit of fingers, the joint should be tight enough to hold together without glue when jammed together endwise.

PREPARATION OF VENEER

As in lumbering, veneering requires prime logs in terms of size, straightness, and freedom from most defects. However, unlike in lumbering, the log must be prepared for veneering. In addition to such treatments as debarking, round-up, and sawing into cants (in the case of slicing), the most important preliminary treatment involves heating the wood to improve its cutting properties.

There is also another important contrast to lumber. In producing lumber, interest centers only on the qualities of the board; the parent piece (the log) and qualities of the chips and sawdust are of less importance. In veneering, the qualities of both the cut piece and the parent piece are important since each sheet of veneer carries cutting results on one side or the other.

Veneer Cutting

The production of veneer from logs is the breakthrough technology that initiated the elevation of wood to an engineering material. It calls on many disciplines to harness the forces, the properties, and the logistics of driving a long knife across the surface of a log in such a way as to derive a thin sheet of wood without breakage.

In veneering, either by lathe (Figure 6-6(a)) or by slicing (b), the knife edge is parallel to the grain of the log, and it is forced across the grain. Cutting forces imposed in cutting veneer are therefore all in the weakest direction of wood: tension (and some compression) perpendicular to the grain. Hence considerable distortion and fracture occur in the veneer, resulting in "lathe checks" (Figure 6-7(b)) and possible rough surfaces. Both conditions increase with increased veneer thickness and virtually disappear in thin veneer.

Lathe checks, since they run parallel to the grain, do not seriously affect strength properties in that direction. Perpendicular to the grain, however, they encourage a "rolling shear" action that reduces cross-wise strength in a panel. They also affect bond formation in allowing greater penetration of the adhesive than an unruptured surface. With severely checked surfaces, allowance for this extra penetration must be made by

(a)

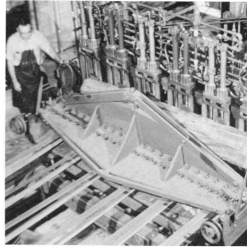

(b)

Figure 6–6   Veneer cutting: (a) lathe, and (b) slicer. (USFPL)

increasing the amount of glue applied. These checks also play a role in test procedures for evaluating bond quality since they affect the amount of wood failure that occurs in destructive testing. In assembling plywood, the face and back veneers are normally placed with the lathe-check side turned toward the glue line. This avoids any unsightliness it might contribute. Even so, long-time exposure to cyclic moisture content conditions can bring lathe checks to the surface.

Two means are used to alleviate these fractures across the grain: heating the wood and pressure. Pressure applied over and just ahead of the knife edge reduces the tendency to break with the wedging action of the knife (Figure 6–7(a)). Pressure is applied through a "nosebar" held so as to maintain an opening between it and the depth of cut that is slightly less than the thickness of the veneer. Varying the size of this opening produces more or less pressure on the veneer as it is being cut. The pressure also produces friction and thus creates another source of distortions, changing the angle of the lathe checks and loosening surface fibers. Too much pressure from the nosebar can cause the lathe checks to merge internally through rolling shear action and thus create a complete plane of weakness at the center of the

veneer. The use of roller nosebars reduces the friction and its consequences.

Rough surfaces on veneer have four main sources: dull knife, instability of the knife edge, properties of the wood, and cutting thick veneer. Some woods are particularly difficult to cut into veneer because they tend to crumble ahead of the knife instead of cutting cleanly. Species with abrupt summerwood transition also tend to force the knife edge out of its path. As with any cutting tool, when the edge wears and becomes dull, its efficiency as a wedge reduces drastically and greater force is necessary to advance it through the wood. Force operates on the knife as well as on the wood, causing both to deform. When the knife deforms out of its prescribed path, it produces different thicknesses. Two kinds of thickness variation ensue: one in which the knife chatters up and down, producing a rough, washboard surface, and one in which the knife deforms only in one direction (sometimes more at the center than at the ends), producing veneer that varies in thickness from side to side or end to end. Rough veneer not only requires a greater amount of glue but also a greater amount of pressure to bring the surfaces together. Both escapes are unsatisfactory, costly in terms of glue use and in terms

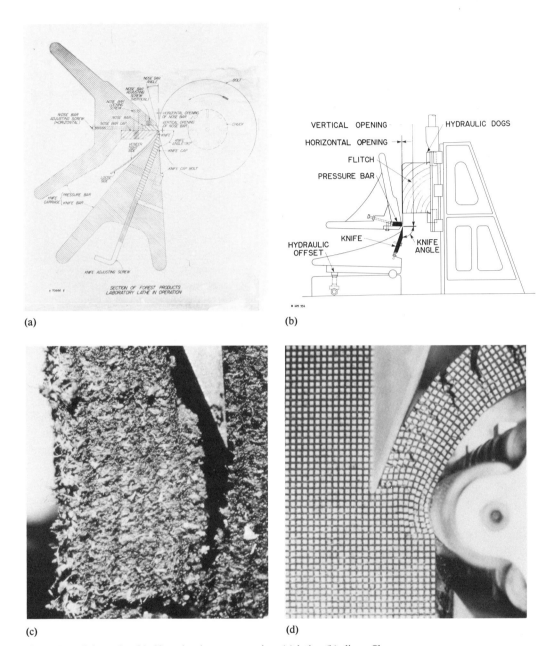

(a)

(b)

(c)

(d)

Figure 6–7   Schematic of knife action in veneer cutting: (a) lathe, (b) slicer. Close-up of knife edge and the production of lathe checks, (c) and (d). (USFPL)

of thickness loss from compression sustained by the veneer.

Thick and thin areas in the veneer produce even more trouble in the glue line because of the direct effect on pressure distribution. High spots are bound to receive a disproportionate amount of the platen pressure, while low spots may receive little or none. Bond formation will be equally spotty. Unfortunately, while rough veneer is easily spotted by eye and can be rejected on line, thick and thin veneer may go unnoticed and their effects may not show up until panels are inspected or have gone to the customer.

One of the older methods of applying glue, the roller spreader, functioned as an inspector for variable-thickness veneer, refusing to apply glue uniformly if the thickness were not uniform. The poorly spread veneer was then discarded with the loss of both the veneer and the glue on it. However, it saved a poorly manufactured panel with even more veneer, more glue, and press time in it. With the advent of more modern means of applying glue—spraying, extruding, and curtain coating—a uniform spread of glue can be applied no matter what the quality of the veneer. Those aware of the consequent loss of monitoring from methods other than the roller spreader have installed thickness-measuring devices to intercept poorly cut veneer.

Many of the cutting defects in the veneer that originate in wood properties can be ameliorated by treatments that reduce the strength of wood: heat and moisture. Consequently, most veneering operations have provision for hot soaking or steaming logs. The temperature varies with species, and the treatment time varies with the size of the log and its thermal diffusion properties. In any case, the temperature is bound to grade from the outside to the inside of the log, with consequent differences in cutting properties as the log is being peeled. Overtreatment with heat either in degrees or in time can result in permanent reduction in strength or a "mushy" response to the knife. In addition to a probable temperature gradient from surface to core of the log, there is certainly a wood quality difference as the knife proceeds from the sapwood to the heartwood regions where knots, decay, and old injuries make for less suitable veneer.

After the veneer is peeled from the log in a continuous ribbon, or in smaller pieces depending upon the quality of the log, it is clipped to remove defects or to reduce to full size sheets if there are no defects. The veneer is then sorted for grade, size, heartwood, and sapwood, and forwarded to the drying operation. In the case of sliced veneer no clipping is done until the ultimate users introduce it into their processes. It is kept in *flitches* (stacks), exactly in the order in which it came from the slicer, is dried sequentially, and restacked in order.

Veneer Drying

Because veneer is a thin piece of wood compared to lumber, it does not develop the steep gradients of moisture and temperature that result in the drying defects experienced by lumber. Therefore, veneer drying can be greatly accelerated without structural damage to the wood. Through use of very high temperatures and rapid air flow, the drying process for veneer is reduced to minutes, compared to the days and months needed to dry lumber properly. Because of this speed, veneer drying is a continuous process rather than a batch process as for lumber. The green veneer is simply passed through a heated tunnel with temperatures to 400°F or more (Figure 6-8). During its transit, the veneer is kept from curling and kept moving by live rolls on top and bottom. For thin veneers, a combination of hot platen pressing and rolls helps ensure flatness.

From this simple scheme, many variations exist in the source of heat, the programming of temperature at different times and locations in the tunnel, the venting of moisture, and the means of applying flattening pressure. Of these, the temperature and the source of heat create the greatest effects on subsequent gluing. As noted in the section on surface properties, heat produces changes at the surface that have a repelling effect on the adhesive, either from inactivation due to contamination from within or from chemical modification of surface molecules. Veneer drying pushes temperature to the burning point and beyond in some cases. The veneer avoids charring by the cooling effect of its own evaporating moisture.

Some dryers are heated with steam, limiting the temperature to that producible by steam, around 350 to 400°F. More modern dryers are heated by direct-fired oil or gas burners; that is, the burning gases are passed directly over the veneer. Thus in addition to being subjected to high temperature well over 400°F, the veneer surfaces are also

(a)

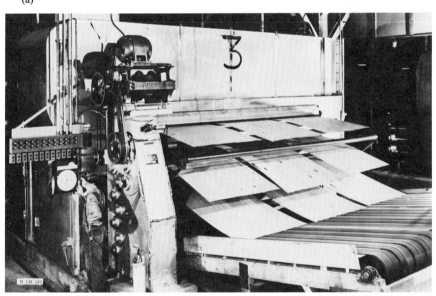

(b)

Figure 6-8   Veneer drying: (a) infeed and (b) outfeed. (USFPL)

swept by the products of combustion, and any incompletely burned hydrocarbons being exhausted. The resulting contamination/ inactivation, when excessive, causes problems in adhesion at links 4 and 5 as a direct effect. Indirectly, link 1 can also be affected. A filtered condition may develop as the result of insufficient water loss during assembly periods due to reduction in hygroscopicity. The consequent higher water content of the glue line increases the mobility of the resin during hot pressing, causing it to diffuse into the wood and leaving the less mobile and less adhesive fillers to form link 1.

As in lumber, final moisture content and uniformity within and between pieces is the major criteria for quality of drying. With target moisture contents between 5 and 10 percent, some veneer can be under 5 or over 10 percent depending upon the moisture content it had initially. While the greatest concern in hot pressing is veneer over 10 percent, overdried veneer also presents gluing problems. Veneer outside either moisture content extreme presents different problems for the adhesive, involving conflicting properties in the adhesive.

On the high-moisture-content side, hot pressing encounters two problems, one dealing with mobility of the resin and the other with the formation of steam blisters between plies. The latter situation is a simple case of moisture being driven by the heat of the press platens toward the center of the assembly where it is trapped. If the temperature is above the boiling point, steam pressure develops, and when the press opens, the steam raises a blister at a weak point between veneers. Blisters have been known to be so large that they filled the opening in the press, making it difficult to eject the panel. While weak bonding is usually associated with blisters because of the first problem—too much mobility—blisters can form even with well-made bonds if the internal steam pressure is high enough. Cutting into such a blister would show wood failure on both veneer surfaces.

When blisters are coming too frequently, suggesting a run of high-moisture veneer, the press operator has an immediate adjustment at hand: slowing down the pressure release part of the cycle, and holding the press closed at zero pressure momentarily. This allows the steam to dissipate, and a blister may not rise. This procedure, unfortunately, will only save the panels that have achieved good bonds. In others the weak bond may still exist, and the blister will show at a later time when the wood is exposed to swelling conditions or is otherwise stressed. Blisters are only the visible evidence of excessively high moisture. There are also borderline cases where there is enough moisture to filter the glue line but not enough to raise the blister. These are the ones that escape notice and become an aggravation to the ultimate user.

The other problem with high-moisture veneer in hot pressing is that it tends to increase the mobility of the resin. While this could be either good or bad, depending upon other factors, in practice it is mostly bad. The good comes with very long assembly times when the adhesive has suffered too much loss of water and is at risk of immobility. In this case, heat from the press drives some of the moisture out of the veneer and into the glue line to restore mobility. The negative aspects of high moisture veneer show at the short end of the assembly time period. At this point, veneer moisture and glue water that might have left the glue line, given more time, combine to produce excess mobility, and the resin filters from the glue line.

Hot pressing veneer that is excessively low in moisture content runs into the converse problems: decreasing the mobility to good or bad levels depending upon other factors. Dry veneer draws water from the glue line relatively quickly. At short assembly times the mobility therefore optimizes quickly, and the best bonds are produced. At the longer assembly times, too much water is drawn from the glue line, immobilizing the adhesive, with no chance of getting help from veneer moisture coming to it during pressing.

The amount and type of fillers and extenders have an effect on the interaction of veneer moisture content with assembly time in hot pressing. In fulfilling their role of reducing flow and penetration, these additives tend to do more harm on low-moisture veneer at long assembly times, and more good on high-moisture veneer at short assembly times. This entire interaction of veneer moisture content and assembly time in hot pressing is further discussed in Chapters 9 and 10.

In cold pressing, the effects of excessively high- or low-moisture-content veneer on mobility are essentially the same as hot pressing, but some of the consequences may be different depending upon the type of adhesive. Some cases in point: Adhesives that harden by loss of water will lose their water and harden faster on veneer of low moisture content than on veneer of high moisture content. This means that they cannot tolerate as long an assembly time on dry veneer as they can on wetter veneer. Adhesives that harden by chemical reaction suffer the same effects. However, they have an additional consequence on dry veneer of losing not only the water that permits their coarser mobility functions, but also the water needed to allow molecular movements of the hardening reaction. This moisture content effect on mobility is so important it merits repeating at every opportunity.

In a well-run veneer drying operation, moisture sensors at various stages in the process monitor the status of the veneer and control the speed of the rollers, the temperature, or the humidity accordingly. Since, as in lumber, heartwood dries at a different rate than sapwood, they are often sorted and dried separately. Pieces having both heartwood and sapwood present a problem as one or the other area could emerge overdried or underdried. Nonuniform moisture distribution can also occur as a result of feeding error when veneers overlap in the dryer. High moisture edges and possibly filtered glue lines in those areas could result.

A problem that occurs especially with thick veneers, and more on hardwood than softwoods, is heavily wrinkled or wavy endgrain edges. This is due to too-rapid drying at the ends, resulting in "set" before shrinkage could take place. It causes problems in applying glue by roller spreaders, in leading to open assembly conditions at edges, and later in the press, flattening causes the veneer to split and overlap itself.

PREPARATION OF FLAKES

The great importance of flakes in the family of primary wood products resides at the very beginning of the process: in the forest. Five factors underlie the importance: the increasing demand for plywood type products, the necessity for high quality logs to make plywood, the increasing scarcity of high-quality logs, the abundance of low quality logs, and the technology of flakes that permits plywood properties to be derived from low-quality logs. The efficient production of flakes from low-quality logs is therefore one key to this breakthrough.

Although logs of any quality can be accepted in the flaking process, economics dictate that only those unsuitable for lumber or veneer be used for flakes. These include short, crooked, or small-diameter trees, with one major exclusion: species of high density. Density is a factor primarily for market reasons; the product would be too heavy. There are also some technical reasons that have their basis in density: compressibility, dimensional stability, and machinability.

Cutting Flakes

Flakes are always produced with knives cutting in the same aspect as for veneer: knife edge along the grain, travel across the grain. Three different cutting mechanisms are in use: drum flakers in which the knives are mounted on the outside of a drum; ring flakers, in which the knives are mounted on the inside of a drum; and disc flakers, in which the knives are mounted on the face of a disc. Each receives the wood (green in all cases) in a different form: drum flakers in a bolt or log form, disc flakers in block form, e.g., 12 inches long; and ring flakers in chip form, 1 to 3 inches long (maxichips).

Drum flakers (Figure 6–9) have knives as wide as the length of the flake to be cut, mounted on the surface of a drum spaced one knife-width apart and in staggered rows. As the drum turns against the log, each row of knives will take its cut, leaving the intervening wood for the next row to cut. The logs are fed broadside to the drum usually by gravity down a hopper with some help by chain lugs moving downward. Thickness of the flakes is controlled at a maximum level by the amount of protrusion of the knife edge beyond the surface of the drum. Minimum thickness is not strictly controlled but is dependent upon how uniformly the log advances with each cut. Length of the flakes (along the grain) is governed by width of the knives and the spacing between them in each row.

The width of the flakes (across the grain) is also not strictly controlled but depends upon how much breakage occurs *after* it is

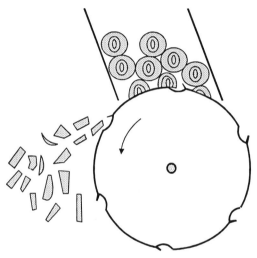

Figure 6-9 Drum flaker. (Bison-werke)

cut. The flake is carried out of the cutting zone in a gullet or pocket under the knife and is flung out when the knife emerges from the cut. The size of the pocket and the amount of wood taken in per cut determines how much packing and how much breakage might occur, all being along the grain and reducing the width accordingly. Sometimes further width reduction is desired, and this is done by mild milling as through a fan.

Disc flakers have knives arranged along the radii on the face of a heavy disc (Figure 6-10). The wood in block form, length no longer than the length of the knives, is held against the disc. As it turns, each knife sweeps across the block, taking off a layer of wood. In front of each knife is a row of sharp spurs, the purpose of which is to score the wood block to the depth of the cut. Their spacing determines the length of the flake along the grain. Width of the flake as it comes off the knife is the width of the block. Disc flakers are capable of producing very large flakes because instead of being curled up in a gullet under the knife, they pass through slots in the disc under the knife and emerge freely on the other side. As they emerge, they may be wafted gently away or impacted by vanes on the back of the disc and reduced to narrower widths.

Ring flakers (Figure 6-11) operate on an entirely different principle from disc and drum flakers. In a ring flaker, the knives are mounted on the *inside* of a short drum, the ring. The wood, in the form of large chips,

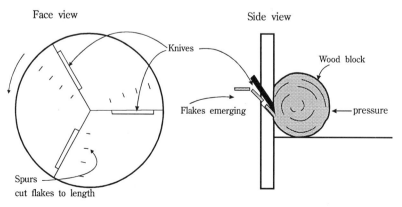

Figure 6-10 Disc flaker.

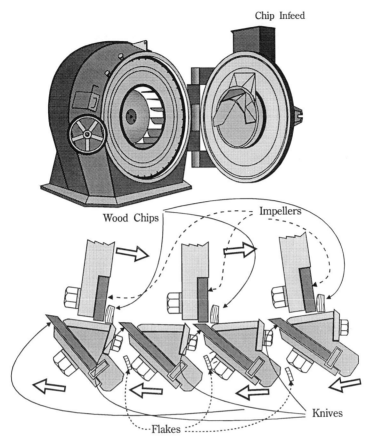

Figure 6-11   Ring flaker shown with front hatch open. Schematic shows arrangement of knives on outer ring and impellers on inner ring. (Pallmann brochure redrawn)

is thrown at the knives running in one direction by impellers running in the opposite direction. The resulting impact provides the cutting forces needed to sever flakes from the surface of the chips. This appears to be a random process with little control over the dimensions of the flakes. However, although the process produces a higher proportion of undersize material, the actions are statistically weighted in favor of clean cuts along the grain, particularly if the chips are elongate with a principle dimension along the grain.

These three processes, with many variations, produce the basic flake as well as the derivative wafer and strand. The ring flaker produces only strands, while the other two produce flakes that can be subsequently cleaved to strands by further milling. Wa-

fers, being essentially thick flakes, can best be made on the disc or drum flakers. However the true wafer requires some modification. A true wafer has tapered ends to provide better nesting and to reduce stress riser effects due to the greater thickness. The tapered ends are obtained by overlapping the paths of knives slightly so they cut part of the next flake.

As far as bond formation is concerned, interest is less on dimensions of the flakes (which mostly affect the properties of the glued product) than on quality of their surfaces. In flake bonding we are dealing with a very small amount of adhesive, too small to allow much penetration. Loose fibers and subsurface damage are therefore not likely to be rebonded, and links 6 and 7 remain weak. Great emphasis must consequently be placed on sharpness of the knives. Hence the ease with which knives can be changed and sharpened becomes a crucial matter in the production of flakes. Thickness of the flakes has a close relationship with quality of the surface. As in veneer, the greater the thickness of flakes, the more difficult it is to obtain a good surface.

Flake Drying

Wood flakes are dried while being airborne or tumbled through a heated drum or cylinder (Figure 6–12). There are many versions of equipment to control the interaction of hot air with the flakes to assure a precise dwell time, and to dry uniformly despite differing moisture contents of entering flakes.

In some the heavier moisture-laden flakes are made to move more slowly, and in some the flakes may take several passes up and down the cylinder before emerging. In all cases, combustion gases are fed to the drum directly from a burner. In this respect, the flakes suffer the same surface inactivation conditions as veneer in direct fired dryers.

Green flakes are fed at one end, the "hot" end where the hot gases also enter, at 900 to 1100°F. The evaporating moisture keeps the surface temperature below the charring point. As they are swept through, the temperature drops to 250 to 400°F, the actual temperature being an indicator of how well the drying is going on. The outgoing temperature thus becomes the monitoring point indicating when or if a change might be neces-

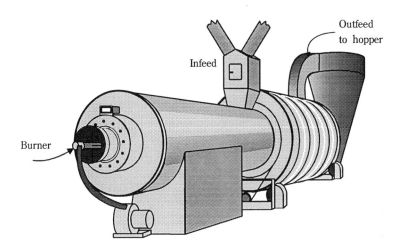

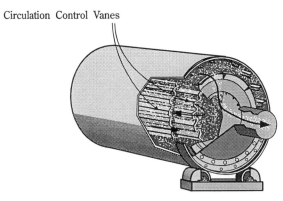

Figure 6–12   Drum dryer for drying flakes and smaller elements. (Heil Dryers brochure redrawn)

sary. A low exit temperature is an indication that insufficient drying is taking place; in other words, a great deal of moisture is being evolved and a great deal of cooling is therefore occurring. Conversely, a high exit temperature is evidence that too much drying has taken place. Compensation is made by varying the infeed rate of green flakes.

Optimum total moisture is 10 to 12 percent going into the press. Hence the flakes must be dried to something less than this to compensate for the water that will be added with the resin if liquid resins are used. In the case of powdered resins, no additional water will be applied, and therefore the flakes can be dried to a higher moisture content.

The moisture content of the flakes has a number of implications in the consolidation of the board. These are briefly discussed here and further amplified in Chapter 9.

Flakes, like veneer, become more fragile as they are dried, despite the fact that wood gets stronger the dryer it is. Overdried flakes, like overdried veneer, are therefore prone to break or split into narrower pieces during handling or transport. In the press, however, their greater compressive strength operates against the forces of compaction, and tends to reduce contact between flakes at a given pressure. Consequently, overdried flakes produce boards that are lower in density or less well bonded. If additional pressure is used to increase contact, more cell wall breakage occurs, and the board becomes weak from that standpoint. Moreover, the board will experience greater instability in the thickness direction as imbibed water induces recovery from compression.

The moisture content of flakes also has an effect on the mobility of the glue during the time between application in the blender and heating in the press. The effect is similar to that in veneer except that the glue line is composed of fine droplets of resin without fillers. Besides there being so very little resin to lose by penetration, there is also no help from fillers. The resin is therefore at risk to mobility influences and must be served with the optimum menu of factors that control mobility. Both extremes of mobility are eas-

ily exceeded by slight changes in time, temperature, and moisture content. Sensitivity to moisture is evidenced by the common observation that the region near the center of the board is its weakest point, coinciding with the point where the greatest amount of moisture accumulates during pressing.

PREPARATION OF PARTICLE ELEMENTS

The low cost availability of large quantities of dry waste wood is the main incentive for producing wood elements of particle size and configuration. Dry waste wood has two major uses: burning for heat and particleboard; animal bedding, packing material, wood flour, and mulch are secondary uses. Particleboard, however, offers the route to the most value added.

Dry waste wood occurs in the form of trim and edgings, and shavings and sawdust from lumber processing. Dry sawdust is not a good candidate for direct reconstitution because of its short fiber length. Shavings, on the other hand, though seemingly not too much better in regard to fiber length, can be hammermilled to a useful element that provides a relatively smooth surface to the board. Edgings and trim are chipped or hogged as a first reduction step, and then hammermilled to the size desired.[3] Different geometries in size and shape can be produced by hammermills depending upon size of grate openings (Figure 6–13).

Following hammermilling, particles are screened to remove material that has been reduced too much or too little, the latter to be reprocessed and the former to be burned, or further subdivided into fillers for plastics. Because the starting materials are already dry, most need no further drying and can be advanced directly to the next operation. Those that need some drying are processed in dryers similar to flake drying.

---

3. A chipper is similar to a disc flaker except the log is advanced end-wise at a slight angle so knives that project further sweep across end grain, severing a cross-sectional area that breaks up into chips. The action of a hog is similar to a hammermill, modified to produce more cutting than grinding.

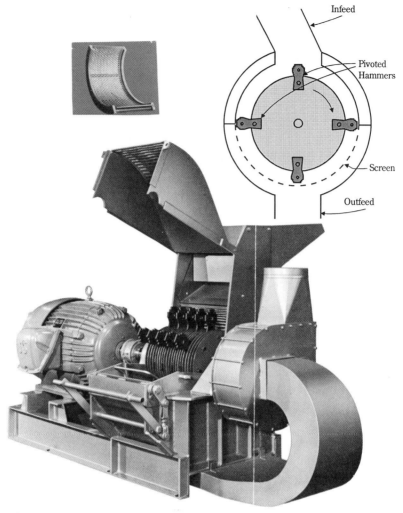

Figure 6-13   Hammermill. (Shutte brochure)

With respect to gluing, one can sense from the manner in which particles are made—impact and cleavage—that their surfaces have all sorts of roughness and that they may be rife with internal damage. They do not have a natural lie and may assume more angles of repose than the previous elements. Contact between elements may be more difficult to achieve and more pressure may be required, further damaging the internal structure of the particle.

PREPARATION OF FIBER ELEMENTS

Wood fiber production derives much of its technology from papermaking, modified in three important respects: The problem of removing lignin is avoided by leaving it as part of the fiber; a wood element larger than an individual wood fiber, a clump of fibers or "fiber bundle" is produced; and no chemicals are used. As in papermaking, the process begins with wood reduced to chip form, wood the size of half a thumb, green, and of selected species. Species selection includes a much broader mix of choices than for paper, but as in paper, the mix must maintain a fairly constant ratio of each species in order to manufacture products with uniformly dependable properties. This is because different species demand different milling conditions or different consolidation conditions.

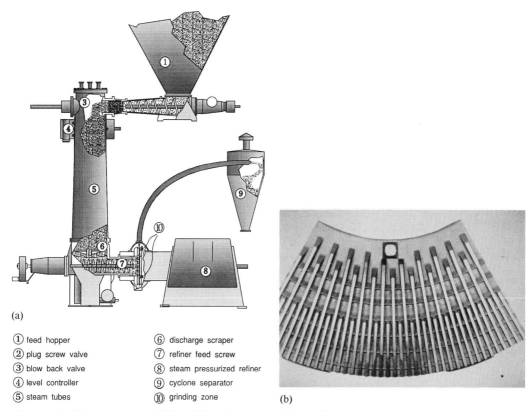

(a)

① feed hopper
② plug screw valve
③ blow back valve
④ level controller
⑤ steam tubes
⑥ discharge scraper
⑦ refiner feed screw
⑧ steam pressurized refiner
⑨ cyclone separator
⑩ grinding zone

(b)

Figure 6-14   Steam-pressure attrition mill grinding system (a), and (b), one of many kinds of grinding plates. (Sprout Waldron brochure redrawn)

The chips are fed into the vortex at the center of a disc in an attrition mill, and receive a severing action of shearing and tearing as they are reduced to fiber bundles (Figure 6-14). The surfaces produced vary greatly in composition and topography depending upon where and how in the fiber wall cleavage took place. Lignin, cellulose, or lumen areas may be exposed on the surface. Also, depending upon the extent of cleavage, the actual surface area of the fibrated material can vary over a wide range. How the wood fiberizes also produces another important variable, bulk density. Unanticipated changes in bulk density of the fiber mass produce confusion in metering systems that introduce material volumetrically. Changes in volume then mean changes in weight delivered, which in turn translate into changes in wood-to-resin ratio, and into the density of the final product.

There are a seemingly infinite number of variations that can be imposed on the chips as they are being milled. These include the nature of the grinding surface; distance between the discs (the smaller the distance the smaller the fiber); speed of rotation; direction of rotation, which can change the action from shearing to rubbing; rate of feed, which determines how fast the wood passes through the grinding region and therefore how much grinding gets done; chip moisture content and temperature, which determines the power needed; and to a certain extent where cleavages occur in the wood. High temperature and high moisture content plasticize lignin, softening it and weakening it so it yields comparatively easily under shearing stress. With cleavage in the lignin region of the cell wall, the fiber should bear a surface rich in lignin. Conversely, low moisture content and low temperature should encourage cleavage between the cellulose layers of the cell wall. The latter then would require more energy to produce.

The features of a good fiber bundle are not easily defined. Much depends upon what product properties are being targeted. Obviously, all properties that depend upon grain have been destroyed, and all future properties will depend totally upon the nature of reconstitution. For low-density products, such as insulation board, fiber length is important to ensure enough strength with a minimum of compaction. For high-density products where very smooth surfaces are demanded, the fibers must be uniform in size with no large pieces, "knots," to produce elevated spots in the board.

Fiber Drying

In two of the three processes for producing fiberboards—wet and semidry—all of the drying is done in the press during consolidation. The boards are pressed on screen cauls to facilitate the escape of water and steam and to avoid blistering. An impression of the screen remains on the back of the board, indicating its method of manufacture.

Dry-process fiberboards begin manufacture as green or steamed chips. When fiberized through the attrition mill, they may be at or above the fiber saturation point. Drying the water out of fiberized wood is the easiest of the drying processes because the small dimensions practically elminate diffusion problems, and there is no need for delicacy to preserve either the wood structure or the element shape or size. The drying process therefore reduces to its simplest form: carrying the fibers in a stream of hot air through a tube that is transporting them to the forming machine. However, the process must be closely controlled because the fibers generally have already been blended with resin,

which needs to be protected from precuring. Drying time is thus reduced to flash drying, a matter of seconds.

Controlling moisture content is similar to that for flakes, monitoring the outgoing air temperature and comparing it with the incoming air temperature. Similar moisture content criteria hold for fibers in "dry process" consolidation as for flakes except that the maximum must be slightly lower to avoid steam build-up in the panel. These boards have a tighter structure than those with larger elements and steam cannot escape as readily. Moreover, since there are no screens to carry away the steam, steam formation must be kept to a minimum by reduced moisture content. This extra caution, however, allows the production of one of the major modern advances in fiberboards, smooth two-sided boards.

In all but the wet process, the important consideration from a gluing standpoint is the very large surface area per pound of wood that needs to be bonded. With economics holding the resin addition to between 2 and 10 percent of the weight of the wood, bond formation enters another realm of near-molecular resin actions rather than mass actions as occur in other composites. Virtually no resin can be given to penetration; it all must function at the very surface. The most favorable surfaces from an adhesion standpoint are those that have cellulose exposed, followed by lignin-rich surfaces. Lumen surfaces, because of their encrustation with cell residue and extractives, may provide no bond at all. However, since lignin is known to rebond to itself as in the original Masonite process, bond formation is more easily obtained and with less resin on lignin-rich surfaces. Under dry-process conditions of low moisture content, lignin does not activate sufficiently to create a durable bond alone; it must be reinforced with resins.

7

Wood Geometry Factors

This chapter presents a further exploration of the wood elements entering the gluing operation, this time in terms of dimensions and configurations: geometry and how it affects the process. If, as indicated in a previous chapter, glue is the common denominator among wood composites, geometry of the wood elements is the main discriminant, the first order characteristic that reflects particular performance of a glued product. Confirmation of this is seen in material specifications where type of wood element often is the first item mentioned—flakeboard, lumber core plywood, and so on. This is because geometry of elements makes a significant contribution to both the strength and the performance of the composite. The adhesive, on the other hand, makes it possible for the elements to deliver and retain their performance. Other factors such as species of wood, kind and amount of glue or additives, and consolidation conditions introduce secondary characteristics that broaden the utility of the product or facilitate its manufacture.

Geometry of the wood element, however, also affects the choice of glue and consolidation procedure. Lumber size elements, for example, are largely assembled in clamps at room temperature, using glues that set at room temperature. Flake and fiber elements, on the other hand, are mostly consolidated at elevated temperature and pressure, using glues that require those conditions.

As has been repeatedly emphasized in the foregoing discussions, three factors underlie what has come to be regarded as the material science of wood. One is the size and shape (the geometry) of the wood elements incorporated into the composite. The second is the inherent properties of wood and glue that ultimately become amalgamated into the composite. The third derives partly from the first: the opportunity to organize these wood elements in the composite in such a way that they contribute their properties in a most effective and cost-efficient manner. The fact that the contributions of these factors depend upon the bond that can be established to hold the elements together, and that some qualities of the bond in turn depend upon size, shape, species, and organization, produces a circle of dependencies with endless permutations. The many peculiarities associated with manufacturing different products is a direct result of these interdependent activities. It is therefore not surprising that the same circle of dependency also often confounds the relationship of cause and effect when it comes to the assessment of product performance. Some of the analytics in this book should aid in sorting out the more useful connections.

In this chapter, the effects of wood element size and shape on product performance are overviewed, together with effects that intersect other aspects of the forest enterprise. This restates the importance of geometry as a primary consideration in producing a particular product. The demands on bond formation and bond performance then become more definable, and more approachable as separate matters. In a following chapter, geometry interacts with consolidation factors to produce a functional integration of the equation of performance. A rather sharp differentiation occurs among industrial processes on the basis of geometry that allows a separation into various representative processes. These are used to illustrate the more important distinctive features of each. But first it seems appropriate to take a general excursion through the various wood elements to discover some of the ways geometry controls properties and processes.

GEOMETRY AS DIMINISHING DIMENSIONS

In the right hand column of Figure 2-1(a) were listed in diminishing order the common wood elements, and in Figure 2-1(b) the type of product into which these elements are converted by gluing. In Table 7-1, the commoner elements have been more specifically categorized by adding dimensions in an attempt to form size classes. The classes can only be defined by a range, since most sizes have already been established by use in the

industry. Although some overlapping exists, the sequential nature of the list is evident. Placing the wood elements in a series of diminishing dimensions suggests a continuum that can be interpolated or extrapolated to achieve properties not yet experienced. Reference will frequently be made to this sequence of wood elements by way of tying in relationships that derive from size and shape.

An example of how the sequence of elements can be manipulated to show certain relationships is given in Figure 7-1. In this arrangement, the wood elements have been recategorized into four strata; within each stratum some continuum of sizes is evident despite the discrete classes suggested. The sequences within strata suggest that, to the extent sizes can be varied at will, the properties of composites can also be varied at will and fine-tuned to rather precise specifications. This arrangement invites the development of other wood elements of different sizes and shapes to fill a potential need. For example, long, narrow strips of veneer reconstituted into heavy structural members (a product recently developed and marketed as PARALLAM by MacMillan Bloedell, Vancouver, B.C., Figure 2-5[d]) derives from reducing the width dimensions of veneer beyond that previously thought useful. Further reductions in width would yield very long, strawlike strands that might result in a more tightly consolidated composite. Where a new element might fit the scheme of utilization

Table 7-1. Diminishing Dimensions of the Wood Elements

ELEMENT	LENGTH (INCHES)	WIDTH (INCHES)	THICKNESS (INCHES)	GLUED PRODUCTS
Lumber	4–20 (ft)	4–12	.5–2	Beams and arches
Veneer	4–8 (ft)	4–48	.02–.5	Plywood and LVL
Wafers	1–3	1–3	.025–.05	Waferboard
Flakes	.5–3	.5–3	.010–.025	Flakeboard
Strands	.5–3	.25–1	.010–.025	Oriented strand board
Splinters (slivers)	.25–3	.005–.025	.005–.025	Splinterboard
Particles	.05–.5	.005–.050	.005–.050	Particleboard
Fiberbundles	.05–1	.005–.020	.005–.020	Fiberboard
Fibers	.04–.25	.001–.003	.001–.003	Paper
Cellulose/Lignin		Molecular dimensions		Plastics, films, filaments

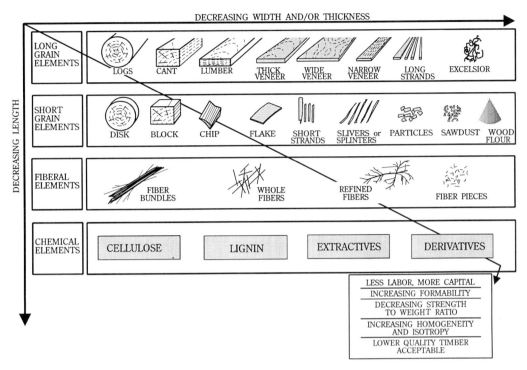

Figure 7-1 Stratified table of primary wood elements.

can be deduced from the trends that sweep vertically and horizontally across the chart. Some deductions include:

1. Greatest strength derives from the upper tier of wood elements declining downward and to the right—with the exception of discs, squares, and chips, which are not normally composited. (Logs and cants are also not composited, but are at maximum strength in their individual state.)
2. Proceeding from left to right and from top to bottom, elements become more formable (able to be bent to shapes) and more conformable (able to achieve proximity with another element).
3. From top to bottom, quality of the timber becomes less decisive in controlling properties.
4. From top to bottom and from left to right, surface area per pound of wood increases dramatically.

These and other trends have been formulated into a number of laws that reflect in a general way how wood element size affects some of the pertinent characteristics of composites and associated activities.

The Laws of Diminishing Dimensions

Some generalizations that emerge from the many effects of wood element size seem reducible to laws that, though exceptions occur, make the relationships stand out. The laws underscore the basic tenet that when a tree is subdivided into smaller and smaller elements, predictable changes occur in products and processes. The changes may be sensed as gains or losses from some norm, which may be taken as a clear, straight-grained piece of lumber.

As in health care, where treatments have benefits as well as undesirable side effects, so in treating wood by subdividing, there are benefits as well as undesirable side effects.

In the case of wood, the benefits and side effects—the gains and losses—are in terms of product properties or economics.

Although the gains and losses are referenced from a lumber norm, each different wood element can be considered to have a norm characteristic of its geometry. The many other behaviors that arise from operational factors associated with gluing—such as heat, pressure, and glue—can then be interpreted as modifications of those ordained by size and shape of the wood elements.

Law 1: Strength. *The strength of reconstituted products tends to decrease with decreasing wood element dimensions when compared at the same density.* In Figure 7–2, strength data from various sources has been averaged and adjusted to a common specific gravity of 0.8. The adjustment was based on the assumption of (1) a straight line relationship between modulus of rupture and specific gravity, and (2) the regression lines of all products have the same slope. Although

these assumptions are not completely valid, the resulting array confirms intuition that reducing wood to smaller and smaller pieces destroys the continuity of fiber structure, which is the basis of strength.

Part of the flexural strength differences among plywood, oriented strandboard, and flakeboard are due to grain alignment at the very surface, where extreme fiber affects register. The values for plywood and for strandboard represent the strength measured along the grain of the faces. In the perpendicular direction the values would be much less. On the other hand, the values for flakeboard would be essentially the same in both directions due to the randomness of grain all the way to the surface. The average of the two directions might then show that strandboard and flakeboard were equal in strength. (Degree of alignment is often measured as the difference in strength between perpendicular face directions expressed as a ratio.)

Within the flakeboard family, length and thickness of flakes produce strength vari-

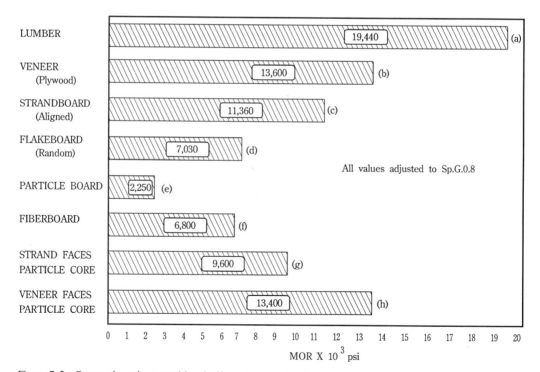

Figure 7–2 Geometric and compositional effects on strength of wood-base composites. (adapted from the literature)

ations that are closely related to dimensions. Figure 7-3(d) shows a relationship of flake length to strength in bending. As the aspect ratio reduces below 100:1, the strength reduces almost proportionately. On the other hand, at aspect ratios above 100:1, the rate of strength increase diminishes and finally levels off at some maximum determined by the strength of the wood and the thickness.

The greater strength offered by longer elements, however, can be offset by a difficulty in handling that may interfere with the formation of a uniform mat.

Thickness of flakes introduces a number of conflicting effects that need to be optimized in each case. Thin flakes generally can be cut with smoother surfaces and less subsurface damage. However, they also have

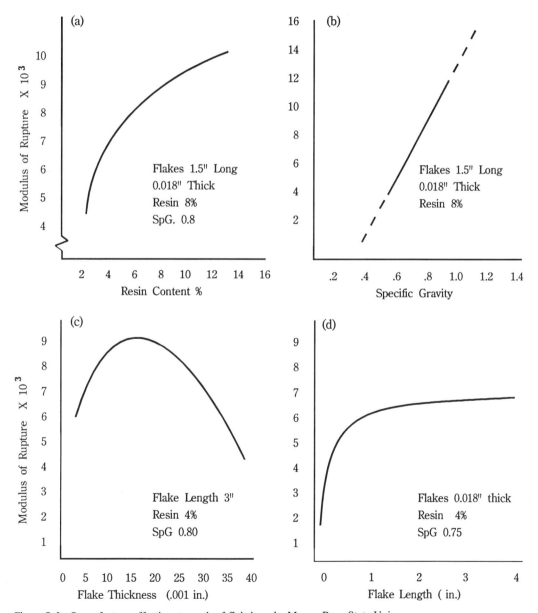

Figure 7-3   Some factors affecting strength of flakeboards. Marra, Penn State Univ. Press. (adapted from the literature)

more surface area per pound of wood than thick flakes, and therefore the amount of adhesive available per unit of surface is less. On the other hand the better surfaces of thin flakes allows a suitable bond to be made with less adhesive. Thick flakes introduce stress riser effects due to the large discontinuities existing at the ends of the flakes. The resulting loss of strength can be reduced somewhat by scarfing—reducing the thickness at the ends of the flakes. Figure 7–3(c) shows a typical relationship of strength with thickness of flake, indicating an optimum at about 0.020 inches. In this case the amount of adhesive per pound of wood was held constant, meaning that the thicker flakes had more adhesive per unit area of surface. Nevertheless they ultimately lost strength due to the combined effect of stress risers and poorer surfaces.

In the case of particles (Figure 7–2[e]) two factors are at play in developing strength: shape and bond quality. Their more or less cubicle shape and granular nature reduces their ability to achieve close contact for best bonding to take place. Moreover, each particle has a greater tendency to orient itself with its grain perpendicular to the surface. All particles so oriented become in effect small knots, contributing little to bending strength. Being also relatively short and blunt, they produce significant stress riser effects. It follows that the more linear the particles can be made, the greater strength they will yield.

Fibers (Figure 7–2[f]) appear to produce an anomaly, being of smaller size but producing strength considerable greater than particles. The same two factors are operating here but the play is different. Being highly linear, fibers, like strands, dispose themselves predominantly in the plane of the panel surface, wasting little strength in the perpendicular direction. Being smaller in cross section, fibers will also conform to each other in a closer association, thus providing greater areas for effective bonding. Unlike other wood elements, fibrous elements produce a degree of interlocking, adding a mechanical component to the bonding action. Also unlike other wood elements, lignin residing on the surfaces of the fibers has a chance to function as additional adhesive. The presence of lignin on these surfaces is, however, less a consequence of size than it is the method of producing the fiber.

The power of wood element combinations to recoup strength is shown in the lower two composites of Figure 7–2. Strategic placement of high strength-giving elements over low strength elements elevates the overall strength several fold. In (g), oriented strands have been added as a surface layer over particles; in (h), veneer is the overlay.

The decrease in strength with decrease in size of wood elements has its basis in several factors that derive from or accompany subdivision; one of the most important is the shortening and consequent breakup of the continuity of grain in the plane of the panel. With increasing subdivision, the multiplication of surfaces also increases the chances of incomplete rebonding, and in any case the increased surfaces all carry a burden of subsurface damage into the composite.

With the smaller wood elements, it is the general practice to use an amount of adhesive gauged against the weight of the wood rather than area of wood surface to be bonded. For a given adhesive-to-wood weight ratio, the smaller the wood element the less adhesive there will be for each unit of surface area (Table 7–2). In the case of fibers, the amount of adhesive appears to be preposterously low in terms of area and the functions adhesives are supposed to perform. Were it not for lignin activity and felting advantages of providing for some interlocking, fibers might not produce boards of the strength they exhibit with the small amount of binder normally used. The original Masonite depended entirely upon lignin and interlocking.

The case of the long grain elements warrants additional consideration. When they are reconstituted with the grain parallel, rather than crossed, they are usually intended for structural purposes, where

Table 7-2. Comparison of Adhesive Usage in Consolidating One Cubic Foot of Douglas Fir, Reduced to Various Elements

WOOD ELEMENT	BOND AREA SQ. FT.	LBS RESIN /MSGL	LBS RESIN /CU. WOOD	LBS RESIN /LB WOOD	GLUE LINE THICKNESS IN.
1″ thick[1] boards	11	39.0	0.43	0.015	0.005
0.1″ thick[2] veneer	119	7.5	0.89	0.032	0.0012
					0.0028 [5]
0.050″ thick[3,4] wafers	240	5.8	1.40	0.050	0.0009
0.010″ thick[3,4] flakes	1,200	1.2	1.40	0.050	0.00019
0.001″ thick[3,4] fibers	24,000	0.05	1.40	0.050	0.000009

1. Lumber spread @ 60 lb/MSGL, adhesive @ 65 percent resin solids.
2. Veneer spread @ 30 lb/MSGL, adhesive @ 25 percent resin solids.
3. Wafers, flakes, and fibers blended with 5 percent resin solids.
4. Assuming no compaction.
5. With 50 percent filler.

strength is a prime consideration. Subdivision provides some distinct advantages in producing structural wood materials: (1) greater load carrying because dry wood strength values, rather than green values, can be used in engineering calculations; (2) higher strength through the selection and strategic placement of highest strength pieces in the highest stress regions of the composite; (3) more uniform strength within and between product entities because no one element dominates the overall properties, and because the multiplicity of pieces assures a more constant average strength; (4) less splitting, not only because the wood is assembled with a low moisture content and therefore has less tendency to shrink, but also because each piece of wood is likely to have a slightly different grain angle. When bonded together the differing grain angles produce an interlocking effect that restrains the propagation of a split.

As we move across the upper stratum in Figure 7-1, where elements become both thinner and narrower, the subdivision process produces some countervailing effects in the consolidation process. The wider wood elements, lumber and veneer, necessitate hand work in assembling the pieces. This results in greater precision in locating each piece, assuring accurate juxtaposition and organization in the composite. With narrow, thin elements, mechanization of the assembly operation is possible and necessary. The resulting imprecision in deposition is both inevitable and beneficial from a structural standpoint because the small cross graining effects tend to reduce splitting. However, from a bonding standpoint the cross graining between elements has negative aspects because it promotes bridging and gaps. Nevertheless, a high degree of uniformity persists, and some strength is recouped by greater densification.

First Corollary to Law 1. The strength lost through subdivision can be recouped by increasing density (with some reservations with regard to stability), (Figure 7-3[b]).

Second Corollary to Law 1. The strength lost through subdivision can be partially recouped by increasing adhesive content (Figure 7-3[a]).

Third Corollary to Law 1. The strength lost through subdivision can be mostly recouped by aligning the wood elements (Figure 7-2[c]).

Fourth Corollary to Law 1. The strength lost through subdivision is often offset by greater uniformity within and between product units, because higher exclusion limits can be set (Figure 7-4). The exclusion limit is a value below which only a predictably small number of pieces will occur.

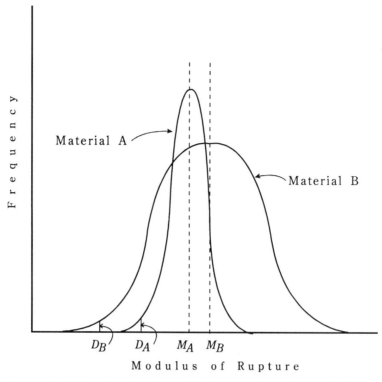

Figure 7–4   Relationships of product uniformity to working values used in design.

Law 2: Stability. *Shape and size stability of reconstituted panels improve with decreasing wood element dimensions.* The basis for this law lies in the reaction of wood to moisture vis-à-vis shrinking and swelling, and the variability of this reaction within and between individual wood elements. The former creates dimension instability; the latter, shape instability. Because a force accompanies dimension change, shape instability is also due to unequal restraint of dimension change, which causes unequal movement. Law 2 takes into account the ability to reorganize the structure of wood such that the directions of dimension change are equalized or neutralized, and such that the forces generated are also equalized or neutralized. The general principle operating here is that as wood is subdivided to smaller dimensions, the capacity of each individual piece for dimension change, for generating force, and for restraining force is proportionally reduced.

The word *force* requires some contextual explanation. If a block of wood containing 5 percent moisture is placed on a table, and its moisture content raised to 12 percent, it will increase in dimensions. Being free to expand, it creates no force. However, if it were enclosed in a tightly fitted steel box, the box would experience a force tending to expand it. The force originates because of the restraint offered by the box. The magnitude of this force is related to: (1) the density of the wood, (2) the amount of moisture change, (3) the mechanical properties of the wood during the moisture change, (4) the amount of restraint, and (5) the size of the wood piece. The greater the density, the moisture change, the size, and the restraint, the greater the force generated. However, the mechanical properties introduce a counteracting influence since at the higher moisture content the wood is weaker and therefore less able to exert as much force against the restraint. Other factors acting to reduce the amount of force generated are creep and re-laxation, which tend to relieve stresses; mois-ture gradients accompanying moisture changes, which means that all of the wood is not undergoing the same amount of swelling at the same time; and swelling internally into the cell cavities.

Nevertheless, the main tenet of the law holds that the force potential introduced by a moisture change is directly related to the size of the wood element. In Figure 7-5, a sequence of panels varying in degree of sub-division may be studied for stability against a moisture change. Beginning with (a), a typ-ical wide, solid board contains both flat- and quarter-cut grain. Given an increase in mois-ture content, the board would become wider by its full swelling potential. The differential swelling between the radial and tangential parts produces in addition a shape distortion across the face of the board. (The reverse would occur with a reduction in moisture.)

If this board is cut into narrow strips and every other strip flipped over so they alter-nate bark side up before gluing them back together, as in 7-5(b), the increase in width with an increase in moisture would remain essentially the same as the solid board. How-ever, its overall shape stability would be im-proved because the direction of the devia-

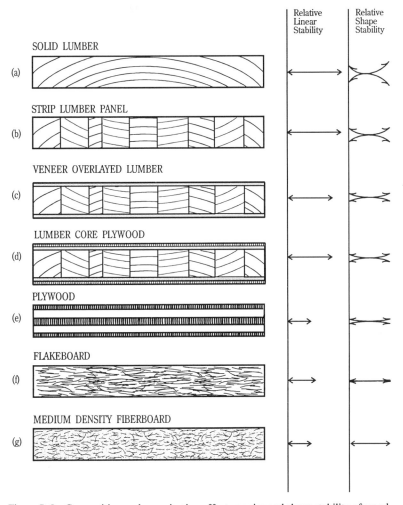

Figure 7-5   Composition and organization effects on size and shape stability of wood-base panels.

tions produced by each strip would now be reversed and cancelled by its neighbor.

While this procedure would produce a panel with the maximum shape stability of solid wood, it is too time-consuming for mass production operations. Most manufacturers therefore rely on random assembly of previously cut strips out of a hopper. However, two other forms of instability may intrude, both due to the difference between radial and tangential dimension change. All boards, except those containing true radial or true tangential grain, will not maintain a perfectly square cross section when moisture content is changed from that at which they were cut. They tend to assume a diamond shape in cross section (see Figure 5–16), and consequently tend to produce little shape deviations of their own.

The second intrusion derives from the dimensional instability in the thickness direction remaining the same in these manipulations. However, small differences between individual strips are now accentuated by the sharp glue line of demarcation between strips. Each strip of lumber retains its own total potential to shrink and swell, and the likelihood of its being different from its neighbor is increased by the multiplicity of pieces and by the dispersion of radial and tangential grain among the strips. This aspect of stability in thickness is dealt with further in Law 3 on surface smoothness. In the meantime, the problem of width stability as well as the problem due to diamonding remains. They are attacked by the next stage of subdivision.

Controlling the stability of solid wood panels in the width direction calls for the use of restraint. The panel must be restrained from shrinking and swelling. Continuing the sense of subdivision, visualize a thin slice (veneer) being removed from both faces of the lumber strip panel shown in Figure 7–5(b). Turning the two slices 90 degrees and gluing creates two improvements by playing the strength and stability of the new faces in one direction against the instability of the solid wood core in the perpendicular direction, (c). Thus the faces restrain the mois-

ture-induced movements of the strip lumber core and reduce the amount of dimensional change that might otherwise occur. The restraint also reduces the differential in movement inevitably existing between the two sides of each lumber strip as the chance distribution of cell structures dictate. Thus the veneered panel will be more stable in both size and shape than the one with solid wood strips alone.

An important proviso: the two slices of veneer on the faces of the lumber core must provide *equal* restraint. Any difference in restraint will result in shape deviations (warping); the greater the strength difference between the two veneers, the greater the warping with moisture changes. This is because the solid wood strips, being still of comparatively large size, can exert a large lateral force. If restrained more on one side than the other, the panel will deform toward the side providing the greater restraint.

There are a number of causes of unequal restraint. They include all those factors that influence the strength of wood such as (1) differing species, (2) differing densities, (3) differing moisture content, (4) differing thickness, and (5) differing grain direction, edge to edge as well as surface to surface. Of course, a differing quality of glue bond on either face would also result in a different restraint. All of these factors combine into a single cause of warping: "unbalanced construction," a most abused aspect of panel assemblies. Freedom from warping in reconstituted panels requires that the properties of the wood on either side of a plane at the center of the thickness dimension be exactly the same. Considering the kind of variables in operation, the balance is probably never fully achieved. Fortunately, the variations have tolerable limits and only extreme cases result in unacceptable products.

The above construction is seldom used today, though it is to be found in older furniture. In modern practice, a second veneer layer is added to both sides with grain perpendicular to the first layer of veneer and parallel to the grain of the lumber strips. This produces what is known as lumber core plywood, in which the solid wood is the core, the first veneer layers are crossbands (signifying their function of crossing and restraining the grain of the core), and the two outside layers are face and back (Figure 7-5(d)). Typically the face veneer is decorative while the back is chosen to provide a balanced construction. The stiffness of the solid wood core helps maintain flatness, though a true balance probably seldom occurs for one or more of the following reasons:

1. Face and back are normally of different species.
2. Face and back are normally of different thicknesses, either originally or because the face is sanded more than the back.
3. The grain in decorative veneers is highly deviant in one way or another (a cause of decorative features), while that in back veneers tends to be considerably straighter.
4. The face side of decorative panels receives finish coatings often omitted from backs, making the two sides different in moisture permeability.
5. The four bond lines can be of different quality, and therefore permit different restraints.

The lumber core plywood construction, however, has a number of advantages, including:

1. High strength, particularly in the direction of core grain
2. Greater stability than solid wood
3. Resistance to splitting
4. Decorative
5. Machineable edges
6. Good anchorage for mechanical fasteners

and a number of disadvantages, including:

1. Potential for "show through" of individual core strips due to differing dimensional stability in the thickness direction
2. Potential for "show through" of glue

lines between core strips, the "sunken glue line" problem
3. Potential for warpage with large swings in moisture content due to the mass of the core and its expanding power.

An escape from these disadvantages lies in continued subdividing of the core to veneer thicknesses, and reaggregating with the grain of alternate layers in perpendicular orientation, the typical plywood construction (Figure 7-5(e)). This degree of subdivision still retains the longitudinal strength of the wood, but in reducing the thickness of each piece, the lateral forces any one piece can exert from a moisture change is greatly reduced. Consequently, adjacent plies with perpendicular grain can easily restrain swelling forces developed through the entire range of moisture change from oven dry to soaked, if the bond between them is adequate. The lumber core plywood can not survive such a moisture change, even if the bonds are adequate, because the expansive power of the thick solid wood in the core would be great enough to rupture the crossbands in tension along their grain. If the crossbands were made stronger by increasing their thickness to lumber dimensions, swelling forces would shatter the wood somewhere back of the glue line.

The maximum thickness that can safely be cross-plied depends, in the first instance, upon species and the swelling forces they can generate through a moisture change. Low-density species would permit greater thicknesses than high-density species, perhaps no thicker than $\frac{3}{4}$ inch in any case. The characteristics of the hardened glue line have a strong bearing on this situation. Since the wood on either side of the joint would be trying to move in perpendicular directions, if the glue were brittle, durable, and unyielding, it would prevent movement through its bond to wood of perpendicular grain, which could have undesirable consequences. Preventing movement at the glue line transfers stress to the region *back* of the glue line. Failure will then occur in that region when the strain limit of the wood is exceeded. On the other hand, if the hardened glue remains fairly flexible, the glue absorbs some of the strain, and the wood on both sides of the bond is somewhat freer to move without building up stress. The strain limit in this case resides in the glue, and when exceeded, failure would occur there. Between these two types of failure, one in the wood and the other in the glue line, there must exist a bond that would permit movement below the rupture point, itself deforming elastically to accommodate the stresses.

An example of such accommodation is shown in Figure 7-6, where wood specimens glued with different adhesives in cross grain aspect with $\frac{3}{4}$-inch-thick wood blocks were subjected to moisture changes, after gluing, of plus and minus 4 percent. The three glues

used—urea, casein, and polyvinylacetate—were all able to resist the moisture content changes and reversals. The polyvinylacetate, however, a fairly flexible glue, appeared to improve with all changes, while the two more rigid glues suffered some loss when tested at a reduced moisture content.

These effects are the result of internally generated stresses of the wood working against itself across the glue line. Externally imposed stresses from loads on the joint are also modulated by the thickness of the wood. A number of identical specimens were made with maple lumber glued with the grain crossed. After the glue had hardened, the amount of wood back of the glue line was varied by sawing away or adding on by gluing. Thickness of the wood then ranged

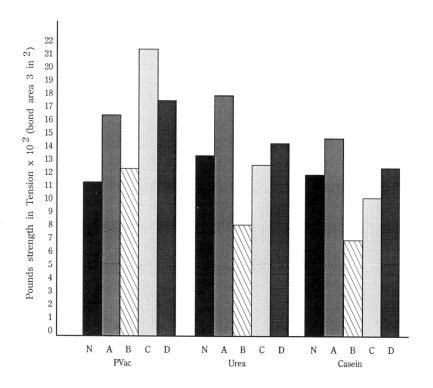

N = Tested as glued, 9% MC

A = Tested after change to 13% MC

B = Tested after change to 5% MC

C = Tested after reduced to 5% MC and returned to 9% MC

D = Tested after increased to 13% MC and returned to 9% MC

Figure 7–6 Response of cross-lap specimens to moisture-content changes before testing. (Marra, For. Prod. Jour.)

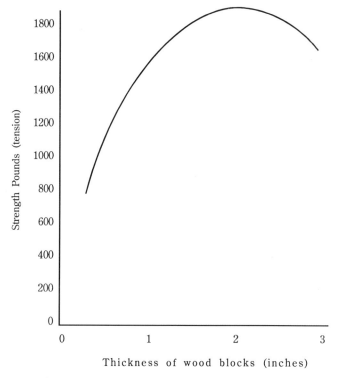

Figure 7–7   Influence of wood thickness back of the glue line on observed strength of the cross-lap specimen. (Marra, For. Prod. Jour.)

from $\frac{3}{8}$ to 3.0 inches. Loading them in tension perpendicular to the glue line produced the data plotted in Figure 7–7. It is evident that the same-quality glue bond can sustain more external load when the wood is thick than when it is thin. The manner of loading of course has a bearing on the results. In this case, the load was applied at the edges of the bond in tension. Stress concentrations developed that were also related to the thickness of the wood, decreasing with increasing thickness. Had the load been applied in shear, thickness of the wood would not have made a difference.

The down-turn in apparent strength at the greatest thickness is also of interest for it reveals another action that accompanies the hardening of water-borne glues—the effect of dispersal of the water into the wood. Immediately after the glue is applied to the wood, the water begins to enter the wood next to the glue line and the wood swells. After the glue has hardened, the moisture diffuses out of the system and the wood tries to shrink to its original preglued state. Now restrained by being glued to wood of 90-degree grain orientation, the restrained shrinkage results in built-in stresses. These interact with the applied load stresses, which change in composition as the wood thickness increases, first neutralizing the internal stresses at the thinner dimensions and then allowing them to be additive at the greatest thickness.

Returning to the all-veneer construction in the search for greater stability, opportunities for warpage still exist. These arise mostly from unbalanced construction, and the prime offender in this case is variation in grain direction. Grain direction in veneer may not be parallel to surfaces or edges because of the way trees grow and the way a veneer knife engages the log. The main recourse against the effects of grain deviations is thinner veneer (i.e., further subdivision), as well as thicker panels. In other words, a $\frac{3}{4}$-inch-thick panel has a greater chance of remaining flat than a $\frac{1}{4}$-inch panel with veneer of the same thickness, if only because its greater thickness makes it considerably stiffer (by the cube of the thickness). Further, a $\frac{3}{4}$-inch-thick panel made with nine thin plies would stay flatter than one made with five plies of thicker veneer because the distortion effect of each ply in any one direction is reduced. Exploiting this thought another step, the nine-ply, $\frac{3}{4}$-inch panel could be improved in shape stability incrementally by even thinner veneer. There is a practical limit in handling, breakage, water intake to the assembly, and glue consumption that precludes going very far in this direction with a veneer element. Hence the next step toward greater stability without these interfering practicalities is a smaller element, the flake.

The flake opens a whole new world of panelized wood products:

1. It accepts a much lower quality log as raw material.
2. It provides for greater mechanization and automation.
3. It uses less adhesive per unit of surface.
4. It gives opportunities to compose panels versus:
   a. organization as random, aligned, or stratified
   b. varying binder kind and distribution within panel
   c. varying moisture distribution within panel
   d. varying density profile within panel.

In terms of panel stability, the dimensions associated with flakes (strands, wafers) further reduce the ability of each single wood element to influence the course of dimension change effects. The shorter dimension along the grain limits the distance a single flake can transmit a stress. Its thinness limits both the amount of force it can develop with a moisture change and the amount of force it can resist. The resulting effect can be sensed, for example, in a $\frac{1}{2}$-inch-thick panel containing 50 plies of .010-inch-thick flakes (in practice, more like 60 plies to allow for compression

during consolidation). When flakes are formed into a mat by a free-fall, random process, varying degrees of grain crossing will occur between adjacent flakes. If grain-angle measurements were taken through the thickness at any point, it would display the entire 360-degree angular orientation. Resolving all grain angles mathematically to the two perpendicular planar directions would show an equal amount each way, at 90 degrees as in plywood. For this reason, one would expect flakeboard panels to be as dimensionally stable as plywood.

Unlike plywood, however, the grain of the wood at the very surface at both face and back is totally random, and therefore the forces that would cause warpage are partially neutralized at the point where they are most effective. In plywood the grain direction is still concentrated in one direction at the surfaces. The full dimensional change potential is therefore free to leverage its swelling and shrinking forces. For this reason, panels made of flakes can be expected to maintain a flat aspect better than panels composed of veneer or larger wood elements. And the thinner the flakes, the more stability can ensue.

In this context, strandboard, with its aligned elements crossed in layers, has essentially the same structural composition as plywood. This should place it in the same shape-stability class as plywood. However, since each strand is not perfectly parallel to its neighbors, some restraint to dimensional change exists within "ply," and therefore each ply does not exert as much expansive force as veneer.

Waferboard, with its thicker, randomly deposited elements, would be similar but slightly less stable than flakeboard. Several other advantages and disadvantages of waferboards are also attributable to element thickness. The lower surface-to-weight ratio of wafers (by a factor of 2 over normal flakeboards) means that there is twice the amount of resin in each glue line if the same weight percentage is applied. If this extra resin in the glue line results in a better bond, greater stability should result. But the poorer surfaces that sometimes accompany the cutting of thick flakes can negate the increased resin.

The above reasoning carries into the final wood elements—particles and fibers—with some anomalous behavior in the case of particles occasioned by the shape of the element, as distinct from its size. Being comparatively thick and stubby, particles have a relatively low slenderness ratio. This means that, due to the short dimension along the grain, the instability of individual elements cannot be as well controlled by bonding to adjacent particles; there is not enough area of contact to provide the needed bond strength. The generality of this statement must, however, give way to the very wide range of sizes and shapes that fall into this category. Since the configuration of the particle is produced by hammermill action, the shape can range from elongate slivers to rather cubicle granules. The more linear the shape, the more stable the product. Fibers present the ultimate potential for stability since each element can only exert a small force in any one direction. Random deposition in a board diffuses the effects and further improves stability.

The preceding discussions have dealt largely with dimensional stability in the plane of the panel. Perpendicular to the face plane, i.e.,—in the thickness direction—the instability (thickness swelling) tends to be greater with decreasing dimensions, except for extruded panels. This instability is affected primarily by consolidation conditions, particularly pressure and the compression set it causes. Heat, pressure, amount of compression, moisture content, and amount of binder also have a bearing on thickness stability. However, thickness of the wood element has an indirect bearing on thickness stability of panels as a reaction to these operational factors. Generally the thicker the element, the more likely it will sustain compression deformations that are later released and observed as instability in the thickness direction. This is particularly evident in poorly

screened furnishes that contain a fraction of oversize particles. The oversize elements tend to swell more than the smaller, resulting in surfaces of uneven smoothness.

Some help in thickness stability from a geometry standpoint comes from organization within a mat. Short, stubby particles compose themselves with a higher percentage in the vertical direction as they are being felted into a mat. The vertically oriented elements thus present their most stable direction in the thickness direction. Extrusion processes accomplish the same effect to an even higher degree because of the manner of felting and pressing. Since pressure is applied parallel to the faces in extrusion, rather than perpendicular to the faces as in platen pressing, springback (release from compression set) does not affect the thickness direction but it does affect the length direction.

Law 3: Smoothness. *Smoothness of the surface of reconstituted products improves with decreasing wood element dimensions.* The surfaces of wood-base panels are controlled by a number of parameters originating in the behavior of wood interacting with the size of the individual wood elements of which they are composed. These begin with the inherent pore structure—lumen sizes—of wood. This has been previously described as a basic roughness characterizing all wood to a degree dependent only upon species. The roughness offered by the pores is compounded by differences in shrinking and swelling between springwood and summerwood, and between radial and tangential directions. It should be noted that the differences in surface elevations attributed to springwood–summerwood instability as a result of moisture change are accentuated by the cutting action that produced the surface.

As we proceed down the dimension scale, we can perceive how more and more of these sources of roughness are brought under control. Referring again to Figure 7–5, the solid wood panel would contain the roughness contributed by the cell structure, that contributed by the springwood and summerwood intersecting the surface, that contributed by the different amounts of radial and tangential material at the surface, and that introduced by the cutting action.

The panel composed of narrow strips of lumber would contain the same sources of roughness and therefore would have the same smoothness as the solid panel. Unfortunately, the reconstitution has made it more likely that some of the differences in elevation that create a sense of unevenness have been made more abrupt and more visible by the straight line demarcation between individual strips of lumber. This is particularly true when adjacent strips are of different species, different densities, or different grain.

The glue line itself also sometimes contributes to the unevenness of the surface. This can happen either by the effect of glue water on swelling and shrinking of wood (sunken glue lines), by the glue line being differently abraded in sanding, or by some dimensional stabilization of the wood by the glue adjacent to the glue line.

The above effects become less visible when the glued strips are overlayed by a heavy crossband as in Figure 7-5(c). However, the unevenness created by glue lines and differential movement of each strip in the lumber composite is not totally eliminated from sight by overlaying. "Show through" or "telegraphing" still occurs, particularly on highly polished surfaces.

A further improvement in smoothness ensues if the lumber strip core is reduced to veneer and reassembled as plywood. This eliminates the differential contributions to unevenness of each strip. All-veneer panels thus can be theorized to have a smoother surface than lumber core plywood. However the springwood–summerwood differentials remain, and can cause show-through effects with overlays. (Plywood underlayment will sometimes show its annual ring pattern through resilient floor tile covering).

When wood elements are reduced to flakes, the unevenness produced by springwood–summerwood differences tend to reduce somewhat, depending upon the species

and the thinness of the flakes. Very thick flakes, as in waferboard, create surfaces in which the pattern of the flakes persists. Very thin flakes made from a soft, even-textured wood like aspen yield the maximum smoothness with this type of element. Heat, pressure, moisture, and resin produce surfaces that are very smooth, even glazed. These factors create smoothness by compressing and collapsing the structure of wood.

Smoothness created by compressing and collapsing the wood structure unfortunately is not durable against moisture. Exposure to moisture invariably tends to restore the thickness dimensions of the flakes (springback), and releases another latent source of unevenness, that due to nonuniform mat formation. The latter leads to nonuniform compression and nonuniform springback. Here again small, thin flakes would result in greater smoothness because their greater numbers allow them to be deposited into a mat more uniformly.

Uniformity of mat formation is enhanced further by reducing the wood elements to small particles; the smaller the particles, the smoother the surface. Mat-forming methods that contrive to place the smallest sizes of a furnish in a layer on the surface of the mat do so to ensure smoothness. (Note that smoothness is achieved at the expense of strength. Strength demands elongate elements.) Note also that at this stage of subdivision, all semblance of the original wood grain is virtually gone. The lumen structure begins collapsing at the flake level of subdivision and continues at the particle level. Both the lumen structure and the grain structure are further reduced at the fiber stage of subdivision, particularly the vessel structure of hardwoods, the more so as the wood element approaches an individual fiber or parts of a fiber. Fibrous elements can therefore be expected to produce the smoothest surfaces with the greatest expectation that they will stay that way. In this case, the smoothness is not only due to the small size of the element, but also to the operational factors—high temperature, high pressure, and in some cases high moisture, all of which induce the mat of fibers to flow and form a tighter structure. (While these same operational factors can be applied to veneer and solid wood products to produce smooth surfaces, such surfaces are often too unstable for most intended uses because of the crushing that occurs and the springback that follows.)

Operational factors also are responsible for apparent exceptions to Law 3. The potential for creating ultimate smoothness is accompanied by the potential for subverting it. The dependence of smoothness upon small size of the wood element means that any larger sizes finding their way into the mat can produce a blemish. The elimination of oversize wood pieces in medium and high density fiberboards is thus a prime quality-control factor.

Law 4: Uniformity. *As the dimensions of wood elements diminish, properties of the composite become more uniform, both within and between pieces.* Uniformity through subdivision is simply a manifestation of the ability to randomize features that create the wide range of variation in wood properties as they come from the tree. Wood normally varies within and between trees of the same species in such properties as density springwood–summerwood amounts, grain direction, sapwood–heartwood amounts, moisture content, compression wood, and tension wood. A single piece of lumber could contain many knots for example, while another piece from the same tree could be free of knots. One board could have compression wood, while another could have none. Under test, these four boards could produce data from both extremes of the strength range for solid wood, all from the same tree.

Reduced to fibers and reconstituted, the four boards would produce products that would vary minimally about a mean value, the highs and lows having been comminuted into individual insignificance and homogenized into reconstituted uniformity.

This law holds with virtual mathematical certainty for the entire range of wood element sizes. However, two major caveats cast their shadows over the practicalities of processes. One is the dependence of product uniformity on mat-forming uniformity or assembly uniformity. The other is the dependence of product uniformity on the uniformity with which bonds can be formed between the wood elements. Both these factors have their own variabilities that demand control.

Assuming good control of bond quality and assembly quality, two major consequences ensue from the operation of this law: (1) the lower strength that comes with subdivision (Law 1) can be partially offset by the greater uniformity (fourth corollary to Law 1), and (2) greater uniformity results in ability to meet narrower tolerances in specifications. The opportunity for uniformity to offset strength loss stems from the necessity for architects and engineers to design with a strength value that has only a 5 percent chance of being lower. With the normally wide range of strength values for solid lumber, the design value must be set rather low to ensure that no more than five boards in a hundred are below that value. With a more uniform product where the variability is less, the design value can be higher without increasing the 5-percent chance of getting a weak board. Referring to Figure 7–4, the board with the lower average strength can have a higher design value because its greater uniformity raises the 5-percent exclusion limit.

The same factor that allows greater design values for boards of lower average strength also provides the ability to meet tighter tolerances because the range within which all values will fall is narrower.

Law 5: Controlled Tropism. *The subdivision of wood into smaller and smaller dimensions allows manipulation of the tropic (directional) properties of composites.* A tree, in its necessity to maintain an upright position while supporting a large crown, concentrates most of its strength in one direction—axially—to function most effectively as a post and as a cantilever anchored to the ground. This, in conjunction with shape and growth characteristics, produces the anisotropic properties of wood, properties differing in each of the three principal directions. The concentration of strength in one direction is one of the reasons wood is so strong for its weight.

Wood retains its anisotropic properties throughout the subdivision series until it is reduced to its molecular constituents. At the molecular level, chemical wood elements are essentially isotropic with the exception of cellulose. Cellulose projects its molecular linearity into crystalline structures that also have directional properties.

It is important to bear in mind that the anisotropy resides within each wood element, and that this property is due to a high state of organization both at the molecular level and at the cellular level. In manipulating this property, the approach is basically one of disorganizing followed by controlled reorganizing. The procedural options include: (1) the shape and size of the wood element created in beginning the disorganizing process, and (2) the degree of alignment and/or stratification of the wood elements achieved in the reorganizing process. Three objectives generally guide the reorganization: strength, stability, and surface smoothness. The tropic options in the reconstituted products include:

1. Anisotropy: strength and stability concentrated in one direction, stability uncontrolled other two directions
2. Diotropy: strength equalized two directions, stability equalized same two directions, no control third direction.
3. Isotropy: Strength equalized three directions, stability equalized three directions—properties the same all directions.

The first two options are more frequently practiced, and with the relatively recent technology for aligning fibers, are possible with all wood elements that possess a linear geometry.

Alignment can be considered to have two extremes, 0 degrees and 90 degrees. With the grain of all elements aligned in the same direction, 0 degrees, the composite approaches the anisotropic properties of the original solid wood. Due to normal variations in felting, the grain direction between elements can only be nominal. The small grain-angle differences between elements nevertheless have important effects in such composites. While tending to reduce the axial strength, they improve stability slightly and reduce splitting in the plane perpendicular to the face and parallel to the elements. To enhance splitting resistance even more, adjacent elements in laminated veneer lumber are sometimes deliberately set with grain angles at, for example, 15 degrees to each other. Manufacturers of laminated airplane propellers have used this alignment procedure to strengthen the leading edge against splitting.

When the grain of adjacent wood elements is at 90 degrees as in the typical plywood construction, strength and stability in the plane of the face are theoretically equalized, and it might be said that the properties in that plane are fairly isotropic but remain anisotropic with respect to the other direction, through the thickness. Because of the need to have a balanced construction to avoid warping in plywood, face and back veneer have to have grain in the same direction. This maintains a balance as far as tensile strength is concerned, but produces an imbalance or degree of anisotropy as far as bending strength is concerned because, depending upon the direction of bending stresses, the extreme fiber will be either parallel or perpendicular to them.

Reducing the wood element from veneer to flakes or wafers tends to remove most of this remaining anisotropy in the plane of the panel. The random alignment of flakes brings to both surfaces of the panel an equal distribution of grain angles.

The above discussions have focused on the two face directions of reconstituted products, along and across. The directions perpendicular to the face are also important, particularly with regard to thickness stability and tensile strength (conventionally called thickness swell, TS, and internal bond, IB). In general, tensile strength perpendicular to the surface is greatest in solid wood and degenerates throughout the subdivision series. On the other hand, thickness stability does not follow the series closely, being more influenced by operational factors than size factors since the instability resides within each element. The degree of compression suffered by the wood elements under the pressures of consolidation sets the stage for later expansion beyond that of normal wood when moisture invades the panel. Resistance to instability is mainly provided by the restraint imposed by neighboring elements to the extent they form a well-glued, interlocking network. The contribution of this plane to the anisotropic properties of reconstituted products remains one of the weaker features that demands research attention.

There are some exceptions, and some manipulations that improve the situation in the thickness direction. Extrusion processes provide not only a high degree of particle alignment perpendicular to the surface, but also, in applying consolidation pressure to the edge of a panel rather than to the face, effectively transfer compression-induced instability from the thickness direction to the face direction of extrusion. The anisotropic properties of an extruded product therefore involve different planes, the weakest being where face-pressured panels are strongest. In mat-formed panels, the degree of verticality achieved by the wood elements as they are deposited offers some means of improving thickness stability. Shorter wood elements are more likely to assume a vertical position when dropped onto a mat. For this reason, particleboards tend to have slightly higher properties in the thickness direction.

While it would seem urgent to improve strength and stability through the thickness of reconstituted panels, the available approaches tend to reduce other values. In a truly isotropic product, not more than one-third of the original strength of the wood would be available in any one direction, an unacceptable loss for structural materials. If the consolidation pressure were reduced so there would be less compression and less springback, interparticle bonding would suffer.

Hence another compositional option is needed that allows reduced pressures without loss of bonding action. The most obvious approach in this respect is an increase in the amount of adhesive to provide bonding across greater gaps between wood elements. This is expensive at current prices of adhesives, but might yield to research on discovering less-expensive adhesives. More-expensive adhesives could be tolerated provided their use resulted in savings in other parts of the process. For example, if higher moisture content could be tolerated, if less heat or no heat were required to consolidate, or if pressure could be lowered, equipment costs could be reduced. Higher production rates could be maintained if the adhesive were fast-setting.

Reconfiguration of the wood element is also a possibility, as mentioned at the end of Chapter 2. An unexpected outcome of this approach was that densities could have a wide range with very little compression of the wood to cause later springback. While high in materials cost because of high resin consumption, this concept demonstrates a technological extrapolation of particle configuration and adhesive properties to achieve a potential solution to one of the more refractive problems in reconstituted wood products.

Organization options. In reassembling wood elements to have prescribed directional properties, also provide the opportunity to combine them in various proportions, orientations, and stratifications. This represents the essence of wood material engineering. Two major categories of product organization exist: *homogenous,* composed of a single type of element, and *heterogenous,* composed of two or more types of element. For example, a laminated beam containing only lumber, and flakeboard containing only flakes, are homogenous. Lumber core plywood containing lumber and veneer or a product that has flakes in the surface layers and particles in the core would be heterogenous.

Both these categories can be further divided into random organization of elements or aligned organization. The heterogenous composites can also be blended mixtures in which the various elements are simply mixed together in a single furnish, or stratified, in which each element is deposited in different layers. Homogeneous composites may also be stratified by quality of the single element, better quality on the surfaces and lower quality in the core. Stratification may be either distinctly layered or gradually distributed size elements. Each layer can also be treated differently with respect to moisture content and resin type and content. Operational factors allow variation of density within panel as well as between panels.

Figure 2–6 presented the options on an element by element basis, suggesting the possible combinations of up to three elements. Each combination contains the further options of different organizational deployment within the composite, combinations of different species, and combinations with different materials such as fiberglass or plastic overlays mentioned above. Organizational options available to the wood materials engineer summarize as follows:

1. Random deposition
2. Oriented deposition
3. Stratified deposition
4. Graded deposition
5. Homogenous
6. Heterogenous
7. Combinations

Product shape options also engage the tropistic behavior of the composite. Such shapes as linear, planar, and volumetric all can be favored with an appropriate selection of elements and organizational structures.

Law 6: Surface Area. *As dimensions of wood elements are reduced, the surface area per pound of wood increases dramatically.* The increase in surface area with increase in degree of subdivision is a simple mathematical outcome of the process. If one begins with a block of wood $12'' \times 12'' \times 12''$, one cubic foot, weighing 28 pounds, it will have

a surface area of 6 square feet, or .21 square feet per pound. Cutting this block into 1-inch-thick boards produces an additional 22 square feet of surface or 1 pound of wood per square foot of surface. Slicing each board into 10 sheets of $\frac{1}{10}$-inch veneer produces an additional surface area of 216 square feet with the same weight of wood, or 7.9 square feet per pound. Slicing each of those sheets further to $\frac{1}{100}$-inch thickness (thin flake dimensions) increases the surface area by 2160 square feet, still with the same weight of wood, or 77.4 square feet per pound. Reducing the thickness to .001 inch increases the surface area by another 21,600 square feet or 771.6 square feet per pound of wood. Finally, if these very thin sheets were then sliced vertically to produce rectangular elements $\frac{1}{1000} \times \frac{1}{1000}$ inch square, fibral dimensions, the total area of surface doubles to 43,200 square feet, or 1543 square feet per pound of wood.

The amount of surface area per pound of wood has meaning both in the energy expended in creating the surfaces and in the amount of adhesive needed to rebond them, giving rise to Laws 7 and 8.

Law 7: Glue Coverage. *As thickness reduces, the amount of glue per pound of wood increases, and the amount of glue per unit surface area decreases.* On the premise that two surfaces produce one glue line, the surface areas calculated resolve themselves into the glue coverage reported in Table 7–2. Although the coverage decreases sharply with reductions in element size, presumably limiting bond formation and bond performance, it is surprising that product performance has found acceptable limits. This is attributed in part to increased knowledge of materials and structures that allows the more accurate measurement of pertinent properties in terms of measurable product performance.

Law 8: Energy Consumption. *As wood is subdivided to smaller and smaller dimensions, more energy is expended in subdividing, and more energy is needed in reconstitu-*

tion. As in Law 6, energy consumption in subdivision is a mathematical consideration, modified in this case by efficiencies that can be engineered into the process. Physically, no matter what the material (even water), a certain amount of energy is required to create a surface where none existed before. The energy is taken up in breaking chemical or physical bonds holding the material together, in a sense the same molecular forces responsible for strength properties.

Without getting into actual energy values, one can derive some interesting comparative information by looking again at the amount of surfaces generated during the various stages of subdivision. Consider a one-foot-long section of a log that after slabbing produces a block of wood one foot on a side—one cubic foot of wood. Discounting the two end-grain cuts because they are constant for the series (discounting also the surface area of the sawdust created by the saw, considerable to be sure) the cutting operation has produced 8 square feet of new surface, 4 on the block and 4 on the slabs. This is the minimum amount of surface for an initial stage in the breakdown operation. If x amount of energy is consumed per square foot of surface, a quantity of energy equal to $8x$ was used in this case.

Following the surface increases as before, reducing this cubic foot block of wood to 12 one-inch-thick boards (in a kerfless operation) would generate 22 square feet of new surfaces, using $22x$ amount of energy. Slicing the 12 boards into veneer $\frac{1}{10}$-inch thick would consume $216x$ more energy. Slicing further to $\frac{1}{100}$-inch-thick veneer (flakes, actually) uses $2160x$ amount of energy (ignoring ends and edges again). If the wood is reduced to fiber dimensions $\frac{1}{1000}$-inch in cross section, $43,200x$ units of energy are consumed, almost 2000 times more energy than needed for cutting lumber.

Depending upon the cost of energy and the equipment for delivering it to the surface being created, there appears to be ample incentive for avoiding any more subdivision than needed to produce a desired effect. One of the advantages of wafers at a thickness of .05 inch over flakes at .025-inch thickness is this saving in energy. The lower surface area per pound of wood also means greater adhesive coverage for the same percentage added.

Law 9: Industrialization. *As wood is subdivided, processes become more capital intensive and less labor intensive.* Wood in the form of lumber and veneer requires considerable handling and judging of qualities (careful handling in the case of veneer to avoid breakage), and human intervention is necessary at almost every operation. In the form of flakes, particles, and fibers, wood elements render hands useless and judgments too fleeting. Consequently, machines and electronic monitors must take over.

Capitalization reaches its peak at the papermaking fiber sizes and drops off toward lumber. With the drop off in capitalization comes a welcome greater flexibility in adjusting to economic conditions. Since massive machines cost money whether they are running or not, a strong imperative exists to keep them running, if only to minimize losses. Operations running on people-power can be, and unfortunately too frequently are, shut down and the people sent home. Highly mechanized or capitalized operations also tend to consume large amounts of wood, and often wood with a narrow quality range such as certain species and certain sizes. This places pressure on local resources, and in the absence of good forestry practices can warp the natural succession, leading to forests containing a high proportion of unwanted trees, or monotypic planted forests vulnerable to catastrophic destruction.

Law 10: Homogenized Resources. *As wood is subdivided into smaller and smaller dimensions, species differences become less crucial.* This should be one of the most important laws of diminishing dimensions for it suggests an approach to utilizing the highly diversified forest so essential to the perpetuation and assurance of forest resources. Forests in which many species exist in the entire range of age classes are deemed to be more resistant to catastrophic destruction by fire, disease, insects, or atmospheric disturbances. Such diversified forests also provide better wildlife habitat and greater control of water run-off than forests made up of single (especially coniferous) species.

One can sense the application of this law by examining the extremes of the wood element size series. In the case of lumber, its utility depends primarily upon strength, which in turn depends upon large trees having long, straight boles and a minimum of limbs. In the case of molecular elements, any woody material can be reduced to chemical constituents irrespective of size, form, or species. In between, varying degrees of size/form/species dependencies exist, being lowest for fiberized materials and highest for veneer.

The opportunity to "homogenize" a highly diverse forest into a single uniform product thus seems in prospect with the smaller wood elements. To a certain extent this has been possible. However, it has been found necessary to maintain uniform *proportions* of each species in order to produce uniform products. This is due in part to the interaction of species properties with the means of subdivision in producing different element configurations, and the interaction of species properties with the means of consolidation.

What is evidently needed are means of subdividing that can override species differences and produce elements having the same configuration. Fibral elements have the best chance here. Flake elements can be produced from most timber qualities, but they retain too much of the original wood properties, such as density and compressibility, which then have to be overcome in the consolidation step before a uniform product can emerge.

Actually, the production of wood elements interacts strongly with assembling and consolidating them. Mechanized operations are not very forgiving of variations; they need to be fed a uniform diet of raw materials. Variables such as changes in bulk density due to slight changes in wood element configuration like curling or fuzziness confuse the electronic monitoring systems, making it difficult to set metering devices. Resin addition becomes uncontrollable, mat formers deliver different weights, dryers produce differing moisture contents, and bond qualities vary.

It is logical to assume that if bond formation could be kept at a high level despite the vagaries of the wood element, more tolerance to species differences might ensue. One of the reasons for poor tolerance is the very small quantity of binder that is allowed by economics in processes using highly comminuted wood. More binder would reduce the need for densification in consolidation, which in turn would reduce the dependence on element uniformity. The obvious solution is lower-cost binders so more could be used without destroying the profit margins.

Law 11: Moldability. *As wood elements reduce in size, they become easier to mold into shapes.* As previously discussed, wood is generally converted to useful shapes by cutting and fastening actions. Most wood structures are in the shape of a box, the chief technological feature of which is the making of corners. With some notable exceptions, such as pallets, making corners is not a process that can be easily mechanized but is left to craftsmen. Shapes other than rectangular are even harder to make. This is because wood, while fairly formable into simple curves, is difficult to form into more complex shapes. Its anisotropic properties mitigate against bending in more than one direction as necessary, for example, in molding a salad bowl.

However, subdivision can dramatically ease the process. The effect of wood element size can be deduced from the organization in Figure 7-1. Beginning in the upper left corner, it is obvious that logs and cants can only be assembled into rectangular structures. Lumber can be formed into curved shapes by steam bending, or into structural arches by gluing. Smaller radii can be formed with wood reduced to veneer thickness and glued in a mold (Figure 7-8(a and b)). The radius of curvature to which wood can be bent is directly related to its thickness; the thinner

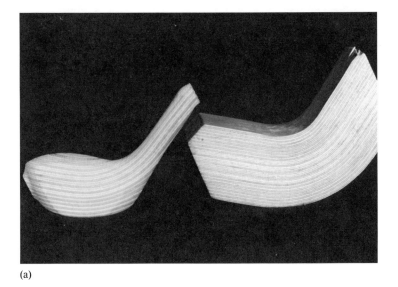

(a)

(b)

(c)

Figure 7-8   Molded veneer shapes: (a) golf club head, (b) Eames chair (c) salad bowl.

the wood, the smaller the radius that can be formed (Figure 7–8(b)). Compound curves with a rather shallow draw can also be formed with veneer. For more severe compound curves the veneer must be reduced further, to narrow strips. The compound curves of salad bowls can be negotiated with narrow veneer, with flakes, and with particles (Figure 7–8(c)). The complex shape of an egg carton requires the formability of wood in fiber sizes (Figure 7–9(b)). For products of intricate shape requiring greater strength than an egg carton, high-pressure molding with a resin-rich fiber furnish yields such items as shown in Figure 7–9(a). Finally the ultimate in formability is reached by wood reduced to its chemical components, cellulose and lignin, molded into plastic parts. The increased moldability that ensues from subdivision is basically a consequence of reducing anisotropy or the effect of anisotropy on conformability to shape.

Law 12: Subdivision Judgments. *Wood should not be subdivided any more than necessary to obtain a desired result.* While sometimes dictated by local economics, it does not make technological sense to reduce to fiber a large straight tree that could produce lumber with structural qualities or veneer with decorative qualities. Since there is an overabundance of wood that can yield everything from flakes to charcoal, reserving the larger trees for the larger wood elements makes for more efficient use of resources.

This is particularly cogent since waste from the lumber and veneer can be used for the smaller elements, thus having it both ways.

The energy needed to break down and reconstitute smaller elements should constrain the economics of oversubdivision. Moreover, although reconstitution can yield not only planar products but also linear products like lumber, the original tree remains the sole source for low-cost, structural lumber and veneer of high value.

Corollary to Law 12: Tree Mandate. In pursuit of biologic efficiency, wood should be used as close as possible in size and function to what the tree intended. A tree is designed to perform a very specific mechanical function and is therefore structured most efficiently as found in the forest. When we begin reducing it to the shapes and sizes we want, we tend to destroy the efficiency of its architecture—round, tapered—with grain in the most favorable, axial direction. Recreating this efficiency is a continuing objective of progress.

GEOMETRY CONSIDERATIONS IN PRIMARY GLUING

The effect of wood element size and shape begins in the forest where it influences the choice of tree from which the wood element is to be produced. The effect continues with the cutting and preparation of the wood elements; the quality of the glue line environ-

(a)

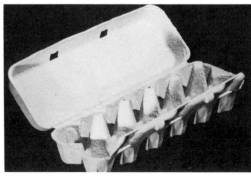

(b)

Figure 7–9 Molded fiber products: (a) pressure molded (Anon), (b) vacuum molded.

ment; the choice of adhesive and its method of application, bond formation, solidification characteristics, and consolidation mechanics; and finally affects the stress the completed glue line must later resist (bond performance).

Tree Choices

The effect of geometry can be broadly perceived through Figure 2–1, which gives an immediate picture of a relationship between size of wood elements and the potential timber resource on one hand, and on the other hand the types of products into which they are composited. Although there are species preferences favoring lumber and veneer with strength, appearance, and convertability, or favoring flakes, particles, and fibers with compressibility and low density, it is clear that the larger elements demand the largest and straightest trees, and the smaller elements can do with virtually any wood material.

Production Methods

As noted in an earlier chapter, methods of production change with wood element size: sawing for lumber, knife cutting for veneer and flakes, hammermilling for particles, and attrition milling for fibers. Each has technologies that speak to the need for efficiency in speed, yield, and quality of product, details of which are beyond the scope of this book. The brief descriptions given in Chapter 6 serve to highlight the essential operating factors that bear on the bonding process.

Drying

Seasoning methods also differ in response to size. Lumber requires stickered piling and a controlled rate of drying at ambient conditions or at a relatively low temperature to avoid degrade. Veneer can be force-dried at a much higher temperature because its thinness permits easy escape of moisture. However, because of its wide expanse of surface area, drying uniformly becomes crucial to

avoid splitting and wrinkling. Also some pressure must be applied during drying to ensure flatness. Flakes, particles, and fibers are completely free of drying restrictions, needing only to avoid burning while they are air-borne or tumbled in currents of very hot air. Some caution is exercised in the case of large flakes and wafers to reduce excessive cleavage into narrower elements.

Glue Line Environment

Wood element size affects the environment in the glue line mainly through the amount of wood back of the glue line. This wood affects the thermal management of that environment as well as the water and glue absorbing capacity. Both the original temperature of the wood and the gradients that generate when heat is introduced reflect through the amount of wood back of the glue line. These effects all increase as wood thickness increases. Greater thicknesses ensure the solidification of water-loss adhesives but operate against adhesives that require heat to cure. The more wood there is between the glue line and the heat source, the slower will be the temperature rise. Also as heat builds up in the wood, forming a temperature gradient, the moisture in the wood responds by forming a gradient of its own. The moisture gradient is reverse to the temperature gradient; the moisture is driven away from the heat source and unfortunately tends to concentrate at the innermost glue line. The additional moisture slows the curing of the glue both by dilution and by slowing the temperature rise. Slower curing coupled with dilution leads to greater mobility of the adhesive, which could be either good or bad depending upon the status of the glue at the time. Other environmental effects involve space or gaps in the glue line and are discussed below.

Surface Qualities

The quality of a surface is related to size in various ways, though some might be considered as much an effect of workmanship as an effect of size. All surfaces prepared by knife cutting have the potential for being as perfect as the wood structure itself, depending mostly upon how well the wood has been preconditioned, and depending upon the care exercised in maintaining and using the equipment. Elements the thickness of lumber encounter rotary knives, which leave a pattern of scallops on the surface. Dimensions of the lumber do not otherwise affect surface quality except as expressed through warpage. In the case of veneer and flakes, thickness has a strong effect on surface quality. As a general rule, the thicker the element, the rougher the surface. Surfaces prepared by sawing or sanding have no characteristics directly affected by the size of the element, although at veneer thicknesses, problems could arise in controlling or feeding the stock.

The smaller elements, particles and fibers, are committed by their size to be prepared by the imposition of random forces, hammermilling and attrition milling. Random forces seek the weaker planes in wood to induce cleavage. The weakest planes in wood are in the springwood zones and in ray tissue. Hence particle surfaces are likely to contain the cell structure associated with these regions, and they will have shapes determined largely by the intersection of these regional boundaries. When steaming is combined with milling, lignified layers are weakened and separations occur there, producing lignin-rich surfaces on elements.

Fit of Joints

The fit or mating of the joint surfaces is rather profoundly affected by size of the wood element. This is the result of mechanical properties of wood interacting with quality of surface. If, because of roughness or mismatch of any kind, pressure must be applied merely to bring surfaces into proximity, stiffness and compressibility become the main issues. In general, it can be said that the larger the wood element being glued, the more the mechanical properties of the

wood will dominate the mating process. One can sense this by imagining how difficult it would be to glue together, and hold together, two railroad crossties, compared to two strips of veneer of the same species and glue line area. Warpage before or after gluing in the crossties would preclude drawing surfaces together for bond formation, or would rupture a bond if made. Warpage in veneer, on the other hand, produces so little stress it can be overcome with finger pressure, and therefore presents little threat to the glue line. On the other hand, high and low spots on the surface would affect veneer and crossties the same in preventing contact, but the situation in crossties would be easier to deal with in this case because the greater stiffness allows for better distribution of clamp pressure.

At the opposite extreme from crossties would be papermaking fibers, actually straplike parts of fibers, which have been processed to a point where they have minimum stiffness and are so limber that they easily conform to one another and achieve bonding proximity merely by evaporation of the water in which they have been suspended. The wood elements between lumber and paper fiber are intermediate in conformability to the extent their thicknesses affect their stiffness.

In summary, one of the major compulsions in wood gluing arises from the relationship of compliance (the ability to deform and achieve proximity) with geometry, and both with the nature of the wood surface. Compliance increases—that is, surfaces mate more easily—as dimensions decrease, particularly in the thickness dimension. Compliance also increases with increasing moisture content and increasing temperature of the wood. The importance of maximizing compliance is that it helps reduce the amount of pressure needed to bring surfaces together. The less pressure used in bringing surfaces together, the less incipient fractures will be generated in the glue line area, and the less stress will be built into the glued structure to reduce the performance of the bond at a later date.

Adhesive Choice

No matter what wood element sizes are involved, the choice of adhesive must first satisfy the anticipated durability requirements of the product. After that, size of the wood element, particularly thickness (and size of the glued product, usually its thickness also), influences choice between cold-setting and hot-setting adhesives. The time and cost of driving heat through wood to reach a glue line removed some distance from the surface motivates the choice. The choice is particularly pertinent with the high-durability structural products where a lower-cost phenolic resin requiring heat must be compared to higher-cost resorcinolic resins requiring no heat. Although curing time is also a cost factor, a breakpoint could be between thicknesses of $\frac{1}{2}$ and 1 inch, thinner favoring phenol and thicker favoring resorcinol.

Beyond heating, additional considerations relate to the mass of wood back of the glue line. Two factors operate here. One is the capacity to absorb glue water and allow hardening to take place. This affects mostly those glues that harden by loss of water, excluding them from operations involving very thin elements; i.e., very high ratio of area to weight. However, even glues that harden chemically, if they are applied as liquid dispersions, need to lose water, especially if heat is also needed to cure them. In general, a first approximation lower limit to wood element thickness for water-borne adhesives applied as a wet film is about .03 inches.

The upper limit of thickness is mandated not by water-absorbing capacity, but by strength of the hardened glue. Internal as well as external stresses tend to rise with increasing wood thickness. Stresses of a continuous nature such as those from dead-loading or from moisture gradients are particularly damaging to bonds made with adhesives that yield under stress. Since density of wood determines to a large extent its ability to create and carry stress, this factor interacts with thickness to produce the upper limit of thickness. In general, the thermo-plastic glues cannot hold together thick, high-density woods through a moisture change, particularly a rapid one when steep gradients develop. Some trials with specific glues, species, and thicknesses are needed to determine limits in particular cases.

Method of Glue Application

The method of glue application loses discretionary options when wood elements reduce to flakes or less in dimensions. Wood elements in veneer or larger sizes can use all the application options—roller coating, spraying, curtain coating, extruding, and sometimes even dipping—with only minor restrictions for size. Narrow boards probably would not be coated by spraying, since overspray would produce too much waste, although stacking a number of boards would reduce the waste.

Flakes and smaller elements are mostly spray-coated. Tumbling with a dry-powder resin is also popular as a means of avoiding the addition of water to the assembly. Fibrous elements destined for wet mats can also be dosed by precipitating resin from solution onto the fibers while they are in suspension.

Consolidation Mechanics

How the assembly is consolidated is somewhat indifferent to size of the wood elements, once past the lumber sizes. From veneer down to fibers, pressing is done through platens or molds, generally heated and hydraulically driven. Lumber, however, participates in almost all forms of pressure devices, from nailing to screw clamps, hose clamps, and friction devices.

Stress Development

Stress on the completed bond is directly related to wood element size, and the direction of the grain on either side of the glue line. A maximum stress-producing extreme (internally derived stress) can be described as one in which flat-sawn lumber of high density is glued face to face with grain at right angles (Figure 7–10(e)). This would produce shear stress on the glue line at the highest level with a given moisture change. The same lumber glued edge to edge (Figure 7–10(a)), would produce less shear stress, but the highest level of tensile stress perpendicular to the glue line during a moisture content change, highest at the ends where the moisture gradient would be the steepest. The amount of stress the wood is able to create on the glue line as a result of moisture change is greater in face gluing lumber thicknesses, (c), than veneer thicknesses, (d). In the case of edge gluing, (a) and (b), the amount of wood between glue lines would also have a bearing on stress development, but it is not clear where the upper limits are.

When grain is crossed on either side of the glue line, the amount of internal stress is directly related to the thickness of the wood. Referring to Figure 7–10, stress in (e) is greater than stress in (f) by virtue of the greater amount of wood creating stress and the greater stiffness that resists warping, a mechanism that relieves stress. In (g), the normal cross-grain plywood construction, stresses exist as in (f) but are balanced, and warpage does not occur.

At the opposite extreme from lumber, fibers consolidated into boards of low density exert the least amount of stress on neighboring fibers during moisture changes. Homosote, a low-density fiberboard with no bonding agent, has been known to survive two decades of exterior service as cladding without protection of paint on a barn door in upstate New York. Although severely eroded and abraded by the weather, it remained intact and in place against the many moisture changes it incurred.

The intermediate wood elements produce stress situations between these two extremes. In all cases, of course, the deleterious effects of internal stress can be minimized by use of moisture excluding treatments or by increasing resin content in the more comminuted composites.

Bond Formation and Bond Performance

Size of wood element has a profound effect on actions on the glue line primarily through two associated factors: (1) the dimensions of the glue line, and (2) the amount of wood that occurs back of the glue line. Each has mechanical consequences in the gluing procedure as well as physical effects on what happens in the glue line. In this section we deal with geometrical dimensions per se on how they influence the bonding process. In Chapter 9, pressure and heat are added to the interaction by way of integrating these and other factors into the final quality of a bond.

For purposes of discussing these two geometry factors, glue line dimensions and wood thickness, we consider only the differences represented by lumber and veneer. Deferring end gluing as a special case for the time being, geometry reduces to two aspects: edge gluing where the relatively narrow edges of wood elements are being joined, and face gluing where the broader surfaces of wood elements are being joined.

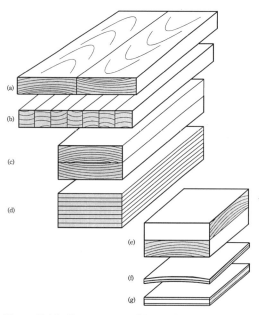

Figure 7–10 Stress generated by moisture-change increases with amount of wood between glue lines both in parallel-grain and cross-grain constructions.

In edge gluing, it can be said that the length of the glue line is infinite, while the width of the glue line is finite and narrow, being defined by the thickness of the wood element. The amount of wood back of the glue line is infinite as far as glue action is concerned, being defined by the width of the wood element.

In face gluing, on the other hand, the amount of wood back of the glue line is finite, being controlled in this case by the thickness of the wood element, while the length and width of the glue line can be considered infinite. Note that as the thickness of the wood element increases, edge gluing and face gluing become essentially similar in geometric aspects.

The extreme effects of these apparently minor differences can be more readily appreciated by considering face gluing versus edge gluing very thin veneer. In face gluing, there is a large glue line area and a relatively small amount of wood back of it. The reverse is true in edge gluing. If each piece of veneer were 10 inches square and .033 inches thick, the following glue line geometries would exist: For the same amount of wood, the glue line area in face gluing would be $10'' \times 10''$ or 100 square inches; while in edge gluing, the glue line area would be $.033'' \times 10''$ or .33 square inches. This creates a number of differences with which the glue must interact:

1. Three hundred times more glue would be applied in face gluing than in edge gluing, assuming the same rate of application.
2. In edge gluing, there is 300 times more wood back of the glue line to absorb glue water than there is in face gluing.
3. Glue water may increase the moisture content of the veneer to an unacceptable level in face gluing while not increasing the moisture content in edge gluing appreciably. For example, if the two veneers were of 30-pound-per-cubic-foot density, spread with an adhesive of 50 percent water content at a rate of 35 lbs/

MSGL, they would receive an increase of 10.6 percent in water content. If the veneer began at 6 percent moisture content, the assembly would enter the press at 16.6 percent moisture. A three-ply assembly with two glue lines and only half as much more wood would go to press at 20 percent moisture. Consequently:

4. Hot pressing the face glued veneer assembly with a water-borne glue would probably be impossible, as blisters would form and overpenetration of the glue would generally occur, giving rise to bleed-through or starving of the glue line, particularly with open-pored woods. However:
5. In face gluing, more unevenness of surfaces can be accommodated because the thin veneer is more compliant, easily deformed, and therefore conforms more readily to its neighboring surfaces.
6. Edge gluing on the other hand predisposes to squeeze-out because of the very short distance the adhesive has to move to escape the bond area. According to the laws of fluid mechanics, pressure in the glue at the very edge is essentially zero, and the glue therefore moves in that direction. (Note that in face gluing, the periphery of the bond area is subject to the same low glue pressure as in edge gluing, one reason for trim allowances in panel products.)
7. In edge gluing, any mismatch in the surfaces represents lost contact that is difficult to overcome, first because very little bending can be imposed to force conformity between the pieces due to the great stiffness offered by the veneers in the edge direction, and second because of the mechanics of applying pressure in that direction on thin substrates particularly.
8. After bonding, the edge-glued veneer would be subjected to greater stress levels, both in bending and as a result of differential moisture change. Bending of edge-glued veneer, usually across the grain during handling, produces tensile

stress on the glue line, its weakest feature. In the case of face-glued veneer, bending produces shear stress in the glue line, its strongest feature.

As thickness of the wood elements increase toward lumber (i.e., more wood back of the glue line in face gluing), some of the effects of thickness on glue activity increase and some decrease.

In face gluing:

9. The capacity to absorb glue water increases, as previously mentioned.

10. Also, as previously mentioned, unevenness of surfaces becomes more difficult to accommodate because of increasing bending resistance in bringing surfaces into closer contact (increasing by the cube of the thickness).

11. As thickness increases, in face gluing it is more difficult to influence the activity of the glue by heating, because the glue line gets farther and farther away from a heat source with a good insulator—dry wood—intervening. The amount of heat reaching the glue line and the rate at which the glue line temperature increases influence moisture effects and curing effects. Since these effects generally have to be optimized rather than maximized, the amount of wood involved must be factored into the gluing procedure. For example, the glue in a 3-ply, $\frac{1}{8}$-inch-thick panel hot pressed at 300°F will encounter different curing conditions than the same glue in the same press in a 3-ply, $\frac{3}{8}$-inch-thick panel. The rate of heat uptake is faster in the $\frac{1}{8}$-inch panel, but the adverse moisture effects on the glue are greater because there is less wood to absorb it away from the glue line.

12. In bond performance, as thickness increases, stress on the cured glue line as a result of moisture gradients increases up to the tensile strength of the wood across the grain (lowered by the amount of sub-surface damage).

In edge gluing, as thickness increases:

13. Stresses due to warping tendencies of individual wood pieces increase.

14. Conformability becomes increasingly difficult, demanding truer surfaces.

15. The low glue line pressure at the boundaries represents a lower percentage of the whole glue line area, and therefore is less of a factor.

These effects of wood geometry on veneer/lumber gluing extend to all other wood elements in proportionally varying degree. In the case of flake gluing, for example, the amount of liquid binder that can be applied is limited technologically by the thinness of the flakes, just as it is in veneer. With very thin flakes, there may not be enough wood to accommodate the water added with the resin and keep the overall moisture content below blistering level. When resin content must be increased, for whatever reason, either powdered resins may be used or some form of redrying must occur after application of the binder.

Conformability, as a geometry factor, is applicable all the way down to paper fibers, increasing with each reduction in size until at

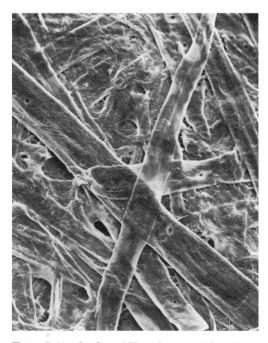

Figure 7-11 Conformability of papermaking fibers. (Wilfred Côté SUNY College of Environmental Science and Forestry)

the fibrillar stage they achieve bond-forming contact merely by surface tension of water evaporating from the fiber suspension (Figure 7–11, see also Figure 9–38).

Surface qualities of consolidated composites are affected by geometry as pointed out under the laws of diminishing dimensions, Law 2. Of particular additional concern is the instability of wood elements after they have been compressed in consolidating. This also varies with size of wood element. Large (i.e., thick) wood pieces tend to produce a more uneven surface when subjected to moisture change after consolidation.

Geometry in End-Grain Joints

As indicated in an earlier chapter, end grain cannot be reliably bonded as a butt joint because the end-grain surface is both too porous to allow optimum adhesive actions and too stress-prone for a bond to resist. The highest stresses in wood structures are carried along the grain. At butt joints, stresses encounter an abrupt discontinuity, forcing the glue line to accept a concentrated stress in tension, its weakest direction.

The obvious recourse is to slope the end-grain surface to produce a scarf. This accomplishes three actions beneficial to the bond. First, it increases the amount of side grain exposed to the glue; the lower the slope, the more side-grain advantage in gluing. A slope of 1 in 12, for example, would be superior to one of 1 in 5. Second, some of the high-tensile stresses traveling along the grain are transformed into shear stresses at the glue line, a better situation from a bond performance standpoint. Third, the stresses entering the glue line do not appear abruptly but are graduated in by the slope of the wood forming a stress gradient. A well-made scarf joint can transmit up to 90 percent of the strength of clear wood.

Scarf joints, however, are costly in terms of wood wasted in cutting the joint surface. In a 1-inch-thick board with a slope of 1 in 12, each joint necessitates the loss of one foot of board length. This loss is avoided in large part by folding the scarf back on itself

several times to form a number of fingers. Four fingers $1\frac{1}{2}$ inches long would produce the same joint surface area as the 12-inch scarf joint, saving $10\frac{1}{2}$ inches of board.

A number of geometric parameters attend the performance of finger joints, including:

1. Number of fingers
2. Slope of surfaces
3. Configuration of tips and gullets
4. Length of fingers
5. Thickness of fingers

Unfortunately, difficulty of producing the fingers also increases with the same parameters that increase strength due to increased rate of wear and increased instability of cutters. A well-made finger joint optimized for geometry and production factors can be expected to capture 60 to 80 percent of the strength of clear wood routinely. Whether or not the slope of the fingers is oriented perpendicular or parallel to the face is more an aesthetic consideration than a strength consideration. For low-strength applications, fingers are generally shorter, of higher slope, lower in number, and with square tips and gullets. For high-strength applications, fingers are longer, of lower slope, higher in number, and with rounded tips and gullets. Of course, the surfaces of the fingers should be smoothly and uniformly cut.

A variation of the finger joint produced by cutters is the *impression finger joint* developed by M.D. Strickler, Washington State University. As the term implies, the finger is produced by an impression process. The fingers are formed by a heated steel die cut to the desired configuration and pressed against the end grain, cut as for a butt joint. The die splits the wood and compresses it into the shape of very fine fingers. Since the wood is split rather than cut, the finger surfaces contain all the side grain of a normal side grain joint. The heat used in making the fingers is then used to cure the glue. Because of side grain on the finger surfaces, bonds are very good and the length of fingers can be reduced to $\frac{1}{4}$ to $\frac{1}{2}$ inch and still produce a joint equivalent in strength to a well-made, cutter-formed finger joint.

GEOMETRY IN SECONDARY GLUING

In secondary gluing, a second element is glued to a product that has previously been assembled and consolidated. Two or more products, such as fiberboard and flakeboard, can also be glued together in secondary gluing. Certain geometries are common to most of these products. With the exception of overlays, secondary gluing involves making bonds some distance from an outside surface. Bonding surfaces may not always be perfectly true and well mated, nor freshly made. Often the parts to be glued are not very compliant and therefore resist being forced into proximity. Most products are in a flat aspect, and usually at least one of the elements is in sheet form.

A partial list of secondary glued products include those given in Figure 2-1c and illustrated in Figure 2-11. Some overlays are in sheet form, overlays, often combining a thin sheet of decorative material with a solid, heavier core; for example, Formica on particleboard for kitchen countertops. Others combine a sheet material with a low-density core to produce a stiff panel of relatively low weight, the so-called sandwich panels. A hollow core door is a common product of such a construction, although a more demanding structure would be floor components in aircraft where high strength with minimum weight is a prime requirement. Still others combine strong panels with a frame of solid wood to produce structural members. Stressed-skin constructions or box beams are examples. Combinations of these also are made when components that are both decorative and structural are desired. Impregnated paper overlays on plywood represent such a component. Post-forming of sheet materials is also a common process to provide structural properties, most notably with paper to produce corrugated board and honeycomb core material. Products of secondary gluing have particular geometries that, together with performance requirements, dictate the glues used and the methods of assembly.

Overlays

As the name suggests, overlays are materials glued to the surfaces of panels or other products to enhance the substrate in decoration, paintability, weathering, and strength, among other features. Although overlays may be composed of many different materials—kraft paper, impregnated paper, veneer, cloth, plastic, metal, fiberglass, aluminum, high-density laminate—they are all sheetlike in geometry and therefore introduce no additional mechanical characteristic to the bonding process, other than perhaps handling in the assembly operation.

However, from a material standpoint, each introduces characteristics that need to be addressed with the help of an adhesive supplier, particularly with regard to bonding dissimilar, impervious, or surface inactivated materials.

Materials in sheet form involve large glue line areas, geometrically similar to veneer gluing. However, both the thickness and the stiffness can vary over a wide range, from a few mils in the case of film overlays to inches for core materials. Nevertheless distance from the surface to the glue line is usually short and therefore easily heated. Stiffness involves the factor of compliance: Some materials, such as film overlays, are very flexible and conform easily to surfaces. Others, such as high-density fiberboard, are rather inflexible, and often also very hard and incompressible. The latter, if not well machined, may produce gaps in the glue line that pressure cannot overcome, particularly if the core material is also hard.

Sandwich Panels

Components comprised of a stiff sheet material such as fiberboard to a low-density core such as honeycomb represent a challenge of producing a bond to a very narrow surface, generally at relatively low pressure (Figure 2-11(e) and (b)). The problem explained in

Chapter 9 is approached primarily through glue formulation.

Structural Components

In structural components, both the surface material and the core material are considerably more massive and correspondingly less compliant. Two major configurations exist; in one the face material is the load-bearing element, as in stressed-skin construction, (Figure 2–11(c)); in the other the core material is the load-bearing element, as in box beams (Figure 2–11(a)). Both employ the I-beam principle of deploying maximum material strength at locations of maximum stress (Figure 2–11(b)).

The I-beam is one of the most revered and widely used structures in the building or fabrication industry. Its popularity is due entirely to a basic engineering principle: Stresses produced by bending forces are highest at the surface of a beam and lowest at the center. Thus the material in the beam can be distributed so as to place more where stresses are greatest (flanges), and less where stresses are lowest (web), achieving a dramatic efficiency in material performance. Most materials that can be converted into an "I" configuration are able to participate in this efficiency, including metals, plastics, and reinforced concrete. The wood I-beam requires gluing in order to connect the web to the flanges, the connection being crucial to the performance of the three parts of the beam. (See Chapter 9.)

Stressed-Skin Structures

In one embodiment of the stressed-skin construction, for a wall component, a framework of 2 × 4's would be built as for a conventional house wall. To both sides of this frame, plywood panels would be glued. In this arrangement, the plywood (the skin) functions as the flange of the I-beam and the 2 × 4's as the web. The bond line situation would be $1\frac{1}{2}$-inch-wide areas of glue line the length of the 2 × 4's. The glue line is the thickness of the plywood away from the surface—diffi-

cult to reach with heat. Surfaces of the 2 × 4's may be neither smooth nor straight, nor fresh if not reprocessed as part of the gluing operation. Hence rather large gaps are likely to occur in the glue lines.

The problem here is the gluing of flat panels to a framework of 2 × 4's on edge, a wide area against narrow edges. A major consideration is calculation of the pressure to be applied if a platen press is used for this purpose. The basis for the calculation is not the total area of the panel but only the surface area of the 2 × 4's. In many cases, expediency will dictate nailing to provide pressure. Nails produce spot pressure with varying pressure between nails, but they provide continuous pressure until the glue hardens and the results are generally adequate. The glues used are necessarily of the resorcinolic type because of the need for durability and the difficulty of applying heat.

I-Beams and Box Beams

In producing an I-beam with solid lumber, the "I" configuration is lost in favor of ease of manufacture. The effect is produced by placing lower-grade lumber in the center areas to represent the web, and high-strength lumber at the flange locations to carry the greatest stresses. The boards are all of the same size and form a rectangle rather than an "I" cross section.

A truer I-beam in terms of configuration and amount of wood is made by using a panel material such as plywood or flakeboard as the web. But the reduced surface for gluing the web to the flange presents a problem. There is not enough glue line area to develop the strength necessary for this high-stress region in the structure. There are two escapes from the problem: (1) creating a groove in the flange member into which the plywood web can be inserted and glued, and (2) using two webs that can then be more easily glued to the edges of the flanges, producing in this case the box beam.

Geometry considerations in box beams are similar to those in stressed-skin structures, being essentially of the same construction and glue line configuration, except that the length is greater and the width is less. The function of the parts, however, is reversed. In this case the solid wood members function as the flanges of the I-beam and the plywood simulates the web. The same gluing situation exists, and pressure, as before, is applied with nails.

Corrugated Board

Two characteristics make the manufacture of corrugated board unique among gluing processes: its geometry and its speed. Speed, in this case, is geometry in motion; the corrugations are made at the same time as the bonds are being made, at high speed.

Corrugated board is a structure of paper in which the center layer is formed to a series of parallel flutes, appearing in end view as a sine wave. High-strength paper is glued to the crests of each wave to form a light weight sandwich of surprising rigidity (Figure 7-12).

The sine wave geometry of the core layer is the seat of the technology. The fluted paper element is made by passing a humidified paper through a pair of fluted steel rolls, nested and heated to form and redry the paper. One facing sheet is glued directly to the corrugations with adequate pressure. The second

Figure 7-12 Glued paper product: corrugated paperboard. (USFPL)

facing sheet is then added, but with only the amount of pressure the flutes can sustain. Since the geometry of the flutes is, by this time, fixed by being glued to one face, the flutes can sustain sufficient compressive force to receive bonding pressure, though much less than that used to bond the first face.

GEOMETRY IN TERTIARY GLUING (ASSEMBLY JOINTS)

The chief objective in assembly joints is to create shapes that provide the ultimate utility for wood as a material. The shape-generating stage in the conversion of wood to useful products has previously been described as one involving craftsmanship, a process of fabricating corners. Corners, contrary to the highly automated, efficient, mass-produced flat panels and linear products we have been discussing, generally require hands-on operations, and at a comparatively slower pace.

Mechanical fasteners are the major means of consolidating corners in the larger scene of building houses and other wood structures. Mechanical fasteners such as nails are fast and easy to use, and they engage more wood than the surface attachment of adhesives. They provide high strength when there is too much stress and too little area for a glue line to be effective. Some disadvantages attend the use of mechanical fasteners: they tend to loosen with moisture changes, they concentrate stresses, they cannot be used on thin members, and they yield or slip to some degree under load. Joints made with adhesives are rigid, distribute stress over an area, and add a sealing action to the joint area.

The geometry of corners is rather simply defined in terms of angles. However, details with respect to wood factors can be complicated. Corners produce one of the great paradoxes of fabricating with wood. The high strength that wood possesses along the grain in the natural state, or conferred by reconstituting, cannot be easily transferred around corners because it inevitably must engage the low across-the-grain strength of the piece it meets at the corner. While corners are mutually supportive in terms of providing rigidity, they also are regions where high stresses accumulate, particularly in torsion and shear. The major mechanical objective of a corner fastening system therefore is to transfer stress around a corner, through the weakest properties of wood. This is true whether mechanical or adhesive systems are used.

Reduced to simplest terms, the corner situations that apply to lumber (and by attribution to other wood-based products), can be represented by only three generic configurations: end grain to end grain, side to side grain, and end grain to side grain (Figure 7–13). Each of these can be studied for the interactions of differing wood stability on either side of the bond line. In (a), for example, the end grain of one piece is interacting with the side grain of the other. Although both act in a similar manner width-wise, in the other direction they act as a cross-grain structure, stable on one side and unstable on the other. In (b) the grain is parallel on both sides of the joint, behaving in an identical and hence stressless manner. Three slightly different configurations are contained in (c), two of which are more prone to difficulty than the other due to the grain orientation on either side of the joint. Board 2 engages its mate with end grain to side grain produc-

ing the same effect as in (a), similar in the width direction but dissimilar in the other direction. Board 3 is in a more difficult position. Even though side grain is engaged across the glue line, it is crossed, producing the maximum in differential movement across the glue line. In boards 4 and 5 the cross grain effect is compounded by end grain on one side of the bond line. Boards 3, 4, and 5 also face the maximum difference in instability across the glue line. In the following discussion, the board 5 configuration shown in (c) is considered representative of the worst-case situation.

Technically, corner joints often involve two geometries that are particularly stressful to bond performance: (1) the usually cross-grained orientation of the wood on either side of the bond line, and (2) the amount of wood back of the glue line, in most constructions quite thick, approaching lumber dimensions, which means maximum forces can develop with moisture changes. Another factor having geometric dimensions is the fit of joints, which along with surface quality has a number of implications in bond formation and performance.

Operationally, two main considerations apply in making wood corners: (1) providing means of increasing the surface area for the glue to operate, and (2) providing a degree of long grain around the corner. This is usually done by inserts of various kinds that reinforce or otherwise span the glue line, such as dowels, tenons, and splines. These inserts also introduce their own element of cross grain that needs to be considered in their performance.

The chief function of the bond in these predominantly cross-grained joints is to create shear resistance, usually against withdrawal of the inserted element. In some instances, the glue line also serves as a gap filler when parts do not fit properly. Bond formation depends primarily upon the quality of the surfaces to be glued (often poor compared to primary gluing), the application of glue, and the amount of pressure that can be introduced to the joint (both uncertain). Bond performance, on the other hand,

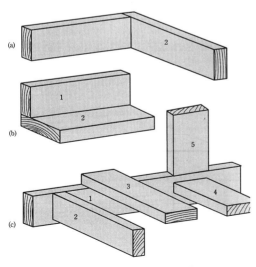

Figure 7–13 Joint geometry options with respect to grain direction.

depends upon grain direction and the mass (dimensions) of wood involved across glue lines. Numbers of joints and their location also have geometric implications, but more with respect to efficiency than performance.

Many adhesive assembly systems are in reality mechanical in nature, using wood instead of steel and glue instead of threads and friction. Various ways of increasing bond area or deploying the strong direction of peg-type wood elements into and across the weak direction of wood are illustrated in Figure 7–14. Different sizes, numbers, and configurations are used depending upon the amount of wood to be joined and the stress expected.

One representative assembly joint system is discussed by way of illustrating the geometric factors operating. Readers may extrapolate to other types of joints where similar considerations apply.

The Dowel Joint

Many of the wood and geometry characteristics operating in end-grain to side-grain joint operate also in dowel joints. Considering first the dowel itself, its strength is primarily a function of its diameter and its length for a given species of wood. The diameter and

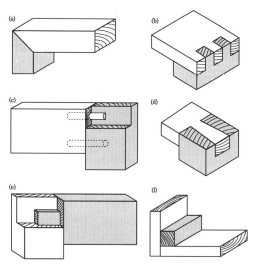

Figure 7–14 Representative assembly joints: (a) miter, (b) slip corner, (c) dowel, (d) dovetail, (e) mortise and tenon, (f) corner block. (USFPL redrawn)

length also relate to the amount of surface area available for bond strength development. Were these the only considerations, the larger and the longer the dowel, the stronger would be the joint formed. However, the hole into which the dowel is inserted produces a weakening effect, much as a knot in a board. A compromise in dimensions is therefore necessary to optimize strength of the corner. At times, two small dowels are preferable to a large one of the same surface area. Although the surfaces of dowels have no specific geometric implications, they are sometimes textured with grooves to facilitate adhesive functions.

The hole into which the dowel is inserted provides its own context of factors, chief among which is its diameter with respect to the diameter of the dowel, both in size and in configuration. While nominally round, holes and dowels tend to be elliptical either as produced or as the result of moisture-content change. The anisotropic properties of wood influence the drill bit as it cuts the hole, tending to deflect the drill more toward the side-grain side of the hole than toward the end-grain side of the hole, and more toward low-density areas than high-density areas. A moisture change further distorts the hole in piece (a) Figure 7–15, since there would be more dimension change across the grain than along the grain of the wood surrounding the hole. The hole in board (b) would also sustain a differential dimension change with a moisture change, but in this case due to the difference between radial and tangential instability and therefore of lesser magnitude. In addition to diameter changes in the holes, the length of the hole also sustains a change; very slight in board (b) because it involves along-the-grain changes, but substantial in board (a). For reasons of variability in both materials and procedures, dowel holes are always made slightly longer than the dowel. The dowel itself would change its circular dimensions only by the difference between radial and tangential instability as its moisture content changes.

The dimension changes as a result of moisture change may seem slight in absolute

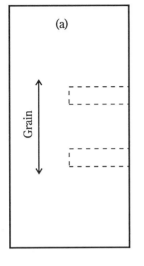

Side view

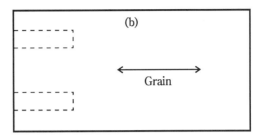

In piece (a), location of holes is constant but shape can become elliptical with moisture content changes.

In piece (b), location of holes varies with moisture content but shape of holes is more constant within the differential between radial and tangential stability.

Under changing conditions, dowels move with the holes in (b) causing misfit or splitting.

Dowels in (a) match the changes in the hole horizontally but not vertically, causing compression of the dowel at top and bottom under swelling conditions, and rupture under shrinking conditions.

Edge and end view

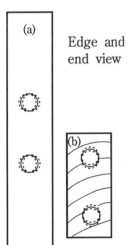

Figure 7–15   Movements in dowel joints with respect to grain direction and moisture change.

terms, but they can pose problems when changes occur from opposite extremes in the dowel and the hole. Greater problems would of course occur with high-density species than low-density species due to their greater instability toward moisture change, and due to their greater strength in creating swelling and shrinking stresses.

When the dowel is inserted into the hole, the same relative dimension changes can be expected with moisture changes, resulting in a complexity of movements only a forgiving material could tolerate. In board (a) the length of the hole will change in a direction that the dowel is most stable. The diameter of the dowel, on the other hand, will change in a direction that the hole is most stable—against end grain. This latter differential dimension change is responsible for some dowel joint failures due to compression fracture in the dowel, which under shrinkage conditions may fail in tension across the grain. In board (b), hole and dowel change together in both length and diameter except as they may differ in radial and tangential orientation, different densities, or different original moisture content. The consequences of this situation are discussed at greater length in Chapter 9 with respect to performance after bonds have been established.

Another differential movement between board (a) and board (b) affects position of the dowel holes if more than one is involved. The width of board (b) may change, carrying the holes with the change. The holes in board (a) however, remain in the same relative location. Dowels may then fail to meet their holes when assembled. If the change occurs after joining, a splitting action can occur.

These differing movements, related primarily to grain and geometry, account for much of the behavior of this type of joint throughout its production and performance. Since differential movement between the various components of a joint are key to performance, it is imperative that all wood elements be at the same equilibrium moisture content at the time of assembly. Although stresses will generate even with equal moisture changes due to differing grain directions, at least the extremes will be minimized by a uniform moisture content initially.

The above discussion on dowel joints applies equally to all joints in which a peg-type wood element is inserted to join solid wood or composites. These include the mortise-and-tenon joint and the turned-pin end joint. The slip joint and the dovetail joint partake of similar characteristics but do not have to contend with end-grain constraints that produce compression failures.

There are several geometric recourses to inevitable adverse effects in dowels and tenons. One of these involves cutting a narrow kerf part way through the center of the dowel or tenon and placing this kerf so it is oriented perpendicular to the long grain of the receiving wood. The kerf provides a space for shrinking and swelling to occur within the dowel without creating compression forces.

Miter Joints

The miter joint must be considered a different structure because of its peculiar configuration. It illustrates dramatically the effect of geometry on performance of bonds. Consider a 45-degree cut across a board. If it is reassembled and glued in the original aspect (Figure 7–16(b)), the resulting joint would be relatively unstressed through a moisture change. On the other hand, if one member is turned 180 degrees and closed, (c), the glue encounters the same surfaces and will form the same quality of bond, but the bond will encounter some of the most severe internal stress situations. The adverse effects are due to the differing amounts of wood back of the glue line from inside to outside of the corner, with resultant differing amounts of deformation that place the bond under tension stress perpendicular to the plane of the joint, (d) and (e).

Fit of Joints

Assembly joints do not generally enjoy good fit. This is not only because of disagreement on what good fit is, but also because good fit is difficult to achieve continuously under high-production conditions. The problem is two-fold: producing accurately sized mating parts, and producing good surfaces. Torn grain is a frequent feature of dowel holes and mortises, a sure progenitor of weakness in the finished joint. Very sharp tools, continuously maintained, are absolutely necessary if smooth, unruptured surfaces are to be prepared.

The problem of fit also affects the application of the adhesive. If the adhesive is placed on the dowel and then slid into the hole, the adhesive will be wiped off if the fit is as tight as it should be. If the adhesive is poured into the hole, a piston effect will keep it from reaching the gluing surfaces, unless the hole is somewhat larger than the dowel (i.e., a poor fit). To circumvent this impasse, flutes are machined on the surfaces of the dowels, either along the grain or spirally, to provide a passageway for glue to rise from the bottom of the hole and reach the sides. While this may not allow complete coverage of the dowel surface, at and near the flutes the glue will be able to form a bond.

A second approach to the fit impasse is the

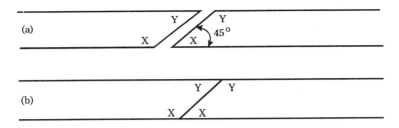

When cut and rebonded Y - Y and X - X, joint is relatively free of internally derived stresses due to moisture induced dimension changes

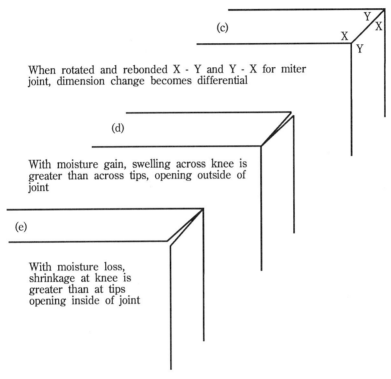

When rotated and rebonded X - Y and Y - X for miter joint, dimension change becomes differential

With moisture gain, swelling across knee is greater than across tips, opening outside of joint

With moisture loss, shrinkage at knee is greater than at tips opening inside of joint

Figure 7-16   Geometry of the miter joint and its effect on bond performance.

use of compressed dowels. In this case the dowel and the hole can be made the same size, but before insertion, the dowel is compressed to a smaller size. With glue in the hole, the compressed dowel can be inserted without interfering with the rise of the glue along the sides. Water from the glue then swells the dowel to form a tight joint.

In the case of slip and dovetail joints, a snug fit is the only option, and the glue that is wiped on must take its chances of being wiped off as the joint is brought together. While these are particularly effective joints for making corners, there is little in their construction that favors the production of strong glue bonds, and due to the crossed-grain structure, much that can mitigate against performance of a bond.

In most assembly joints, the application of pressure to the glue line is at best haphazard. While pressure is applied to close the joints, the fit determines whether any reaches the glue. Hence one cannot expect all adhesive actions to occur, except those that ensue spontaneously as properties of a liquid system. Often the solidified glue acts merely as a mechanical interlock, filling space, preventing movement, and stiffening the joint; the primary source of strength being the configuration of the system.

Attempts to reduce fit of joints to a measurable quantity are somewhat inconclusive due to the variables in wood properties, in machining qualities, in joint configuration, and in glue qualities. As a general starting point, in peg-type and interlocking joints, the fit should be snug enough to hold the structure together without glue and loose enough to allow assembly by hand, without pressure of clamps. From this point, allowances can be made for modifications, such as a snugger fit if the peg element is grooved or fluted to permit better distribution and anchorage of the glue, or the hole or mortise has a poorly machined surface; or a looser fit if the glue has gap-filling properties, or the peg element is precompressed and will swell with glue water after insertion. From the standpoint of the glue, most can tolerate gaps of five to ten thousandths of an inch. It should be remembered that gaps represent, in this case especially, a situation where the glue receives no pressure and must function entirely on its own mobility properties. Glues that contain no solvents—100 percent solids such as epoxy, polyester, and urethane—can sustain greater gaps since they do not shrink appreciably on curing.

Choice of Glues

Since strength resides primarily in the configuration of the joint, the main criteria in choice of glues are speed of cure and ease of use. Instant holding power is desired in order to have very short dwell times in clamps. The configuration also makes it impractical to introduce heat to assembly joints, therefore requiring glues that develop strength fast at room temperature. Because joints vary in fit and quality of surfaces, the glue must have a tolerance for such variables, particularly gap-filling. Durability is generally not a consideration since most products assembled in this manner are used indoors and not normally exposed to severe conditions.

GEOMETRY IN COMBINED PRIMARY AND TERTIARY GLUING

When wood elements are bonded in final shapes as in molding, a number of geometric considerations come into play. Molding processes are in a sense trying to bypass the separate making of corners in assembly gluing. They therefore are processes that develop a measure of end-grain strength around corners at the same time that consolidation is taking place. Basically, wood is being asked to bend around the corner. Factors that affect bending (without breaking, of course) are therefore under consideration. These include thickness of the wood as the primary concern because resistance to bending (i.e., stiffness) is directly related to the third power of the cross-sectional depth, as given by the flexural formula for Young's Modulus. This accords with Law 10, diminishing dimension, which stipulates that as dimensions decrease (thickness in this case), formability increases.

Since forming long-grain elements around corners involves bending, the radius to which a piece of wood may be bent, other factors being equal, can be smaller and smaller (representing a greater degree of formability) as its thickness becomes smaller. A 2×4, for example, may be bent to a radius of perhaps sixteen feet without exceeding the breaking strain. The same wood reduced to $\frac{1}{4}$-inch thickness might take a two-foot bend, and if reduced to $\frac{1}{40}$ inch would form to a two-inch radius.

Formability also increases as the grain of the wood becomes more perpendicular to the direction of the bend. This is because E, the modulus of elasticity, becomes minimum across the grain and allows more deflection before breaking. Complete parallelism would defeat the purpose of bending for structural uses, though not for decorative uses. In many applications, a 45-degree angle of grain to direction of bend produces a compromise in stiffness that allows for smaller radii with greater thicknesses, such as for making tubular products.

A further increase in formability as might

be needed for compound curves is achieved by reducing both the thickness and the width. This progression continues to the finest wood elements producible, although eventually strength must suffer according to Law 1. Figure 7–1 shows the progression in stratified detail down to molecular dimensions.

Aside from providing wood elements of sufficient conformability, perhaps the main consideration in molding parts to shape is the application of pressure. This factor comes down to design of the mold through which the pressure is delivered to the assembly. A number of devices for delivering pressure, and sometimes heat, are available. These may be rigid or flexible, and may deliver pressure in one direction, two directions, or omnidirectionally. They are discussed in Chapter 9.

8

Characteristics Conferred in Applying Adhesives

Adhesives, as purchased from the manufacturer, have basic properties that are ordained by their chemical composition. These properties, however, are delivered to the glue line only if the adhesives are properly used or handled prior to meeting wood. The word *proper,* seen throughout the literature on adhesive use, covers a wide range of cautions in applying adhesives and is generally used loosely and without sufficient specificity to aid the user. In this chapter the various factors involved in "proper" use are discussed in relation to how they affect performance so users can determine for themselves that "proper" is achieved in their particular operation. The terms *handling properties, use properties,* and *working properties* are often used to describe particular characteristics of adhesives that pertain to the manner in which they are treated in the interval between manufacture and the glue line. Use instructions normally include this kind of information by way of instructing the user on methods to obtain the most efficient and most reliable bonding.

We have seen that gluing factors have positive and negative effects on performance, and that they operate with a number of different mechanisms. However, since their effects all register through the actions glues must perform in bond formation, the common denominator approach to understanding can be used in this case also. The key to "proper" that fits all factors and resolves their effects into a common denominator is again *mobility.* The effect on mobility of the adhesive before, during, and after hardening determines to a profound degree how a particular use factor influences the performance of the ultimate bond. It is important to bear in mind that mobility covers a wide range of movement, from mass flow of the adhesive during application and bonding to atomic adjustments and orientations. With this in mind, each factor is discussed below in the approximate order in which it occurs in a gluing operation. In this way, one can perceive a course of mobility changes with each step in the process.

STORAGE

After manufacture, glues remain viable for a length of time (shelf life), which is partly dependent upon storage conditions. Between manufacture and actual deposition on the glue line, adhesives are subjected in storage or transportation to varying conditions and lengths of time. Some deleterious effects are traceable to this interval. The main factors operating in storage conditions are time, temperature, and moisture or humidity, usually the lower the better, though there are exceptions. In the case of latexes, for example, some formulations should not be frozen because an irreversible separation of ingredients may occur. Moisture in storage affects mostly those adhesives in powder or film form. At best, moisture causes lumping and difficulty in mixing; at worst, it activates some glues to begin curing, especially those incorporating catalytic agents. The effect is first a reduction in solubility, which reflects itself as a higher viscosity in the glue mix. Second, the effect reduces the intrinsic mobility of the glue on the glue line with all the consequences this entails, usually inhibition of the bond-forming motions the glue must perform. In some glues such as protein or starch, the opposite effect may occur due to breakdown of the particulate structure of the glue.

Heat usually compounds the effects of moisture. High temperature can completely destroy some adhesives in a short time. Liquid resins of the cross-linking type are particularly sensitive to high storage temperature. In the case of resorcinol resins, the liquid portion is quite stable in storage but the potency of hardeners used to cure them diminishes slightly over time.

Storage-life changes are evidenced by viscosity aberrations either in the original liquid or in the prepared mixture being applied. Damage in storage is usually spotted as a viscosity change observed during mixing. A viscosity higher or lower than normal should raise suspicions. Effects of storage on catalysts will not be shown as a viscosity change, however, but as a change in the rate of cure. This is observed by measuring the time necessary for the prepared glue to reach a gel state (see page 65) at a certain temperature. Manufacturers use this method in their own quality control and record the results against lot numbers. They thus have records of the starting condition of their glues. Recommended storage conditions and time limits are usually stated on the package when any cautions should be observed.

MIXING

Some glues are sold ready for use out of the container. Most, especially those used in high volume, require the addition of various chemicals to complete the formulation and prepare it for use. The five most common additives to the base adhesive are catalytic agents, solvents or liquid carriers, fillers, extenders, and fortifiers (see Chapter 4 for types and functions). Of these, solvent or carrier in the form of water is the most often used, and catalytic agents the most critical. These additions, done by the user, provide flexibility in adjusting an adhesive to particular needs. In doing so they minimize the number of different adhesives that must be bought and inventoried; they allow a longer shelf life for the more reactive formulations since they are not catalyzed until used; and they reduce shipping costs (water in most cases). However, they also introduce chances for mixing errors to occur, some of which are difficult to trace, and some of which may produce delayed consequences—the ultimate dismay in glued products.

In conducting the mixing operation, a number of subfactors influence the outcome and demand attention. These include the type of mixer, nature and amount of the ingredients, order of addition, interval between additions, duration of mixing, speed of mixing, and—most important—temperature of the mix. Most of these subfactors are covered in instructions accompanying every container originating with the manufacturer. It is advisable to know the implications attending each in order to sense sources of potential problems associated with the mixing operation.

There are many means for mixing glues, beginning with a wooden spatula and ending with impingement mixing of highly reactive liquid resins. The latter are used with very fast-setting formulations that harden so quickly they cannot be transferred from a mixer but have to be delivered directly to the surface during mixing. Spatulas should be used only for small batches in a cup, although it is preferable to use a power drill driving an appropriately sized stirrer. Power stirring can be scaled up to pail sizes. For larger batches, mixers with multiple or counterrotating paddles are used (Figure 8-1). Some of these can be water cooled to reduce the temperature and ensure a longer mix life, or heated to improve solubility or to cook ingredients such as extenders.

Mixing becomes a particularly crucial operation for those formulations requiring multiple and sequential additions. Often a definite procedure for mixing is part of the use instructions. For the more complicated formulas, a checklist aids the process. The operation is made more foolproof by using containers sized and marked to hold the exact amount of material for the formulation being mixed. A universal caution is that

whereas liquids may be measured and added by volume, powders are always measured by weight since their volume is not constant but varies according to past handling.

A lumpy mix is a most distressing sight, frustrating to correct but easy to avoid. One particularly vexing situation is partially dissolved, gelatinous spheres floating in a liquid, seemingly unresponsive to the stirring action and defying further solution. Several courses of action are available: (1) Discard and start over. (2) Continue stirring for a long time and risk a shorter working life. (3) Enlarge the mix by adding only the powder fraction first, thereby stiffening the mix and making the stirring more effective. In starting over, the lumps may be avoided by first placing only a portion—half to two

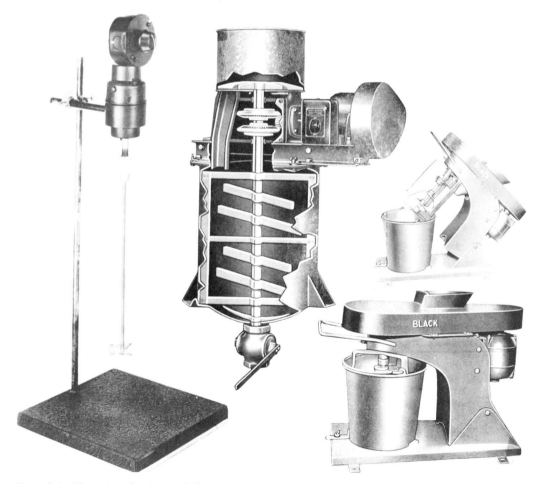

Figure 8-1 Glue mixers for shop and plant.

thirds—of the liquid in the mixer or container and then adding all the powder. This results in a much stiffer mix in which lumps do not form or are easily rolled out. The surest way of avoiding lumps is to sift the powder into the liquid with the mixer running. The mixing thereby occurs progressively and lumps do not form.

For those formulations that require the addition of catalytic agents, this step may be the most crucial. Because they control the speed of hardening, these agents directly control the rate of production. They also control degree of hardening. Through the effect of speed of hardening on the mobility functions of the adhesive, and degree of hardening on the strength and durability of the bond, catalysts exert a profound influence on performance. Most of the problems associated with catalyst addition stem from the small amounts needed. These are usually powerful chemicals in the sense of ability to alter properties, and small differences in the amount added can produce significant changes in performance. For this reason, the means and manner of measuring becomes important. Liquid catalysts can be measured volumetrically using a vessel of just the right size and shape to hold the quantity needed. The preferred shape is one that is tall and narrow rather than short and squat as this provides the most accuracy in judgment. As with other powders, catalysts in powder form do not have a constant density and therefore are always weighed.

Both methods have sources of possible error, the first being calculation of the amount needed. Instructions usually state the amount of catalyst to be so many parts per part of resin—5 parts catalyst per 100 parts resin, for example—or 5 percent catalyst based upon the weight of the resin. Either method will produce the same answer, but only if the arithmetic is correct. If the weight of resin is 25 pounds, 5 percent of that would be 1.25 pounds of catalyst. If dealing in parts, for every part (pounds, ounces, grams) of resin there is 5/100 or .05 parts of catalyst. Then 25 parts times .05 parts equals 1.25 parts as before.

In weighing catalysts, several errors can arise. One comes from using the same scale to weigh the catalyst as used to weigh the resin. A scale with the capacity to handle the poundage of the resin does not usually have the sensitivity to weigh catalyst precisely. Moreover, resin scales tend to accumulate spillage, creating downside errors in weighing that may be tolerable for the resin but intolerable for the catalyst. Finally, the most insidious error is in failing to deduct the weight of the container from the total weight, thereby leading to an undercatalyzed mixture. In any case, catalytic agents are normally added last in order to prolong as long as possible the working life of the mix.

Because of the crucial nature of catalyst addition, this factor is a point for close control in a gluing operation. The worst-case situation is forgetting to add the catalyst. Only slightly less worse is adding it twice, as can happen when more than one person is involved. Recognizing this source of human error, manufacturers often incorporate a strong dye in the catalyst; the intensity of the color change in the mix signifies the presence of the catalyst.

Solvents or diluents are added as a dispersion medium (most wood adhesives are not true solutions but suspensions of particulate materials in a liquid) to facilitate application and to permit adhesive actions dependent upon fluidity or mobility to take place. The latter includes, in the case of adhesives that harden by polymerization, the mobility of individual molecules in achieving atomic proximities and orientations for chemical couplings to be made. This role of solvent in bond formation is a frequently overlooked consideration in the development of strength and durability. Sometimes the solvent is adjusted to suit the application equipment being used. Spraying requires a lower viscosity than roller coaters, for example, and sometimes extra diluent is added.

After the adhesive has been placed on the glue line, the solvent has a different and continually changing role to play. At first it facilitates the motions necessary for bond formation, and later it must disappear to allow full strength to develop. Programming the effects of solvents on the glue line is one of the subtle technologies that spell success or failure in a gluing operation since both good and bad effects occur. (This is discussed at greater length below.)

Solvents (usually water in the case of most wood adhesives since they are polar in nature) are added in amounts as prescribed in the recipe for the formulation and, as in the case of catalysts, are stated as a percentage of the base adhesive or as so many parts per part adhesive. Again a major caution is the arithmetic in scaling a formulation to size.

The process of adding the solvent poses some problems, especially where very large amounts are involved. The aids used to insure accuracy can themselves become culprits. For example, when water is gravity-fed from a holding tank, even though fitted with a sight gauge marked to indicate the amount of water to be let in, faulty operation of the let-down valve can add more or less water to the mix. The use of a dip stick also has its gremlins. If it is referenced from the top rim of the mixer, an overage of water can result if glue accumulates and raises the level of the rim. A tank or vessel holding only the correct amount of water, perhaps guarded by an overflow outlet, provides some assurance of correct water addition.

Improper mixing that could arise from complicated or difficult mixing procedures can destroy all subsequent good work. Consequently, some means is needed for checking whether mixing is in accord with instructions and the mix is in condition to be used. A target viscosity, monitored with suitable means of measuring it, provides clues to many of the prior factors that may have gone wayward. For those glues that set by chemical reaction, measuring reactivity on a regular schedule provides a check on whether catalysts have been added correctly. Methods for carrying on tests of viscosity and reactivity are given in Chapter 10, Quality Control.

The act of mixing introduces another variable that often is neglected but that has the potential for causing trouble, particularly in bonds that are intended to carry high stress. Stirring or churning at high speed sometimes creates foam or bubbles in the mixture. Two problems can ensue. In applying the glue, the increase in volume due to bubbles can give a false impression of the amount being placed on the surface. Also, bubbles in the hardened glue line weaken it. The weakening effect is engendered in several ways. In the first place, a bubble represents a void, a spot of zero strength. Second, the stress that the spot might have helped resist shifts to adjacent areas, increasing the stress there and producing stress concentrations. In the latter case, even the shape of the bubble has significance—elliptical bubbles are more stress-sensitive than round ones. The orientation of the elliptical bubbles also has a bearing. When they are produced by the pressure of the press squashing round bubbles, their main axes are in the plane of the glue line, the worst direction for most joints under shear, cleavage, or tension stress. Third, while the air in a bubble can be squeezed out into the wood, it is conceivable that small bubbles can be pushed into the pore structure of the wood and block the penetration of the adhesive.

Problems with bubbles are more likely to occur with cold-setting glues than with hot-setting glues since heat, pressure, and lower viscosity in the latter give the bubbles a better chance to dissipate. Although some glues are intended to be applied as foams, they usually are appropriately formulated to discount the effects of bubbles. Glues that are especially prone to foam during mixing or spreading, such as the proteins and latexes, have antifoaming agents as part of the formulation.

Excessive foaming due to too much or too rapid mixing can be observed by letting the mixture rest a few minutes. Bubbles will rise to the surface and form a layer of frothy material on top. In less pronounced cases, the presence of air bubbles in the mixture can best be assessed by spreading a small amount of it on a piece of glass and observing it against a light. One can use varying amounts of magnification to obtain a closer view;

bubbles will invariably be found. The question then arises of how much is too much. Specifications generally do not stipulate the tolerable level of bubbles in a glue mixture. It depends upon a number of factors that vary with applications, such as the ultimate strength needed and the opportunity for bubbles to escape during pressing; the latter being dependent upon the wood being glued, the pressure applied when applied, and the temperature. In most cases, except as noted above, bubbles that do not rise to the top can be disregarded.

A topic of related interest is the study of the morphology and disposition of these mobile voids in the glue line as they are affected by viscosity and pressure. In addition to shape and size, one can visualize a distribution effect of pressure. If pressure forces them toward boundaries (or low spots) in joints, a concentration of weaknesses can occur at critical regions. Some idea of the translocation of bubbles in a joint can be recreated by placing a spot of glue between two glass plates. Applying pressure will cause the mass to flow, dispersing the bubbles in all directions.

WORKING LIFE

After mixing, glues remain useful for a period of time that varies with the formulation and surrounding conditions. The time and conditions are usually stipulated by the manufacturer of the glue and presented as part of the instructions for use. Three main factors affect the working life (*mix life, pot life*): (1) temperature, (2) evaporation, and (3) chemical activity. All operate to change the viscosity of the glue, in all cases unfavorably since the initial viscosity is by design the optimum. Depending upon the composition and mechanism of hardening, the factors operating may increase or decrease the viscosity. For example, hot animal glue tends to break down under prolonged heating in the pot, and although it appears to be thickening, this is due to loss of water; if water content is restored, the viscosity would be lower than originally, confirming the breakdown. This

is the reason for discarding leftover glues after a day in the molten state. On the other hand, a glue like resorcinol increases in viscosity during its pot life due to continual molecular growth, which it is designed to carry into the glue line. Resorcinols, like other highly reactive glues, release heat as they polymerize (an exothermic reaction), shortening the pot life further and at an exponential rate. The glue may be good one minute and gone the next, particularly if in large quantities where the heat generated is less able to dissipate. Hence the recommendation to mix smaller quantities, or to provide a cooling jacket around the pot. Ready-to-use glues have pot lives that equal their storage lives shortened only by loss of solvent, usually evidenced by a skinning over of the surface.

In general, most glues will forecast the end of their pot life by being too thick to apply. By keeping an eye on the spreading operation one can observe when glue has gone past the point of usability.

APPLYING THE ADHESIVE

Placing the glue on the joint surface has two objectives: placing the right amount, and placing it uniformly. (Some applicators contribute a slight pressure to initiate penetration.) The means for applying adhesive depend upon the amount to be spread, the speed of hardening, and the viscosity of the glue. In addition, the quality and the area of the surface have a bearing on the spreading operation.

For small areas with high-viscosity glues, one would use a spatula or a trowel. Under production conditions such a glue might be extruded as a bead from a nozzle under pressure. Glues that set by cooling also would be extruded but through a heated nozzle; alternatively the wood piece (a dowel, for example) might be dipped in the heated glue.

The more fluid glues may be applied by brush on small areas. On large areas, a paint roller might be used if only a few parts are involved. Under production conditions, large stationary rollers transfer glue to wood surfaces in a manner similar to a printing operation. Texture and resilience of the roll surfaces accord with the viscosity of the glue and roughness of the wood surface. On large surfaces the adhesive can be applied by spraying or by extruding beads of liquid or foamed glue through multiple openings in a pressurized header spanning the joint surface.

On very small wood pieces such as flakes, the adhesive is applied by spraying them while being tumbled or stirred. In the case of powder adhesives, the powder is simply added to flakes in a tumbler and the amount that clings to the surface is sufficient to create the desired bond.

Each of the methods of applying glue that involves power equipment is vested in its own set of mechanics, but the end product is the same: a uniform deposition of glue in the specified amount. They are discussed more specifically later in this chapter.

SPREAD AMOUNT

Glues vary considerably in the amount needed to form a good bond. Theoretically, the correct amount is predicated upon the ratio of solids content to solvent content of a formulation. The reasoning here is that only the solid material functions in the final bond. Wood factors influence the amount of glue one might apply in a given instance. An important factor is roughness of the surfaces; obviously the rougher a surface is, the more volume there will be between two surfaces, and therefore the more glue solids will be needed to fill it. Porosity is also a factor; highly porous woods require more adhesive solids to allow for greater penetration while leaving enough on the glue line to form link 1.

Other factors of an operational nature also influence the amount of adhesive applied; two are of particular importance. The first is assembly time, which harbors factors that project themselves into the functions that occur after pressure is applied. The factors collectively influence the mobility of the glue. Those that are affecting mobility through solvent loss can sometimes be over-ridden by the amount of glue applied. For example, a light spread will engender a more rapid solvent loss and enforce a shorter assembly time. Conversely, a heavy spread reduces the percentage rate of solvent loss and lengthens the assembly time.

A second effect of amount of glue spread is the amount of solvent that is thereby added to the assembly. Those adhesives that harden by loss of solvent are directly affected in their rate of hardening—low spreads, faster; high spreads, slower—all controlled by the rate at which wood can absorb water from the glue. For those glues that require heat to harden, the amount of solvent that enters the assembly can become a liability when the temperature exceeds the boiling point. When solvents vaporize they do two harmful things: absorb heat (heat of vaporization), thereby reducing the rate of temperature rise; and they create internal pressure, causing blisters or separations between laminations.

The amount of glue applied to a surface is expressed in two ways: in weight of adhesive per unit joint area, or in the case of comminuted wood, in weight of adhesive per unit weight of wood. The latter is somewhat surprising because the important parameter of area does not enter the equation. However, this latter unit does lend itself to easier measurement and it does enter the cost calculations more precisely. In the latter case the weight of the glue is always the weight of resin solids alone, not the total weight of the mixture, and the weight of the wood is also always the oven dry weight (OD weight). The ratio is usually stated as a percentage:

$$\frac{\text{Percent}}{\text{adhesive}} = \frac{\text{OD weight adhesive}}{\text{OD weight of wood}} \times 100$$

For example, 12 pounds of OD adhesive on 200 pounds of OD wood represents an application rate of 6 percent.

Spread amounts in the other case are always stated as pounds of glue mix per thousand square feet of single glue line, abbreviated to lbs/MSGL. Sometimes the glue is spread on both sides of the joint, half on each side, and is then referred to as double spreading. Nevertheless, when the two surfaces are brought together they form a single glue line and the two amounts become additive. (The softwood plywood industry has its own convention for expressing spread rate. In this special case, the spread is expressed as pounds per thousand square feet of double glue line or lbs/MDGL in order to reflect the amount of glue used in a thousand square feet of 3-ply plywood, which has two glue lines.

It is often useful to consider the amount of *total glue solids* per thousand square feet of single glue line. In this case it is necessary to learn from the formulation what percentage of the total weight is solids and calculate accordingly. Similarly, if it is desired to know only the amount of *resin* solids alone that are being spread, i.e., separate from fillers and extenders, the amount derives from the percent of resin solids in the applied wet mix.

ASSEMBLY TIME

Assembly time is a simple statement of how long a film of glue can stand on a surface before pressure is applied. Some glues tolerate only a few seconds; others can go days or weeks without injurious effects. The assembly time is often divided into two distinct periods: one in which the surface spread with glue is exposed to the atmosphere, designated as *open assembly time;* and one in which the joint surfaces are in nominal contact but without pressure, called *closed assembly time.*

The assembly time period is one of the most important operational variables affecting the quality of a bond. Of the two periods, open assembly time is the most influential because of the rapid solvent loss that can occur during that time. The solvent lost at this point can have beneficial as well as detrimental effects, depending upon the nature of the operation and the type of glue being used. Consequently both a maximum and a minimum time are specified in some cases, although it is normal to specify only

the maximum permissible time with the implication that there is no minimum. In the latter case, it is permissible to apply glue and press immediately. In general, assembly time produces a thickening of the glue film, the main consequence being a reduction in adhesive mobility. Some glues are too fluid when applied and need a minimum assembly time to firm up before pressure is applied, which might force it out of the glue line. Some glues also need to lose solvent before pressing if the adherends are unable to absorb it.

PRESSURE

All use instructions carry a stipulation on the amount of pressure that should be applied to the joint. Theoretically this is not so much a function of the glue as it is the quality of the joint surfaces. Poor fitting and rough joints require more pressure. Also high-density (stronger) woods require more pressure than low-density woods. The main function of pressure is to bring the joint surfaces into reasonable contact; most adhesives will take it from there. However, considering the mobility status of the adhesive on the glue line at the time pressure is applied, pressure must also provide the impetus for the mass movements the glue must perform in forming the links of a bond. The mobility status of the glue at the time pressure is applied is a function of the composition of the glue but also its history up to that point—storage, mixing, assembly time. With regard to a particular glue, the basic requirement of pressure is presumed to be due to its composition or fluidity. However the actual pressure needed in a particular instance must be derived by experience in the context of the materials being glued. Use instructions generally recommend 100 to 150 psi on low density woods and 200 to 250 psi on high density woods.

PRESSURE PERIOD

The length of time the assembly must remain under pressure is primarily a function of adhesive composition and the construction being glued. When there is likely to be any stress on the glue line coming out of the press, as might exist in gluing curved assemblies, or where high pressure is needed to close a joint, pressure must remain until the glue is fully cured. The environment in, and next to, the glue line—particularly moisture and temperature—also strongly affect the pressure period by influencing the rate at which glues harden. Consequently one or both of these two factors are usually included in statements of pressure period. In general, high moisture content in the wood will reduce the rate of moisture loss from the adhesive and vice versa. Consequently, adhesives that harden by moisture loss will require a longer time under pressure when wood is at the top of the acceptable moisture-content range. Pressure period must also be lengthened when the amount of glue applied has been increased for some reason, since there is more moisture to contend with. With respect to temperature, the pressure period is generally shorter as temperature increases; the major exception being the hot melts in which the reverse is true.

It should be remembered that conditions calling for a longer pressure period also invite greater motion of the adhesive since each action will have a longer time in which to act. Conversely, conditions that prompt shorter pressure periods may result in curtailed adhesive motions. A situation of the latter kind can occur in the gluing process referred to as the separate application catalyst system. The catalyst is applied to one surface to be glued, and the resin is applied to the mating surface. When the two surfaces are brought together, the catalyst diffuses into the resin and curing takes place. The purpose of the process is to allow the use of very fast catalysts that otherwise would produce too short a working life. A problem can arise at the surface receiving the catalyst. The concentration of catalyst produces a very rapid progression of rheological effects on adhesive motions, too fast for them to optimize, and they can fall short. Moreover, since such catalysts tend to be acidic in nature, concentrating them risks later deterioration of the wood as well as of the adhesive.

A similar situation can occur when surfaces are deliberately preheated to speed adhesive curing without having to drive heat in after pressure is applied. Both processes have nevertheless been successfully used with proper compounding of the adhesive.

CONDITIONING PERIOD

When an assembly comes out of the press, there are still a few remaining motions to be accomplished. One of these is final cure or hardening. Since the pressure period is the chief determinant of rate of production, it is tempting to shorten it as much as possible. Many products, particularly those that are flat or straight, have a minimum of stress between the wood pieces, as they lie fully supported. These are often removed from the press before the glue is fully hardened and are then set aside to complete the hardening while the press is used for the next assembly. This is particularly advantageous in hot-press operations. The assemblies come out of the press at press temperature and are immediately stacked. Wood, being a good insulator, retains the heat and allows the glue to finish hardening.

In the case of cold-press operations a similar situation exists, except that after a pressure period sufficient to achieve holding strength, the assembly is set aside for the glue line to lose more moisture or solvent, to complete chemical reactions; or in the case of hot melts, for the glue to cool further.

A major need for a conditioning period is due to the redistribution of moisture that goes on during pressing. In hot pressing particularly, moisture, in migrating away from the heat source, tends to concentrate in various locations in the assembly. The resulting moisture gradients cause stress between wood pieces and warping of the assembly. Moreover, the surfaces of the assembly that have been next to the heat source become exceedingly dry. These surfaces will later pick up moisture and swell, increasing the dimension of the piece and causing buckling. If the later moisture pick-up is uneven, the assembly will warp.

A particular situation occurs in edge gluing lumber where the moisture from the glue enters the wood on both sides of the glue line during pressing. The resulting effect leads to one of the more distressing consequences in panelized lumber products: sunken glue lines, if a conditioning period of sufficient duration is not allowed. This phenomenon is explained in Chapter 9.

SPREADING AND ASSEMBLY TIME IN SPECIFIC CONSTRUCTIONS

The manner in which adhesive is applied to wood surfaces and the events that transpire before pressure is applied are dictated in large part by the geometry of the wood element and the hardening properties of the adhesive. Since it is the beginning of wood/adhesive interactions and includes actions that are crucial to bond formation, this phase is one of great importance to the success of the operation. It is discussed here in wood element sequence from largest (lumber) to smallest (fibers).

Lumber

This one element has three different gluing processes according to which surface is being glued: edge, face, or end.

Edge to Edge. Following preparation of the wood, the first step in gluing lumber edge to edge in a production operation is selecting and gathering together all the boards destined to become one panel. Usually the boards vary in width but are calculated to add up to the final width of the panel, sometimes in a definite order to enhance decorative features. The group of boards are then placed on edge and moved forward to glue application.

Glue Application. Typically, the boards are passed as a group over a roller turning in a hopper of glue (Figure 8–2). In slower production operations, the glue is applied with a squeeze bottle or brush, or a bead may be extruded from a pressurized gun. In all cases, each board receives a layer of glue on only one edge. The roller applicator places the glue on the bottom edges, while the other means of application place it on the top edges. This minor difference can be a source of potential problems for the roller application. Depending upon how steadily the boards ride over the roll, skips may occur in transferring glue from roll to edge. In some installations, the operator will shuffle the boards into a clamp without an opportunity to see whether the application was uniform. Skips or wipes can also occur in any case due to careless handling of the boards after glue application.

Assembly Time. Normally the time between applying glue and applying pressure in the clamps is very short, perhaps too short

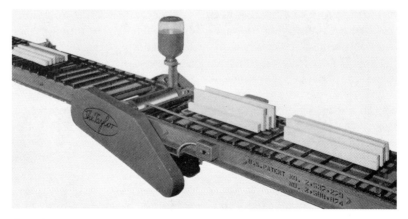

Figure 8–2 Preselected and serially grouped boards are passed over a glue roller to receive glue on lower edges.

in some cases. During this time, both desirable and undesirable events can happen. One of the more insidious occurs after the boards have been reassembled in a horizontal position and are awaiting the clamps. During this time, the glue can drain away from the upper edge to the lower edge, starving the former and overloading the latter. A properly formulated glue would be thixotropic enough to hold itself in place against the pull of gravity, yet fluid enough to perform adhesive actions necessary in bond formation.

During assembly times, glues gradually thicken with loss of water, loss of heat, or chemical reaction, as the case may be. This time period therefore has limits, mostly on the long side, but often enough on the short side to be a source of concern both ways. Long assembly times that result in severe loss of mobility have all the consequences associated with failure of the transfer action (recall that the glue was placed on only one side of the joint). All the factors that affect the rate of moisture loss from the spread adhesive—such as temperature, moisture content of the wood, and relative humidity—directly affect the assembly time. Short assembly times invite vulnerability to squeeze-out and starved glue lines due to the short distance to an edge and high pressures used to ensure conformation of surfaces. Starving is particularly apt to occur with glues that are fluid enough to be brushed on easily.

Face to Face. Due to the long assembly times often associated with large beam constructions, the adhesive is preferably applied to both surfaces of a bond line. This is done with a double roll glue spreader (Figure 8-3 (a and b)). Alternatively, an extruder is used, laying down ribbons of glue along the length of the board, but in this case only on one side Figure 8-3(c). In both cases, the glue is in bead form, creating a stripe pattern on the surface. Concentrating the glue in bead form rather than in a thin film prolongs the assembly time. In bead form there is less surface area losing water to the surroundings, and therefore less viscosity increase. An advantage of spreading both surfaces of a bond

line as in double roll spreading is that it avoids the transfer action by the glue, and thereby allows an even longer assembly period. A disadvantage of roll spreading is that, over time, the glue tends to thicken on the rolls particularly at the ends, changing its viscosity-dependent characteristics. The amount of glue that is applied by the rolls, for example, increases as viscosity increases, but with its lower water content would have a shorter assembly time than fresh glue. Also penetration activity of the glue decreases as viscosity increases. Two recourses to these disadvantages exist: (1) constant addition of fresh glue to the rolls, and (2) removal of the end plates from the rolls so that glue flows out the ends. Constant addition of fresh glue thus provides a flushing action; The glue falls into a pan from which it is recirculated to the rolls.

A second problem that sometimes occurs with roll spreading is due to a propensity to feed lumber through the rolls constantly at the center. The rolls then wear unevenly and deposit glue unevenly. The recourse is to alternate the feed from one end to the other to promote more even wear.

Assembly Time. The main assembly-time challenge in lumber laminating is bringing together the many long pieces of glue-coated lumber for a large arch (Figure 8-4). Even after clamping begins, considerable time can elapse between tightening the first clamp at the middle and the last clamp at the ends. One problem is staying within the maximum allowable assembly time for the first board that has been spread with glue, and the minimum allowable assembly time for the last board passing through the spreader. For example, some use instructions may specify a maximum assembly time of 90 minutes and a minimum of 15 minutes. This means that after the last board has been spread, it must stand for 15 minutes, during which time the first board should not go beyond 90 minutes before pressure is applied. These limits are of course modified according to ambient conditions, shortening in dry, warm atmospheres and lengthening in humid or cool sit-

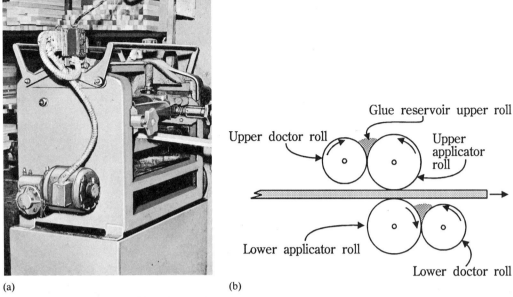

(a)   (b)

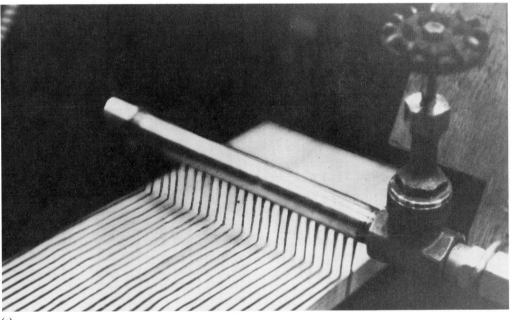

(c)

Figure 8-3 Spreading equipment for applying glue to boards: (a) double-roll spreader
(USFPL), (b) schematic, (c) extruder (USFPL).

uations. The time limits are also shortened if too much open assembly is involved.

During the assembly time both detrimental and beneficial actions are going on, hence the limits at both ends. At the short end, the expectation is that some thickening of the glue will occur so it can accept pressure without squeezing out of the joint. At the long end, the concern is that the glue may lose too much of its mobility to respond sufficiently

Figure 8–4   Laminated arch being placed under pressure (USFPL).

to pressure. The assembly period also harbors the effect of gravity on the liquid glue. If the laminations rest on edges awaiting pressure, the glue can drain from the upper edges to the lower edges, starving one and flooding the other.

It is well to keep these time and place events in mind should there be occasion to review performance at a later date. Effects of long assembly times are to be found in the first lamination spread with glue, the ends slightly more than the middle. Effects of short assembly time are to be found in the last lamination. Effects due to drainage from one side to the other will register most strongly in the first lamination, at what was the upper edge during the assembly period.

End to End.   On scarf joints, the adhesive is usually applied by brush when there is only a small number to be made, or when it is an operation separate from face gluing. When it is part of the face-gluing operation, the scarf is given its coat of glue as the rest of the board passes through the spreader.

A means of dealing with the porosity problem of end grain surfaces presents itself at this time—sizing. Both surfaces of the joint can be given a thin coat of lightly thinned-down adhesive and allowed to dry partially. This seals the end grain, at the same time achieving some of the penetration and wetting actions. The joint then proceeds normally through the process.

In the case of finger joints, this procedure would also be beneficial, but it would disrupt the efficiency of high-speed continuous processes. The compromise is to apply a relatively high-viscosity adhesive and assemble immediately. Although the glue can be brushed on, under production conditions it is wiped on by a wheel configured to the size and spacing of the fingers and provided with a continuous flow of adhesive (Figure 8–5).

As with other types of joining operations, it is always better technically, though not always economically, to apply the glue to both surfaces of the joint. This insures the transfer action and promotes the wetting and penetration actions.

Assembly Time.   In the scarf joint, except for the sizing step, the assembly time depends upon the type of operation, whether it is separate from or part of beam assembly. In the latter, the assembly time for the scarf is the same as for the lamination. When scarf joining is a separate operation, assembly

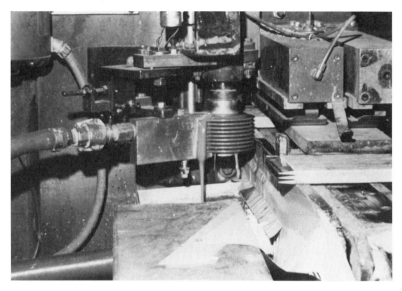

Figure 8-5   Applying adhesive to finger joints by use of an applicator wheel (USFPL).

time can be controlled more favorably for adhesive actions on end grain, generally profiting by longer assembly times to allow for additional thickening beyond what would be good for face gluing.

Finger joining typically suffers a very short assembly time, since production is often a continuous operation from cutting the fingers to pressing. In batch type or job shop operations, the assembly time can be controlled to favor adhesive actions as in separate scarf-joining operations.

Veneer

Because of the large glue line areas to be spread in veneer gluing, virtually all application methods can and are being used, including dusting on of powders, dry films, wet films, double roll coating, spraying, extruding, curtain coating, brush, scraper, and hand roller application. Although more than one option often exists in any one situation, factors influencing choice often come down to moisture content of veneer, quality of surface, thinness of the veneer, speed of production, and method of layup. Some accommodations include the use of:

1. Powdered resins or dry films on high-moisture-content veneer in hot-press operations where the moisture in the veneer, aided by heat, causes the resin to flow, wet, and bond

2. Dry adhesive films on very thin decorative veneers for hot pressing; because of their thinness these adhesive films are unable to absorb glue water, or because steep grain angles are vulnerable to bleed-through

3. Wet films on veneer that is rough and uneven and requires more "body" in the glue than can be delivered by other means

4. Roller coating when both sides of a sheet of veneer need to be spread (the traditional method)

5. Spraying, extruding, or curtain coating for applying glue on rough and uneven veneer, but only on one (the upper) side

6. Brush, serrated trowel, or hand roller are used on small jobs. On rough and uneven veneer, the trowel and hand roller are less effective.

In all cases, the ability to place the glue on the veneer surface in some sort of a pattern, typically a ribbon or bead pattern, enhances the assembly time tolerance of the spread glue. At the same time, it provides a bench mark for judging mobility of the glue in later post-mortem observations.

It should also be noted that application methods that are sensitive to the quality of the veneer in terms of roughness, waviness, or uneven thickness, notably double roll spreaders, are most likely to signal the presence of such quality and alert personnel to remove it from the line or make allowances.

The sensitivity of roll spreaders to quality of the veneer is due partly to the need to apply pressure on the veneer both to draw the veneer through and to aid in transfer of glue from roll to veneer. Uneven veneer surfaces influence this roll pressure with the result that the amount of glue deposited varies; high spots or thick veneer receive less glue, and low spots or thin veneer get more glue than targeted.

To minimize the effect of veneer variation, the rolls are sometimes covered with a resilient material, or air pressured so they have some "give" to conform with the changing surface elevations and still provide "draw" on the veneer. Grooves on the rolls are specially designed for low or high spreads, and for low- or high-consistency glues.

Another factor applying chiefly to roll spreaders is the opportunity for volatile fractions in the glue to evaporate off the rolls. While not a serious problem in steady operations, when the roll is left running during interruptions, thickening of the glue can occur with the result that more glue is applied when work resumes. Thickening can also occur due to continued pickup of fiber and dust from the surfaces of the veneer, or from foam formation as the turning rolls pull air into the nip roll hopper. Many of these effects are alleviated by feeding fresh glue constantly to the rolls, drawing off the overflow, straining, and recirculating.

Most of the other methods are not sensitive to veneer variation, applying the set amount of glue no matter what the condition of the veneer. The chief objective of all ap-plication methods is to place the glue on the surface in the prescribed quantity and uniformity. It is common practice to increase the amount of glue on rough surfaces or in operations where assembly time is inclined to be long or the veneer on the too-dry side. This amounts to paying for rough veneer with glue, or buying assembly time with glue. Glue spread is also higher in cold pressing than hot pressing for a number of reasons, including the need to maintain a greater degree of fluidity throughout the assembly time because there will be no heat to lower the viscosity. Also there will be no heat to lower the crushing resistance of the wood, and therefore there will be less compliance and less proximity of the mating surfaces; all of which means there will need to be more glue solids on the bond line in cold pressing to fill the added space between surfaces.

In a typical softwood plywood operation (Figure 8–6), glue is applied to both sides of individual pieces of crossband veneer, which is then assembled and interleaved with unspread faces and backs to form assemblies in the number needed for a single press load. In conventional processes this is a time-consuming, labor-intensive area, hence it is a target for development of automated lay-up systems and mechanized handling of the veneer.

Assembly Time. In veneer gluing as in lumber gluing, assembly time produces good effects as well as bad effects depending upon the glue, the condition of the veneer, and the process. The action in all cases is the same: an increase in viscosity of the glue as it lies on the wood surface. Optimum times occur when the consistency of the glue on the glue line is in accord with mobility requirements when pressure arrives. So many factors are involved in this optimum that it is difficult to prescribe it except through experience.

In most instances, assembly time limits, both minimum and maximum, have a stated practical range from which deviations may be made as experience teaches. Limits on the long end are usually more crucial than limits

Figure 8-6   Glue spreading in plywood manufacture (USFPL).

on the short end. The following special cases are examples of situations where assembly times other than normal should be considered.

- With thin, highly figured veneer and liquid glues, a longer minimum assembly time tends to reduce bleed-through, particularly in hot pressing, but also to a lesser extent in cold pressing.
- Rough veneer, and veneer that has deep and profuse lathe checks, is usually given a higher than normal spread of glue. Increasing the minimum assembly time may then be necessary to allow the additional water to be dissipated before it has a chance of causing excess mobility in a hot press.
- Because of its effect on rate of water absorption, moisture content of the veneer strongly influences assembly time. Low moisture content shortens assembly time, while high moisture lengthens it. How much to shorten or lengthen the assembly time depends upon the construction, species, amount of glue spread, formulation, and method of pressing. Therefore only an on-site experiment with assembly time as the main variable will reveal the permissible time period in a particular case. It should be noted that changing the assembly time is one of the on-line maneuvers that can be instituted when there is an unexpected change in moisture content of incoming veneer.

- Sapwood and heartwood as well as springwood and summerwood also affect differently the rate of water absorption from the glue, and allowances in assembly time would seem prudent. Unfortunately, the more absorptive springwood and the less absorptive summerwood occur in alternating bands on veneer, and optimizing both at the same time is impossible except within a narrow interval of time where the maximum for springwood might overlap the minimum for summerwood. Otherwise the problem remains in the lap of the adhesive formulator to devise the most appropriate adhesive properties for both wood types.

With regard to the rate of water absorption of sapwood and heartwood, since heartwood dries (desorbs) at a slower rate than sapwood, it can be assumed to absorb at a slower rate. Consequently on veneer at the same moisture content, a glue film can be expected to dry faster on sapwood than on heartwood. Therefore a shorter maximum assembly time might be necessary for sapwood, especially in cold pressing where mobility actions of the glue are more dependent upon fluidity. In hot pressing, the concern shifts to heartwood where a longer assembly time may be needed to reduce the water content of the glue line and prepare it for the additional mobility inducements of heat.

Surface activation treatments have the potential for producing very absorptive veneer due to the increase in surface energy they are designed to produce. Consequently, surface-activated veneer may require considerations similar to sapwood veneer.

• Temperature has three different effects on the mobility of glue during assembly times. One is due to the higher rate of evaporation and diffusion at higher temperatures, another is due to the higher rate of polymerization at higher temperature, and the third is the general effect of high temperature on lowering viscosity. Converse situations produce converse effects. Temperature effects can arise either from the atmosphere or from the veneer, from hot or cold days, or from hot or cold veneer. Adhesive manufacturers often account for seasonal temperature differences by offering different formulations for winter and summer use. Veneer that is used too soon after drying will often contain heat that can affect the condition of the glue during assembly times.

• Low relative humidity in the workplace—which mostly affects operations needing long open assembly times—occurs in northern indoor climates during the winter. Two recourses are available: shorter assembly times or humidification.

In cold pressing, the upper limit of assembly time is always determined by the loss of inherent fluidity of the adhesive. Since only one surface of a bond line is spread with glue, the worry is that transfer to the unspread surface may not occur, and if it does, that penetration and wetting may be inhibited.

In hot pressing, the inherent fluidity is supplemented by plasticization due to heat. This means that longer assembly times can theoretically be tolerated in hot pressing. In practice, this is the case only with neat resins, and only when they are not heavily catalyzed. When no extender is present, a properly formulated resin can tolerate days or months in assembly time (resins that might be used to produce dry film glues, for example). Extenders severely limit assembly times. Since they generally have poor flow properties themselves, they restrict the mass actions of the glue; increasingly so as glue water is lost to the surroundings. Special additives are available to prolong assembly times of flour-extended formulas over a weekend.

Assembly time has an interesting effect on hot-press glues comprised of mixtures of resin with added fortifiers or extenders. Differential motions of the ingredients may occur during pressing that are related to assembly time. For example, when a urea or phenolic resin is highly extended with various kinds of flours in hot-press formulations, the bond is more resistant to soaking if made at longer assembly times than at short. Presumably, within the mass movements of the glue, the resin fraction has experienced some independent movement, migrating toward and into the wood surface, thus impoverishing the glue line. In a sense the glue line ends up with a higher proportion of extender than intended. At best there remains a resin gradient in the glue line—rich in resin at links 4, 5, 6, and 7, and poor in resin at link 1, which then becomes the weak link. At worst the extender is left pretty much alone on the glue line, the resin having filtered into the wood—the classic example of a filtered glue line. As assembly time increases, the reduced mobility of the glue encourages the ingredients to remain together and move together. Of course, the remaining mobility must sustain all the motions necessary in bond formation.

A similar but less noticeable effect occurs when a urea resin is fortified with melamine resin to improve its durability. Better results are obtained at long assembly times than at short; boil resistance improves with assembly time, while dry strength remains essentially the same. The implication is that the loss of boil resistance at short assembly times can be attributed to reduction in melamine content on the glue line due to its more facile mobility.

Assembly time also creates a problem around the edges of a lay-up that is both annoying and detrimental to bond quality. Water from the glue is absorbed on the inside surfaces of the outermost plies. These inside surfaces then swell rather rapidly, causing the veneer to curl away from the core. This action not only exposes the edges of the glue line to the more rapid moisture loss of open assembly time, but also makes it difficult to place the assembly into the narrow opening of the press. If not handled carefully, splits and overlapping can occur. Various aids are used to facilitate insertion into the press, the most common of which (particularly in hardwood plywood) is the placement of steel or aluminum cauls over and under the assemblies. These keep the veneer from curling and also aid in handling multiple panels into an opening. Another advantage of cauls is protection from precuring of the first assemblies into the press while the rest are being loaded.

In hot pressing softwood plywood, another recourse taken during assembly time is prepressing, where an entire bundle of assemblies scheduled for one press load is placed under pressure in a manner similar to cold pressing. During this time under pressure, the glue line achieves sufficient tack to keep outside veneers from curling. At the same time an important adhesive action—transfer—is being assured, and some penetration and wetting are getting an early start.

In veneer gluing, the question of open assembly time normally does not arise because, unless there is a sudden shortage of faces and backs, the assembly is completed almost instantly. However, portions of a glue line may not be completely closed due to wrinkles and bulges in the veneer. Particularly around edges, the glue line may suffer premature drying or skinning over during long but otherwise tolerable assembly times. The use of cauls or prepressing avoids this problem also.

Edge Gluing Veneer. Veneers are spliced either with tape, with string, or by edge gluing. In the latter case machining edges and applying glue often occurs in the same operation. The first step is to produce a straight, square edge. This is done on a jointer specially designed to hold veneer while a cutter is dressing the edge. Two approaches are in use: one where a bundle of veneers travel on edge over a stationary cutter; the other where the stacks of veneer are stationary, held under pressure while a cutter passes, dressing all edges at once on one side of the bundle. The bundle is then reversed and the edges on the other side are dressed. In both, glue is applied by rolls following the cutters. Glue can also be applied by a thin disc positioned so as to wipe glue onto the edges of the veneer as they are being fed into the pressing system.

The glue in both cases is of a type that can be allowed to dry before the edges are brought together for bonding. Generally this will be a urea or melamine type that produces an invisible glue line. They are compounded so they can be fused with heat at a later time. Since glue will usually have been applied to both surfaces of the joint, most of the bond forming functions will already have occurred, and there remains only the fusing and curing of the films to form the completed bond.

Flakes

In the early development stages of flake gluing, the adhesive was applied by roller coater. One can imagine the problems that had to be overcome: uneven application, high moisture addition, and clumping of the flakes as they were being distributed into a mat. Through an ingenious combination and organization of wood elements, however, suitable panels were nevertheless made with remarkable strength, smooth surfaces, and low density.

The glue line in this early case was more of a continuous film. It produced the ultimate in bonding between flakes, but the cost in excess resin compelled the introduction of alternative means of applying the binder.

Spraying was the obvious answer, adopted and adapted from particleboard, the precursor of the true flakeboard. In particleboard, resin drop size and distribution was not a critical factor because the granular furnish could be tumbled in a mixer until the resin was wiped from one particle to another and finally became uniformly distributed. In flakeboard, on the other hand, only limited tumbling can be permitted because of the need to avoid breakage of the fragile flakes. Hence drop size and distribution had to occur "on the fly."

Many methods have been developed to accomplish the atomization of the adhesive and its placement on the flake surface in a uniform manner. Three general methods are in use: air spraying, airless spraying, and disc spraying. In air spraying—the process commonly used in paint and finish applications—high pressure air jets impinge upon a stream of resin emerging from a nozzle and break it into droplets. Air and resin then are propelled toward the target. In airless spraying—such as that used by garden sprayers—the liquid is forced from a small orifice under high pressure. Upon emerging to atmospheric pressure, the stream breaks up into small drops spontaneously. The latter process eliminates some problems that air creates: turbulence, disposal of the air without loss of resin, and overspray. Disc sprayers employ a high-speed spinning disc upon which the resin is fed. Centrifugal force moves the resin to the periphery of the disc, from where it is spun off in small drops.

In the first method, atomization depends primarily upon pressure of the air, in the second upon pressure of the resin; both are also influenced by the size and configuration of the orifices through which air and resin emerge. In the disc, speed of rotation controls drop size. Viscosity of the resin is, of course, an important factor in each case; the lower the viscosity, the finer the atomization. Hence all factors that control viscosity—dilution, molecular size and configuration, solvents, and temperature—play in the reduction of the resin to droplets.

Once the resin drops have been formed and are airborne, the flakes must be properly presented to receive them. This is done in *blenders*, whose function it is to provide movement of flakes (or other furnish) past or through the spray pattern. One of the simplest types (Figure 8–7) is a drum with vanes deployed around its inside periphery, and a boom carrying the spray nozzles extending into the central areas. As the drum turns, flakes are carried up the sides until they fall off the vanes in a cascade to be hit by a shower of resin drops. In another type the drum is stationary and the flakes are propelled past the resin spray by revolving paddles. In both cases the objective is to give each flake an opportunity to be engaged by the spray.

Blenders can be either batch type or con-

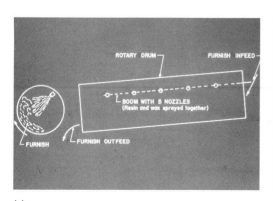

(a)

(b)

Figure 8–7 (a) Drum blender for applying resin to flakes; (b) short-retention-time blender.

tinuous. In batch blenders, the resin and flakes are weighed separately and charged to the blender. This assures an accurate ratio of components. In continuous blending, wood and resin are metered into the blender in a continuous flow. Accuracy depends upon proper functioning and calibration of the meters to control the amounts of each ingredient entering the blender.

Two factors control the quality of the application procedure: the amount of flakes being processed at one time, and the length of time the flakes are exposed to the spray. The more flakes there are in the blender, the greater the likelihood that some flakes will be in the "shadow" of others as they pass in front of the spray. This will increase the number of flakes that receive no resin. Prolonging the blending time to overcome the shadow effect risks more breakage, and also may incur overloading with resin.

In the case of applying resin in powder form, the process is greatly simplified. Powder and flakes are charged to a blender in the amounts calculated, and after a brief churning are ready to be composed into a mat. After blending, the furnish is delivered to a hopper where it can reside for a relatively long time before assembly time limits are exceeded. However, because the resin is rather weakly held to the surfaces of the flakes, the amount of handling must be kept to a minimum to avoid losing binder.

At this stage in the manufacturing process, it is possible to incorporate various chemical additives for the purpose of increasing some property of the composite. Wax is the most common additive since it controls to a significant degree the ingress of water into the panel while in service, and thereby improves bond and panel performance. Paradoxically, wax also aids the bond-forming process; in the case of liquid resins by promoting retention of the drop form, and in the case of powder resins by holding the powder on the surface during application and handling. Other additives include fire retardants, preservatives, and dimensional stabilizers.

Figure 8-8 shows qualities of resin cover-

age on the surface of manufactured panels. In (a) globs of resin appear with normal resin spots, a sign of inadequate cleaning and maintenance of the blender, and there are flakes without resin, suggesting that the blender was overloaded or the dwell time too short. Panel (b) exhibits the ideal uniformly distributed droplets of resin.

It should be noted that flake gluing is the equivalent of "double spreading," where both surfaces in a bond can carry glue into the glue line. Hence the likelihood of a glue line having no glue at all is rather small.

Particles

Blenders appear to be similar to those used for flakeboards, but much more energy is expended in tumbling and mixing the furnish to ensure uniformity. Since particles have already been subjected to severe punishment in a hammermill, further punishment in a blender can do no more harm. Consequently, considerable force can be used in distributing resin binder on the particles (Figure 8-7(b)). Both the speed of mixing and the duration can be increased to effect a proper distribution of resin. As in flakeboards, the application of a liquid resin does not cause clumping because the moisture is instantly absorbed into the wood.

The uses to which particleboards are put do not demand the ultimate in binder durability as do flakeboards. Therefore it is logical to use the lower cost and faster setting amino resin, urea formaldehyde. The other amino resin, melamine formaldehyde, as well as phenol formaldehyde are also used but to a less extent and only when greater durability is needed.

Fibers

Since fibers are converted to products either in wet processes or dry processes, differences exist in the manner of applying binder and other additives. In the case of wet processes binders (if any), sizing agents, and other chemicals are added to the slurry and simply stirred. They are introduced as solutions or emulsions, and then caused to precipitate

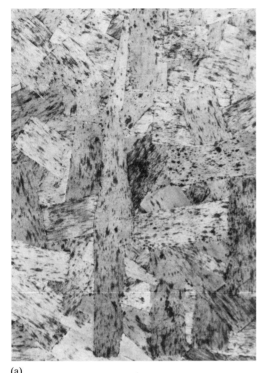 

(a)   (b)

Figure 8–8   Resin coverage as observed on the surface of flakeboards: (a) well-blended furnish, (b) poor blending—note resin glob at left center and light coverage on flake at upper left.

onto the fibers by changing the pH. Although some is lost in the water that is later drained away, additives become well distributed on the fibers.

In dry-process fiberboard the problem of adding a very small amount of resin to a very large area of comminuted wood generated a number of solutions. One of the simplest and most novel is the use of the attrition mill as both a fiberizer and a blender. The resin is either applied to the chips before they enter the mill or the resin is dribbled in simultaneously. In either case, the intense activity between the grinding plates insures good distribution on the fiber.

In a second approach, the resin is applied in a blender similar to that used in particleboard; the fibers are churned vigorously by paddles while the resin is sprayed on.

In a third process, the resin is sprayed on the fibers while they are being transported in an air stream to the felter—"blow pipe" application. In all three cases, other additives are incorporated in like manner. Good and frequent cleaning are requirements in each case to avoid the formation of clumps of fibers or chunks of resin that would produce blemishes in the finished panel.

SECONDARY AND TERTIARY GLUING

In the assembly and subassembly of wood parts, application procedures are often adapted to the construction being glued. Depending on the size of the operation, the mechanics may range from powered rolls to brushes, hand rollers, scrapers, and even dipping. Assembly times tend to be very short except when some predrying is needed. Since the construction controls this part of the process, it is addressed more fully in Chapters 7 and 9.

9

Consolidation

In previous chapters we have described and analyzed most of the factors in the equation of performance. Here they are summed through the crucial interactions that consummate the bond and establish the performance of the product. The latter awaits the final test of service conditions to determine conclusively how well the factors have expressed themselves on the glue line in joining two pieces of wood. At this time the concern is mostly with consolidation, the process where the surfaces are brought together and the glue caused to harden. The act primarily involves applying pressure and sometimes heat.

As before, the context for this stage is the basic wood elements and their assembled structures, many of which, as we have seen, invoke unique technologies. Again, as before, the objective is to show how principles apply, rather than show the specifics of how to conduct the operation. (The latter encompasses so many forms and involves so many different mechanical approaches that separate books are needed to cover each class of product at the production level.) The fact that process mechanics derive from the proper application of principles implies that though processes differ, principles remain the same; though circumstances vary, a bond must be formed between two pieces of wood. In the final analysis, the bond formed with whatever adhesive and whatever process is dependent primarily upon the environment existing between two wood surfaces.

In all processes, after the wood has been prepared for size, surface, and moisture content, and the adhesive chosen, prepared and applied, the same three operations ensue:

1. Application of pressure to bring surfaces together
2. Application of heat (when required)
3. Postconditioning to equalize any gradients and restore equilibrium conditions

Within these three operations the entire gamut of adhesive actions occur in all the many configurations of wood elements and structures that exist.

Since the quality of the bond at this point is dependent upon the environment between two wood surfaces, our attention is focused there for each product and process, noting the differences that exist and observing the way they play on the adhesive. We are dealing primarily with a space or volume between two surfaces, characterized by an area and a height or thickness (gap). Area dimensions of the glue line have a bearing on adhesive actions, both in terms of edge effects and in terms of heat and moisture gradients that develop in hot pressing. The area may have one or both dimensions short or long, but fairly constant within a product item. The thickness of a glue line introduces one of the more fractious factors in bond formation because it directly affects how the glue functions.

Size and shape of the glue line enters the equation of performance indirectly through edge effects and center effects. Edge effects are always at play, not only in bond formation where pressure, surface qualities, and assembly time actions are most likely to be unfavorable, but also in bond performance where stresses are most likely to develop or concentrate, and where environmental factors most readily attack. For example, in the edge gluing of veneer, the glue line is so narrow that it may be considered dominated by edge effects. On the other hand, in consolidating large flakeboards (such as 8 × 16 feet), the center region dominates bonding conditions because of the extremes in moisture and temperature effects that concentrate there during hot pressing. In this case, unfavorable edge effects become part of the trim and are discarded, representing more of a cost (waste) than performance of the product. Size and shape of the glue line is one of the principle reasons that results from laboratory scale experiments sometimes do not match industrial scale experience.

While size and shape are constants within a product line, gaps vary not only within a product but also within and between glue lines. Because the gap varies, pressure that arrives on the glue varies and pressure-dependent actions therefore also vary. This gap, which really defines how close two surfaces come together, is an enigma. There are small gaps and large gaps, positive gaps and even negative gaps, all related to how well the surfaces are prepared, how much pressure is applied, and how compressible the wood is. Compressibility may either be inherent in the wood species or modified by moisture and temperature. Negative gaps are produced when so much pressure is applied that elevations on one surface are pushed into the opposite surface, breaking through the boundary in between. One never knows how big the gap is unless it is near an edge,

nor what the effects are unless they are so large as to cause early failure of the bond.

Adhesives differ as to the gap or thickness of glue line they can tolerate. True gap-filling adhesives are those with 100-percent solids, such as epoxies, polyesters, and urethanes that have no solvents to evaporate after hardening to cause shrinking within link 1. All waterborne adhesives undergo some shrinkage after hardening. The degree of shrinkage determines to a large extent the magnitude of the gap they will tolerate. How they shrink and how plastic the bond is also has a bearing on gap filling. Adhesive functions that are dependent upon pressure are the main victims of gaps. In general, a gap of about .005 inches or less is considered optimum for most glues, whether gap-filling or not.

Pressure has two main functions in bond formation: to bring surfaces together, and to aid in penetration and wetting of the adhesive. Other functions include expressing air, and sometimes water, from the glue line, and promoting flow and transfer of the adhesive. Pressure on the glue film during hardening also aids in developing maximum cohesiveness in link 1, particularly when large amounts of fillers or extenders are incorporated. Since the adhesive on the glue line loses volume as it hardens due to shrinking and penetrating, pressure is needed to compensate by reducing the distance between surfaces. The actual pressure needed by the adhesive to perform its functions varies with its chemical and physical properties. With most glues that are applied in liquid form and pressed while in a fairly mobile state, the amount of pressure needed is relatively small, on the order of 10 to 20 psi.

Although the strength of joints generally increases with increasing pressure, there are limits, first on the amount of adhesive that might get squeezed from the glue line, and second on the amount of crushing and deformation that is acceptable in links 6, 7, 8, and 9. The latter limit deals with the weakening effect on the wood, as well as the recovery from compression effect. Except for gaps that are caused by bending or warping deformations, the closing of all gaps involves crushing wood at the high spots. Bending and warping problems occur mostly with lumber gluing, and to some extent in secondary and tertiary gluing. The problem in the case of lumber must be attacked at the wood-preparation stage, either to eliminate warp or to produce true surfaces despite it. The latter is not the best solution because the potential for warp remains and becomes built into the glued product. When warp deformation enters the gluing operation, pressure becomes an even greater variable because some unknown amount is consumed in straightening out the lumber. This amount is withheld from that part of the glue line that resisted closure, and added to that part where closure occurred easily.

Moisture and temperature are the two other major glue line environmental factors. Both can vary before, during, and after bond formation. In the preparation of the wood before bond formation, variations are produced in the course of seasoning or storage. During bond formation, glue water adds to water already in the wood, and when heat is involved, gradients develop that affect glue actions accordingly. Moisture and temperature also affect wood properties, mostly in a transient manner but some with lingering effects. During bond formation the effect of moisture and temperature on wood properties is beneficial because it aids conformation of the surfaces. After bond formation, moisture and temperature comprise the two most influential factors controlling the performance of a glue line or the product it is in. For this reason these two factors are the main elements of evaluation procedures for determining bond performance where high durability is expected.

The wood bounding the glue line plays a commanding role throughout the gluing process, not only in terms of its properties (Chapter 5) but also in terms of its dimensions (Chapter 7). How the glue is applied is partially determined by the dimensions of the wood element, but beyond that, the wood back of the glue line acts like a sink for absorption not only of glue water but also of glue, heat, and pressure. At the same time it acts like an insulator of heat and a barrier to the rapid transport of both heat and water.

In this varying, unsteady environment, a liquid glue is asked to anchor itself to the wood, chemically and/or physically, bond itself to itself, harden and form link 1, and then hang on as stresses of various kinds and sources course along and across its thin domain.

The important thing to remember is that the environment we have been discussing is the central difference between gluing processes, and this difference must be clearly visualized if adhesive performance is to be understood. In all the processes, it is also well to bear in mind that the requirements of the adhesive dictate the procedures. The consequences on the wood are of secondary importance in current practice, though penalties may follow, such as compression losses and springback. Accordingly, as the various processes are discussed, the emphasis is on the glue line environment and the consequences that arise there.

PRIMARY BONDING

The reaggregation of the primary wood elements into useful products, chiefly linear or planar in nature, represents one of the major avenues for adding value to wood. For many products, this is the first step in the conversion chain toward ultimate utility as a functioning assembled product. It is also the first time that glue meets wood; presumably there is something pristine about it, but this is generally lost in commercial operations. Seldom does the glue see the real wood, but rather some version modified by the methods of preparing the wood elements. Only the craftsman preparing a solid piece of wood for gluing by using a plane or a shaving tool has a chance of offering true undamaged wood for the glue to embrace. In exploring the consolidation of the wood elements, it will be seen again how differences have entered the gluing process through their geometric configurations. The exploration proceeds down the path of diminishing dimensions, beginning with lumber and ending with fibers. The processes are covered generically as representative of many embodiments where size of the operation, equipment, and species may differ but the principles remain the same. The focus remains as much as possible on the glue line and its immediate environment, influenced primarily by the geometry of the wood element.

Lumber Gluing Processes

Within the family of glued lumber products, three separate configurations occur, as noted in Chapter 2: edge to edge, face to face, and end to end. (Special cases of edge to face, and end to face or edge, also occur and are included in a later section as part of assembly gluing processes.) Each configuration presents a different glue line situation and generates its own context of factors affecting bond formation and bond performance. The differences are emphasized by the fact that the three processes are sometimes carried out as independent operations, although end gluing is often incorporated into face gluing when that product needs to be lengthened.

In all cases a dominant consideration prevails throughout: the very large mass of wood engaged with each glue line, the largest among wood elements. The properties of wood therefore control outcomes to a greater degree than with other wood elements because of the relatively poor conformability of lumber elements. This is due primarily to their greater dimensions, which adds the mechanical property of stiffness to that of compressibility in achieving proximity between mating surfaces. The greater difficulty of gluing oak compared to gluing yellow poplar is partly a consequence of the influence of these properties. Operationally, this means greater attention must be paid to preparation of the element, particularly its surface. The degree to which these factors are attended determines a large share of success in bond formation. In some respects lumber is the most difficult of all the elements to glue together.

The interplay of these and other factors is brought out in the discussion that follows as the essentials of each process are scrutinized.

Edge-to-Edge Lumber Gluing. Considering first the dimensional aspects of the situation, the width of the glue line is relatively short, commonly one inch going up to two inches, rarely more, and seldom under one-half inch. This short dimension magnifies edge effects not only because they represent a greater percentage of the total glue line area, but also because the short dimension makes the effects more likely to be negative since fluid pressure that exists on the glue at all edges is low. What happens at the edges can also have an effect on what happens at the center; excessive squeeze-out, for example, can drain glue from the center regions of the glue line. On the other hand, the short dimension makes the glue line more accessible to external influence—heat to accelerate bond formation, for example, or moisture to degrade it.

The amount of wood back of (perpendicular to) the glue line may be considered infinite in edge gluing as far as glue line action is concerned. The capacity to absorb glue water is ample. Because of the high stiffness that accompanies this dimension, achieving conformity of surfaces against major deviations from the plane of the glue line may be difficult. However, this same stiffness insures that pressure received by the boards from individual clamps will be more uniformly distributed to areas between clamps.

The dimension of the wood perpendicular to the glue line also figures strongly in the *performance* of the bond under service conditions due to differential dimension changes at the ends. It is not known what the maximum limit of this dimension is from a stress standpoint, but a practical limit seems to be four inches, with some consideration of species, decreasing for high density and increasing for low density. The same limit holds for shape stability; the tendency to cup increases with width of the individual board.

The dimension of the wood perpendicular to the glue line has an indirect effect on bond formation that arises often enough to warrant mention. The mass of wood represented by this dimension can be either a heat sink or a heat bank as dictated by its temperature. Depending upon the mode of hardening, some glues would have trouble on wood that is too warm or too cold as it enters the gluing process. Adjustments can sometimes be made in other parts of the operation, for example by shortening the assembly time to take advantage of warm wood in speeding the hardening of the glue. Double preheating (preheating both surfaces in order to speed

curing of the glue) is an extreme example of this modification.

Pressure Application. Clamping devices to apply pressure while the glue forms a bond are many and varied, as shown in Figure 9-1. They include pipe or bar clamps, bench clamps, clamp carriers, continuous presses, and dielectric presses. Each provides pressure perpendicular to the glue line, and some provide additional pressure to maintain the panel in a flat aspect during the pressure cycle. Bar clamps are the simplest and most

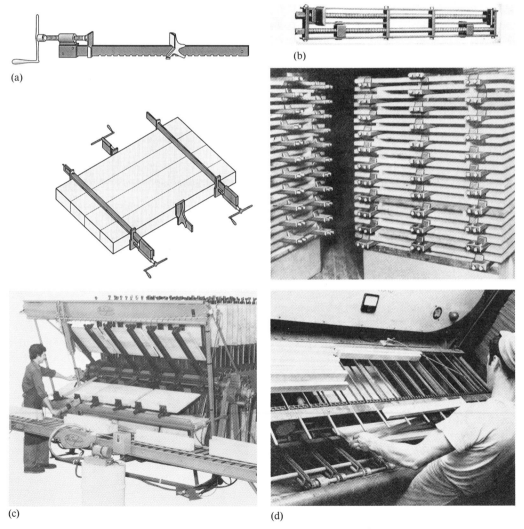

Figure 9-1 Clamping devices for edge gluing lumber into panels: (a) bar clamp, (b) bench clamp, (c) clamp carrier, (d) continuous clamp. (Taylor brochure)

versatile but also the most easily misused. Because pressure is necessarily applied eccentrically by each clamp, they tend to bow the panel unless they are distributed equally on both sides and tightened equally. Bench clamps are made to clamp first to a bench to ensure flatness, and then to successive layers of panels. Bar and bench clamps are used mostly for small shop work.

A clamp carrier employs the same mechanics as bench clamps, but the clamps are arranged like the spokes on a wheel. They are meant to mechanize the operation and eliminate the handling of many clamps. As a clamp is filled it indexes forward, bringing the next set of clamps into position bearing a completed panel. Unloading and reloading thus proceeds at a more scheduled pace, and with appropriate timing the wheel can keep going without stoppage; solidification of the adhesive is achieved during one turn. The carrier is sometimes enclosed and heated to speed the cycle time. A danger in heated enclosures is possible drying at the ends of the panels, with shrinkage and open joints greeting the operator at the completion of the run.

Since the pressure in these clamps is developed through a screw, the exact amount delivered depends upon the torque applied to the screw. Often this is done by hand through a lever, sometimes aided by blows with a maul. In the more sophisticated operations, torque is produced with a torque wrench or an air wrench, both of which can be monitored.

Continuous edge gluing presses have a different approach to the application of pressure. By substituting friction on the faces of the panel for the restraining end of the clamp, boards can be fed in one end and pressed through the friction system while advancing the entire panel through the press. The friction maintains pressure while the panel is progressing through. To speed the process, hot air is circulated around the panel as it moves forward. Under high-production schedules, heat has time to penetrate only a small distance in from the edges of the glue lines. For appropriately catalyzed glues this is sufficient to create a bond near the edges strong enough to hold the panel together while it is being set aside to complete its cure at the center. Since the plane of the glue line is always perpendicular to the direction of travel, panels can be made endless for width, or individualized by withholding glue from certain edges.

Dielectric presses are essentially batch presses producing individual panels. Their chief characteristic is very fast curing of the glue, less than a minute in most cases. Consequently they are geared for fast loading and unloading. Contrary to the continuous press, the plane of the glue line is parallel to the direction of feed travel, making it possible to bring pressure in from the sides. In operation, panels are assembled out of the press and pushed into position. A top pressure is applied to prevent buckling, and then the full bonding pressure is applied from the sides.

All of these pressure devices have only one primary function: to deliver a force perpendicular to the edges of the panels. This they do with virtually equal efficiency. However, the quality of the output can differ markedly. One reason lies in the differing interactions of machine with board variables. For example, a board with an out-of-square edge, when pressed against a board with a square edge, will have a tendency to jackknife as the two surfaces accommodate to each other without lateral restraint as in bar clamps (Figure 9-2 (a) and (b)). A good bond will form but the panel will not be flat. If jackknifing is prevented by applying top pressure, or is beaten down with a maul, the boards will lie flat but the joint will be open, (c), on the lower side (out of sight of the operator, and not discovered until resurfacing operations). If the open joint is on top, (d), and observed by the operator, he is inclined to increase the edge pressure until the joint closes, (e). In doing so he will have crushed the wood at the lower edge of the joint, and perhaps achieved a functioning bond. However, compressive stresses have now been built into the joint along with a starved condition. These may release at a later time and

(a)

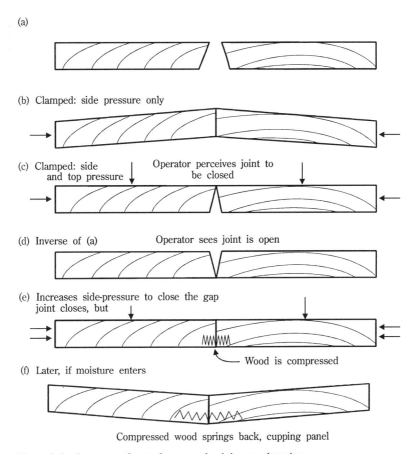

(b) Clamped: side pressure only

(c) Clamped: side   Operator perceives joint to
    and top pressure   be closed

(d) Inverse of (a)   Operator sees joint is open

(e) Increases side-pressure to close the gap
    joint closes, but

    —— Wood is compressed

(f) Later, if moisture enters

Compressed wood springs back, cupping panel

Figure 9-2   Response of out-of-square edge joints to clamping.

cause the panel to warp, (f). Most high-production presses provide top pressure to keep panels flat while being pressed. In these processes open joints are inevitable with stock that is jointed out-of-square. Increased attention at the jointer is needed to avoid the problem.

Gaps due to inaccurate machining, warping, or moisture-induced dimensional changes are the bane of lumber gluing in general, and edge gluing in particular. The stiffness of lumber in the width direction (perpendicular to the glue lines) means that gaps due to crook require a very high force to close. The high force becomes concentrated at both ends of the gap, as in the footings of a bridge, with undesirable consequences both for the wood and for the glue at those points. The use of pressure to compensate for poor-fitting joints creates more problems

than it solves. First, wood that has been crushed on the glue line has expansive tendencies, inducing tension perpendicular to the glue line. Second, wood that has been bent edgewise retains the bending force within the elements. (See Figure 5-35). Like a spring, its restoring energy translates later into more tension stresses across the glue line. Tension perpendicular to the glue line, it should be recalled, is the weakest direction of a joint. While the high clamping forces are building stresses or compression failures into the joint, the extraordinarily high pressure at high spots is weakening the glue line by causing too much glue to flow away from those regions, starving them of adhesive. Thus high spots receive too much pressure and low spots not enough.

In a high-frequency dielectric press, these gaps, plus an entirely different kind, interact to produce low-strength bonds. The schematic in Figure 9-3 shows an assembly of boards being pressed with electrodes top and bottom. The electrodes are intended to touch the glue lines or the squeeze-out and provide a higher conducting path for the current than the wood, thus heating the glue line but not the wood. Where an air gap occurs between the electrode and the glue line due to a thin board, the flow of current into that glue line is reduced and it will remain undercured at the end of the press cycle. Gaps within the glue line suffer for the opposite reason: too much current. The result is "arcing" or burning in the glue line. The resulting char is even more conductive, and the process becomes self-feeding—the worst that can happen. The glue in the gap can also boil and form into a weak froth as it cures.

One might wonder how a glue line that under normal conditions would take four hours to cure differs when it is caused to cure in less than a minute. One difference can arise from the very fast temperature rise: boiling temperature in a few seconds. The possibility exists for drastic effects on the five actions bond formation requires. Mobility experiences a sudden, short burst upward due to the lowered viscosity accompanying the temperature increase, augmented by the expansive pressures generated in the glue line by the boiling and volatilizing glue water. On porous woods some of the glue may be blown into the pores, resulting in increased penetration; increased squeeze-out may also occur. Bubbles invariably form in thick glue lines as in gaps, leaving link 1 in a fragile condition. Wetting action is unaffected and may actually be improved. Recourse to the negative effects is to be found in factors that reduce mobility. These include: (1) reducing the water content of the glue, which reduces the entering viscosity as well as the boiling vapor pressure; (2) lowering the electrical current so as to reduce the rate of temperature rise; and (3) increasing the assembly time to allow some glue water to leave the glue line.

In addition to these grosser motions of the adhesive, an adverse effect on quality of solidification can be postulated. With the virtually instantaneous loss of glue water, reactions between adhesive molecules that lead to a high-quality cure may be curtailed, leav-

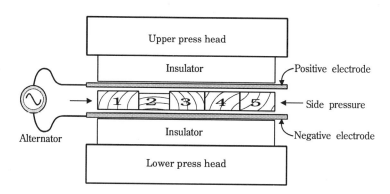

Because electrodes make direct contact with conductive glue lines, most of the electrical energy concentrates there and heats only the glue.

Glue line between boards 2 and 3, however, will be undercured because of electrical gap caused by thin board. (Only one edge of each board receives glue.)

Boards of high moisture content and higher density draw more power, reducing energy to glue lines, and causing undercure.

Figure 9-3   Wood factors in a dielectric edge-gluing press.

ing the glue less resistant to degradation though perhaps sufficiently strong to function as a bond.

Another wood/machine interaction that occurs with dielectric presses and not with cold clamp pressing is the effect of wood moisture content. Boards with high moisture content draw more power from the electrodes than wood of low moisture content. Power that goes into the wood is power lost to the glue line. Consequently a panel containing high-moisture boards will have undercured glue lines at the end of the press cycle. This condition is easily observed by a sweep of the hand across the panel immediately out of the press. Boards that are warm to the touch are high in moisture. In this sense, the dielectric press is serving a quality-control function because there could be trouble ahead in the form of warping or splitting even if the glue had completely cured.

Post-Conditioning. As panels come out of the pressure cycle there may be residual moisture gradients or temperature gradients that ought to be stabilized before further processing occurs. In high-speed operations where the press controls production, there is a strong temptation to remove panels from the press before the bond is fully formed. This is done with the understanding that a partial bond will hold the wood together if it is not highly stressed. The glue can then complete its cure outside the press. As a general rule, it is often assumed that three-fourths of the total cure should occur in the press, and one-fourth can occur out of the press. Parenthetically, with some glues, maximum strength occurs before final cure; additional curing mostly enhances durability. Accordingly, panels are placed in a stack and given a period of two to four days to further cure and recondition. This reconditioning time also allows moisture added with the glue to diffuse, thereby reducing the incidence of sunken joints. Stickers between the panels improve air circulation and prevent panels from bonding together at points of squeeze-out.

The sunken joint (Figure 9-4), one of the more common complaints with edge-glued lumber, is a depression that sometimes forms over the length of a glue line and is especially noticeable in highly polished surfaces, even through a crossband and face ply of a lumber core panel. The problem arises from the effect of glue water on the swelling and subsequent shrinking of the wood adjacent to the glue line (Figure 9-5). Although other causes exist, such as differential abrasion of the glue line during sanding and the possible stabilization of the wood by the glue in the vicinity of the bond, the majority of sunken joint problems disappear when panels are given a period of reconditioning before final surfacing.

Glues that harden by loss of water engender an additional reason for postconditioning. The squeeze-out, being a mass of glue, does not lose its water as fast as the thin film inside the joint. It thus remains in a rubbery state longer. In this condition, it fouls up planer knives and loads sand paper.

Dielectric heating with thermosetting resins generally produces fully cured glue lines while still in the press due to the intense heat generated. With latex adhesives an additional caution arises due to the plasticity induced by heat. If glue lines are under stress as they emerge from the press, some immediate opening of joints may occur. Otherwise, strength increases as cooling takes place.

General Considerations. Pervading all edge-gluing operations is an interaction between basic compositional and structural properties of wood and the processes of preparation and gluing. The hygroscopicity of wood, and the way moisture moves through wood, cause differential dimensional changes that can subvert or override the best machining or gluing practice. The subversion begins in the seasoning stage and arises from variation in moisture content between boards. Variation in moisture content within boards also results in problems but the simpler between-board variation is dealt with first.

Open joints are a dismaying sight in any panel operation. We will trace the genesis of some of these. The moisture content in a

Figure 9-4 Above, sunken joints in lumber core plywood as seen against low-angle
light. Below, lumber core plywood with properly reconditioned core. (USFPL)

charge of lumber coming out of a dry kiln may have an average moisture content of 9 percent. Assuming the distribution of moisture content to be normal between boards, 66 out of 100 boards might be between 7.5-percent and 10.5-percent moisture content. Seventeen out of the hundred boards might be below 7.5 percent, and seventeen above 10.5 percent. Of the latter group, a small percentage could be at 6 percent or less, and an equal number at 12 percent or more. The range between high and low moisture content boards and the number of boards at each level of moisture content depends upon how well the kilns have been run. Three to five boards out of the 100 may occur at each extreme.

Boards at the extremes of the range are

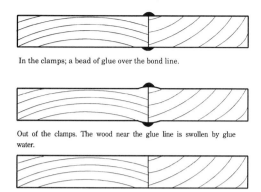

In the clamps; a bead of glue over the bond line.

Out of the clamps. The wood near the glue line is swollen by glue water.

When the panel is planed too soon, swollen wood is removed over the glue line area, and unswollen wood is removed from the rest of the surface.

As glue water later diffuses out of the wood at the bond line, this wood shrinks more than the rest, producing a depression along the length of the glue line.

Figure 9–5   The development of sunken joints in edge-glued lumber panels.

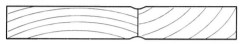

one source of later trouble. What happens in the glue room is an interaction of moisture content with time and relative humidity as the gluing process proceeds. The interactions operate through the hygroscopicity of wood substance and its tubelike structure. Relative humidity changes that are different from those that produced the EMC of any one board cause the wood to take up or lose moisture. This in turn causes changes in dimension. Because of the tube-like structure of wood, the changes in moisture content and dimension occur faster at end grain surfaces than elsewhere in the board, on the order of 10 times faster.

The speed of change at the ends of boards is surprising. The relative humidity of a rainy weekend can raise the moisture content at the ends of kiln-dried boards, neatly piled on a shop truck, by several percent, and the dimensions will change accordingly. Conversely, the same truckload in a dry atmosphere or parked near a heater could dry the ends of boards by several percent. This apparently insignificant change can cause havoc at several stages in the gluing process, particularly with those boards that begin at

an opposite extreme in moisture content from the changes taking place, for example a board at 6 percent subjected to a relative humidity that would result in 12-percent moisture content is set for trouble.

We will follow the effects on two boards from each extreme as they go through an edge-gluing process and into service. Consider first two boards from the dry extreme (Figure 9–6) left side. They begin, let us say, at a fairly uniform 6-percent moisture content stacked on a truck with the rest of the kiln charge, (a). While in the storage shed awaiting the next operation, some humid weather sets in. This drives the moisture content at the very ends up toward 12 percent or more, (b), grading inward down to 6 percent, the length of the gradient depending upon time of exposure to the high humidity. The moisture gradient produces a swelling gradient at both ends of the board, giving a fish-tail configuration, perhaps not noticeable but there nevertheless. Edge jointing the

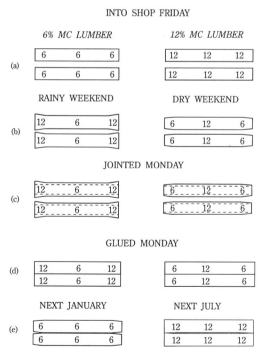

Figure 9–6   Changes experienced by boards from high and low moisture content extremes as they might proceed through an edge-gluing operation and into service.

boards in this condition produces a straight edge for gluing, (c), which is accomplished without apparent problem, (d). The problem develops sometime later under drying conditions such as the atmosphere in a northern home in the wintertime. Because the boards were edged while the ends were swollen, the ends—having lost more wood—in effect begin shrinking from a reduced dimension compared to the interior of the boards. The dimensions at the ends after reequilibrating to 6-percent moisture content will therefore be less, producing tension perpendicular to the glue line, and open joints if the shrinkage is severe enough, (e). If the glued panel equilibrates at 12 percent joints may still open due to the differential swelling of the center of the boards.

Similar but opposite effects can result from boards at the upper extreme of moisture content, Figure 9–6, right side. If, while waiting to be edged, they are subjected to dry conditions such as might exist in a heated building in the wintertime, they will incur a moisture gradient, and a dimension change at the ends as shown in (b). Edge jointing then removes more of the original wood from the center region than from the ends. After gluing and later equilibrating to 6-percent moisture content in a heated home in winter, the ends of the boards will remain the same dimension, but the middle will want to shrink to a lesser dimension, creating tension perpendicular to the glue line in the middle and opening the joint if severe enough. It is further disheartening to realize that there is no single equilibrium moisture content that will result in a stress-free state in panels with either the low or the high MC boards. Stresses will course along the panel at all moisture contents from this time on.

The cases where moisture varies within boards pose the same sequence of events and the same potential for unfavorable results since much of the variation occurs at the ends of boards.

Sometimes a waiting period occurs *after* the boards have been edged. In this case the same moisture and dimension changes can occur. They then would have gone into the press varying in width from middle to end and would have shown open joints *immediately*. This sequence is depicted in Figure 9–7.

A common procedure for circumventing problems due to moisture variation and change is to edge and glue the same day (Figure 9–8(a)). This will produce good joints. However the later differential movement of each board in a glued panel due to each undergoing different changes in moisture content and dimensions assures a constant source of stress on various parts of the glue line.

It should be noted in this connection that it only takes a few boards at either moisture extreme to ruin a large number of panels. Consider 5 such boards 8 feet long among a cutting being processed for end tables 24 inches square. The 5 boards could yield 40 or more pieces after cutting to length and ripping to suitable widths. It is possible, though not likely, that 40 panels could be affected, if each piece ended up in a different panel. A sure sign that the condition for

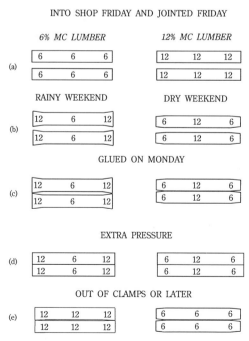

Figure 9–7   Same as Figure 9–6 with machining the same day and delayed gluing.

(a) INTO SHOP FRIDAY; JOINTED FRIDAY; GLUED FRIDAY

(c)

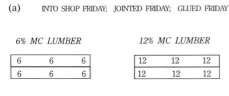

6% MC LUMBER

6	6	6
6	6	6

12% MC LUMBER

12	12	12
12	12	12

NEXT JANUARY

6	6	6
6	6	6

6	12	6
6	12	6

NEXT JULY

12	6	12
12	6	12

12	12	12
12	12	12

Severity depends on species, quality of bond, and moisture-content change. Conclusion: There is no alternative to lumber uniformly seasoned to the average moisture content the product will experience in service.

(b)

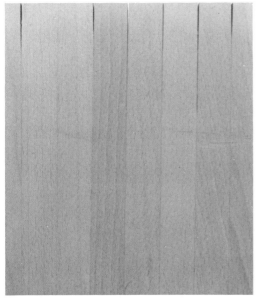

Figure 9-8 (a) Even under the best gluing schedule—the entire process in the same day—boards at extreme moisture contents can cause trouble. (b) Open joints and prehardened glue in glue line (c) confirms extreme moisture change or dubbing in jointer before gluing.

open joints existed before gluing is illustrated in Figure 9-8(b). Cutting off a piece shows a prehardened glue line (c)—one that never received pressure.

Superimposed on the moisture difference problem is the difference in dimension change between quarter- and flat-cut boards. Since flat-cut boards tend to swell and shrink about twice as much as quarter-cut boards in the edge-gluing context we have been describing, flat-cut boards would have a greater tendency to produce open joints. Also, in this interplay of board moisture variation, one would expect to find a difference between heartwood and sapwood effects. Since heartwood dries more slowly, it could enter the gluing process at the high side of the moisture content range, while sapwood could be at the lower extreme. Local grain deviations, as in the vicinity of knots, further contribute to the instability of surfaces before, during, and after gluing.

The conclusion one draws from this is that there is no alternative to well-seasoned lumber entering the gluing process. After gluing it would seem logical to protect end-grain edges from rapid moisture-content changes that produce damaging gradients.

Sunken Boards. A common but more accepted problem with edge-glued lumber panels is the sunken (or raised) board. This condition has less to do with gluing, although it partakes of some of the same dimensional-change effects mentioned in the previous paragraphs. The condition and causes are illustrated in Figure 9–9. A mixture of species, (a), will produce varying thickness boards with any change in moisture content. Combinations of quarter-cut and flat-cut boards, (b), no matter how well seasoned and glued, will also develop sunken or raised boards with any change in moisture content other than that at the time of gluing. When brought back to the original moisture content all boards will return to the same thickness. On the other hand, when boards varying in moisture content are glued into panels, (c), there is no future equilibrium moisture content where all boards will be the same thickness. Mixtures of moisture contents and planes of cut compound the effects of either.

Cupped Panels. Edge-glued panels that do not lie flat have a number of causes arising from wood properties, the operation, or subsequent treatment (Figure 9–10). Two of these were mentioned previously, the out-of-square edges (b) that naturally buckle when pressed together, and the poorly surfaced edges (d) that receive excessive pressure in attempts to close the joint in the clamps. Springback from the resulting compression of the wood is a latent warping mechanism triggered by later moisture increases.

A third mechanism that causes warpage of

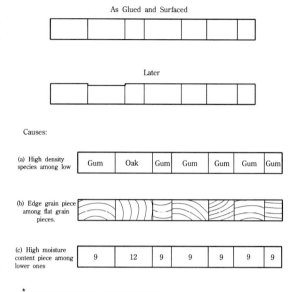

Figure 9–9 Causes of the sunken (or raised) board in edge-glued lumber panels.

edge-glued lumber panels is more subtle and perhaps inevitable since it arises in the differential dimension changes between tangential and radial directions (c). Perfectly flat-cut or quarter-cut boards will remain rectangular

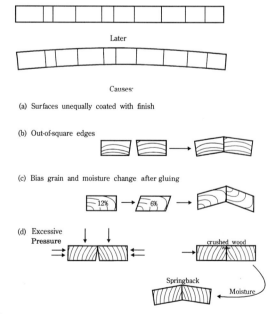

Figure 9–10 Causes of cupped lumber panels.

after a moisture change. However, boards that contain combinations of each become rhomboid at any moisture content other than the original one (Figure 5-16). While this may be slight in each board, a number of boards of like propensity glued together can accumulate their individual deviations into a perceptible warp.

Edge-glued tabletops sometimes receive several coats of finish on the upper side and none on the lower side. The lower side is thus free to take on or lose moisture as the surrounding relative humidity permits, and shrinks or swells accordingly. The upper side, being protected by the finish film, remains at a fairly constant moisture content and cannot counterbalance the dimension changes going on at the other side; the resulting imbalance causes the warpage. The obvious recourse is to apply equal coats of finish to both sides.

Glues Used. Since edge-glued lumber panels are most always used in protected interior environments, a wide range of glues are available and the choice often comes down to working properties and sometimes color. Ureas and catalyzed polyvinyls, for example, might be chosen for their colorless glue lines and high strength on high-density decorative hardwoods. The aliphatics and conventional polyvinyls are easier to use and are appropriate for lower-density woods. Animal glues and casein are also used, although the latter may stain certain woods. Starch glues also find usage, especially when the product is to be used as a core material. Resorcinols would perform adequately but would present a red glue line.

Face-to-Face Lumber Gluing. The same boards used for edge-glued panels, when turned face to face, present a different context for the gluing process. The change in dimensions of the glue line affects the process of preparing the surfaces from the beginning. Applying the glue, assembly time, pressing, and postconditioning also incur changes.

Whereas in edge gluing the objective is to convert lumber elements into panel form, in face gluing the objective is to convert lumber elements into beam form. In panel form, appearance factors generally control, even though the panel may later be overlaid with another element. In beam form, strength factors control since structural or load-carrying applications are of prime importance. With load-carrying come integrity and safety considerations and greater emphasis on control of operations. Organization of the laminations within a beam—along with their thickness, width, and length—becomes an engineering matter. By placing the highest-strength laminations at the top and bottom of the beam and lower-strength laminations in the middle, an I-Beam effect is produced even though the beam itself is rectangular.

Because of the greater bond area and the lesser amount of wood back of the glue line, the basic operations in face gluing differ significantly from edge gluing. Although when laminations are relatively narrow and short, edge gluing processes can be used to make short posts or billets for turnings, attention here is focused on the production of large beams with the understanding that the characteristics apply to all face gluing.

Application of Pressure. After the boards have been prepared and the glue applied, pressure is invariably applied by gangs of screw clamps in face gluing lumber elements (Figure 9-11). While such clamping systems are very effective and have made possible the fabrication of large structural members in well controlled processes, it is of interest to note an interaction they have with wood and glue. All screw clamps have one characteristic in common: once tightened, they maintain a fixed distance across the load, rather than a fixed pressure as with a hydraulic system. Pressure is maintained only as the compression and relaxation properties of the wood allow. This means that the clamps do not accommodate to glue movements or wood movements that occur gradually during the press cycle, and therefore the actual pressure varies over time.

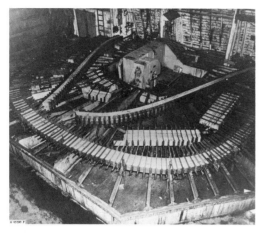

Figure 9–11 Clamping systems for laminated beams and arches. (USFPL)

Glue movements during the press cycle include penetration and squeeze-out, as well as the subsequent shrinkage that accompanies hardening. These actions reduce the volume of glue on the glue line. The remaining glue may not have sufficient mass to form a complete bridge between surfaces, placing link 1 in jeopardy. One might therefore sense the need for a pressure system that provides "follow-up" pressure to assure close proximity of the surfaces as the glue changes volume. Inflated hoses are sometimes placed between the clamp head and the wood to equalize pressure.

Wood movements further complicate the problem. In addition to the relaxation of stress (built up by the clamps) in the wood that occurs over time, the wood swells in response to the diffusion of glue water, which increases clamp pressure. However, the restraint of the clamps causes compression set in the wood (like in a hammer handle), which ultimately results in shrinkage to something less than the original thickness of the lamination. One can visualize an undulating pressure on the glue line as these actions play out their effects during the pressure period. At the end of a long pressure cycle, 24 hours or more, it is not unusual to find the clamps loose, suggesting that the pressure has dropped to zero as a result of shrinkages that have occurred. It would be of interest to place strain gauges on the clamps and obtain a record of pressure history over a 24-hour period.

Amount of Pressure. Pressure requirements are usually prescribed by the adhesive manufacturer and are dictated primarily by the density or species of wood being glued. High-density species require more pressure than low-density species, based upon the compressibility of the wood and the ease with which it can be deformed to force conformation to a mating surface. A range of 100 to 250 psi satisfies most applications.

Obtaining the correct amount of pressure is a question of torque on the screw advancing the head of the clamp. This can be delivered accurately using a torque wrench driven pneumatically with known air pressure or powered by hand. In addition to these devices, reference can be made to Table 10–2 which gives an indication of the amount of force that can be generated by different hand clamps.

Spacing of clamps is not only calculated on the basis of the load each must deliver, but also on how uniformly the load can be delivered to all parts of every glue line. Rocker heads (Figure 9–12) help distribute pressure across the width of an assembly. To distribute the load between clamps, bolsters—thick blocks of wood—are placed adjacent to the last laminations on both sides of the assembly.

Figure 9–12   Rocker head clamps help distribute pressure across width of beam assembly. (USFPL)

Clamping a stack of boards coated with a slippery liquid poses problems in trying to keep them from sliding out of line or squirting out of the clamps altogether. Some lateral clamping may sometimes be necessary. It is desirable that these be in place before face pressure is applied. Hammering the boards into line after face pressure has been applied disturbs the pressure-enforced point-to-point mating that has preceded, and can only disrupt the bond forming actions that have already occurred.

Application of Heat.   Due to the large mass of wood involved in gluing lumber, it is generally impractical to accelerate curing by applying heat from a source outside the assembly. The rate at which the glue line can usefully be heated from an outside source in most cases does not result in sufficient press-time reduction to warrant the extra cost. Many situations are geared by number of clamps, size of screw, and space to operate on a 24-hour cycle. Even cutting the cycle in half by some form of heating would not in itself result in greater production on an 8-hour shift schedule.

When World War II urgencies spurred production, a form of heating was used that reduced clamp time. This involved covering the clamped assembly with a tarpaulin and feeding steam under it. The steam not only provided heat but also kept the wood from drying out, which would have been the case had hot air been used.

There are other methods of accelerating the hardening of the glue that do not involve driving heat in from an outside source. Two of these use the principle of stored heat.

Single Preheating. In single preheating, one surface of a joint is heated and the other surface is spread with glue. When the surfaces are brought together, the heat diffuses into the glue line and cure ensues according to the chemical and temperature relationships of the system. Two cautions apply. One is that assembly time has to be very short; hence spreading, assembly, and clamping have to be engineered almost into a single continuous operation. No time can be lost after the first two boards come into contact. One way of accomplishing this is to heat both sides of every other board in the assembly. While they are heating, the intervening boards can be spread with glue and assembled on edge with space in between. When all is ready, the heated boards are shuffled into their respective spaces and the entire assembly swept into a hydraulic or pneumatic press.

A second caution deals with bond-forming actions of the glue. The glue now faces two different environments in the same glue line: the one provided by the heated surface, and the one provided by the unheated surface. Adjusting the rheological properties of the adhesive to respond appropriately to both situations is the province of the adhesive technologist. Rate of hardening must be more closely coordinated with penetration so as to provide an adequate action in both the heated and the unheated directions.

Double Preheating. Extending the single preheating one step further, preheating both surfaces of a joint should allow even more heat to be introduced, and an even faster hardening produced. Despite the cautions of single preheating carried to a more severe degree, an operation of this kind has proven to be successful. Two products were made, one requiring two boards to be glued and another requiring three boards to be glued. The boards are passed laterally through an infrared heater and linearly through a spreader, assembled, and then linearly through a continuous roller pressing system. The greater heat input allowed a lower-cost glue to be used. With assembly time approaching zero, and a very hot surface, the adhesive had to have precise physical and chemical characteristics—an example of very fine tuning to fit a unique situation.

Dielectric Heating. The geometries of the product and the clamping system make it difficult to harness the internal heat-generating potential of dielectric energy. With dielectric current traveling perpendicular to the plane of the glue lines, too much of the energy will be absorbed by the intervening wood, i.e., the entire package must be heated to curing temperature. In order for the current to be directed parallel to the plane of the glue lines and thus heat only the glue lines, some means must be devised to insulate the electrodes from the connecting rods of the clamps. The difficulty of doing this controls the economics of this approach.

Microwave Heating. The factors that apply to dielectric heating also apply to microwave heating, although because this type of energy can be more controlled directionally, applications to the production of thick members such as beams is increasing.

Separate Application Catalyst.   Some adhesives can be catalyzed to the point where they will cure to a solid almost instantly without heating. Such an adhesive would never get out of a mixer or off a spreader. This situation has been circumvented with more success than logic would predict. By applying the catalyst as a liquid onto one of the mating joint surfaces, and the uncatalyzed adhesive onto the other, a situation similar to single preheating is engineered. After mating the surfaces, the catalyst diffuses into the glue and effects the cure.

Logic suggests (1) that penetration of the adhesive into the surface containing the catalyst might be limited by the rapid curing at that point, (2) that the catalyzed surface is in a sense contaminated, (3) that the catalyst may not diffuse completely into the opposite surface to reach the penetrated uncatalyzed glue, and (4) that even if a bond formed, the high acidity of the catalyst would in time corrode the wood on the side receiving it. Despite these negative predictors, the successful gluing of sugar maple lumber into billets for bowling pins attests to the viability of this approach. Nevertheless, all such operations should not be approached without experimentation to assure performance in any given instance.

Modern meter mixing processes now permit catalyst and resin to be mixed while being delivered to the surface by spraying or extruding. This avoids both the mixing time and the dwell time on a spreader. In operations utilizing this procedure, the emphasis shifts to the speed with which the assembly can be made and transferred to press.

Post Conditioning.   Little deliberate reconditioning of face-glued products normally takes place, due to the rather minor amount of water that has been introduced with the glue compared to the amount of wood involved. However, laminated beams are usu-

ally kept at the mill for a period of time to undergo machining, patching, and fitting operations, and to await tests for quality assurance.

Because of the large cross section of glued lumber products, end-grain surfaces are particularly vulnerable to transient moisture gradients arising from atmospheric events. These result in stresses on the glue line as well as in the wood. It is good practice to protect end-grain surfaces from rapid moisture changes throughout the service life of the product.

Glues Used.   For structural purposes, only two glues have traditionally met the requirements of the industry: high-grade casein for interior or protected environments, and resorcinol or phenol resorcinol for exterior exposures. (Urea glues and all glues of a thermoplastic nature are specifically excluded.) Both the casein and the resorcinolic glues have tests standardized by the American Society for Testing and Materials. As noted earlier, this grade of casein glue is no longer marketed in the United States. Criteria for suitability are covered in Commercial Standards, promulgated by the Department of Commerce. The American Institute of Timber Construction monitors the quality of the products manufactured by its members and certifies conformance to standards of each piece destined for market. Recently urethanes specifically formulated for this purpose have met the standards for exterior durability and are being increasingly accepted by the industry. Their formaldehyde-free composition and colorless glue lines provide market advantages. Due to restrictions in the use of toxic substances to enhance durability of casein against biologic attack, a suitable casein glue for structural purposes is currently not available in the USA.

For nonstructural purposes such as turnings, furniture parts, and other interior uses, ureas, aliphatics, and catalyzed polyvinyls are the most appropriate. For curved parts where residual straightening stresses remain in the finished piece, ureas or any of the cross-linked, high-modulus, (i.e., rigid,) adhesives are recommended.

End-to-End Gluing.   Two characteristics of end gluing dominate the technology: (1) surface composition, being highly porous due to the plane of the cut across the ends of cells, and (2) the angular aspect of the gluing surface with respect to the main surface directions of the lumber element. The angular configuration of the joint is dictated by the need to avoid butting the ends squarely to each other. A square butt joint, as noted earlier, presents the maximum open tubular structure of wood to the liquid glue, encouraging overpenetration and a weak bond. At the same time, it produces a sharp discontinuity in the path of the high stresses that course along the grain, stresses that easily exceed the strength of even a good bond. Apart from the problems of producing suitable joint surfaces, the angularity of these joints influences both bond formation and bond performance. The greater the angle of the bond surface to the grain of the wood, the more pronounced the end grain effect in promoting overpenetration. Hence there is technological motivation for designing joints with as low an angle as possible consistent with the mechanics of cutting and the economics of wood cost.

From a bond-performance standpoint, the motivation is also toward low angles of joint surfaces because of the nature of the stresses that might intercept the glue line. As the angle decreases from 90 degrees of the square butt joint to zero degrees of a face-to-face joint, stresses change from relatively pure tension to relatively pure shear. Since glue joints handle shear stress better than tension stress, the lower the angle the more it will favor the kind of stress it can best resist.

Although the production of the two kinds of end joints—the scarf joint and the finger joint—differ markedly in some respects, the actions of the adhesive are similar. Scarf joining is usually done as part of a laminating operation, producing long beams and arches. Finger joining may also be part of a laminating operation, but often it is a sepa-

rate enterprise producing long lumber, as for full-length siding, door frames, or for salvaging short lengths of lumber. Preparation of the joint and application of glue was covered in Chapters 6 and 8.

Application of Pressure.   Assuming that clamps or pressure systems deliver the prescribed loads, the pressure on the glue line can be affected by a number of operational factors. In the case of scarf joints the chief source of variation is horizontal slippage as face pressure is applied (Figure 9–13(a)) resulting in little if any pressure on the glue line. When the scarf joint is glued as a separate operation, two procedures insure pressure on the glue line. One (Figure 9–13(b)) is to provide a slight overlap of the scarfed surfaces. This in effect produces a greater overall thickness at the scarf joint, insuring that clamp pressure will arrive there. A restraint is then applied at points X to prevent endwise slippage, followed by the closing pressure at point Y, over the scarf. (This calls for a subsequent surfacing operation to restore the board to a uniform thickness; as a result the lamination will be thinner than the starting board.)

When the scarf joint is assembled in the beam along with the laminations, some other means of preventing endwise slippage must be used. This usually takes the form of a

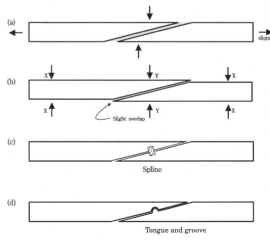

Figure 9–13   Glue line pressure control in scarf gluing.

tongue and groove machined into the surfaces of the scarf, or grooves fitted with a spline (Figure 9–13(c) and (d)).

In the case of finger joints, pressure is delivered not as one might presume by pressing in a direction perpendicular to the glue line, but by pressing in an endwise direction. This jams the fingers together, the pressure on the glue line generating a sliding, wedging action. Applying pressure endwise adapts easily to a continuous process. Running the assembled pieces through sets of drive rollers, the first set operating at a faster speed than the second, creates the jamming action on the fly.

Several interactions between pressure and previous operations occur at this point:

1. If the board did not receive a square cut prior to finger jointing, fingers and gullets will vary in length, and the assembled joint cannot close at the short edge. A forced closure will result in a finished product with crook.
2. If the fingers are longer than the gullets, a situation that develops with knife wear, the sloped surfaces cannot mate and will receive no pressure.
3. If fingers are thicker than gullets, a situation that also occurs with knife wear, the fingers cannot engage fully and bond area will be reduced. Also the gap at the bottom of the gullets will be greater.
4. If slopes of fingers are different from slopes of gullets, no mating and no pressure develops.
5. If outer fingers are too thin, they will spring outward, subverting the wedging action and failing to develop pressure on the outermost glue line. This then becomes the weakest part of the joint, a part that often carries the greatest stress in bending. To compensate, outer fingers can be made thicker. But this creates a wider gullet and places a greater grain discontinuity at the worst location. The discontinuity can be reduced by resurfacing down to the bottom of the notch, adding cost in lost wood and machining time.

An alternative is to cut outer fingers and gullets as thin as machines allow and then provide some face pressure as the stock moves through the process. A slight offsetting of the two parts of the finger joint would produce a high spot as in scarf gluing, and would focus side pressure to the precise point needed with a minimum of additional force.

Application of Heat. When the scarf joint is assembled along with the laminations, heat is normally not introduced, hardening of the glue proceeding in concert with the rest of the assembly. On the other hand, when scarf joining is a separate operation, it may be desirable to accelerate the hardening. Heat can be introduced in three different ways: hot platens, conductive rubber blankets, and dielectric heating. Of these, only the last requires special attention, particularly with respect to the glue chosen. Rather than resorcinol or phenol resorcinol—normally used in laminating—the choice reduces to melamine, urea, or combinations of the two, since they can be formulated to receive dielectric current. There is of course the possibility that the boiling that goes on in the glue line with dielectric heating may aggravate the added penetrability of the end-grain surfaces and increase the potential for driving the glue out of the joint.

Finger joints intended for interior uses are usually not heated. The joints hold together well enough after pressurizing to allow cold-setting glues such as urea, catalyzed polyvinyls, and aliphatics to harden at their own rate. For more severe exposures, resorcinolics may be used, or melamine or melamine-ureas with dielectric heating. Since joints hold together after brief jamming in a pressure device, finger-joined boards can also be subjected to a surface heating treatment such as infrared or hot platen later in the process.

The impression finger joint enjoys the advantage of being heated in advance by the impression operation. It can then proceed with the double preheat method to achieve very rapid curing, using virtually any glue suitable for the intended exposures.

Postconditioning. End-grain joints are generally not given a postconditioning treatment other than that necessary to cure the glue more fully. It should be noted that because the glue line surface contains end-grain effects, glue water will be absorbed rather quickly into the fibers and swelling actions will occur in a short time. Dressing the end-glued board may then remove swollen wood as in edge gluing and invite a slight depression later on.

Veneer Gluing

In gluing wood elements of veneer dimensions, the motivating objective is to obtain a sheet product, and to endow it with greater dimensional stability as well as more equalized strength properties. Increased decorative properties also ensue from the manner in which the wood element is prepared. As mentioned in Chapter 2, there are two distinct processes for gluing veneer, hot press and cold press; two constructions, cross plied and parallel plied; and two species-based divisions, hardwood plywood and softwood plywood. There are also mixed composition constructions: the lumber core and particleboard core panel used extensively in the furniture industry and, increasingly, the inclusion of hardwood in softwood plywood. As one would expect, these all have direct and indirect interactions with nearly every factor in the equation of performance, producing an exponential maze of technological detail. Nevertheless, the principles of bond formation and bond performance apply, and can be reasoned into every situation capable of definition. We will explore some of the major interactions to demonstrate the application of principles peculiar to most of them. In doing so, we will again focus on the glue line to examine the environment produced by various aspects of the gluing process and to note the consequences on behavior of the glue.

As far as adhesive action is concerned, the wood in sheet form is no different from wood in lumber form except for its thinness, its very large surface area, and the characteristics that have been conferred in preparation. The thinness makes it more compliant than lumber and therefore more conformable with adjacent surfaces. In some processes only atmospheric pressure in the form of a vacuum is needed to bring surfaces together. In others, tack of the adhesive alone can hold veneers together.

The thinness also makes it feasible to introduce heat from a platen to speed the hardening of the adhesive. However, the interaction of heat with whatever water there is in the panel becomes intense because of the reduced theater of activity in the smaller amount of wood back of the glue line. Moisture in the veneer and water added with the glue therefore become crucial matters in hot pressing veneer. Moisture and temperature gradients develop that complicate the entire bond-forming process by creating a constantly changing environment for relatively fixed adhesive properties. Hence veneer must be dried to a very low moisture content, an escape that leads to other problems: increasing fragility of the veneer, more surface inactivation, and less conformability.

Contributing to the problems associated with thickness of the veneer is that the thickness is deliberately varied in many constructions in accord with specifications for face, back, core, and inner plies. Lumber core plywood is the most variable in this respect, having relatively thin faces and backs, somewhat thicker crossbands, and a core 10 times thicker yet. The thick core can act as a heat sink in hot pressing, requiring longer press cycles than the depth to the glue line otherwise demands.

The small dimension in the thickness direction of veneer also brings grain angle to a more crucial point. The greater the angle with respect to the face plane, the greater the tendency for the glue to overpenetrate, particularly on hardwood veneer with its larger pores. In extreme cases, this leads to bleed-through and the blemishes that result during finishing operations.

The greater surface area of veneer strongly affects the way it is handled and transported from operation to operation to avoid breakage, a contributor to higher costs. The major effect from an adhesive standpoint is the large amount of adhesive compared to the amount of wood being glued (see Table 7–2). From face gluing lumber of one-inch thickness to gluing veneer of one-tenth-inch thickness, there would be a ten-fold increase in the amount of glue needed to bond a pound of wood if the spread rate were the same per glue line. Resin consumption, however, has only a two-fold increase due to extenders and lower spreads on veneer.

The gluing of veneer provides the first opportunity to vary the grain direction in a product deliberately. With the same species and veneer thickness, the two grain directions, parallel and perpendicular, produce different glue line environments apart from that due to the different thicknesses of the products. Recalling that the structure of wood likens to a system of tubes, the cross orientation can only produce point contacts where crests intersect. Add to this the irregularities in surface topography that also tend to run parallel to the grain, and a further reduction in contact points can occur. On the contrary, when veneer is oriented with grain parallel, its surface irregularities have a better chance to nest and allow more points of contact, which can then result in more bonds when pressure is applied.

Application of Pressure. Two chief means of applying pressure in producing veneer panels characterize this part of the process:

hot platen presses pressurized with hydraulic cylinders, and cold presses in which the initial pressure is provided hydraulically and then screw clamps are attached to the load to retain the pressure for the duration of the curing period (Figure 9–14). In hot pressing, whether single opening or multiple opening, panels are placed one per opening on the hot platen and pressured as soon as possible. Sometimes with very thin panels, two and even three panels are loaded one over another into each opening. With very short panels, one or more may be placed side by side in each opening.

In cold pressing, panels are assembled one over another until a stack of some height is reached. The stack is then placed in a press with an opening wide enough to receive it. The designated pressure is applied, retaining clamps are tightened, and the clamped stack is set aside for the glue to cure while other stacks are being assembled. I-beams and thick headboards distribute pressure across the load. Wooden cauls interleaved at intervals throughout the stack also help even the pressure both vertically and horizontally and ensure the production of flatter panels.

The chief points of difference between hot and cold pressing, as far as adhesive action is concerned, derive from the heat factor; the effects of pressure, however, are also significant. In platen pressing, the assembly of plies is pressed against rigid steel platens on

(a) (b)

Figure 9–14 Presses for producing plywood: (a) multiopening hot press (USFPL), and (b) cold press. (Black Bros. brochure)

both sides. Any thickness irregularities in the veneers translate into areas of high and low pressure on the glue line since the unyielding platens cannot conform. Compression of the wood at thick areas must therefore occur before any pressure can arrive at thinner areas. The resulting unevenness of pressure can result in uneven bonding quality. Depending upon the mobility status of the adhesive when pressure arrives, either the thick areas or the thin areas may suffer. The thin spots that receive less pressure will suffer at the long end of the permissible assembly time when mobility is likely to be low and penetration therefore curtailed. The high spots that receive more pressure may suffer at the short end of the assembly time when the adhesive is most fluid and subject to excess flow.

In cold pressing, where assemblies are stacked in a bundle several feet thick, panels bear against each other rather than against a rigid platen. Adjacent panels provide a more resilient bearing that can deform locally and accommodate to point-to-point veneer-thickness variations. Moreover, the multiplicity of plies tends to even out statistically the high and low spots as far as their cumulative effects vertically are concerned. Ply to ply throughout the stack, however, deformations of a bending nature are going on as high spots are pressed into low spots across the panel surfaces.

Thus in the two processes, conformation between glue line surfaces is promoted by compression deformations in platen pressing and by bending deformations in cold pressing. Consequently, one might deduce that platen pressing would produce panels with a flatter surface but spotty bonding, whereas in cold pressing the surfaces of the panels might deviate from flat but the bonding would be more uniform. Two countering mechanisms tend to level out these differences. Heat in hot platen pressing plasticizes the wood so that it compresses more easily. This makes the high spots deform more and allows pressure to reach the low spots. In cold pressing, the more liberal use of cauls between panels helps prevent major defor-

mations from propagating through every panel in the stack. Flatter panels can thus be made depending upon the extent to which a bundle of panels in the press is pay load or filler load.

Another source of pressure variation arises from how well assemblies are loaded into the press. In multiple opening presses, it is necessary that the same-size panel be placed in each opening and that each panel be placed in the same vertical line both endwise and sidewise. The same holds for cold pressing. Otherwise, three consequences ensue:

1. Portions of panels out of line will receive less pressure than calculated.
2. Portions of panels in line will receive more pressure than calculated (because the effective load-carrying area is reduced).
3. Platens and cauls become distorted.

A source of pressure variation that seems obvious but is too often overlooked—particularly in job-shop operations—is changing the size of panels without changing the gauge reading on the press or without changing the number of rams operating. If, for example, a press is designed with eight rams to produce 4-by-8-foot panels, loads consisting of 3-by-5-foot panels would result in too much pressure on the glue line when the same gauge reading is maintained. Moreover, with the smaller panels, some of the rams would have nothing to work against and could distort the press heads. With short loads, it is necessary to inactivate the unopposed rams and to recalculate the gauge reading accordingly. A similar—and seemingly equally obvious—situation occurs when several panels of *different thickness* are placed in the same opening. This is most likely to happen in making lumber core panels where the thickness of the core, the crossband, or the face and back may differ when more than one order is processed at the same time.

The amount of pressure to be delivered to the glue line is conditioned by the species being glued. Recalling that one of the two purposes of pressure is to bring surfaces into contact or close proximity (the other is to aid adhesive actions), the amount of pressure required to do this depends upon two factors: how well the surfaces are prepared to mate, and how deformable the wood is in compression and bending. Use instructions normally recommend 100 psi for low-density species and 200 psi for high-density species. When a panel contains both low- and high-density species, the low-density controls. The supposition is that the low-density veneer will do most of the accommodating in responding to pressure, since low- and high-density veneers would be in different layers.

Duration of pressure is always controlled by the rate of hardening of the adhesive under the conditions in the glue line, as in all gluing processes. In hot pressing, distance from the platen to the farthest glue line is the determining factor. This glue line must reach curing temperature and be held for the time needed to achieve most of its cure, normally a matter of minutes depending upon thickness and rate of cure. In cold pressing, after clamps are applied to the pressurized load, the bundles are set aside and allowed to remain, usually until the next day. It should be realized, of course, that the temperature of the veneer will have a bearing on the rate of cure in cold pressing. If the veneer is much below room temperature it can overwhelm chemical mechanisms of curing, while those hardening by loss of water would be only slightly affected.

Speed of press closing and pressure buildup has a bearing primarily in multiopening hot presses, hand loaded one panel at a time. The first panel in must rest on a hot platen while the remainder are loaded and closing commenced. The first panel is therefore undergoing the early stages of curing before pressure can help the glue execute its mobility-dependent actions. In general, the faster a press can be closed and brought up to pressure, the better for uniform bond quality. Glues will differ in tolerance to closing time, the more reactive ones being the least tolerant.

Application of Heat. When panels are heated in a hot press, two dynamic events are set in motion: a temperature gradient and a moisture gradient. Both have profound and differing effects on the glue in the various glue lines. All bond-forming effects register through the mobility of the glue during the pressure and heating period. Mobility is affected in the following ways:

1. Increased as a direct effect of heat on viscosity
2. Increased in the inner plies as a direct effect of dilution by migrating water (driven by the temperature gradient)
3. Decreased in the outermost glue line due to loss of water by migration inward; also due to the fast heat uptake and consequent shorter mobility period
4. Decreased due to onset of curing of the resin
5. Prolonged in the inner plies due to decreased rate of curing because of (a) dilution, and (b) slower temperature rise.

These effects transpire in all hot-press cases. The cumulative effect is the quality of the bond that is expected to deliver the designed performance. One can sense that the cumulative effect will vary according to all the factors that affect heat and moisture transport. These include the following.

Moisture content of the veneer has multiple effects. Since dry wood is a better insulator than wet wood, we would expect the glue line in high-moisture-content veneer to heat up faster and cure faster than in low-moisture-content veneer. This is not necessarily the case, especially if the glue requires temperatures in excess of 212°F. To rise above this temperature water must take on heat of vaporization, and until it does so there will be no temperature rise above this point. During this period of no temperature rise, the glue, with the combined presence of some heat and excess moisture, is experiencing high-mobility effects. It is during such periods that the resin is most likely to flee the glue line and leave a filtered condition.

In the development of glues for high-moisture-content veneer, it is primarily this effect that must be circumvented.

The *water content of the glue* and the *amount of glue applied* have a direct bearing on the total amount of water in the assembly. Both variables can sometimes be used to counteract or balance effects due to too much or too little moisture.

A common occurrence in hot pressing at high-moisture content is the formation of blisters between plies. At temperatures substantially above 212°F steam is formed, and pressure develops if it has no place to go. When the press is opened, the steam, if it is trapped inside, expands and raises a blister. Sometimes blisters are obvious, sometimes not; when not obvious they reveal themselves later as moisture changes swell or shrink the loose veneer over the blister.

Blisters are normally associated with starved or filtered glue lines, and therefore one escape would be to reduce the mobility of the glue. However, blisters can also occur over bonds that are otherwise well formed. In this case, the steam pressure is great enough to rupture the wood back of the bond, exposing wood on both sides of the break. The escape in this case is to release the press pressure gradually to allow steam pressure to dissipate through the veneer. This maneuver is common practice in most hot-press operations as a precautionary measure, though it must be realized it cannot save the panels that blister due to too much mobility, nor those that blister due to the opposite extreme, too little mobility, as might occur with excessive assembly times.

Reactivity of the glue affects the duration of the high mobility period. A fast-setting glue that arrives at the hardened state quickly will therefore sustain less movement than a slow-setting glue merely because it has less time to move. Speed of hardening thus becomes one of the options for controlling mobility when other factors such as high moisture content are invading the process.

Steepness of the temperature gradient (rate of temperature rise) affects mobility in the same manner as adhesive reactivity. A slow temperature rise prolongs the period of high-resin mobility, while a fast temperature rise shortens it. One can sense the compound effect of a fast temperature rise with a highly reactive resin, producing a very short mobility period and perhaps inadequate penetration and shallow bonding. On the other hand, a slow temperature rise with under-reactive resin sets the stage for too much mobility, overpenetration, and starved or filtered glue lines.

Temperature of the platens and their ability to maintain temperature under the draining influence of the load determine both the ultimate temperature of the glue lines and the rate of temperature rise, with the attendant effects on mobility. Platens heated with steam, hot oil, hot water, or electricity all can be expected to achieve a prescribed temperature. However, they may differ substantially in uniformity of heat from point to point on the platen, and in maintaining the target temperature throughout the press cycle. Water logging of steam-heated platens due to dysfunctioning of steam traps is sometimes a cause of low temperatures; also inadequate flow of steam from the boilers, which can happen when fuel supplies, particularly waste wood, are low.

Steepness of the moisture gradient (rate of moisture movement within the panel both vertically and horizontally) may compound as well as counteract the effect of temperature. The fact that the two events may not be in phase, moisture advancing ahead of the temperature wave, suggests that a dilution and mobility increasing effect of moisture can begin before heat arrives first to compound the effect and then stop it by promoting cure.

In addition to the effect on bond formation, moisture gradients affect the subsequent stability of the panel. The completion of the bond locks the dimensional status of the plies in the condition they were in when the surfaces on both sides became immobilized against the perpendicular grain of adjacent plies. The times during which plies change dimension as the result of glue-water imbibition, and the time surfaces become locked, are open to some conjecture. Certainly during the assembly time, plies are absorbing glue water, the spread side more than the unspread side, and the unspread side picking up differing amounts from spot to spot depending upon contact with the spread side. When full pressure is applied, the veneers are locked together by friction and can no longer change dimensions externally. However, dimension changes can also take place internally within the cell structure. The degree to which compression set or tension set occurs under the influence of heat and moisture change then determines future consequences. The curing of the glue line establishes a fulcrum of conflict between movements of opposing veneers and produces stress or warping.

The effects of moisture gradients on stability of hot-pressed panels have differences in the number of panels that are placed in each opening. In the case where there is only one panel being pressed per opening, the outer plies will emerge very dry and the centers at some higher moisture content, unevenly distributed but balanced between the two sides. When two panels are pressed per opening, the sides against the hot platens will be very dry, while the sides toward each other will be higher in moisture, setting up an imbalance that results in immediate warp out of the press. In this case it is unlikely that equilibrium moisture contents established in the future will result in a completely stress-free condition in the panel.

In this connection it should be noted that one of the advantages of microwave and dielectric heating is that these methods, rather than concentrating moisture, tend to disperse moisture or equalize its distribution. This is because their energy is drawn to areas of high moisture.

Thickness of the veneer contributes effects in several ways. A thin face veneer will of course transmit heat to the first glue line very rapidly, conversely for a thick veneer. However, the main effect of veneer thickness is on the capacity to accept glue water and to provide void space for steam to expand harmlessly. Consider the situation of gluing veneer $\frac{1}{32}$-inch thick versus veneer $\frac{1}{8}$-inch thick. As much as 20 pounds or more of water can be added for each thousand square feet of glue line. The thicker veneer would have about 4 times the capacity to deal with the added glue water.

Number of plies has the overall effect of increasing the total amount of water in a panel, since every ply carries another watery glue line into the assembly. Because the water will be driven by heat and will concentrate at some inner point, the more plies, the more accounting must be provided to avoid the effects of water on glue mobility and curing.

Location in the panel becomes a point of concern both from the standpoint of rate of temperature rise and amount of water accumulation. Glue lines furthest from the heat source will experience the slowest heat rise and the greatest amount of water, with consequences as mentioned above.

Assembly time has less effect on heat transport than on moisture. During assembly time, glue water disperses into the wood or is lost to the atmosphere. Portions of it remain as liquid in the glue or in adjacent cell cavities, or as vapor wherever space allows. This water is relatively free to move as heat compels. Water that condenses and becomes adsorbed to cell-wall constituents will require additional energy to desorb and vaporize. In any case water that is lost from the glue line by whatever mechanism during the assembly time has the overall effect of reducing the mobility of the glue on that glue line.

Pressure also has several effects on heat transport. In the first place, pressure improves the contact between the heat source, the platens, and the wood surface, thereby reducing air gaps and improving the diffusion of heat across the wood-steel interface. Compacting the entire veneer assembly also improves heat flow since diffusion of heat increases with increasing density.

As heat flow improves with pressure, moisture flow decreases due to the closing or constricting of passageways within and between veneers. Extreme cases of high pressure, high moisture, and "tight" assemblies combine to favor the formation of steam blisters as mentioned earlier. So-called "breathing" of the press is a procedure for relieving steam pressure inside the panel before it has a chance to destroy glue bonds. This procedure is standard practice in many of the panel processes where comminuted wood is used. Relieving internal steam pressure by slow decrease of press pressure at the end of the cycle has been discussed previously.

The direct effect of pressure on bond formation is part of the fundamental necessities motivating adhesive actions. In the case of veneer gluing, there is the added necessity of repairing the subsurface damage—lathe checks—created by the veneer knife on the loose side of the veneer. The glue needs some pressure to penetrate into these checks. It also needs pressure to eliminate air bubbles trapped in the glue line, especially when the glue is applied as a foam.

Heat, moisture, and pressure have a combined effect of increasing the compliance and therefore the conformance of veneer. Heat and moisture plasticize the wood, making it easier to deform under pressure. High spots thus flatten out, allowing pressure to reach low spots. While this is going on, of course, the assembly is becoming tighter, and impervious to egress of steam, and more disposed to blistering. When inner plies of a plywood panel are assembled as individual pieces, there is often a space between them which provides a groove through which steam can escape during pressing. With the increasing use of one-piece inner plies for automated lay-up operations, these grooves are disappearing and the potential for blistering is therefore rising.

While each of the above factors follows physical laws in a rather predictable manner and most are measurable, it has been impossible to sum their effects in advance for the purpose of predicting the outcome of a particular set of circumstances. However, first approximations are possible using the reasoning offered above. Trial and error in an experimental context then must take over to fine-tune an operation.

Postconditioning. In normal practice, post conditioning of either hot press or cold press plywood is seldom done as a deliberate part of the process. Other operations, such as trimming, sanding, inspection, and repairing, follow as scheduling permits and usually involve sufficient time for any equalizations to occur. However, hot-pressed urea bonded panels are sometimes cooled as quickly as possible to prevent overcuring and hot pressed phenolic bonded panels are "hot stacked" to preserve the heat of the press as long as possible and ensure a more complete cure.

Some uses of plywood, such as underlayment, require panels to be laid tightly with no gaps. As they come from the hot press, panels are very dry and subject to moisture uptake and swelling. If laid tight while still dry, they would eventually expand and buckle. Postconditioning to ambient humidities either by the manufacturer or the user is recommended.

Patching and Repair. Veneer may contain defects or blemishes that detract from the appearance of the finished panel. Sometimes these are corrected before gluing and sometimes after the panel has been consolidated. Veneer that is destined to become a face may have its knots or knot holes patched before gluing. The procedure involves stamping out the knot with a standard-shape die. A standard-shape piece of veneer is pressed in place. Friction holds the patch until the glue line can fasten it in permanently.

After the panel has been made, defects that originate as a result of handling or pressing are patched by routing out the defect to the depth of the face and applying a patch of the correct size. In this case, glue must be applied and set by pressure from a hot anvil. Patching that avoids routing is also widely practiced. This involves filling the voids in the surface with a patching compound that adheres and hardens without shrinking, becoming relatively indistinct as it does so.

Edge Joining Veneer (Splicing). As veneer logs become smaller and contain more defects, the likelihood of producing sheets of veneer to the full size of panels decreases. To build up the width from narrower pieces, some form of edge joining must be undertaken. In decorative plywood especially, edge gluing is the chief means of producing interesting grain patterns such as book matching and diamond matching.

Depending upon the use of the panel, edge joints may be tight or loose. For example, in decorative plywood, the joints need to be tight and as invisible as possible, whereas in structural panels and interior plies the joints may be rather loose, the main reason for joining in this case being to facilitate handling and to avoid large gaps between veneers.

The geometry of the system—glue line and wood—presents an extreme in configuration. The glue line is long and narrow, but there is more wood back of the glue line than in any other construction. Aside from the difficulty in machining straight and square joint surfaces, the act of applying pressure to the joint also becomes a problem because of the thinness of the sheet. Pressure cannot be applied directly to the glue line as is done in most cases, but must be applied by friction shoes working on the surface of the veneer. Despite the apparent awkwardness of the mechanics for bringing joint surfaces together, ingenious machinery has been developed for this purpose. Two approaches exist: One in which the veneer travels under heated shoes parallel to the glue line, and the other in which the veneer travels perpendicular to the glue line while being crowded together. The latter produces a continuous sheet that is then clipped to size. The former is used to assemble a predetermined number of pieces to build up a specific size sheet.

The large section modulus of the veneer sheet in the direction perpendicular to the glue line means that very little bending can occur to encourage conformation of the gluing surfaces. Since the amount of pressure actually delivered to the glue line may be small, any inaccuracies in machining generally remain as visible openings on the surface of the sheet. The narrow glue line makes it possible to introduce heat rapidly to cure the glue and provide fast and continuous production.

In structural panels where appearance is not a factor, veneers can be edge joined simply by tying them together with string. The veneers are crowded together as desired, and the string is attached across the grain either by stitching or by gluing to the surface with a hot-melt adhesive. In this case the trueness and squareness of the veneer edges are not critical since they are not being glued. A continuous ribbon of edge-fastened veneer can be made in this way and later cut to full size sheets (particularly desirable in automated lay-up systems). The inner plies of structural plywood are usually not spliced but merely assembled edge to edge in the lay-up.

The glue used in edge gluing veneer must have properties that would seem to be mutually exclusive. Because squeeze-out comes in contact with hot metal surfaces, it must not adhere or it will foul up the pressure shoes, yet it must adhere to the veneer. Formulating an adhesive system that will function under these circumstances is one of the most difficult in the gluing art. Much laboratory time is spent on these adhesives that provide an essential link in veneer operations but with relatively low sales volume to pay for it.

Parallel-Laminated Veneer. When veneer is assembled with the grain all in one direction, the main objective is to produce a structural material of high strength and piece-to-piece uniformity (Figure 2–5). In parallel-laminated veneer (PLV), or laminated-veneer lumber (LVL as it is also called, the directional properties of solid wood remain. Improvements over solid wood ensue because of the ability to select the strongest veneers for the outside layers, and because of the multiplicity of plies that statistically even out the variations normal to wood. There is also an unexpected improvement in splitting resistance due to the slight ply-to-ply differences in grain direction that produce an interlocked grain effect.

The thinness of the veneer element provides a secondary advantage in that it can easily be bent lengthwise to rather sharp curves without breaking. This allows the fabrication of angular components having far greater strength than anything that can be constructed of solid wood; the grain follows the curve around the corner instead of being interrupted at some point (Figures 2–13 and 7–8). Chair legs and braces are familiar examples of such laminating. The process of creating shaped articles from veneer is considered at greater length in the section on molding.

At this point we are concerned only with the production of structural sizes of laminated veneer. Two main dimensional features distinguish it from plywood production: the thickness of the product, usually 1.5 inches in deference to the standard thickness of structural lumber; and the length of the product, theoretically endless but practically up to 72 feet in response to a demand for lengths greater than normally available from sawmills. Width is variable but is chosen in multiples of structural lumber so it can be resawn as needed for other purposes. Width is not a major factor in the process, but thickness and length are.

Thickness invokes the same considerations as in plywood, to a more extreme degree. It brings into the panel more water contributed by the many glue lines, greater distance of the farthest glue line from the heat source, and slower temperature rise in the farthest glue line. The length calls for joining 8-foot-long veneer end to end in a manner that does not reduce the lengthwise strength properties. It means the spreading and exact placement of nearly a hundred pieces of veneer sequentially over the 72-foot length, and it means a long press. The pieces of veneer are joined endwise by scarfing or by a short overlapping during assembly, staggering them so they occur at different places throughout the thickness. Pressing in a continuous press would shorten the assembly time to acceptable limits since the assembly operation could then also be continuous and in pace with the press.

The structural nature of the product demands the durability of a phenolic bond. But we have previously seen the problems arising from having to drive heat through many plies and glue lines, and from contending with temperature and moisture gradients. An Alternative is the use of a combination phenol and resorcinol resin that reduces the heat demand and therefore the problems that come with high heat. The extra cost of the glue would be offset by the problem it eliminates.

Parallel-Laminated Veneer Strands. The logic of cutting veneer to very narrow strips or strands does not emerge as much from the sequence of size reductions shown in the upper tier of elements in Figure 7–1 as from the desire to utilize veneer more efficiently to manufacture high-value structural materials in the format of laminated veneer lumber Figure 2–5(c) and (d). It is, in the scheme of wood utilization, a rather ingenious approach, and a bold engineering exercise. This reduction in width of veneer to $\frac{1}{2}$-inch permits a more mechanized operation than LVL. A number of unique operations are built into the process:

1. Glue is applied by dipping.
2. The strips are redried to tack-free state.
3. Assembling and aligning is accomplished by cascading onto a moving belt.
4. Pressing is continuous.
5. Heating is by microwave.

Applying glue by dipping means that all surfaces are covered, hence the transfer function is assured. In order to avoid too much glue pickup, the formulation is fairly fluid and surplus can drain off. Fluidity also permits good penetration and wetting to be accomplished during a brief transition through a dryer. Final bonding thus requires only the fusion of resin coated surfaces, and curing.

Microwave heating avoids temperature gradients and consequent moisture gradients, resulting in a uniform, stable product. Microwave heating also allows a thick section to be made with durable phenolic resins. A cross section of 4 inches by 12 inches can be used in full dimension or resawn to smaller sizes.

Glues Used.   Two glues provide most of the bonding for veneer products; both are synthetic resins and both are waterborne. Phenol formaldehyde is used almost exclusively in the softwood plywood industry for plywood destined for outdoor or otherwise severe exposure. For indoor and less severe exposures, the phenolic resin is extended with various powdered materials such as sander dust, milled bark, lignin, or grain husks. The hardwood plywood industry relies mostly on urea formaldehyde resins, with fillers and/or extenders, depending upon the product. For severe exposures the hardwood plywood manufacturers turn to melamine formaldehyde, or combinations of urea and melamine resins, and for some purposes may use phenolic resins. Protein glues, casein and soybean, find usage in some grades of plywood, the latter for its low cost and enough resistance to moisture to withstand occasional wetting. It is expected that the isocyanates, due to their freedom from formaldehyde, will make an impact on the plywood scene as their handling properties come under better control and costs are reduced.

Flake Gluing

Geometrically, a flake is a wood element that is relatively square, not more than three inches on a side, and rather thin, 0.010 to 0.025 inches. Variations of the flake include the "wafer," which is thicker (up to 0.050 inches), and the "strand," which is narrower. Although each element produces a different panel—flakeboard, waferboard, or strandboard—production conditions for each are essentially the same.

Two characteristics of flakes dominate the technology of making flakeboards: the small, thin pieces of wood, and the large cumulative surface area. The small size of these elements precludes any handling by hand as is done in veneering processes. Instead the process is almost entirely automated; operations are electronically controlled and largely continuous. The thinness of the elements lends greater conformability in mating surfaces and contributes to uniformity of the product by the greater numbers needed to produce a given thickness. In addition, because of the small thickness dimension in flakes, there is not much wood back of the glue line to absorb glue water. However, this means there is also less tendency to create stress after bonding.

The large area of flakes results in much less adhesive per unit area of surface in order to maintain economic feasibility; this factor places a severe limitation on all bond forming and bond performance factors. The dramatic difference in adhesive coverage between veneer and flakes was shown in Table 7-2. It seems incredible that one sixth the amount of adhesive used per unit glue line area in veneer gluing can produce flakeboard strength suitable for plywood applications. The process of subdividing plays a part in this phenomenon. Theoretically, the amount of adhesive needed is only that which will create a bond equal in shear strength to the tensile strength of the flake along the grain. Since the cross section of flakes is small, their strength is relatively small. Experience has shown that a resin addition of 4 to 6 percent based on the weight of oven dry wood will provide adequate strength if it all functions as intended. Higher resin levels increase strength (Figure 7-3(a)) as well as other properties, but the increased cost must be judged against the values desired by the market.

The small amount of resin per unit area of surface opens the door to the need for compromises between "good enough" and "better" if more resin is provided. An upper practical (if not economical) limit is reached with liquid resins when too much water would be added to the furnish with the resin, raising the potential for blisters. Powder resins reach an upper limit by the amount of powder that can adhere to the wood surface in the blending operation. Using both types of resins would allow the limits of either to be exceeded, if the resulting better panel could be justified.

The technological objective in flakeboard manufacture is to make that small amount of adhesive do its job. Unfortunately the gluing action with flakes is almost totally blind compared to the visibilities of lumber and veneer gluing. There is no discrete glue line to provide evidence of the different adhesive actions, yet they all occur. The small amount of adhesive must undertake its functions in a most efficient manner if it is to form a useful bond. Efficiency is achieved by: (1) bringing surfaces closer together, thereby reducing the thickness of the glue line, link 1; (2) reducing the glue line area; (3) reducing the amount of flow and penetration of the adhesive.

Bringing surfaces closer together produces one of the major distinctions between the gluing of lumber and veneer previously discussed, and the gluing of wood elements subdivided to small dimensions. In order to obtain the proximity necessary for a small amount of adhesive to form a bond, pressure is applied to the point where substantial compression of the wood occurs. This compression, while aiding the bond formation aspect of the process, also creates a bond performance debit that plays out over time and service exposure, reducing the general integrity of the panel.

Compression in this context means breakage and deformation of cell walls, as well as building internal stresses that remain in the panel and are later released. The slow recovery from compression that inevitably occurs with uptake of moisture lowers strength and increases the thickness dimension. The need to compress the wood also constrains the process to the use of low-density species. This helps produce panels that are more acceptable from a weight standpoint, although typically the density of panels is always

higher by ten to twenty percent than the wood from which they are made. High-density species require higher pressure to achieve proximity, further increasing both product density and instability.

Reducing the glue line area is essentially a means of gathering into discrete drops a film of resin that would otherwise be too thin if it were applied uniformly. By applying the adhesive in drop form the glue line is discontinuous, rather than continuous as it is in lumber and veneer gluing. With the adhesive in drop form rather than film form, its mass becomes sufficiently concentrated to allow most of the bond-forming functions in a sort of spot-welding fashion. A good enough bond is produced, even though there remains a high percentage of unbonded interfacial areas.

This raises the issue of optimum drop size, and how this varies with species, surface properties (in terms of roughness and inactivation), moisture content, distance from the heat source, temperature, pressure, and mobility properties of the resin. These factors have not been studied quantitatively as interactants. However, each can be experimentally observed, and reasoned conclusions drawn for specific cases.

Reducing the flow and penetration of the resin in order to leave more on the surface in one small spot produces what appears to be mutually exclusive conditions: reduction in these fundamental actions of bond formation without interfering with equally fundamental actions of adhesion and transfer. The problem is resolved to a satisfactory degree by resin chemistry. Normal surface inactivation that occurs under the severe drying conditions used to remove moisture from flakes, and the addition of wax to reduce water absorption, also contribute. Both tend to keep the resin in drop form and size, and help insure that enough resin exists to form a surface-to-surface bond.

It should be noted that without penetration, there is very little if any allowance for doing anything about links 4 and 5—subsurface damage—which can be severe in flakes, especially as their thickness increases. A technological escape from this bind is smoother and tighter surfaces. Frequent fitting of the knives is a prerequisite for good surface preparation. Better surfaces are also more likely to be made on thin flakes than on thick flakes, introducing another point of compromise: thin flakes, smooth surfaces, and more surface area per pound of wood, versus thick flakes, rougher surfaces, and less surface area per pound of wood. With the same percentage of resin addition, the thin flakes will receive less resin coverage per unit surface area, and the thick flakes will receive more resin per surface area. Therefore there is some built-in compromise to the situation, though a closer study might reveal an advantage of one over the other in a particular operation.

Assembly: Felting, Forming, Composing, Deposition. Between the storage hopper and the press, the resin-covered flakes undergo what is perhaps one of the more crucial steps in the manufacture not only of flakeboards but also of all reconstituted products: the formation of a mat. This is where uniformity—a prime objective in such products—is established. Uniformity greater than in lumber or veneer panel products is possible because of the higher degree of subdivision represented by flakes, (and successively greater as dimensions decrease further). Whereas in a $\frac{3}{4}$-inch-thick lumber panel a single element is involved across the thickness and in a veneer panel 3 to 7 elements are involved, in flakeboard 40 to 60 elements average their properties through the same thickness. However, the averaging is not automatic; it must be achieved.

The mechanical objective in producing a uniform mat is to move flakes from the hopper to a tray or "caul" so that every point of area is covered, for example, with 60 flakes. Points that have less than 60 will receive less pressure during consolidation and may suffer insufficient contact for proper bond formation. These points will also be lower in density and proportionately lower in strength. Conversely, points that have more than 60 flakes will be higher in density,

higher in strength and also, unfortunately, higher in springback potential. The range above and below 60 flakes represents a theoretical departure from uniformity, assuming all flakes have the same thickness and original strengths. An uneven surface that develops after exposure to moisture is an indication of the uniformity with which the mat was made.

Felting systems do not count flakes and deposit them precisely in prescribed locations. They depend upon randomness and large numbers to achieve point-to-point, within-panel uniformity. Between-panel uniformity is approached in two ways. First the thickness of the mat must be controlled to a predetermined height. For a $\frac{3}{4}$-inch-thick panel the mat height might be 6 to 8 inches. Unfortunately, the thickness of the mat can vary due to an apparently minor consideration: the amount of curling present in the flakes, which affects their bulk density. Consequently a second control is also necessary: weighing each mat and discarding those that are under or over the calculated weight to produce the required density at the specified panel thickness. For example, the calculated weight for a $\frac{1}{2}$-inch-thick panel measuring 8 feet by 4 feet, having a density of 42 pounds per cubic foot, would be its volume—1.33 cubic feet $\times$ 42 or 52 pounds—plus the moisture in the furnish, and plus what is added for trim losses.

The actual mechanics of producing a mat of flakes involves means for dispersing the flakes at a constant rate and allowing them to cascade down in a free-fall onto a caul passing at a constant speed underneath (Figure 9-15(a)). Machinery for doing this varies in many respects all striving for the maximum uniformity. However, in addition to uniformity, the machinery also addresses a common pitfall in mat formation: the slope of the flakes after they are deposited on the caul (Figure 9-16). The relative movements of falling flakes and advancing caul produces an unavoidable angle of repose between the first flakes deposited on the caul and the last flakes deposited on the top of the mat. This angle produces the same type

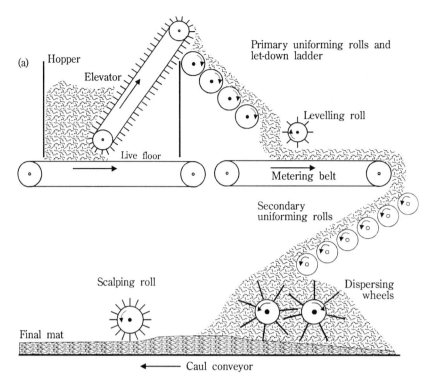

Figure 9-15 (a) Mat formation of flakes in random orientation, and (b) Elmendorf strand orienter using oscillating baffles.

of weakness as slope of grain in solid wood. Moreover, an effect can occur during compaction that is detrimental to the press. With all of the wood elements lying on the same slope, a horizontal component of displacement accumulates along the length of all panels in the press, and all in the same direction. Platens experience a horizontal shift as they close, causing an imbalance in the pres-

sure system and possible warping of the press structure. If the entire mat is formed in one pass, the advancing edge has a very steep angle. Consequently the process is usually accomplished with as many passes as possible in order to obtain the lowest slope.

Rather than passing the caul back and forth under the forming box to pick up its full load, several (usually three) forming

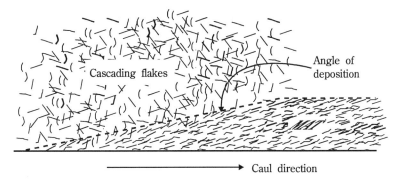

Figure 9–16   Slope of wood elements produced in a panel by the mechanics of deposition onto a moving caul.

heads are provided in sequence so the caul can proceed uninterruptedly down the line without incurring excessive slopes in the deposited flakes.

The three forming heads then also provide means for introducing a number of different compositional variables into the panel, such as producing a three-layer board by placing special furnish for face and back in the first and third forming heads, and a core furnish in the center head. Some of the variables include a different moisture content than core material, different species, different-size wood elements, different resin content, and even a different resin. When the heads are designed to orient the elements (Figure 9–15(b)) (strands in this case), they may be set to orient all in the same direction for maximum strength in one direction; or, as is usually the case, they may be organized to distribute the strands oriented in the long direction from the first head, in the perpendicular direction from the second head, and in the long direction again from the third head. This organization produces the plywood properties attributed to oriented strandboard (OSB).

The infinite variations that can be introduced by a succession of forming heads provide one of the most compelling arguments for subdividing wood to small dimensions and reconstituting: the ability to predetermine and produce engineered wood materials to design specifications. Means for orienting strands include the simplest—dropping them through slots—to the more mechanical grooved rolls, or electrostatically through a magnetic field. Each has advantages and disadvantages in uniformity and efficiency, but their influence on bond formation and bond performance lies primarily on how well they deposit flakes from point to point in the mat.

Prepressing. Prepressing in flakeboard has a different purpose than prepressing in veneer gluing. In the latter, prepressing served to ensure transfer of the glue to the unspread surface, while at the same time generating some tack to hold veneers together while they are being transported into the press. In flakeboard, the main purpose of prepressing is to reduce the thickness of the mat so it can fit more easily into the space (daylight) between the platens. Were the mat to be inserted into the press at the thickness it had from the felter, so much daylight would be needed for a group of panels that the press closure time would be extended beyond tolerance of the resin for dwelling on the surface of a hot platen. The result would be precure of the surface layers of the panel.

Other advantages of prepressing accrue. Edges of the mat become firmer and better withstand the ride to the press without loosening outward or sluffing off. A certain amount of "tack" in the resin also tends to firm up the mat for better handling. The operation also squeezes air out of the mat,

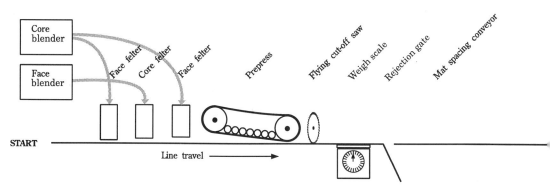

Figure 9-17 Board-forming press line.

gradually, so it does not rush out in the press and disturb the distribution of the flakes. Additionally, since mats are made in a continuous manner, they need to be cut to the length of the caul before proceeding toward the press. Compaction aids this cutting process.

Prepressing may be accomplished in the same manner as normal pressing, but the most efficient is the belt press, since it performs its task on line at the same speed as the other operations.

Pressing (Consolidation). From this point on, the operation of consolidating the mat is essentially the same as pressing hot-press plywood. The press itself is the same: multiple openings, loaders and unloaders, and simultaneous closing to shorten dwell time on hot platens (Figure 9-17). The main difference is in programming the pressure on the panel. In plywood, pressure is built up to a maximum of 200 psi as fast as possible and held there until the resin is sufficiently cured, but in flakeboard, pressure requires additional considerations. Density of the panel as a whole, thickness of the panel, and density variation throughout the thickness of the panel are all controlled in the press at the same time that forming the bond should be the primary objective. Temperature of the platens is another difference from plywood. Whereas in hot-press plywood, temperature ranges from 250 to 350°F, in flakeboard the temperature may be as high as the wood will tolerate, 400°F or more.

The best and most uniform bonding occurs if the mat is pressed in the same manner as in plywood—applying a prescribed amount of pressure and holding for a prescribed amount of time. In such a process, variation in panel-to-panel density is minimal, since density is primarily a function of pressure for a given species, temperature, and moisture content. However thickness would vary out of the press and would require sanding for uniformity between panels.

In order to provide thickness uniformity without the necessity of sanding, flakeboards are normally pressed to "stops"—metal bars the thickness of the panel placed along the edges of the platens in each opening. When platens close and seat on these stops, no further compaction occurs, providing a positive end point for thickness control. In this case, pressure is applied in an amount sufficient to bring the panels down to the stops in a prescribed length of time. Once closed to the stops, pressure is relaxed to that needed to maintain closure. At some point in the press cycle, pressure is reduced to near zero in order to allow moisture, in the form of steam, to escape in a harmless manner, an operation called "breathing the press" or "degassing."

Within this apparently simple context of pressure and temperature, a swirl of changing events sweeps through the thickness and width of the panel during the short time it is in the press. This swiftly changing atmosphere produces a different bond-forming

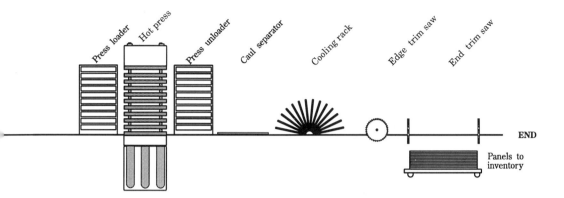

END

Panels to inventory

situation for each flake from the surface inward to the center and from the center outward to the edges. The conditions can span the entire range from too much mobility to too little mobility. The resin therefore must be tolerant to a wide range of conditions. Figure 9–18 presents an intuitive representation of the possible conditions in various parts of the panel during the press cycle. A number of conditions can be generalized from the depicted scenario for a typical consolidation cycle.

The regions nearest the surface reach platen temperature almost immediately.

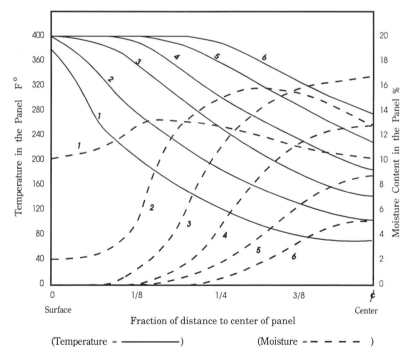

(Temperature = ——————)   (Moisture = – – – –)

Figure 9–18 Representation of the sweep of temperature and moisture gradients through the thickness of a panel beginning with a mat moisture of 10% and a platen temperature of 400°F. Numbers on curves indicate successive time periods in the press cycle. Every point in the thickness has a different combination of temperature, moisture, and time to which the adhesive must accommodate.

They also dry out with only slightly less speed. This means that the resin at this outermost layer may be fully cured during the first few seconds. Moreover, it is unlikely that pressure has reached maximum during this time. Meanwhile the flakes, drying rapidly, can become less compliant, and less able to achieve necessary proximity. Bonding actions can therefore be curtailed especially with a fast curing resin. When maximum pressure is later exerted, surface flakes have become more fragile. Instead of the more plastic deformation that would have occurred at higher moisture content, compression causes fractures within the flakes. Thus in addition to relatively less bonding, surface flakes also incur some internal damage.

Flakes in the region halfway between the surface and the center encounter a similar environmental change, but on a different time scale. For them the pressure is the same, but the rate of temperature rise is slower, and the moisture content, driven up by moisture advancing ahead of the heat wave, becomes momentarily higher than it originally was. It then gradually declines to the end of the cycle. The combination of conditions is such as to make this region the best consolidated in the panel, and normally the highest in density.

At the center of the panel, the wave of moisture, heat, and pressure changes creates still another set of circumstances surrounding the resin and the flakes. Here, most of the moisture contained in the mat at the start of the press cycle begins to accumulate. This moisture has both desirable and undesirable consequences. If it arrives as a vapor, it carries heat of vaporization and therefore helps raise the temperature of the region. If it is condensed, it helps plasticize the wood, but without the proper level of temperature may not improve the compliance enough to be beneficial in this regard. The high level of moisture produces serious negative effects on bond-forming action: by diluting the resin it reduces the rate of curing, and it increases mobility. Besides promoting an undercured condition, these two actions also combine to cause overpenetration and loss of

link 1. The cumulative effect is that the center of the panel is usually the weakest and most unstable. This instability creates one of the negative features of reconstituted products: delayed springback, which increases thickness permanently and also causes a loss of strength. The effect is exacerbated by the generally lower quality of flakes that are placed in the center.

The accumulation of moisture in the center of the panel has the potential of creating one final assault on the integrity of the product. Continued heating vaporizes the moisture to steam, which, if not allowed to escape, creates pressure and raises a blister by rupturing a plane through the weakened center. In practice, this steam is vented during the press cycle by reducing the pressure momentarily, and again at the end by releasing the pressure slowly. However, it should be noted that this practice only avoids the raising of a blister out of the press; a weak plane may still remain. The blister will rise at a later day when moisture prompts it, or when resawing reveals it.

The sweep of moisture, and to a lesser extent temperature, also have another dimension, horizontally, across the width and length of the panel. Because the only escape for moisture during pressing is out of the edges of the panel as it is driven by heat, it tends to move from the center regions toward the four edges. This action, occurring later in the press cycle, adds to that from the surfaces inward, and creates an even higher moisture gradient from the center to the outer regions. The consequences of high moisture would therefore seem to be greatest at the edges of panels, and even more so at the corners.

The needs of the adhesive in forming a useful bond seem to have created a restrictive circle of technological necessities. The small amount that is affordable requires high compression of the wood to bring flake surfaces close enough together to allow linkages to form. Because the adhesive requires high heat to cure in a reasonable press time, moisture content must be kept low in order to avoid steam blisters. Low moisture in the wood adds to the pressure needed to achieve conformability. The added pressure increases internal fractures and dimensional instability in the thickness direction (springback). Several methods of breaching the circle are in current practice.

In combating the inexorable movement of moisture during hot pressing, use is made of the three forming heads. Flakes in the first and third heads may contain a higher moisture content to protect them from too rapid drying, and they may be sprayed with a slower-curing resin to protect it from precuring during pressure buildup. The center head could deliver flakes at a lower moisture content coated with resin that cures at a somewhat faster rate. Thus moisture has a place to go in the dryer flakes, and the resin has a chance to cure before it overpenetrates into the wood.

Moisture is a key player in the conflicting interactions noted above. It is needed to improve conformability of the flakes and to encourage plastic deformation during consolidation. However, it is a detriment to the efficient performance of a minute amount of resin. The ability to consolidate flakes containing a relatively high amount of moisture—18 to 20 percent—would greatly simplify the process. Capacity of flake dryers would be increased by not having to dry down to 5 percent or less, breakage of flakes would be reduced, compressibility and conformability would be improved, less pressure would be needed to consolidate, and panels would come out of the press at the moisture content they would experience in service. The principles of gluing demand only two changes: (1) a resin that can tolerate the higher moisture, being one with a faster curing speed and a slightly lower mobility, and (2) a resin that will cure at a lower temperature. A lower temperature is needed to reduce steep moisture gradients, preventing its accumulation to disrupt adhesive action or to form blisters. Isocyanate adhesives seem to have characteristics to support this approach.

A rather ingenious method of increasing rate of temperature rise to the center of the panel without engendering steep moisture gradients has been developed at the U.S. Forest Products Laboratory. This involves injecting steam into the mat through holes in the platen immediately before compaction, while it is still relatively loose and bulky. The procedure not only provides rapid temperature buildup to the center of the panel, but also promotes greater plasticization of the wood, greater conformation, faster curing, and increased stability of the board.

Postconditioning. After panels come out of the press, they are *dead piled*—piled solid without stickers—to retain heat and complete the curing of the resin to its ultimate durability. In time the panels are trimmed, edge treated with a moisture barrier, and sometimes surface treated with a nonskid material to reduce slippage on steep roofs. Boards are rarely sanded except for special uses; they come out of the press at the specified thickness and with reasonably smooth surfaces.

Following inspection, panels are stamped for intended use and for spacing of studs or joists, strapped in bundles, and otherwise prepared for shipment. During this time also, quality-control personnel conduct routine tests of product properties, particularly internal bond, flexure strength, and stability to moisture. Commercial standards also exist for beams and plywood.

Glues Used. Phenol formaldehyde resins are used almost exclusively in the production of flakeboards. They are used either in powder form or liquid form diluted to an easily sprayable viscosity. Extenders are not normally incorporated due to difficulty in spraying. However, economics and improved extenders prompt increasing use to lower the cost. In order to reduce as much as possible the hygroscopicity of the glue line and the consequent water absorbency of the composite, the pH of the resin is kept as low as possible commensurate with the desired high speed of curing, which reduces at the same time. Some tack is also built into the resin in order to help maintain a firmer mat between the felter and the press.

Combinations of resins are sometimes used in attempts to speed the cure or lower the cost. These include the addition to phenol of varying amounts of melamine or urea resins, or phenolic extractives from lignin and from bark (tannins).

Because of their fast cure at lower temperatures, isocyanates are increasingly being introduced, particularly in the central layers of a panel, where their properties respond better to the unfavorable conditions normally found there.

Research. The excellent performance of flakeboards as a replacement for plywood in construction prompts research to improve them further. Research objectives include:

1. Reducing the high cost of the adhesive component, perhaps by use of natural products, particularly those that might be derived from trees, extractives, waste wood, or waste plastics. Reducing the amount of binder used is not the best option for reducing cost because this approach leads to the use of greater pressure in consolidation, which already causes excess swelling in the thickness direction, one of the problems with performance of flakeboards.

2. Behavior of wood under compression and following compression. The additional instability conferred on wood as a result of pressure in consolidation represents a latent mechanism that reduces the performance of the panel. Compression of the wood is needed in order to ensure close proximity of surfaces and maximize the efficiency of the small amount of adhesive it is economically viable to use. Means of compressing wood so it remains permanently compressed would solve a major shortcoming in this type of product.

3. Methods of cutting flakes to reduce surface and subsurface damage would improve the bond-forming potential of the small amount of binder used.

Particle Gluing

This, the largest and oldest segment of the reconstituted wood industry that pioneered dry-process consolidation, was born out of the necessity to utilize dry mill waste. Previously burned either to dispose of it or to produce heat for plant operations, it became the raw material for an entirely new industry. The nature of this raw material had a strong bearing on the technology of its conversion and on the characteristics of the products manufactured. The original, and the easiest to use, was and still is dry shavings from planing mills. It needed only to be made more uniform by milling and screening before being blended with resin and cast into a panel. Green shavings are only a drying operation away from serving the same purpose. Edgings and trimmings are only a hogging or chipping operation further back. Chips from forest residues are technically acceptable but economically cannot usually bear harvesting costs.

Two features characterize the furnish derived from these types of mill waste. One is the short grain that is inevitably created in shavings by the rotary action of planer knives. This means that strength can never be a major consideration in products reconstituted from this material. However, other advantages exist, including low cost of the furnish, smoothness of surface, and stiffness. Major uses of particleboard are as core materials in furniture panels and underlayment in flooring. Since these are interior uses, the adhesive can be of lower durability and consequently of lower cost. Urea resins used almost exclusively in these products also cure faster and at lower temperature.

The second feature of the furnish made from mill waste is the geometry of the particle and the nature of its surface. The particles, especially those made from chips, are less linear and their surfaces are neither smooth nor sound. Two consequences of these characteristics are that proximity of surfaces is more difficult to achieve, requiring high pressure, and that more binder may be necessary to obtain promotable properties. Some of the worst-case particles—those of a blunter nature—have six sides that can receive resin, but only two of the surfaces will receive pressure to form a useful bond. Other particles of this class that may be more linear and in which the width approaches that of the thickness, such as the sliver, have four sides that receive resin, and again only two sides forming the bond.

The process of converting this type of furnish into products is essentially the same as for flakeboards, perhaps simpler because the particles are somewhat easier to handle; they flow and cascade more freely and breakage is not a problem, permitting them to be driven more forcefully and efficiently through equipment. Since high strength is not an objective as it is in flakeboards, mat formation is less demanding, although uniformity is still a prime target.

Mat Formation. Particles that are cubical or granular in nature can be cascaded onto a moving caul, leveled off, and passed on to the next operation (Figure 9-19). Any linearity, however, demands that they be laid in the mat as horizontally as possible. Otherwise a constant slope of the particles in the mat will cause a further reduction in strength adding to that due to poor geometry, and the same effect on the press can occur during compaction as indicated for flakeboard. Consequently, mats are formed by passing the caul under a sequence of forming heads as in flakeboard. This also allows the deposi-

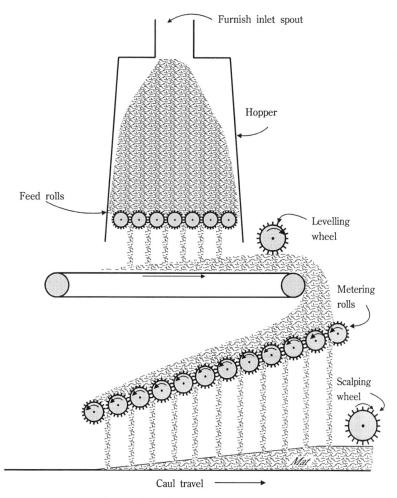

Figure 9-19 Schematic essentials of a particleboard mat former.

tion of different kinds of furnish in definite layers in the board.

An ingenious forming system combines the felting process into a single head while producing a layered effect. The layers are not in distinct stratifications in this case, but in a gradually changing composition across the thickness of the mat. In this system (Figure 9–20), the furnish containing the entire range of particle sizes is cascaded in front of an air stream blowing in both directions over a moving caul. At this point, aerodynamics comes into play. The finer particles are blown the farthest in both directions, and the coarser particles achieve an increasingly shorter trajectory in both directions as dictated by their mass and surface area. The process therefore serves a classifying as well as a felting function, separating the various size fractions and distributing them from top to bottom in a consistent manner, fine particles on top and bottom, producing smooth surfaces with coarse particles in the middle. Batch-to-batch uniformity is determined by how well the proportions of each size particle can be maintained in the feed furnish. The range in particle sizes introduces some consideration for resin distribution in the blender, since the finer particles absorb a greater share of resin in proportion to their weight. However, since the finer particles end up on the surface of the panel, the greater resin coverage enhances surface quality.

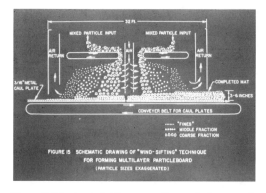

Figure 9–20 A wind sifter system for producing a graduated layer mat from an unclassified furnish. (Maloney, Wash. State Univ., Bison-Werke)

Products made with wood in particle form tend to have greater dimensional stability in the thickness direction than other reconstituted materials. One reason for this lies in the way short particles dispose themselves in the mat, free-falling from the forming heads. Because of their reduced length, they have less inclination to lie horizontally when they hit the mat. Consequently a higher percentage of them remain in a vertical position, placing their most stable dimension in the thickness direction.

Prepressing. Prepressing of particle mats is accomplished in the same manner and for the same reasons as for flakeboards: to reduce the thickness and thereby reduce the amount of "daylight" needed in the press. This in turn reduces the distance the rams have to travel to close the press, which reduces the time needed to close the press. This is precious time not only from a production point of view but also from the standpoint of bond formation. Precure of the resin at the surface is reduced and more of the panel can be properly consolidated.

The compaction that is achieved in prepressing also helps establish enough integrity in the mat to move into the press without edges sluffing off in transit.

Pressing. Consolidating mats of particles is also accomplished in the same manner as for flakeboards, but with two basic differences in process conditions due to the urea resin binder. One has to do with the propensity toward precure, and the other toward overcure. Because urea resins cure faster, they arc easily precured in the presence of heat. Consequently lower press temperatures and fast press loading and closing systems are mandatory to avoid precuring the surface layers. Even so, the surfaces of consolidated panels are frequently sanded not only to establish the final surface smoothness but also to remove poorly adhered surface particles. The lower temperature is also needed to avoid overcuring and decomposing the bond.

Curing speed of urea resins is strongly affected by pH, being faster at low pH and lower at high pH. Because of the small amount of adhesive relative to the amount of wood with which it is in contact, the pH of the wood has a major effect on curing speed of the resin. Hence the catalyst incorporated in the resin must be adjusted for species. Precuring at one extreme or over-penetration at the other extreme are consequences of improper adjustment to variation in wood pH.

Compounding the effect of pH on adhesive actions is the effect of the temperature of the wood particles at the time resin is applied. When particles are delivered to the blender directly from the dryer, the heat retained in the wood can advance the resin to precure status before the mat can reach the press.

Extrusion Pressing. The desire for improved dimensional stability in the thickness direction of the panel prompts an alternative to platen pressing. Since in platen pressing the most unstable dimension of the particles is in the thickness direction of the panel, and since instability is aggravated by the pressure needed to consolidate the particles, also in the thickness direction of the panel, some means of improvement are to be found in re-orienting these two factors. The application of pressure in a direction perpendicular to the edges rather than perpendicular to the faces, and the orientation of the particles in a plane perpendicular to the faces rather than parallel to the faces, would accomplish the desired result. Extrusion processes accomplish this change (Figures 9–21 and 2–14(a)). They not only achieve dimensional stability in the thickness direction, but also markedly reduce the cost of equipment, though disadvantages as well as advantages ensue.

The most notable disadvantage is the loss of strength in the linear direction. Also the instability now transferred to the length direction of the panel accumulates to cause major dimensional changes lengthwise. Both of these disadvantages are partially overcome by overlays that restrain the dimension

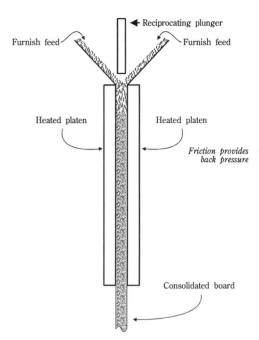

Figure 9–21 Schematic of an extrusion press, edge view.

change and at the same time contribute strength. Thus the panel becomes primarily a core material with properties controlled by surface reinforcement.

Mechanically, the extrusion process combines mat formation with consolidation. Geometry of the particle contributes to the operation. After being coated with resin, particles are cascaded down into a vortex leading to the end of a die, the cavity of which is the rectangular shape of the panel in thickness and width. As they fall into the narrow space between the die faces, they lie horizontally as they would when felted in a normal manner. The horizontal direction, however, is now the plane perpendicular to the face. If the length of the particles is less than the die opening (thickness of the panel), a reasonably random orientation occurs in that plane, creating a measure of stability both in the thickness and in the width dimension of the panel.

Pressure is delivered by a ram reciprocating up and down with each charge of particles shuffled into the opening, a charge being an amount sufficient to produce about $\frac{1}{4}$ inch of panel length. Particles sense pressure only to the extent that there is friction between them and the die faces. If there is not friction there can be no pressure, since there would be no reacting force. Friction therefore controls the actual pressure imposed on the particles. With each stroke of the ram, particles experience two actions: compression, and being pushed farther down the die, eventually emerging at the opposite end. Heat is delivered by maintaining the faces of the die at the proper temperature.

The action of the ram introduces an additional characteristic in the consolidated panel: each charge of particles becomes a discrete layer with surfaces created by the ram face. These surfaces form a continuous plane across the width of the panel, and if not well bonded, constitute a plane of weakness that reflects in the bending strength cf the product in the length direction. Moreover the constant impacts of the ram can cause displacements that may rupture bonds at critical stages in their formation.

An interesting effect of this process develops by lowering the die ninety degrees, to a horizontal position. Particles cascade as before, but now they fall onto an extension of the lower die, and therefore assume a horizontal aspect also. When the ram engages a charge of particles, some are upended, some are driven into the previous charge, and some are turned to conform to the surface of the ram. The result is a better fusion between charges. However, another characteristic is introduced: finer particles tend to gravitate to the bottom of the charge as they hit the die, creating a panel with differing faces.

Postconditioning. Unlike flakeboard and hot-press plywood, particleboards bonded with urea resins are not hot stacked to complete the cure, but are open stacked, Figure 9-22, to allow rapid cooling. This arrests the curing and avoids overcure.

Since most particleboards are destined to be covered with flooring materials and overlays of various kinds, smoothness of surface is a necessary attribute, and this is accom-

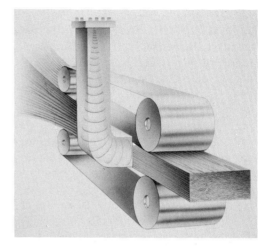

Figure 9-22 Stacking system for cooling of panels that have been consolidated with urea resins. (Maloney, Wash. State Univ.)

plished with a final sanding to produce the finest surface that the size of the surface particles allow.

The Problem of Springback. In these and other composites formed of discrete elements of wood and consolidated with high pressure, the problem of springback—recovery from compression—represents a property that subtracts from the otherwise favorable performance they exhibit. It is therefore one of the major remaining problems that demands basic research on the behavior of compressed wood. Many means of alleviating this problem at the practical level have been proposed, including stabilizing the wood elements chemically, thermally, or by adding polymers before consolidation; adding wax or other materials to the furnish; or treating after consolidation. Since compression and the stresses it fixes inside the wood elements is the basic cause of springback, an approach should include means of compressing without creating the stress that remains in the panel. This would seem to lead to means of plasticizing the wood. Moisture is the cheapest way to accomplish this, especially with the prompting of heat. But moisture also leads to blisters in the panel. A compromise that balances the demands of heat, moisture, pressure, and adhesive, to optimize the stressless deformation of wood, would improve this aspect of board performance.

It should perhaps be noted that much of the moisture that exists in a typical mat during consolidation may be either "free water," water vapor, or condensed on surfaces and may not be located where the plasticization needs to occur. In order for useful action by water to take place, it must be deep in the cell wall. The water in green wood is in the right place. When wood is dried to 5 percent or less as it is in these processes, moisture cannot reenter the lignocellulosic structure of the cell wall except over long periods of time. Therefore water added with the glue or sprayed on separately cannot be expected to help to the degree necessary in plasticizing the wood. Although it aids in heat transfer it also contributes negative effects of promoting overpenetration or creating blisters. By this reasoning, an appropriate approach is to avoid drying the wood in the first place and leave the original moisture where it was in the green condition.

The Problem of Formaldehyde. The release of formaldehyde from wood products consolidated with urea resins, including lumber and veneer, has long been noted. It is due in part to the presence of free formaldehyde in the original resin that is not used up in the curing process. Some of this formaldehyde is trapped as a gas in the cured film, and some is trapped in the wood cells surrounding the glue line. In addition, formaldehyde can be formed by hydrolysis of the cured resin, prompted and accelerated by moisture and heat, and to a certain extent by acid conditions that may prevail in the glue line due to low wood pH or low resin pH. This gas slowly diffuses out of the product and becomes part of the neighboring atmosphere.

In closed spaces having little or no air interchange, the amount of formaldehyde in the atmosphere can accumulate to a noticeable extent. People sensitive to the odor will experience a noxious effect, with watering of the eyes and perhaps a skin rash after a short exposure to a very small amount of formaldehyde vapor. Because of the immediate affront to the senses, and because of the threat of cellular disturbance, restrictions have been placed on the amount of formaldehyde that may be present in the atmosphere inhabited by people. Since a very small amount is easily detectable by eye or nose, the acceptable limit may be set lower than is feasibly attainable.

The resulting formaldehyde concerns have energized a search for glues that are as easy to use and as low in cost as urea resins. Nevertheless, urea resins remain the mainstay of wood gluing processes. It is hoped that methods will be found to control formaldehyde release from glued products. Approaches include lowering the amount of free formaldehyde in the resin, incorporating formaldehyde scavengers in the resin, treating the product with formaldehyde scavengers, and coating the product with formaldehyde-excluding films. Adhesives that contain no formaldehyde are also receiving increased attention.

Fiber Gluing

The consolidation of fibrous materials into sheet products is an ancient art, steeped in papermaking techniques. The art has been modernized mostly by machinery, but also by adhesives and chemical additives. Wood, reduced to fiber form, becomes essentially homogenous in the sense that its original structure is lost, and with it the mechanical properties that depend upon structure, notably its high ratio of strength to weight and its ability to generate and sustain internal stress. Many physical properties remain in full force including: hygroscopicity, anisotropy, and dimensional instability against moisture. Reconstitution provides a means of restructuring, and to a certain extent modifying properties.

Generically speaking, a wood fiber element has, like other elements, a particular configuration and a range of dimensions. It is highly linear, small in cross section, with a high length-to-width ratio. Dimensions at the upper end begin as a subdivision of a sliver, which morphologically is a subdivision of a flake. The subdivision that produces a sliver can be thought of as a longitudinal cut through the flake, producing an element that is as wide as it is thick, and as long as the flake. Subdividing the sliver—which can be visualized as chopping to shorter lengths and cleaving to narrower widths—would produce elements composed of aggregates of wood fibers, or *fiber bundles*, not single fibers, as the nomenclature suggests. (Single fibers as well as parts of fibers accompany the production of fiber bundles, but these are incidental at this state of breakdown. They fall into the next smaller subdivision that leads to papermaking.)

The essential demarcation between a sliver and its adjacent subdivision as a fiber bundle element is not only one of size but also one of properties. The distinguishing characteristic is stiffness. Slivers are stiff because they still retain wood structure. Fiber elements are flexible, the structure having been destroyed in their preparation. The difference may be sensed by squeezing a handful of the material as if making a snowball. Slivers will prick the flesh and remain fairly loose, whereas fibers will collapse pliantly without pricking and will form a coherent ball.

The basic importance of flexibility in this case is that it provides conformability. Imagine a pile of wooden matchsticks compared to a pile of short pieces of string, both of the same dimensions. The strings will have a much greater area of contact than the matchsticks because the strings have a tendency to drape or wrap over one another rather than make point contacts as do the matchsticks. Moreover, the strings will have a greater degree of intertwining. The greater contact and the greater intertwining produce a measure of cohesive strength in the mat as a result of felting alone, without consolidating actions. This is fundamental to fiber bonding, part of the reason fibers can be bonded with little or no additional bonding agent. Another part of the reason is that some processes are carried out in such a way as to reactivate lignin to perform bonding functions.

After fiber bundle elements have been made (Chapter 6) they are consolidated by three processes. Characterized primarily by the amount of water involved in forming and consolidating the mat, the processes are appropriately referred to as "wet," "dry," and "semidry." Three classes of products are made, based solely upon density: (1) low-density fiberboard, with a density range under 37 pounds per cubic foot, (2) medium-density fiberboard, with a density range of 37 to 50 pounds per cubic foot, and (3) high-density fiberboard, with densities over 50 pounds per cubic foot. The properties of each are conferred primarily by their densities, although various treatments can be applied to enhance certain features. Density is conferred principally by the amount of pressure used during consolidation. Uses of fiberboards are also related to density, hard and smooth at the upper end of the density scale for overlays and regraining, soft and porous at the low density end for insulation and sound absorption, firm and stable at the middle densities for furniture and siding. Those products that have smoothness as a major attribute share a common problem: The occasional presence of oversize wood elements in the furnish. After consolidation these oversize pieces tend to swell and produce a raised point on the surface.

Wet Processes. Emulating closely the mechanics of papermaking, wet processes for producing fiberboards therefore represent the earliest attempts to produce boards from fiber. The fiber is dispersed in water at a very low concentration, forming a *slurry*. The water has several functions in preparing the furnish and forming the mat, after which the water becomes a problem of removal and disposal. Basically, water is a carrier for the fiber, allowing it to be handled easily and uniformly in the various operations.

Felting. The most important function of the water is in producing a uniform mat. For sheet products, the slurry is dispersed on a screen, and as much water as possible is drained away before further consolidation by heat and pressure. The operation of transferring the slurry from a tank to the screen is critical in terms of uniformity. Two distinct methods are in use: continuous and batch. In continuous felting, the slurry is flooded onto a screen belt moving at a constant speed. A slot at the bottom of the box holding the slurry allows a metered flow across the width of the screen (Figure 9–23). Flow speed and screen speed combine with the concentration of fiber in the slurry to control the amount of fiber deposited per unit area. The movement of the screen and the flow of the slurry as it engages the screen tend to align the fibers in the direction of motion, producing a slight difference in properties across and along the resulting board.

In the batch process, the screen is stationary, in what is called a *deckle box*. The slurry is flooded in, and since there is no directional movement the fibers retain the randomness they had in the slurry. No "machine direction" ensues.

Both the continuous felting and the batch felting contend with a common problem: ensuring that fibers remain in a horizontal plane. Fibers that assume a vertical aspect, especially if they are in clumps, tend to destroy the smooth appearance desired in the finished product, particularly in the high-density fiberboards. Areas that contain vertically oriented fibers are more absorptive and respond differently to finishes.

Consolidating. Wet processes are capable of producing the entire range of density boards simply by controlling pressure. Low-density boards receive only a drying treatment after felting when the product is to be used for acoustical or insulation purposes. Some compaction is introduced for sheathing applications where more rigidity is required. In either case, the strength resides as much in the natural frictional resistance between intertwined fibers as in actual inter-fiber bonding.

With increased pressure via hot-platen presses, boards achieve medium density and ultimately high density with a maximum approaching that of cell wall material (90 pounds per cubic foot). The high pressures coupled with high heat and moisture cause lignin, as well as hemicellulosic material, to flow and act as a bonding agent between fibers. They are sufficiently activated to perform the entire adhesive function without further binder addition. (However, increased durability is imparted to the board by incorporating small amounts of phenolic resin.)

Water has a varied role throughout the process. After facilitating the felting operation, its drainage from the mat tends to draw fibers together due to capillary forces exerted through receding meniscuses among the fibers. The water also increases the flexibility of individual fibers, allowing more drape and conformability during consolidation. At the higher board densities, water aids in plasticizing the lignin and improving its flow properties for bonding.

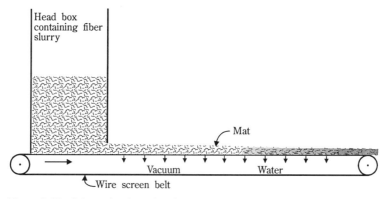

Figure 9–23 Schematic of wet-forming a fiber mat.

Having accomplished all this, water must be removed from the board without causing problems of steam blisters. This is done by using a screen as a caul to carry the mat into the press. The screen provides passageways for moisture to escape through the back of the panel while it is being pressed. The screen then leaves its impression on the back of the panel, evidence later that the panel was made by a wet process. The use of a screen to facilitate moisture removal creates an operational as well as a performance problem. Fibers must be washed before felting to remove fines and sugars that would clog the openings in the screen and reduce their efficiency for moisture egress. Also the screen impression on the back of each panel produces a roughness that must be planed away if a smooth surface is desired on both sides.

The production of high-density fiberboards by the wet process—involving as it does high heat, high pressure, high moisture, and high communition—encounters extreme conditions in both bond formation and bond performance. Specific effects, however, are difficult if not impossible to sort out because the high degree of comminution has blurred interfaces, and the components are too well homogenized to allow separation of consequential actions.

There is, to be sure, a great deal of surface area to be bonded. But with lignin as the main bonding agent, its presence in large and well distributed amounts insures ample adhesive. When fibers are prepared at high temperature the rupturing action that creates them occurs primarily in the middle lamella. Consequently the fibers have lignin-rich surfaces and need only to be fused back together. The bond made by fusing lignin to lignin is strong but less resistant to moisture. Means of improving the durability of the bond seem warranted. The addition of small amounts of phenolic resin has been mentioned.

A more challenging and perhaps more effective approach that is being researched involves the "activation" of ligno-cellulosic surfaces by chemical treatment so they become in a sense self-adhesive. Actions by various oxidants produce chemically reactive sites that can engage in bonding activity. Because lignin is already well attached and diffused into the fiber, three adhesive actions have already been accomplished: transfer, penetration, and wetting. The high pressures that are normally used in high-density fiberboards play to the advantage of such a bonding mechanism since very close proximities are needed to engage the bonding sites across an interface.

The consolidation conditions of high temperature necessary to achieve lignin bonding with a wet furnish constrain the product to thin panels. Otherwise deterioration of surface regions occurs due to heat before the center can attain bond-forming temperature. In high-density fiberboards, thicknesses of $\frac{1}{8}$ to $\frac{1}{4}$ inch out of the press are the normal dimensions. Greater thicknesses are made by laminating in a secondary gluing operation. Secondary gluing is also involved in adding decorative overlays of films or veneers. Decorative features can be produced by other means, such as embossing during pressing or scoring after pressing.

Postconditioning. Low-density boards require no conditioning after pressing since they are relatively stable. Their stability derives from their porosity, which allows dimension changes that do occur to take place internally without accumulating to affect the exterior dimensions seriously.

The higher the density, the lower the porosity, and the less this effect can take place. Therefore high-density boards receive some moisture reconditioning to bring the moisture content nearer to that encountered in service. It is important to realize in reconditioning that the equilibrium moisture content achieved by fiberboards at a given relative humidity is lower than that of the wood from which they are made. For example, pine lumber that might equilibrate at 8-percent moisture content under a relative humidity of 40 percent might not rise above 5 percent under the same conditions after conversion to fiberboard.

Postconditioning sometimes involves treatment with oxidizing oils. Such treatments increase strength as well as size the surface, resulting in so-called "tempered" fiberboard. Other post treatments impart water-excluding properties. The many pretreatments and posttreatments, when added to the pressing treatments, provide a wide selection of characteristics to serve a variety of end uses.

Dry Processes. The problems of dealing with water in the previous system led to development of processes eliminating it. The substitution of air for water to carry and deposit fibers, and the addition of resin to take over the bonding function, greatly simplified the process. An important result was the elimination of the screen in pressing. This allows panels to be made smooth on two sides. Other adaptations and innovations contributed to the success of waterless fiberboard processes.

Felting. Mat forming is similar in appearance to that of particleboards in having multiple forming heads and a continuous belt drawing the mat along. The major mechanics in this case are to provide a constant cloud of fibers that drift gently and uniformly down onto the belt (Figure 9-24). The drift, however, is guided by unseen air currents to deposit fiber where it is most needed and to ensure a uniform mat. Like in wet-process felting, the belt is a screen, but serving a different function. A vacuum under the screen draws away the air in which the fibers are fed to the felter. At the same time, because the vacuum is always greatest where there is the least amount of fibers deposited on the screen, more fiber will always be drawn to those scant locations to even out the mat. Means are also available for depositing more fibers along the edges in order to compensate for outward squashing movement later in the press.

Because of the very low bulk density of dry-fiber mats, less than 2 pounds per cubic foot, prepressing is mandatory to reduce the thickness of the mat. This provides the same benefits as for particle and flake boards: (1) It firms up the mat so edges do not sluff off

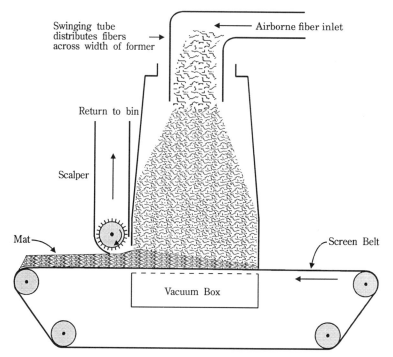

Figure 9-24 Schematic of an air-lay system for producing a mat of fiber by the dry process.

during transport to the press; (2) it reduces "daylight" in the press, thereby reducing the amount of travel the rams must undertake to close the press; (3) the press can close faster, thereby reducing chances of precuring surface regions; and (4) it provides a denser mat for easier cutting to length.

Consolidating. Dry-formed mats are consolidated as in particleboard at temperatures according to the resin used, pressures according to the density desired, and cycle time according to the thickness of the panel. Pressing systems are also similar with the multiple-opening hot-platen press as the mainstay of the industry. Unusual departures include single-opening presses, some of which are very long and heated dielectrically, while others may be short, pressing a continuous mat in a progressive manner. A totally different approach uses a steel belt stretched over a large heated drum rotating at a slow speed. The furnish is squeezed between the belt and the drum, producing a continuous, thin panel.

In all cases, bonding conditions are similar to those in particleboards except that there is considerably less binder per unit of surface area to be bonded. However, some important counterbalancing distinctions exist. One is the better conformation of the surfaces of the elements, which together with intertwining produces more and closer interfacial contacts. It could also be postulated that there is so little wood back of the glue line—a few cells thick—that no penetration of the adhesive is needed and the resin can therefore be formulated to remain on the surface and assure formation of link 1. With no allowance for penetration, less adhesive would be needed. This logic is disturbed by reasons emanating from the sequential nature of adhesive motions. In order for an adhesive to stay on the surface, it has to be deprived of its mobility to flow and penetrate. If it cannot flow and penetrate, it cannot wet either. Hence any bond that forms can theoretically be no better than unanchored, and largely mechanical in nature. Actually, since the binder is applied in liquid form, it has a

chance to wet at least the first surface it encounters.

Because there is no water to contend with during pressing other than that present in the fibers or added with the binder, and because the resin binder is the main dictator of consolidation conditions (rather than lignin in this case), products having thicker cross sections can be made. For furniture paneling, thicknesses of $\frac{3}{4}$ inch are common. Additionally, mats of low moisture content allow pressing without screen backing, producing panels smooth on both sides.

The elimination of water means that consolidation has no ally in lignin but is totally dependant upon the added binder. Two resins carry most of the production: urea formaldehyde and phenol formaldehyde. Melamine formaldehyde is sometimes added to improve the durability of urea. As in particleboards, phenolic binders require a longer press time and a higher temperature than ureas. Although not experimentally proven, the reaction of wood to high heat suggests that this may be a partial reason for the slightly greater stability of phenolic-bonded panels.

The elimination of water also makes pressure more crucial because of the lowered flexibility of the fibers. This limits the product density range to medium- and high-density fiberboards, it being impossible to achieve panel integrity at lower pressures.

Postconditioning. As with particleboards, panels emerging from the press are placed on edge for rapid cooling if they have urea binders that must be protected from overcuring. Those that contain phenolic binders are dead-piled to retain heat and complete the cure. Since most dry-process panels are destined, because of their smooth surfaces, to be overlayed with decorative treatments—paint, films, paper—inspection for defects, particularly those that cause telegraphing blemishes, is a necessary component of final processing.

Semidry Process. In some respects, the semidry process is an amalgamation of the wet and the dry processes. As the name indicates, the moisture in the furnish (approximately 30 percent into the press) in the semidry process is higher than in the dry process (10 to 12 percent) and much lower than in the wet process. The moisture is sufficiently high to necessitate provision for eliminating it harmlessly from the mat during pressing. This is done, as in the wet process, by the use of a screen backing. The fibers are prepared from steamed chips but are not dried as in the dry process. Instead they are blended with resin (as in the dry process) and fed directly to the felter from the fiberizer. Drying occurs in the press as in the wet process. Thus water aids the conformability of the fibers but does not present as much a removal and disposal problem as in wet processes.

Operationally, the press is functioning as a dryer. This eliminates a separate operation from the process, and the fibers do not have to be heated twice, once to dry them and a second time to cure the binder. This is more efficient in terms of energy, but may not be more efficient in terms of time since the two operations are no longer conducted in parallel.

In these three fiber processes, with their different moisture contents during consolidation, the binder must have different mobility characteristics in each case. In the wet process, at one extreme, the resin must have the least mobility in order to stay in place on the surface against the motivation of water. In the dry process, the other extreme, the resin must have the maximum mobility in order to execute its motions without the aid of water. In the semidry process, an intermediate mobility must be struck to balance that contributed by the moisture in the fiber.

Resin Efficiency

In bonding comminuted wood, and to a less extent, lumber and veneer, the issue of adhesive efficiency is a consideration. Efficiency is a ratio indicating the amount of adhesive needed to produce a given result. However, different kinds of decisions are made in the name of efficiency. Both the "amount" and

the "result" therefore need to be defined. A business manager may only want to know how many pounds of adhesive came out of inventory against how many square feet or pounds of board were shipped; in other words, pounds resin per pound product. A researcher may want to know the percent of resin needed to achieve a certain strength with the wood element being consolidated. The business manager includes waste and rejects in the calculation; the researcher may be more interested in learning how much resin participates in bond formation and how much is performing some other function.

Both are confounded by the efficiency of adhesive application in respect to the amount delivered from the container as opposed to the amount actually put on the wood. Both also depend on the uniformity with which resin coats all particles; 0 percent on some particles, x percent on some, and $2x$ percent on others will produce a poorer board than x percent on all particles. The researcher may also want to discount resin that is applied to surfaces, such as edges, that engage minimally if at all in consolidation.

Generally, efficiency is used as a criteria in comparing adhesives. For example, phenolic resins are considered more efficient than ureas, powdered phenolics more efficient than liquid phenolics, and isocyanates more efficient than either. Reasons for these differences must lie in the molecular structure of each, and the basic adhesion each has for the wood substrate being bonded.

However, it must be remembered that all seven links of the bond chain are still operative and all five actions of the adhesive are still viable and in control of them. A true measure of efficiency demands that all actions be optimized and all links maximized. This is seldom the case because of the transient and variable nature of bonding conditions inside a mat during consolidation. Consequently optimization can occur, if at all, in only one region of the mat where temperature, moisture, pressure, and porosity combine to produce the correct mobilities for total link formation. Unfortunately both extremes of mobility often occur in the same mat, giving rise to precure at surface regions and overpenetration at interior regions. Both lead to some degree of inefficiency. In light of this, it is probable that resins have different "apparent" efficiencies only because their composition may have demanded a different sweep of bonding conditions. In other words, bonding conditions have as much to do with efficiency as resin properties, and should be factored into the estimate. A mental integration of Figure 9-18 with Figure 12-1 of Chapter 12 suggests the intricate interactions of the factors affecting resin efficiency.

Papermaking

Strictly speaking, from the standpoint of adhesive bond formation, papermaking belongs in a class by itself, since in the conventional process no adhesive is needed; the elements provide their own adhesive actions. However, from the standpoint of wood element geometry, it represents a theoretical endpoint to the concept of diminishing dimensions. The fibers in this case are not bundles of fibers, nor even single fibers, but parts of fibers unraveled from fiber walls as ribbonlike appendages—fibrillated, to use a technical term.

Further subdivision beyond these fibrillations does occur. Some of this is due to incidental "overmilling" and is usually separated out and discarded as "fines." Fines interfere with water drainage in wet processes, as well as reduce strength properties in the final product. Fines are also produced in generating most wood elements and are generally regarded as undesirable for reconstitution because of their lower contribution to strength and their large surface-to-weight ratio, which increases adhesive consumption. However, in accord with one of the laws of diminishing dimensions, which promises increasing smoothness of surface with decreasing dimensions, fines are sometimes used or added as a surface layer when smoothness is the prime requisite. Wood flour, a powdery subdivision of wood not considered in the dimensional series of wood elements because it does not contribute structural properties, nevertheless has many uses of its own, being an additive in plastics and adhesives to modify strength and flow properties and to lower cost.

Although the process of making paper does not involve gluing per se, it does involve adhesion. Adhesion is achieved by contriving the utmost in proximity between wood elements; the utmost being interparticle distances close enough for their hydroxyl groups to share each others' hydrogens and form hydrogen bonds. The proximity is achieved by creating an element that has maximum conformability (flexibility) and hydroxyl-rich surfaces. Separated by chemical and mechanical means from the wood cell wall, these fibrillated fiber elements are flat, straplike, and linear, and stripped of lignin and other encrusting or stiffening materials (Figure 7-11). In its highest grade, the element's chemical nature is virtually all cellulose. As such it has a very strong attraction for water, and in fact when dispersed in water is referred to as "hydrated cellulose."

The strong attraction (adhesion) of cellulose for water is used as a precisely directed force in bringing element surfaces together. As explained in wet-process fiberboard, as water is drained or otherwise driven from the mat (a sheet in this case), the film of water on each element finally forms a meniscus that naturally retreats to the junctions of elements due to capillary action (Figure 9-25). As the meniscus recedes due to drying, the tension forces it contains become more intense (concentrated) and it draws the elements together. In this progression, the attraction of water to water, together with the attraction of water to cellulose, culminates into an attraction of cellulose for cellulose. This bond formation involves only one of the five actions normally associated with adhesives: that of wetting or adhesion, links 4 and 5.

One can sense that the strength of the resulting paper depends upon the number of hydrogen bonds that are formed. This in turn depends upon the number of hydroxyls

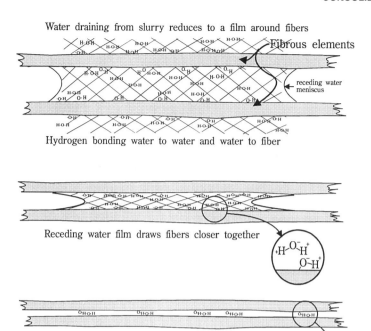

Figure 9-25   In papermaking, surface tension of water draws conformable fiberal elements together as drying causes meniscus to recede and contract, allowing hydrogen bonds to form between surfaces.

present per unit of surface, and the number that actually engage across the surfaces. Many factors affect these numbers, including the cross-sectional thickness of the element that influences its strength and conformability, the amount of lignin present (which affects flexibility, conformability, and hydroxyl content inversely), the pH of the system, heat, and pressure. The length of the straplike elements also affects strength of the product; the longer the stronger, up to a point where length begins to interfere with the production of a uniform mat. This system of factors contains conflicts and compromises. Increasing flexibility, for example, involves more mechanical action to produce finer elements but also results in shorter elements. It can also involve more chemical action to remove additional lignin, which at the same time increases cost and reduces yield.

Hydrogen bonds are susceptible to the same water that formed them; they are easily disrupted by moisture. Hence means of improving moisture resistance of paper are important to the technology of papermaking. The addition of sizing agents to exclude water, or the inclusion of resins such as urea, melamine, or phenolic, impart moisture resistance and even wet strength in the case of the resins. Products that depend upon absorbency, such as paper towels, napkins, and toilet and facial tissues, contain no sizing agents and leave the cellulose with its natural affinity for water.

Hundreds of different kinds of papers exist. They multiply through the species of trees used, the manner of fiber production, the degree of comminution, blends with other fibers and additives, and the mechanics of felting, drying, and calendering.

Although paper is not itself regarded as a

structural material, it does enter the structural field through secondary and tertiary gluing processes. Such products as corrugated board, honeycomb, overlays, and boxes take advantage of paper as a raw material in further manufacturing. Paper can also be combined with plastics and densified to make decorative, long-wearing surfaces such as Formica, or to improve weathering and paintability of normal wood.

The susceptibility of paper to moisture, while considered a negative factor, is also an advantage in one important respect: it can be recycled simply by wetting and stirring. Ink and additives must of course be removed in order to reconstitute it into a white paper with reasonable strength.

SECONDARY BONDING PROCESSES

Most of the products produced in primary gluing, essentially flat panels, proceed directly by various mechanical fastening systems, to their ultimate destination in useful, boxtype structures such as buildings and furnishings. But a large percentage first receives a second gluing operation to attach reinforcements and decorative overlays, or to produce subassemblies for componentized structures. These second gluing operations are more easily performed in a flat aspect before assembly into structures because of the greater ease of handling and applying pressure. In general, flat, smooth, freshly made surfaces are the most easily glued. The more one departs from this ideal, toward contoured or otherwise shaped products with old or variously contaminated surfaces, the more one may encounter difficulties of one kind or another.

Secondary gluing is characterized by the diversity of bond forming and bond performance conditions that may exist in a single assembly. Each component may have one or more of the following features: (1) Surfaces to be bonded may not be freshly made, and may in fact have been partially inactivated by contact with a heated surface such as a hot press platen; (2) available pressure might be quite low, and variable; (3) the ma-

terial on opposite sides of the glue line may be different and therefore require greater tolerance in the glue; (4) the material on opposite sides of the panel may be different, producing an unbalanced construction and a tendency to warp; (5) heating the glue line may be difficult due to the distance from a surface.

There are an endless number of materials, combinations, and configurations that can be considered for secondary gluing. Fortunately, many have similar circumstances with respect to bond formation and bond performance. Three types of products are selected to illustrate pertinent features operating in this kind of gluing: overlays, panel and lumber components, and paper products.

Overlayed Panels

Overlayed panels represent a class of materials characterized by a base core to which a face or surfacing layer is added. A familiar example is kitchen countertops where a sheet of high-density, decorative fiber composite has been glued to a particleboard core. They are usually designated by both the face and the core, such as veneer-overlayed particleboard or vinyl-overlayed fiberboard.

The geometries of the materials, mainly large sheet forms, suggest that the mechanics of the process may be similar to that for plywood. Many overlays could be applied in that manner and some are, but the need for speed discourages the use of long pressure cycles, driving the technology toward fast assembly-line operations.

Techniques of application are controlled mainly by the properties of the overlay. Two properties are particularly dominant: stiffness and porosity. With regard to stiffness, the bonding process differs between overlays that are flexible and those that are rigid. Overlays that are porous demand a different process from those that are nonporous. Their purpose—whether decorative, protective, or structural—has a bearing mostly on the type of adhesive used.

The base panel, or core material (the other element in overlayed products), also contributes both to bond performance and to bonding conditions. One of the most important properties of the core material is surface smoothness. Stability is a close second, although an argument can be made that it is first, since latent deformations can affect the quality of the surface. Means are available for satisfying both to a useful extent if the ultimate product demands improvement. Strength, when it is a necessity, can usually be achieved by proper choice of composite for the core. For example, lumber core might be selected instead of particleboard if greater strength is needed in one direction. In some cases the overlay itself becomes the strength-giving element.

From a bond formation standpoint, the density and porosity of the core panel play important roles. Many of the core materials are very dense. With density comes high compression strength, which means that high spots will remain as high spots under pressures normally used in gluing. Consequently variations in surface topography must be dealt with in the gluing operation. Even so they may persist or reappear in the final product, subtracting from decorative qualities. Moreover, the surfaces of core panels presented to the glue are likely to be rather hydrophobic as a result of having been hot pressed or treated to reduce water uptake and improve dimensional stability, for example wax in particleboard. Sanding removes the effects of heat-induced surface inactivation but not that due to stability treatments. Many panels, however, particularly those with the highest density, are not generally sanded because they emerge from the press with a high degree of smoothness.

One of the more contentious conditions for bond formation arises from the sometimes grossly different properties of the materials on either side of the glue line. For example, when veneer is being glued to a high-density fiberboard, the glue senses a porous material on one side and a nonporous, inactivated material on the other. The application of plastic and metallic films to wood substrates also presents dissimilarities at the glue line.

The binder used in some core boards, especially urea formaldehyde in particleboards and medium-density fiberboards, presents a restriction on the use of heat-cured glues. Since heat has a deleterious effect on urea bonds, more so when glue water is present, press temperatures and press times must be curtailed and closely monitored in order to prevent damage to the substrate. In addition to lower temperatures and shorter press times, panels can be force-cooled to prevent further breakdown.

Given the above context, the environment for bond formation would seem to indicate less reliance on waterborne adhesives and more on solvent-borne or solventless adhesives. Heat-cured adhesives can be used in most cases but must be controlled on urea bonded boards. Because of the rather low surface energies on most surfaces (both overlays and cores), adhesives with even lower surface energies have the best chance of producing an adequate bond.

As noted earlier, the cost of glue at this stage of conversion is relatively low compared to the value added to the product. This is converse to primary gluing, where the bonding agent in some cases represents the highest-cost item in the composite.

A few specific examples will illustrate the functional parameters that play in overlaying processes as different materials are used.

Application of Veneer to Lumber Core. The instability of wood across the grain, coupled with the parallel-grain orientation of the wood pieces in lumber cores (which accumulates and maximizes dimension change in one direction), makes this construction unique among core materials. Other core materials are more stable and more uniform in properties. In order to control and also to mask the instability of the solid wood core, one layer of veneer must be applied with its grain perpendicular to the core grain on both sides (crossbands) (Figures 7–5(d) and 2–5(b)). This layer has the function of restraining the dimensional changes of the core in the width direction. The face veneer, along with its balancing back veneer, is then applied with the grain parallel to the core grain, making a five-ply panel. The gluing operation is carried out as in normal plywood manufacturing, with the glue generally being urea, applied to both sides of the crossbands by roller spreaders. Consolidation can occur in either hot press or cold press.

Some considerations peculiar to lumber core plywood may be noted:

1. Due to the heat-sink effect of the thick lumber core, somewhat longer press times are needed in hot pressing than would normally be indicated for the thickness of veneer being used.

2. Face veneers are usually thin and highly figured. They often also have a large pore structure, sometimes open and angled toward the surface, an accompaniment of decorative properties. This encourages mobility of the glue from the glue line, moving through the veneer the short distance to the outside surface. The resulting "bleed-through," more probable in hot pressing than cold pressing, causes some problems. In cold pressing where panels are often stacked one on top of another, bleed-through can cause panels to stick together. To avoid this, panels are sometimes interleaved with paper. However, bleed-through also interferes with subsequent finishing operations. Urea glues, even though themselves colorless, prevent wood stains from adhering. Even if the resin is sanded off the surface, penetration of the stain is inhibited by the glue in the pores. Adhesion of other finish-coating materials may also be affected.

When bleed-through is a problem, measures for reducing mobility may need to be instituted. These include use of a faster-setting adhesive, longer assembly time, increased filler content, "drier" glue mix (i.e., less water), and reduced spread rate.

3. Crossbands are sometimes high-density fiberboard to provide a smoother backing for the face veneer and to aid in masking surface irregularities that might be generated by the lumber core. While accomplishing these purposes, they introduce an environment in the glue line different from veneer crossbands. The difference is due to: (a) the impervious nature of fiberboards, which restricts the dispersal of glue water, slowing the hardening of cold press glues and (in the case of hot pressing) trapping moisture in the face veneer; (b) the rigid and incompressible nature of fiberboards, which means that while irregularities between the core and the crossband will be bridged over, those between the face veneer and the crossband will be left unattended except for whatever compressibility can be induced in the thin face veneer; (c) the fact that fiberboards, besides being impervious, have low surface energies (due to the heat exposure they have undergone in manufacture and due to moisture-excluding ingredients) and therefore the formation of links 4 and 5—adhesion—may be adversely affected.

The situation in (a) and (b), hot pressing, can be alleviated by use of a fabric sheet placed over the face veneer. The sheet not only absorbs moisture but also helps distribute pressure over the incompressible fiberboard. With respect to the poorer adhesion-ability (wetability) of fiberboard, two aids are available: light sanding or scuffing, and the addition of an adhesion promoter such as furfuryl alcohol to the urea mix. The latter is usually used as an additive to improve gap filling, but in this case also increases the adhesion power of the adhesive, possibly through solvent action on surface contaminants.

4. Product performance, and to a certain extent bond performance, is dominated by the properties of the lumber core, first by providing maximum bending strength in the direction of core grain; and second in being a source of high stress with moisture-content changes. Paradoxically, the stronger the bond between core and crossband, the greater the stress generated by the shrinking and swelling core working against the more stable longitudinal grain direction of the crossbands. When the crossbands are unequal in restraining power—as they might be if they differed in thickness, species, density, or grain direction—warpage of the panel occurs. More seriously, if the crossbands are

unable to sustain the tensile force imposed by a swelling core due to being too thin or too weak, they break across the grain, splitting the face grain as well. Weakness or failure in the bond avoids breakage failure in the wood but incurs the more embarrassing delamination failure.

Two other product performance characteristics should be mentioned, both dealing with *telegraphing:* one due to glue line activity and the other to lumber activity. Sunken glue lines (Figures 9–5 and 9–4) are the result of glue water interacting with wood instability and both interacting with planing schedules, resulting in a depression over the glue line that can be seen through the face and crossband. Sunken boards or raised boards (Figure 9–9) are the result of individual boards in the lumber core equilibrating from a different moisture content, from a different density, or from a different ring orientation.

Application of High-Density Laminates.
This construction, familiar as countertops in kitchens and desktops in offices, combines a laminated paper sheet material with particleboard. The sheet material begins as paper, porous and absorptive, but after processing to form the high-density laminate (see below under secondary gluing of paper) it becomes very hard, stiff, impervious, smooth of surface, and to a substantial degree not very inviting to glue. However, four adhesive systems have been shown to form a useful bond: rubber cements, urea formaldehyde (both hot- and cold-pressed), vinyls (both catalyzed and uncatalyzed), and hot melts.

The rubber cements are the easiest to use, and preferred by craftspeople or for limited production runs since very little equipment is needed. The adhesive is applied to both surfaces of the bond line by brush or spray. An open-assembly time period, such as 30 minutes, is allowed to transpire. During this time adhesion of adhesive to substrate takes place, more or less spontaneously, in response to attractive forces interacting among the solvent, the adhesive molecules and the solid surface. The solvent or water, as the case may be, no longer needed as dispersant or as facilitator of adhesion, evaporates from the adhesive film, leaving it tack-free to the touch. The two adhesive-coated surfaces are then brought together and pressed by hand or roller (sometimes a block of wood struck with a hammer) to achieve close contact and engage the two films. The bond forms on contact, adhesive to adhesive, and seems virtually assured of success since all other adhesive motions have already been accomplished.

Nevertheless, weak bonds can develop from several sources. One is the timing of the open assembly period: too short, and insufficient solvent will be eliminated, leaving a weak link 1; too long, and the attractiveness of the adhesive surfaces for each other will diminish, again a weak link 1. Perhaps the most common problem is applying an insufficient amount of adhesive, particularly on porous surfaces, predisposing to a starved glue line. Two coats on the core are sometimes needed. Applying the adhesive in an uneven film is almost tantamount to a rough surface and cannot result in a uniform bond or a smooth surface to the laminate. Finally, while applying pressure is a simple process, applying it uniformly is not. Careful attention is required to achieve uniformity.

Since the bond forms on contact, it is imperative that the two pieces be exactly placed at first contact because they cannot be readjusted afterward. To aid in exact placement, two fences can be set at right angles, producing a corner into which one piece and then the other can be thrust so it only has to drop vertically into proper register. Alternatively, a so-called slip sheet, plain paper, can be placed over the core glue line after it has dried to the tack-free state. The overlay can then be dropped into place and adjusted, whereupon the slip sheet can be withdrawn without disturbing the position of the overlay.

Hot melts are also considered high speed since they only require cooling to consolidate the bond. However, the speed that comes with cooling mandates speed in handling the materials. The adhesive is applied in the molten state, hot spray or curtain coater, assembled immediately, and put under pressure. Bond formation can only take place while the adhesive is in the fluid, molten state. Solidification begins the instant the adhesive leaves the nozzle. On the way to the surface, further cooling occurs; and after contacting the surface, cooling accelerates due to conduction of heat into the substrate. This is a crucial time in bond formation because wetting and penetration demands the maximum in mobility, which is limited to begin with and rapidly declining. The quality of the bond depends almost entirely on the temperature of the adhesive at the time surfaces are brought together. One means of prolonging the assembly time and assuring wetting and penetration is to preheat the surfaces before adhesive application. Since there are no solvents in hot melts to aid penetration, the adhesive material must do it alone. However, the lack of solvents means there are no problems with effects of residual liquids on performance of the bond or performance of the product.

The other two glues are applied and pressed as in plywood manufacture, hot- or cold-pressed. Some attention is needed to prevent excessive slippage of the overlay. Generally the panel is made slightly larger in order to allow trimming to final size while removing any mismatching of the two pieces.

Depending upon the thickness of the particleboard and the uses to which the composite may be put, a backing sheet having properties similar to the face but without the decorative layer is normally added to balance the construction and forestall warping should moisture contents change.

Application of Flexible Films. The application of various types of films to manufactured boards has been reduced to a rather simple process. Adhesive is applied either to the back of the film as it passes over a roll or to the substrate. The assembly then proceeds through a nip roll to apply pressure (Figure 9–26). Pressure can also be applied by means of vacuum, particularly when the substrate is highly textured or contoured. Sufficient tack is provided by the adhesive to hold the film in place until it hardens further. Since films are not usually structural, high strength is not required in the bond other

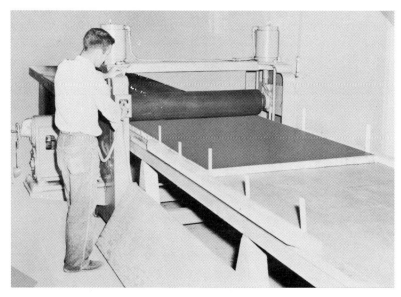

Figure 9–26   Roll press for applying pressure in bonding flexible films to particleboard substrate. (Black Bros. brochure)

than to resist creep or shrinking away from edges. Polyvinyl acetate is the preferred adhesive both for its water base and its lower cost. Solvent-based polyester or urethanes are used when higher strength is necessary, and epoxy resins when it is desirable to avoid solvents altogether. In addition to the usual cautions of using and applying adhesives, the chief operational imperative in this process is avoiding the formation of bubbles in the glue line as the material speeds down the assembly line.

Since films are very thin, two to seven mils thick, they have no hiding power over surface irregularities. Consequently they are at the total mercy of the core surface quality in providing a smooth surface. Nevertheless they are used to impart decorative characteristics to virtually any panel material. By printing or embossing, wood grain patterns can be imparted to the film and made to appear to be that of the substrate, thus achieving the ultimate in conversion technology, the transformation of low-grade or waste wood to the highest value panel material (Figure 9–27).

Figure 9–27 Decorative film over flakeboard substrate for wall paneling, a quantum leap in value added.

Warping. One of the concerns in the performance of overlayed materials is the potential for creating a source for warpage to occur. It is usually the product of instability and moisture content change. (Temperature change is also a factor with some composites, particularly when large density differences occur within a structure.) In virtually all cases, warping is due to imbalances within the composite. The balance point or neutral plane can be assumed to lie midway between the face and back of a panel. Flatness demands that all strength and stability properties be equal at equal distances from the neutral plane, both directions. Imbalances can be of several kinds and sources. They begin with unequal shrinking and swelling responses to moisture change due to (1) unequal densities through the thickness of the base panel, (2) unequal moisture change, or (3) unequal restraint of shrinkage and swelling on either side of the composite. While these three factors appear to be easily controllable in the context of a particular composite, inequalities frequently creep in, which, as predicted in the discussion of diminishing dimension Law 2, Stability, in Chapter 7 include:

1. Different materials or thicknesses in overlays on face and back of panel.
2. In the case of veneer overlays, different species, grain direction, thicknesses, moisture content, density, and sapwood/heartwood on face or back.
3. Sanding more from one side of the core than the other. Casehardening in the case of solid lumber, and density differences from surface to center in the case of reconstituted panels, are sources of imbalance if surfacing is not equal on both sides.
4. Many panel materials that are formed in a continuous fashion tend to have properties that are different in the direction of travel than in the cross direction. This difference is due primarily to a tendency of elongated wood elements to assume a direction in the mat parallel to the direction of travel. When two such materials

are included in the same construction, care must be taken that the machine direction is the same in both. This is particularly important when fiberboards are used as cross bands.

5. Uneven moisture change surface to surface due to different finish treatment or one-directional source of moisture such as might occur in house walls.
6. Uneven temperature change side to side.

These produce either permanent warp or temporary warp depending upon the original situation at time of assembly. If there are temperature or moisture gradients in the panel at the time of manufacture, equilibrium later on has the same effect as a differential change. This can cause a permanent warp since the same gradients are unlikely ever to occur again. On the other hand, if all elements of the composite were at equilibrium with the same relative humidity at time of manufacture, moisture-content changes that develop in service will produce warp that is temporary, coming and going with the changes as they interact with structural imbalances. If the panel as originally produced is both well balanced and at uniform equilibrium, moisture-content changes will not cause warping because the resulting stresses will be in balance.

Smoothness. A second concern in performance is the stability of smoothness. Smoothness has several enemies, all having to do with moisture and density differences. Reconstituted panels from flakes to fibers, besides having a stratified density profile from surface to center, also have density variations from point to point across the surface. The behavior of wood varies not only according to its own inherent density, but also according to the density to which it has been compressed in consolidation. The latter becomes a variable to the extent that the process of felting departs from absolute uniformity in forming a mat, an eventuality that is more closely approached by the finer wood elements but perhaps never completely achieved.

The whole problem of telegraphing in reconstituted panels is one of density variation responding to a moisture change. Unfortunately it takes only a very small dimension change in a spot on a surface to be visible to the unaided eye looking toward a light source reflecting off the surface. Furniture manufacturers have learned that telegraphing is less evident on vertical surfaces than on horizontal surfaces, and less evident under a patina finish than under a high shine.

Panel and Lumber Components

Several different types of structures can be fabricated by adding lumber to previously composited panels. They vary as to characteristics and function of the panel, and characteristics and function of a lumber frame. Structures include the furniture component, panel on frame, in which a thin decorative panel is glued to a frame of lumber; the stressed-skin construction in which the skin carries load; the box beam in which the frame carries load; and the sandwich structure, in which a low density core and a panel material has been consolidated to produce a lightweight structure.

Panel on Frame. Thin panels, $\frac{1}{8}$ to $\frac{1}{2}$ inch, that have decorative characteristics are useful in furniture- and cabinetmaking because of their light weight and lower cost. However, they do not have sufficient warp stability or strength to form cabinets on their own, and because of their thinness they cannot be fastened securely to produce a rigid structure. In order to make such panels useful, they are reinforced along the edges with solid lumber. This not only provides rigidity to the panel, but also provides thickness at the edges for anchoring fasteners in final assembly. The added thickness also permits additional decorative effects such as sculptured edges and recessed or raised borders (Figure 2–11(d)).

The glue line situation is a relatively narrow band, 2 to 4 inches wide around the periphery of the panel. The thickness of the panel and the thickness of the lumber locate the glue line with respect to distance from a surface, a distance that may be too great to drive heat from an external source to speed the cure of a glue.

Surface preparation generally includes planing of the lumber to thickness. The panel on the other hand may be sanded on the face side but not on the back side.

Glue is applied by brush or single roll spreader to one side of the lumber frame because it represents the actual bond location. The panel is positioned, and the assembly pressed according to a predetermined schedule.

A caution applies to this and similar assemblies in calculating the amount of load to be delivered by the press in order to impose the correct amount of pressure on the glue line. While the entire panel will be going into the press, the area to be pressured is not the total area of the panel, but only the area of the lumber comprising the frame, as only this area would be resisting the load delivered by the pressing system.

Speed, in terms of parts per minute, is the criteria for an efficient operation under production conditions. Accelerating the cure of the adhesive by direct application of heat—as by the platen delivering the pressure—is not a preferred method due to depth to the glue line, and due to the danger of warpage as a result of uneven heating.

Dielectric heating does offer the maximum in speed: The short width of the glue line allows high-frequency current to be fed directly into and across it, effecting cures in about 30 seconds. With another 30 seconds for infeeding and outfeeding, production speed would reach one part per minute. Because of their ability to carry a high-frequency current, and because exceedingly high durability is not essential, urea or catalyzed vinyl resins are used. Glue lines cured in this manner must perform all five adhesive motions in a highly accelerated fashion. Conditions ensue as in edge gluing lumber with high-frequency current, and similar cautions apply.

The simplest embodiment of dielectric gluing of panels on frame is the so-called "clam shell" press. Hinged along one edge, the two shells carry the electrodes, hot on the top and ground on the bottom shell. Pressure is applied by closing and clamping the two shells. The size of the press and the location of the electrodes are fixed, determined by the configuration of the parts being glued. Blocking, appropriately placed, keeps the parts in register during the cycle. Different panel sizes require changing location of the electrodes to minimize air gaps and insure uniform heating. Changing sizes also requires changing the wattage input to accord with the different electrical load.

In a cold-press alternative, a number of assemblies are stacked on a headboard, placed in the press, and allowed to remain under pressure until bond formation is complete. This process is considerably slower than dielectric, but cheaper in terms of equipment. A major necessity is a means of avoiding slippage during pressing so that parts stay where they belong. Clamps along the sides of the stack accomplish this purpose.

Sandwich Construction. In a similar but somewhat more intricate construction, a hollow core door (Figure 2–11(e)) involves the gluing of thin panels to both sides of a frame. Because these panels, "door skins" in this case, can be very thin (about $\frac{1}{8}$ inch), and the span between sides of the frame may be large, some type of low-density core is included within the frame to support the skins against buckling actions. All degrees of "hollow" exist, as the nature of the core varies in weight, structure, and cost.

The proper functioning of the core depends upon its compressive strength to prevent inward deflections, and upon its tensile strength plus a good bond to the skins to prevent outward deflections. Sufficient material strength is rather easily obtained. Paper, either in the form of honeycomb (Figure 7–12) or corrugated board, has been found suitable in resisting ordinary stresses. Solid wood cores, or plywood, and other materials arranged in grid form are also used when higher strength is necessary.

The main technological problem is forming a bond between the narrow edge of the paper and the broad surface of the skins. The situation involves conflicting properties in the glue: high viscosity to keep it on the edge of the paper and hopefully to form fillets of glue between the two materials, versus low viscosity to perform bonding functions to the skin at the very low pressure the edgewise paper can withstand. The same glue must also form a bond between the frame and the skins where a more conventional situation exists.

Glues used include casein, polyvinyl, and urea, all cold setting. Hot-setting glues are avoided because of the danger that heat may induce buckling. Geometry of the assembly makes the efficient use of dielectric heating more difficult to engineer. However, it is possible to arrange a system of electrodes with positive and negative bars spaced in alternating fashion such that a portion of the current "strays" down into the glue line, heating and curing it. Other means of accelerating the cure are restricted to catalysis and some form of radiant heating, particularly at the more vulnerable corners of the structure.

The glue is applied by roller spreader to both sides of the frame and core. Skins are positioned and the assembly is added to a stack being prepared for the press. Assembly time is the time needed to assemble 10 or 20 doors to fill the opening of a press. Thus assembly time will be different for each door in a press load, maximum for the first door assembled and minimum for the last door. The extremes in assembly time are therefore at the top and at the bottom of the stack. Inspection should logically focus on those two doors as they are the ones that may suffer from too long or too short an assembly time.

Pressing is done by applying pressure to a stack of assemblies, as in cold-press plywood, using retaining clamps to maintain pressure in the long press cycles that cold-setting glues demand. Pressing systems made up of inflated hoses are efficient and low-cost, and are especially useful when the amount of travel to close the press is very small. Again, as in panel on frame, the load needed to deliver the correct amount of pressure is calculated on the basis of area of the lumber comprising the frame, not on the entire area of the door. In calculating load delivered by an inflated hose it is important to multiply the gauge pressure by the area of actual contact between the flattened hose and the press head. Partially flattened hoses deliver less load for the same gauge pressure. Some load above that needed for the frame may be added when the core is of a material that has substantial surface area and can sustain some compression. Otherwise pressure on the core is minimal, the bond being formed primarily by adhesion forces acting more or less spontaneously. To the extent that gravity affects how well the glue stays on the edges of the core and forms fillets, it might be expected that the down-side faces in the press would develop a better bond than the up-side faces.

In product performance, warpage is sometimes a problem. Twisting of the lumber in the frame as moisture changes occur is responsible for some warping. Uniformity in the skins on either side of the door also has a bearing. Being firmly glued to the frame, the skins transfer their unequal deformations to the entire door. Some elastoplasticity in the bond helps reduce stress transfer between elements and lessens the overall warpage. Skins that are more stable or more uniform—fiberboard, for example—tend to produce doors that are more warp-free. Vents in the door frame also help reduce unequal dimensional effects.

Another problem that sometimes befalls hollow-core doors is the vulnerability of skins to moisture. Even though hollow-core doors are used mostly indoors, delamination (particularly at lower edges) can sometimes be seen in basements and in areas where floors are mopped frequently. Plywood skins are normally bonded with glues that are only water-resistant (as opposed to waterproof), principally ureas, and often with a degree of wheat flour extension that makes them more sensitive to moisture. While these can resist occasional wetting, they are inadequate against prolonged or repeated wetting. Because the glues used to bond the skin to the frame may be no better in water resistance, failure can also occur at this point when excessive moisture exists.

Stressed-Skin Components. In the previous two constructions, the frame was the load-bearing member, and the skins performed decorative or closure functions. Here the reverse holds, where the skins carry the load and the frame serves mostly to hold the skins in place and provide section modulus as well as horizontal shear. Structures in which the outer layers or skins perform the major load-carrying function are basically extensions of the I-beam principle. In this case the flanges merely become continuous laterally, and the webs remain essentially the same (Figure 2–11(c)). They are usually designed for structural applications, bearing either bending or axial loads or both. In a visual sense such structures can be considered scale-ups of the hollow-core door construction where individual components become longer, wider, and thicker. The product can become part of a building—walls, floors, roofs—providing efficiencies in both material use and installation procedures.

Because of their load-carrying function, the skins are chosen for their strength, traditionally $\frac{1}{2}$-inch plywood in their normal 4-by-8-foot sizes. The opportunity to use longer and wider structural panels is a comparatively recent development with the availability of flakeboard and oriented strandboard in lengths of 16 to 24 feet. In the original concept, the webs would be solid wood 2 × 4's, 2 × 6's, and up depending upon the section modulus needed to carry the intended loads. These would be organized as in a stud wall frame and the skins attached. Modern technology has also offered an alternative to the solid wood frame in the form of plastic foams that occupy the entire space between the skins and serve as both insulation and web. Specially made honeycomb treated with phenolic resin has also been used for this purpose.

In all cases, the performance of the product as a stressed skin, load-carrying member depends entirely upon the quality of the bond between web and flange (i.e., between skin and core frame). The gluing situation appears to be similar to that of the hollow-core door, but the element geometries are sufficiently different to warrant a wholly different procedure. Several peculiarities may be noted in the case of plywood on framing lumber:

1. Framing lumber does not have precise dimensions nor ideal surfaces due to the method of manufacture and moisture history before gluing. Therefore thickness of the web section, (the core), can vary. This prohibits the use of rigid platens to deliver bonding pressure since only the high areas would receive it.

2. Geometry of the structure, mainly thickness of the plywood or depth to the glue line, discourages the use of direct heat to accelerate the cure of glue lines.

3. The structural nature of the product demands ultimate durability in the bond. Since heating is difficult, phenolic resins cannot be used. This leaves resorcinol-based glues and urethanes which are cold setting as well as durable, as the most appropriate. Casein glues could be used if the product is to serve in a protected environment.

4. The product is large and cannot be allowed to tie up a pressing device while a cold-setting adhesive cures. Therefore nail gluing is the preferred method of applying pressure; it applies pressure permanently and is able to draw the surfaces together despite local inaccuracies.

5. Because the frame is cumbersome to handle through a spreader, glue is applied by gun in the form of a bead on the edges of the framing members.

Although the procedure is rather straightforward, some special conditions develop:

1. Driving nails through the plywood with repeated blows of a hammer produces a squeegee or pulsing effect on the glue in the glue line, dispersing it more with each blow. Most of the dispersal occurs near the nail, in some cases to the point of starving the glue line. If the nail is not fully driven, or if some

loosening occurs, the bond cannot form completely.

2. A nail gun that drives with one shot is preferable to a hammer.

3. Pressure is bound to vary between nails, maximum at the nail and minimum at midpoint.

4. Because of the extra mobility induced by driving the nail it may be desirable to withhold some mobility from the glue, for example by increased viscosity.

5. Although pressure delivered by the nails is considered permanent, curing conditions (temperature) for cold-setting glues must still be observed. Temperature of the lumber and of the plywood is as important, if not more important, than ambient temperature during assembly and storage. Cold wood becomes a heat sink and deprives the adhesive of curing energy when it needs it most—before it loses its water base.

Once the bonds have been fully formed, the structure becomes unitized and behaves as one piece in monocoque fashion. Because the plywood on both sides is internally balanced, and because of the rather massive frame, the structure is stable for shape except as moisture differences may occur in service. As in the hollow-door construction, vent holes are desirable to help maintain equilibrium between the inside and the outside of the structure.

I-Beams and Box Beams. These structures have a configuration similar to the stressed-skin construction, but the roles of the elements are reversed. Here the lumber carries the maximum stresses acting as the flanges of the I-beam, and the skins perform the functions of the web (Figure 2–11(a and c)). The gluing situation for the box beam is similar to that for the stressed skin construction, and essentially the same conditions prevail.

However, the I-beam configuration presents special problems due to the very narrow engagement between the web and the flange. Gluing the narrow edge of the plywood web to the face of the flange element does not provide enough bond area to carry the shear forces it is called on to transmit.

Consequently additional bond area must be contrived. This is done by rabbeting out a groove down the center of the flange elements, sized to accept the thickness of the plywood web, Figure 2–11(a). Glue, normally of the resorcinolic type, is applied in the groove and the plywood edge inserted. In order to assure good bonding, both the sides of the groove and the plywood edges are sometimes tapered so that insertion can occur without wiping away the glue, and a positive pressure can be applied to properly close the joint. A thin kerf the depth of the groove may also be cut into the edge of the plywood web. This allows a springlike deformation to occur as the web is pressured into place, assuring continued pressure on the glue line and avoiding a splitting action on the flange.

Paper Products. The secondary gluing of paper has a number of interesting applications: applying wall paper; bookbinding; producing high-density laminates; applying impregnated paper on lumber, plywood, and reconstituted products; and manufacturing corrugated board and honeycomb, tubes, bags, envelopes, and the like. Each has its own unique technology and its own world of industry. Some representative processes are briefly overviewed by way of illustrating the adaptability of adhesive systems.

Wallpaper. In the case of wallpaper, no more than hand-roller pressure is available. The glue must therefore have instant tack to hold the paper in place. Two systems are used, one using a water-base starch that is brushed on and the paper applied immediately with roller pressure. The paste-like glue holds while the water evaporates or diffuses to harden the glue. The other involves a tacky film applied by the manufacturer to the back of the paper. The paper is protected from sticking to itself in the roll by being overlayed with a release paper of very low adhesion. This paper is stripped off by the user at the time of application, and the film can then perform its bonding function to the wall. Labels, shelf papers, and other decorative papers can also be prepared for application in this manner. The age-old process of licking stamps and envelopes involves the prior application of a gum or dextrin film that is dried immediately to prevent sticking. They are prepared for use by rewetting with the tongue or other means.

Bookbinding. The art of bookbinding is traditionally based on the use of elastomeric adhesives, solvent-based in order to avoid wrinkling the paper. Evaporation of the solvent forms the bond. Hot melt adhesives also have properties useful in paper bonding. Being solventless and fast hardening, they permit rapid assembly. Heat-activated films that can be applied with a hot iron have similar qualities.

Bonds normally do not have to be any stronger than the paper substrate. Hence strength of the adhesive is not an important criteria. In a contrary sense, low or reversible strength may be more important. This permits undoing a bond when repairs are necessary. Heating, steaming, or wetting will usually reverse a paper bond.

High-Density Laminates. Kitchen countertops are often covered with decorative scratch- and heat-resistant material known generically as high-density laminates (Formica, for example). They are comprised of kraft paper steeped in low-molecular-weight phenolic resin, redried in such a way as to remove the water without curing the resin. The resin, because of its low molecular size, is able to penetrate the fiber wall as easily as water, and in fact replaces it. A number of these redried sheets are then assembled into a pack, and pressed at high pressure (1000 psi or more), and temperatures of about 300°F against mirror-smooth or embossed steel cauls. Unlike plywood manufacture, this laminate must be cooled in the press to about 200°F in order to avoid blisters and delamination, a major drawback to the process not only for the extra time in the press but also for the strain on the press as it goes through successive heating and cooling

cycles. The resin both stabilizes the fiber against moisture and provides the bond holding the sheets together.

The product emerges from the press with a dark brown to reddish color, and as such probably would be used as the back of a panel to balance the construction against warp. To provide decorative properties, the top sheets, imprinted with decorative patterns, are steeped in melamine resin, which imparts a clear, glasslike finish when cured. A thin, plain, tissue-type sheet provides the final protective layer.

Paper Overlays. A paper prepared in a manner similar to that in high-density laminates, phenolic-resin-impregnated, is used as an overlay on softwood plywood, flakeboard, and particleboard, and even on lumber, to improve weather resistance and paintability. The paper is considered a high-density overlay when it contains not less than 45 percent resin, and because of the high resin content provides the maximum degree of protection for the substrate. A paper with less resin content, about 20 percent, designated a medium-density overlay, would be used for less demanding conditions. Papers with lower resin content are also in use. These overlays serve a number of purposes: They reduce surface checking and splintering, increase stability, and provide a smoother surface. The paper with the highest percentage of resin requires no additional adhesive to bond it to a substrate. For the others an adhesive must be used as in applying veneer.

In the case of plywood and reconstituted panels, the question arises as to whether to apply the impregnated paper at the same time that the panel is consolidated or at a later time. Efficiency favors the former because it saves one more cycle through a hot press. With presanding veneers, and care in assembly, one-shot pressing is feasible for paper-overlayed plywood. The other panel types can also be overlayed at the same time that they are consolidated when surface elements are uniformly felted, and especially when they are of the smallest sizes.

There is, however, a case for overlaying as a secondary gluing process. Panels often have rough or defective surfaces out of the press. By sanding and repairing, these panels can be reclaimed and upgraded to a higher value by overlaying. Since there is only a relatively thin paper element between the glue line and the platen heat source, the press cycle for the second gluing can be very short.

Corrugated Board. The production of corrugated paper presents a special challenge: speed. The very high speed at which corrugating machines travel, up to 1600 feet per minute, means that bonds must form in a fraction of a second. In a typical operation, three sheets of paper unwind from rolls and are conveyed to the corrugator, the first and third being high-strength paper that will form the faces and represent the strength-giving element of the board. The center sheet is the corrugating medium, specially composed to be absorptive and formable. It feeds between nested fluted cylinders and is formed to the corrugated configuration. While still engaged, glue is kissed onto the crests of the flutes and pressed against one of the faces as it flies by over a steel roll. At this point, a corrugated core has been glued to one face sheet, already a useful material for wrapping and protecting products for handling and shipment. However it may also proceed through the machine, receive another kiss of glue at the crests, and the other sheet pressed on to form the finished corrugated board. Multiple layers can be added to produce thicker and stronger boards. Since these boards are stronger in the direction of the flutes than across the flutes, this fact must be taken into consideration when fabricating boxes and other load-carrying uses.

It is of interest to consider the pressure factor in the corrugation process. While pressure to form the bond is high in attaching the corrugating medium onto one face, the first step, it is very low in attaching the other face because only the sidewise resistance of the flutes is available to receive pressure.

Aside from the machinery to move the paper and apply heat and pressure in forming and attaching the corrugations, the essential ingredient is an adhesive that can convert from a liquid to a solid in the very short time available. Actually the adhesive is neither completely liquid nor does it need to completely solidify in the machine. It is closer to a paste in consistency, and it only needs to convert to a firm gel to hold the paper together against the low strength of the paper, links 8 and 9 in this case. Two adhesive systems have been found useful, starch-based, and silicate. Both are waterborne and therefore bond formation depends only upon removal of water from the glue line. Besides the high absorptivity of the substrate, heat is applied to accelerate the dispersal of water from the glue line and to restore the dry strength of the paper.

Paper Honeycomb. The manufacture of honeycomb appears at first glance to be a complicated gluing operation, but it is really rather simple. Sheets of paper are passed through a glue spreader that is designed to apply glue in thin lines a fixed distance apart. The lines of glue on alternate sheets are offset midway between those of the previous sheet. A stack of sheets is then pressed until the glue hardens, making a composite held together by lines of glue. This is sliced crosswise of the glue lines in thicknesses of the desired honeycomb. Pulling in a direction perpendicular to the glue lines opens the cell structure. Glues used can be any that provide the durability and economics of the intended use. The paper itself can be treated in various ways to improve stiffness and add moisture and fire resistance.

Many different configurations can be made that allow the honeycomb to be formed to simple and double curves when the design of the sandwich panels calls for such conformation.

TERTIARY BONDING PROCESSES

The bonding processes discussed in the previous sections produce primarily flat or linear materials. Considerable value has been added but relatively little worth; worth being defined as the ability to deliver functions of the final product. A board, for example, has value, but it has limited ability to deliver a useful function as a consumer product. Not until it is fastened to other boards to form a structure such as a house or a piece of furniture does it attain ultimate value. This is the shape-generation stage, the formation of volumetric or three-dimensional structures that enclose and create useful space. It is, in short, the process of making corners, the realm of the craftsman, where joints become crucial and a relatively insignificant amount of glue assumes the total responsibility for performance of the product. The nature of this and other fabrication operations—the cutting, fitting, and fastening of parts to form structures—makes it the most costly of all the intervening processes in the conversion chain from tree to market.

Bond formation and bond performance in assembly joints may be considered to be the most difficult in the entire scheme of wood gluing. Many of the critical factors in the equation of performance are at their worst:

1. Surfaces may be ill prepared.
2. Fit varies and can be either too loose or too tight.
3. Pressure is indeterminate.
4. Crossed-grain effects are at a maximum.
5. Considerable wood occurs back of the glue line.
6. There is great potential for high stress.
7. Different species may occur across glue lines.
8. The disposition of glue is somewhat haphazard.

On the positive side, the glue line in assembly joints has a simple job: developing shear resistance, sometimes merely mechanical in nature, between the components of the joint, its easiest task. Some gap-filling action by the glue line is sometimes also needed when joints are too loose. Tensile stresses perpendicular to the glue line may also develop in some joints when moisture-content changes occur.

In making assembly joints, Figure 7–14, the essential considerations are strength and ease and speed of assembly. Durability is seldom considered since these joints are mostly for indoor applications. However, there is need for a durability that resists high and low humidity cycling such as might exist in atmospheric extremes. Normally assembly joints are not expected to resist soaking or boiling in water, though adhesives with that kind of durability may sometimes be used.

From a bond-forming and performance point of view, factors reduce primarily to amount and direction of grain on either side of the glue line, quality of surfaces, and fit. Much of this has already been considered in Chapter 7 since it is geometric in nature. The gluing procedure, once the joint configuration has been executed, is comparatively simple: The glue is brushed, wiped, dipped, or squirted onto the joint surfaces and the structure assembled. Clamps are applied, briefly, to draw joints together and to assure squareness or other alignment necessities (Figure 9–28).

The simplicity of execution belies the difficulties experienced by the glue in trying to form a bond. In most cases, the act of bringing joint surfaces together has a tendency to wipe the glue from the surface. Unfortunately this tendency is greatest when joints are best fitted, an argument for looser joints. The action compounds when surfaces contain torn or loose fibers. The small amount of adhesive can only bond to these fibers, producing a bond no stronger than already broken wood. Other problems in obtaining good bonds in dowel type joints were discussed under fit of joints in Chapter 7.

Since dimensions and surface locations are fixed by the geometry of the joint, very little if any pressure can be delivered to the glue line to aid it in carrying out its mobility-dependent functions. Clamping serves mostly to rigidize the structure while the glue solidifies, rather than providing pressure to the glue line, a major difference from other gluing processes.

Because the period during which the product must remain in clamps often controls the

Figure 9-28   Case clamp for drawing together the assembly joints in a dresser. (Black Bros. brochure)

rate of production, glues are desired that develop strength fairly rapidly. The gap-filling action is not well served by the adhesives normally chosen for assembly gluing. The best for gap filling would of course be those with 100-percent solids content such as epoxy, polyester, and urethanes that exhibit little shrinkage in curing. These suffer on the ease-of-use requirement, since they are two-component systems and need to be mixed before using. Moreover, they have such a short pot life that frequent mixing of small quantities would take up too much operator time.

The choice reduces to one-component, or ready-to-use, systems. These—primarily the hide glues, polyvinyls, and aliphatics—harden by loss of water. They form a firm enough gel in a short time to permit unclamping in about the same amount of time it takes to prepare the next assembly for clamping. The chemically curing resin glues in water suspension are not generally used in assembly gluing due to their long cure times or the difficulty of heating the glue line to accelerate cure. However, when extreme service conditions are anticipated, they can be

compounded to form good bonds in these types of joints.

After the bond has been made, performance of the joint under service conditions then depends to a large extent upon the grain direction on either side of the glue line and the amount of wood engaged by the glue line. As mentioned previously, the grain is usually crossed at 90-degree angles and with considerable thickness. The significance of this geometry lies in the effect of shrinking and swelling in creating stress on the wood when moisture content changes.

M. L. Selbo (1953 Forest Products Journal) in a comparison of the performance of various glues in different types of assembly joints subjected to moisture cycling, found distinct and sometimes unexpected differences between glues. He observed, for example, that none of the glues were able to hold the corner block in place, confirming the power of cross grain when there is a substantial amount of solid wood back of the glue line. In contrast, in what may be considered an opposite extreme, the responses of the different glues to parallel grain were relatively imperceptible. Nevertheless, as noted in primary gluing of lumber, the stresses at end-grain edges are particularly destructive when generated by steep moisture gradients because they operate in tension perpendicular to the glue line, its weakest direction. Steep moisture gradients occur due to the fast rate at which moisture is adsorbed or desorbed through end grain, compared to that through side grain. (Steep moisture gradients can be reduced considerably by suitable coatings at end-grain surfaces.)

The effect of steep moisture gradients operates at all assembly joints that include end grain. Slip, or multiple-tenon joints and open-dovetail joints, experience a double effect at the bond line. In these joints, shrinkage due to moisture loss also generates shear stress due to the crossed grain orientation, which adds to the tension stress to produce a combined stress that is even more destructive. In the case of swelling, another effect cuts in. The glues normally used in these joints are sensitive to moisture. This means that when the wood is swelling due to moisture pickup and creating stress, the glue bond is also picking up moisture and losing strength. Although increased stress with decreased strength of bond can lead to permanent loosening of the joint under extreme moisture conditions, under milder conditions the slight softening of the adhesive may allow it to accommodate to wood movement without breaking.

Dowel and Mortise-and-Tenon Joints

A special problem generates in dowel and tenon-type joints that is quite independent of the bond. This is the interaction of the grain of the dowel or tenon with the grain around the hole or mortise. Whereas the dowel and the tenon engage with side grain, the holes engage both with side grain and end grain (Figure 7–15). In the hole, the grain on either side of the glue line is crossed. Because the dimension of the hole in the end grain direction does not change with moisture change while the dimension of the dowel does, an event peculiar to wood takes place. Under

swelling conditions, the dowel attempts to increase its dimension, but in the vertical direction, encounters the stable end grain area of the hole. Unable to expand dimensionally, the dowel expands internally, its cell wall material swelling into cell cavities. This creates a new swollen dimension, something less than it would have been if swelling had occurred freely. When moisture content subsequently declines, shrinkage begins from this new dimension, and proceeds to a new shrunken dimension, which is something less than it would have been without the restricted swelling. The action is known as "compression set," a situation that causes loosening of axe handles. The result is a tension stress across the glue line at that point.

Several consequences can emerge from this state of affairs: (1) The bond can fail, which is likely considering that it may have been weak in the first place due to the end-grain porosity and poor attachment possibilities at end grain surfaces; (2) the wood can fail either in the dowel or the hole depending upon which was weaker; if the hole were cut with dull cutters, its surface would be the weaker; (3) under the best case (a well-made bond and normal moisture cycling), the bond would hold, with the caveat that repeated cycling could cause fatigue failure in the bond or in the wood.

Ninety degrees around the dowel another stress situation is going on with moisture changes: the crossed-grain effect, creating shear stress during both moisture increase and decrease cycles. Here the bond is likely to be better, side grain to side grain, and will normally hold through all but the most severe changes.

Meanwhile, at the other end of the dowel, the situation is generally much better. Not only is there side grain to side grain, but the grain is parallel on either side of the bondline. Consequently, moisture changes produce very little if any stress on the glue line, except as the wood produces differential movements due to density variations, radial versus tangential orientations, or widely varying original moisture contents.

These same interactions between the glue

line and the wood on either side of the bond occur in all assembly joints more or less depending upon their geometries. They may be analyzed in like manner to reveal the factors that may be operating in the thousands of joint configurations the mind can conjure. The picture of assembly joints has been drawn to show the forces at work. This is not to imply that they are inadequate for the purpose of holding parts together for most uses, but to suggest that they are vulnerable to mismanufacture and misuse, and that cautions are necessary.

Miter Joints

The miter joint encounters special bond-forming and performance factors because of its geometry, which places a high degree of end grain at the glue line and a differing amount of wood back of it. The end grain, as has been mentioned repeatedly, is difficult to bond because of its high porosity, which encourages excessive penetration with consequent starving of the glue line. End grain also provides limited surface area for suitable anchorage of the glue. The fact that the structure is a corner where large racking stresses concentrate makes the joint more vulnerable than a scarf joint. Both the internal stresses and the external stresses place the bond line under tension stress.

Two situations apply to miter joints, and to a certain extent to other end grain joints. Irregularities in the surface of end grain joints are virtually impossible to compensate by increasing clamp pressure. The high strength of wood in compression parallel to the grain comes into play here in proportion to the angle of the grain with respect to the pressure against it. This high component of compressive strength means that high spots on the glue line may not compress with clamping pressure, and any gaps they create will remain as gaps in the completed joint.

A second point stems from the general purpose of a miter joint, which is to produce a rectangular structure. Clamps are therefore often set up to produce the structure with precise angles (Figure 9-29). This

means that if angles are not cut precisely, the surfaces cannot close completely, and the joints will remain open for the glue to cope with.

Several procedures can be used to improve bonding to such surfaces. One is to size the joint surface with a thinned-down mixture of the same glue to be used in later final bonding. This thinned film is allowed to dry or cure to a semisolid state, but not to the fully hardened state. The joint is then completed as normally done. Alternatively, the mobility of the glue may be reduced either by withholding water or by adding fillers. Another procedure, which is even better from a bond-formation standpoint, is double spreading and prolonging the assembly time. This allows thickening of the glue on the surface where it can start to form molecular attachments while it is waiting to be assembled.

Choice of glues for miter joints is essen-

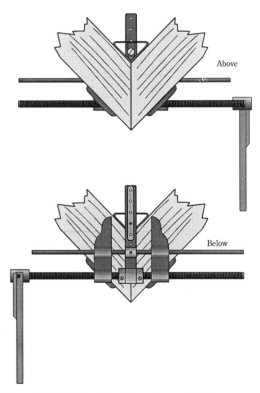

Figure 9-29 Miter joint clamps (note the 90° pressure directions)

tially the same as for other types of assembly joints. The aliphatics have an edge because they are ready to use and have a slightly higher viscosity than other assembly glues. The catalyzed vinyls of high-solids content may be even better, particularly in high-speed processes where dielectric heating is used to accelerate the cure. Both produce a hardened film that is tough and able to withstand the workings of this joint through moisture changes and racking stresses. Since the geometry of the miter joint allows dielectric heating, most of the other resin glues can also be used for rapid production when properly formulated for end-grain application. The 100-percent solids resins—epoxy, polyester, and urethanes—also provide a special advantage. Since they contain no water, the joint does not have to begin its life with swollen surfaces. Their high cost and their mixing needs, however, would preclude their use in any but the most demanding situations. In any case, an adhesive with some flexibility rather than brittleness is desirable.

Once the bond has been formed, its performance faces some of the most virulent of stresses, all in tension across the glue line. In the first place, rectangular designs produce the most unstable structures, racking and deforming at very little loads, as may be noted by the ease with which a box without end walls can be squashed. Racking tends to open corner joints through a high moment lever action with one edge of the joint pivoting against the other. Moisture changes produce a similar effect but with a different mechanism. The action here is generated through the different amount of wood back of the glue line from the inside corner to the outside and was discussed in Chapter 7.

Because of the vulnerability of miter joints, a number of methods for reinforcing them are in use. The most common of these introduce the peg or slip-joint effect to help carry stress around the corner. When material thickness allows, dowels are inserted across the glue lines, otherwise a slot is cut across the corner and a veneerlike slip of wood is glued in the slot with its grain across the glue line. Alternatively, the joint can be redesigned to include a lap effect providing more gluing surface and placing some of the joint in shear (Figure 9-30).

The above has dealt with the formation of miter joints in producing a frame. Similar considerations apply to the production of a box structure. The longer corner dimension, however, demands greater accuracy in cutting since both the miter cut and the cross cut must be true to a finer degree. Although box shapes usually have backs or bottoms that reduce racking stresses, open miter joints are as unsightly as they are weak.

The previous discussion has centered on one aspect of forming the box shape. It is instructive to contemplate the comparative ease with which corrugated boxes are made, and to sense the simplicity of the folding miter joint.

In making the corrugated board box, a sheet of predetermined size is scored (stamped) along lines where folding will occur, and slits are made to allow folding top and bottom. The board is folded onto itself, and one joint is made to fasten the sides together at a corner. The fastening is made as a butt joint held with adhesive tape or as a glued lap joint using a fast-setting glue such as hot melt, starch, or silicate. At this point the box can be collapsed to a flat structure for compact shipping. At destination, the bottom flaps are folded in place and secured with tape or fast-setting glue. After packing, the box is closed with the remaining flaps

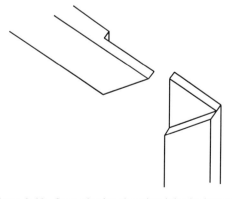

Figure 9-30 Strengthening the miter joint by increasing bond area and including side-grain the miter lap.

and taped or glued. Very little time elapses between applying the glue and completing the bond; the process is virtually continuous.

One might well wonder how such efficiency can be brought to the assembly of other wood-based products. In the case of corrugated board, success bears on two factors: the relatively low strength of the substrate that allows folding to shape, and the use of low-strength, low-durability, but very fast bond-forming glues. For stronger substrates, stronger glues are needed that can perform like the welding of metal, and structures need to be designed where corners are less stressed, perhaps rounded or pre-formed.

A step in producing a box shape from stronger material is taken by a method of recent origin that involves the use of film-overlayed particleboard. After cutting to size, the material is scored with 45-degree V-grooves where the four corners are located. The cuts are made from the back side down to the film but not touching it. When ready for assembly, glue is poured in the groove and the box folded to its square shape. The film functions as a hinge and as a clamp holding the corners together while the glue hardens. The film also functions as a decorative overlay, resulting in a maximum in added value with a relatively simple operation.

and floor and wall coverings, structural glues that amend to the vagaries of on-site conditions have improved the performance of subflooring and the installation of wall paneling.

The squeaking of floors is a familiar sound in many houses. It is due to the rubbing of wood against wood and against nails when they have not been driven tightly or have become loosened in service. Less familiar but equally discomfiting is the deflection or bounce of some floors as live or dead loads impose their effects. Both are improved by the use of glues bonding the subfloor to the floor joists. Glues for this purpose are designated "construction adhesives" and are closely specified by the American Plywood Association to have the properties for on-site application. They are applied as a bead extruded from a cartridge or gun along the top edge of the joist (Figure 9–31). The subfloor is then nailed down as its normally done. After hardening, the glue performs a number of functions as it bonds floor to joist. One is to level low spots and fill any gaps that might exist, thus eliminating areas where small deflections might occur to cause squeaking. Another is to form a good enough bond to provide an I-beam effect (T-beam in this case) between the subfloor and the joist, thus boosting the stiffness of the total system.

House Construction

Gluing has not participated to a large degree in the fabrication stage of large structures such as houses and other buildings. There are several reasons for this, chiefly that on-site conditions do not provide a consistent environment for the needed durable adhesives to form reliable bonds. Temperature, pressure, and moisture conditions sometimes reach extremes that mitigate against bond formation. Moreover, should a bond be formed, the surface attachment it achieves may not be able to carry the loads imposed by the larger components. But there are important applications for smaller members. In addition to the traditional gluing of shingles

Figure 9–31 Applying a bead of glue to bond subfloor and joists for greater stiffness and less squeak. (USFPL)

Decorative wall paneling is normally nailed to studding or furring. This is satisfactory in many cases. But the more finely detailed surfaces may be disfigured by nail heads, missed hammer blows, nail popping, pull-through, or discoloration. The use of glues formulated for this purpose eliminates these problems, and with experience makes for a neater application.

The glues are applied in bead form, as for subfloor gluing, and the panel pressed in place. In this case the weight of the panel works against the process and it must be supported. This is done by driving a few nails along the top and bottom of the panel where they will be hidden by moldings. Warp in the panel may also interfere with bond formation at the early stages, pulling away from the wall before the bond has a chance to develop holding power. If there is a buckle in the center areas of the panel, some means of temporary bracing needs to be contrived. One such brace might consist of a board wedged against a block fastened to the floor and angled to the wall to press against a plank placed across the panel. If the warp is near an edge, a block of wood may be temporarily fastened at that point by driving a brad or other thin nail through it and into the crack between two panels. The brace and the block can be withdrawn when the glue has hardened sufficiently.

COMBINED PRIMARY AND TERTIARY PROCESSES

In these combined gluing processes, the motivation primarily rests on the desire to shortcut the expensive craft assembly stage in producing the shaped products needed in commerce—avoiding the production of corners as a separate operation. The processes involve molding to shape rather than fabricating to shape. By combining primary and tertiary (as well as secondary) gluing, advantage is taken of the formability of individual wood elements, as opposed to their consolidated state when they become more rigid and act as solid wood. As individual pieces, their thinner cross section allows them to be bent

more easily around corners, thus achieving the continuity of grain essential to providing strength in perpendicular directions. Corners in molded products are seldom sharp as in fabricated systems, but usually rounded to accommodate to the limitations both in the bending properties of wood and in the delivery of pressure by molds.

While molding produces shaped articles in one shot from wood element to finished part, some disadvantages exist. Molds themselves are costly because they require high skill in carving or otherwise machining mold contours and surfaces to exactly replicate the shape desired. More importantly, the design of the mold must assure that the internal volume exactly matches the cross-sectional dimensions of the part to be formed. Hence molding processes favor the production of large quantities of the same configuration. Many costly molds are needed for mass production and for inventorying different-shaped articles.

The gluing operation in molding processes is much the same as in primary gluing processes except for the manner of delivering pressure to the glue line. In some cases differences exist that compel additional considerations for the glue line. A brief survey of representative products made with each of the wood elements will illustrate some of the pertinent technologies that apply.

Lumber

Perhaps two of the most spectacular products in gluing lumber are the laminated arch, and the boat keel (Figures 2-4(a) and 9-32). The lumber is prepared and surfaced, and the glue applied as in laminated beams. Clamps are also the same, except that they are deployed and rigidly fixed into the shape desired plus a little extra bend to account for springback that inevitably occurs after release from the clamps. The amount of bend that can safely be made without breaking is calculable from the properties of the wood species, its moisture content, and its thickness. This is obtained by solving for y in the equation for modulus of elasticity, using the

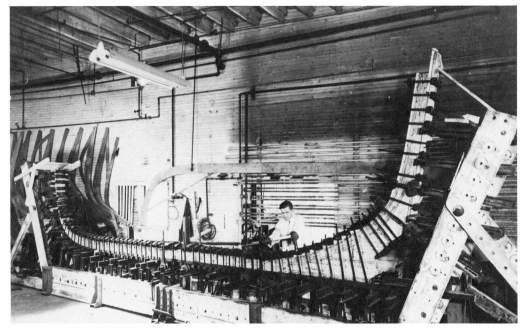

Figure 9–32   A laminated boat keel clamped to its designed curvature. (USFPL)

value for the proportional limit and for E at the given moisture content.

After the assembly is placed in the clamps, pressure is applied in sequence beginning at the center of the arch and proceeding toward both ends. The reason for this procedure is that some slippage must occur between laminations as bending takes place, producing the stepped appearance of the lumber at the ends. If the ends were clamped first, this type of deformation could not occur, and it would be more difficult if not impossible to make the assembly conform to shape.

The slippage necessary to allow bending to shape before bonding must be arrested by the bond so that reverse slippage does not occur, or else the boards will straighten. To do this, the bond must not only be strong but also totally unyielding. This places a constraint on the type of glues that can perform satisfactorily. They can only be those that are highly cross-linked and rigid. Since the laminations had to be forced into their curved shape, the stresses thus imposed linger in the system and provide a constant force tending to restore the wood to its origi-

nal shape. Two types of stresses are thus created that the glue line must resist: tension perpendicular to the glue line as the laminations try to spring back to a straight aspect, and shear as the laminations try to slip upon each other.

Since heating the glue line is difficult the choice of glues is narrow; resorcinolic or urethane if for exterior uses, and casein (assuming structural quality) if for interior dry uses. Other chemically cross-linking, cold-setting adhesives would be suitable if the cost and durability were acceptable. Ureas are specifically excluded because of their potential for deterioration under some use conditions, such as high humidity and heat. All glues of a thermoplastic nature are also excluded since they deform under continuous stress.

Veneer

The molding of veneer into shapes has many ramifications, although divided into only two main classes on the basis of whether the veneers are assembled with grain all parallel or crossed as in plywood. The plywood

types, using veneer in sheet form, are preferably of simple curvature, or of mild double
curvature (Figure 7–8(b)). Because wood is
anisotropic, wide sheets cannot easily be
bent in more than one direction at the same
time without breaking in one of the weak
planes. In order to form the plywood structure into more severe double curvature, such
as salad bowls (Figure 2–4(d)), boat hulls, or
airplane fuselages wide sheets must be reduced to narrow strips. This reduces the
amount of cross-grain deformation that
must occur with each element, and therefore
the tendency to break is lessened.

In laminating veneer with all the grain in
the same direction, the object is to recre-
ate the strength of solid wood in a shaped
product such as the chair legs in Figures
2–13(b) and 7–8(b) and the golf club head in
7–8(a). This construction is designed to carry
around a corner the stresses normally running along the grain. Because the grain in
this case is continuous around the corner,
there is very little loss of strength, if any. In
this respect, this type of construction is far
superior to dowels or mortise-and-tenon
joints, since there are no holes in either
member at the corner, and no cross-grain
differential movement is possible.

Simple Curvature.   Plywood formed to a
simple curve begins as a normal plywood operation, in which a stack of assemblies is first
formed. These are then placed in a press in
which the headers are shaped to the desired
curve (Figure 9–33). Pressing then continues
as in cold pressing plywood. Simple curved
plywood shapes (as well as mild compound
curves) can also be made in hot presses by
heating the molds with steam, hot oil, or
electricity. Molds may be nested to provide
multiple openings as for flat plywood.

A reduction in mold costs is achieved by
using dielectric current for heating. In this
application, the molds can be made of wood.
The mold surfaces are fitted with sheet cop-
per to form the electrodes. Pressure is applied as for cold pressing, with care to avoid
shorting across the electrodes. In this configuration of the electrodes, the entire assembly

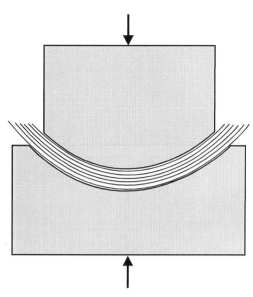

Figure 9–33   Form for molding veneer to simple curvature.

of wood and glue is heated to the curing temperature. (This contrasts with edge gluing of
lumber where only the glue line is heated.)
As far as the glue line is concerned, its atmosphere is similar to normal hot pressing except that there is less temperature gradient.
As a consequence, a moisture gradient is also
less since the temperature rise is relatively
uniform through the thickness of the panel.
The nature of dielectric heating in fact tends
to dissipate moisture concentrations because
areas of higher moisture draw more current;
more heat is generated there, energizing the
moisture and causing it to diffuse. This phenomenon is the basis for using dielectric current to equalize moisture in lumber during
the last stages of seasoning.

In these methods of forming plywood constructions to shape, the adhesive functions
essentially as in flat plywood manufacture,
and therefore the same glues and gluing procedures obtain.

Products in which the grain is parallel in
all plies are made in a similar fashion—rigid
molds, hot- or cold-pressed. They are usually of simple curvature, narrower, and therefore somewhat easier to handle. Because the
grain is parallel, the plies tend to conform

more readily than cross plies where bridging across cell walls and irregularities reduces proximity and requires higher bonding pressures.

Multiple Curvature. The making of cross-grained veneer products having multiple or compound curves, such as boat hulls or salad bowls, involves gluing technology that is totally different from flat plywood in veneer geometry, type of glue, method of glue application, and heat and pressure system. Although the process is not as widely used since its wartime heyday, a brief discussion of the procedure will illustrate how factors can be dramatically modified to fit an unconventional situation.

The wood element is cut to narrow strips, one to four inches wide depending upon severity of the curve. For a salad bowl where radii are small, the narrower strips are necessary to allow conformation around the curves. On the other hand, the milder curves of a boat shell can be fitted with wider strips. The amount of fitting necessary can be illustrated by trying to form a sheet of paper over the curved surface. In order to remove wrinkles that will form, the paper must be slit where the wrinkles occur. The paper then overlaps and conforms. The amount and frequency of the overlap provides an indication of the severity of the curvature; the more overlapping, the severer the curves.

Strips are then coated with an adhesive, redried to a tack-free state, and set aside for assembly. Because of the nature of the resin, a phenolic, assembly time can be indefinite. In the case of a boat hull, the form may be a solid block of wood carved to the inside shape of the boat (the so-called male mold) and provided with a release agent over the surface. The strips are stapled to this form one at a time, and fitted tightly edge to edge until a complete layer has been formed. A second layer is then added with grain at an angle to that in the first layer. A third layer parallel to the first follows, with subsequent layers repeating until the desired thickness has been formed, all stapled in place.

When the assembly is complete, the entire structure is sealed in a rubber bag (hence the term "bag molding") and a vacuum drawn (Figure 9–34). The resulting atmospheric pressure draws the rubber tight to the surface, fixing the strips in place and stabilizing the assembly. The bag with its contents still under vacuum is then rolled into a cylindrical retort. A well-sealed door is locked and steam is admitted to the retort. The steam provides both pressure to bring surfaces together and heat to fuse the glue lines.

Without getting into the mechanical details, it is of interest to note the special events that have occurred. In the first place, the adhesive must be one that can be applied as a wet film, and dried without curing. At this stage penetration and wetting have taken place as liquid actions. No flow has taken place except that accompanying the spreading operation, and transfer is not needed since the adhesive was applied to all surfaces. Fusing of the resin-coated surfaces resin to resin and curing completes the bond. During the fusing period, however,

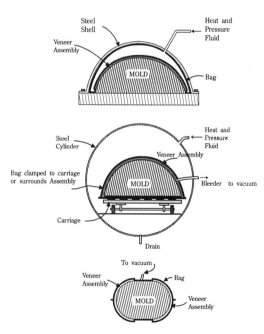

Figure 9–34   Three methods of bag molding to produce curved plywood. (USFPL)

the resin must maintain a fairly fluid state for a period of time, not only to allow flow but also to allow veneers to slip and settle into their final position. The fluidity of the resin during the fusing period must be controlled so as not to encourage too much penetration. This rather intricate scenario of glue line activity is played out with phenolic resins appropriately condensed and configured.

Also noteworthy is the pressure, which—being of a gaseous nature—is omnidirectional, meaning that it is always perpendicular to the surface of the veneers, no matter what their direction. Moreover, it is uniform from point to point across the surface, no matter whether there might be high and low spots. Veneers are brought together merely by bending rather than by compression of high spots as in platen pressing or rigid molds. As a consequence, pressure delivered can be much lower; in some cases as low as that provided by vacuum alone. A caveat here is that with pressure the same over high and low spots, producing equal bonding everywhere, these spots remain at their respective elevations and become part of the visible topography in the product out of the mold. In other words, the price of uniform bonding is an uneven surface.

Nevertheless, this is a good illustration of the desirability of some type of fluid pressure in all assemblies that contain variabilities in thickness of wood elements or their uniformity of deposition as in flakeboard. In the latter instance, a rubber blanket placed over or under the mat during pressing markedly improves horizontal density variation and with it strength variations, but with a rough surface that will need sanding.

The vacuum inside the bag performs another valuable function, absent from other forms of hot pressing: It continually removes moisture and other volatiles from the system, reducing the incidence of blisters.

In making smaller products with multiple curvature such as bowls, heated steel dies provide faster production. Otherwise the process is similar except that strips are interleaved instead of laid edge to edge.

Flakes and Particles

As the wood element reduces in size, conformability improves. Flakes eliminate the constraint imposed by length and width in forming to a compound curve. Finer particles contribute a small but significant measure of flowability such that some movement between wood elements vertically and laterally occurs in response to pressure and topography of the mold.

The process is the same as for making flat panels except that the mat is transferred to a mold instead of to a platen press. The transfer requires that the mat have a somewhat greater integrity. This is accomplished in the first place by using resins that have a little tack. Prepressing to achieve some compaction also helps, especially if it is also preshaped so as to better engage the mold without having to do all the conforming as the mold closes.

Both phenolic resins and urea resins are the primary binders used. Isocyanate formulations have a strong appeal due to their lower temperature requirement and higher moisture tolerance. Their higher cost is offset by their greater efficiency in the bond line, more forgiving of operational variables, and more stability in the consolidated product. The higher moisture content they tolerate is particularly beneficial in molding because it results in greater formability of the flakes. Figure 9–35 shows a chair part formed with strands and an isocyanate binder.

The finer particles cannot be consolidated with binders applied in liquid form not only because of balling up during blending, but also because too much moisture would be added to the mat. The use of powdered resins removes the threat of moisture problems in creating blows and blisters. To remove the threat further, molds are opened periodically, "breathed," to allow volatiles to escape.

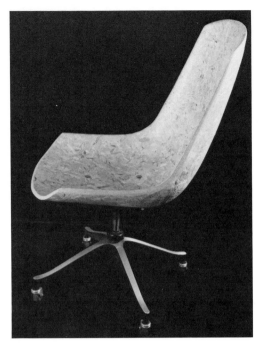

Figure 9–35 Strandboard chair shell consolidated with isocyanate binder in a heated concrete mold. (Courtesy Bruce A. Haataja, Institute of Wood Research, Michigan Technological University)

Fibers

Being the most formable of the wood elements, fibers enjoy the maximum in shape generation (Figure 7–9). Again, the bonding technology derives from that for flat panels where both wet processes and dry processes are in use. However, blending with other materials to form virtual alloys has greatly widened the property range of fiber-based products. From the apparently complex egg carton of low strength, low durability, and highly configured structure, to the highly engineered speaker cone, the furnish and the process follows that of low-density fiberboards. A slurry is made of fiber and water alone. A mold formed from a rigid screen is dipped into the slurry, and a vacuum is drawn through the screen. Fibers are drawn to the mold, forming a mat to the exact contours of the screen and fairly uniformly since vacuum is greatest where fibers are least. This formed mat is dried briefly and then removed from the screen to be further dried unsupported. Enough strength ensues for the intended uses.

Additives such as phenolic resin and sizing agents, precipitated onto the fibers while in the slurry, improve strength, surface smoothness, and durability against water when hot pressed in molds, making products for more severe uses, such as interior car parts. Instead of going directly to molds, these mats can be dried in flat aspect and sold as molding stock for later forming in customer's molds.

In the dry-process version, flat mats are made, sometimes needled, (a process of driving prong-tipped needles through a mat), to improve integrity for handling and shaping, and then molded as in the wet process. Ingredients, blended dry, can include phenolic resin, urethane resin, glass fiber, and synthetic fiber in varying proportion depending on the product. Factors such as strength, stiffness, surface smoothness, stability, and hardness that meet requirements for automobile exterior parts such as fenders and doors are possible with the proper mixture. This is the ultimate in performance of a wood-based composite, and it is fitting to end the chapter on this note.

10

Quality Control

GENERAL PHILOSOPHY

Throughout this book we have stressed the importance of many factors that influence bond formation and bond performance. Unfortunately, it is difficult, if not impossible, to determine how well the factors have operated in a given bond without destroying it by testing. Although some nondestructive methods exist for observing the quality of certain glued products, such as proof-loading of laminated beams, by and large one is left with the decision of whether to sacrifice a product for information or to sell it for profit. The best means of avoiding such a dilemma is to make it right in the first place. To insure this, it is necessary to assure that each factor is at the level stipulated for it, or within permissible limits while the bond is being formed.

If we have been successful in emphasizing the crucial nature of factors in the equation of performance, the glue user or researcher will not want to embark on a gluing project without knowing with some degree of certainty that everything is "proper." What is "proper" depends upon the job and should have been established from manufacturers' instructional bulletins, with perhaps further fine-tuning through the tenets proposed herein, and the particular conditions existing.

Responsibility

Despite strong protestations to the contrary, no factor can be assumed to be ipso facto on target. The safer assumption is that everything varies. Machines vary or drift; people tire or become complacent; wood properties not only vary inherently but by processing, they change with ambient conditions; glue properties change with time. A gluing operation cannot be made foolproof except by constant vigilance. The vigilance must be supported by "eyes" that can see changes and detect wayward behavior. Moreover, the observation that a wayward change is beyond acceptability should carry with it *the authority to make corrections and/or shut down the operation.* Failure-prone products in particular need the protection of tight control. In practice, this means that the person in charge of control must be able to override whoever is in charge of production, authority that has to be delegated from the highest levels of administration. Otherwise, quality control becomes quality observation, and the impetus for correction comes from the marketplace, too late to save embarassment.

Statistical Considerations

Wayward behavior comes in various cloaks. One, for example, comes as a trend, a gradual change in the same direction, unnoticed until products begin to fall apart. Another is sporadic, varying hapazardly on either side of an average value, occasionally swinging beyond the limits set for the operation and causing spotty failures. A third concerns abrupt changes to different levels but remaining constant for periods of time, causing a run of bad parts. Each type of behavior contains information that is valuable in tracking down the cause and instituting the appropriate corrective action as indicated in Chapter 12 on troubleshooting. These three types of behavior can be perceived only if checks are made frequently and at definite intervals. The observations can then be plotted over time to sense on a current basis the

nature of any potential adverse changes that may be occurring.

In monitoring the behavior of moisture content of veneer for a period of time, the plotted data would appear as one or another of the four situations shown in Figure 10-1. These plots show a target moisture content of 8 percent, with an upper limit line at 10 percent, and a lower limit line at 6 percent. In (c) the gradual trend situation is evident, going off limits by the end of the period. The use of this veneer must then be stopped at once, at the spreader if necessary. The continuous upward trend suggests something like a gradual loss of steam pressure for heating the dryers, perhaps running out of waste to feed the boilers. In (d), changes in

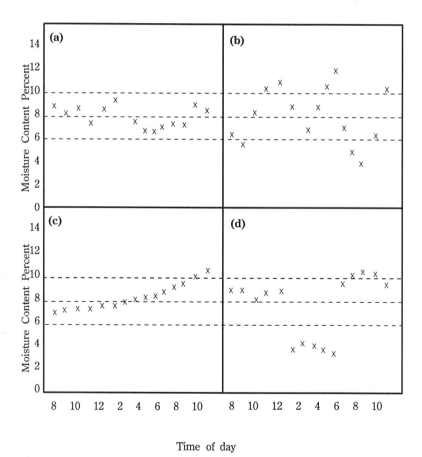

Figure 10-1 Recorded patterns of moisture-content variation in veneer: (a) operation under good control, (b) operation out of control, (c) operation losing control, (d) shifting controls.

moisture seem to occur in batches, though fairly uniform within batch. Such behavior may be due to changes in dryer settings, changes in personnel, change of shifts, or changes in material source. In (b), the drying is totally out of control, producing veneer that is both too wet and too dry and creating complete frustration at press. In (a) the process is in complete control with all observations falling within the predetermined limits.

Methods for setting up such charts on a statistical basis are available in most quality-control textbooks. They also provide instructions for assuring that tests are representative, for determining how many observations to make, how often to make them, and how to detect a spurious result. Means for separating out different kinds of variabilities from the same readings are sometimes useful where the method of measurement superimposes its own variability onto that of the object being measured, or when the material has a natural variability to be distinguished from that added by processing. This requires statistical analysis of well-gathered data, and reference should be made to texts on that subject.

Records

Despite all efforts, occasional glue-line failures occur, sometimes due to production error, sometimes due to customer mistreatment. In such cases, it may be comforting legally and personally to have records on file that tie the product to its conditions of manufacture. This means having the product carry a number that identifies its time of manufacture. Time in turn provides connections with possible material changes, personnel changes, machine changes, and so on.

In this chapter we present the various procedures and equipment used in monitoring gluing operations. They include methods for observing wood properties, glue properties, operational factors, and glued-product inspection. Some of the quality-control procedures use observations that are largely qualitative rather than quantitative, what is felt with fingers or observed by eye. In these in-stances, the translation of the observation to a go/no-go decision depends on experience and the knowledge of what an adhesive needs in order to form a bond. It sometimes needs what appear to be mutually exclusive actions: both penetration and no penetration; molecular conformation and gap filing; high mobility and low mobility. On a porous and naturally variable surface such as wood, striking the mobility balance that provides both extremes is the essence of the gluing art.

As far as bond formation is concerned, an ideal situation to serve as a benchmark would be one in which the glue was applied to both surfaces in a high state of mobility, but pressed at a reduced state of mobility. Thus the important action of penetrating and wetting have an opportunity to occur spontaneously as much as possible. If the adhesive can then firm up a bit, the applied pressure will complete the mobility-based actions without losing glue solids needed to form link 1, and at the same time provide any gap filling that might be necessary. The practicalities of production and economics, however, force a reduced regimen, often one that operates near the extremes of permissible ranges for individual factors. In such cases, even normal variabilities can at times compound to defeat the mobility requirements of the adhesive.

Finally, in reaching for the ideal, it would be technologically appropraite to have a systematic study of the entire operation with the view of determining the magnitude and ranges of all factors and their interactions. A context of known variables is thus formed that can serve as a base from which to judge effectiveness of adjustments. The operation can then be stabilized and controlled with a minimum of spot checks.

MONITORING WOOD PROPERTIES

One of the immutable laws of the glue room that should by now have been deeply inscribed on the walls of the mind is that the properties of the wood presented to the glue are as important as the glue itself in the actions of creating a bond. Moreover, the wood element carries a greater burden of variability and therefore a greater requirement for vigilance in assuring good bonding.

As wood elements approach the gluing operation, the most important and least obvious characteristic is its moisture content. Also important, though more obvious, is the topographic quality of the surface. Roughness and fit of joints, for example, are qualitites that can be observed visually and, with experience can easily be separated into the two groups of "acceptable" and "nonacceptable." Similarly, trueness and warp, affecting mostly lumber and edge gluing of veneer, are relatively easy to observe. Moisture content, on the other hand, requires measurement techniques, and because of its propensity to vary with environment, the measurements should continue to be made up to the time of gluing.

Subsurface damage in most cases would seem to be a constant of a particular surfacing operation. However, in veneer and flakes, and to a lesser extent in lumber, the damage can vary with the preconditioning of the wood, with the settings of the machinery, and with the sharpness of knives. There are no formal tests for this property of a surface prior to gluing, except as a qualitative observation. After gluing, the condition evidences itself as low strength and shallow wood failure.

Surface inactivation—or one of its net results, adhesion failure—is a more subtle characteristic that can vary within the same piece of wood and can get worse with time. Here we are at the mercy of physiochemical properties, and more sophisticated procedures to get even a tentative observation. The inherent roughness of wood created by its cell structure tends to interfere with the accurate measurement of this property. Consequently we will be content with simple tests that only provide approximations.

The temperature of the wood as it approaches the glue applicators is often ignored in the rush of getting the job done. However, since it has a strong bearing on what happens to the glue during the assembly time and later in the press, this factor merits close attention.

Wood Moisture Content

Of the many ways of measuring moisture content of wood, only one is absolute, accurate, and repeatable. This is the oven-dry method. The others employ various electrical or dielectrical properties of wood that vary with moisture content. The latter, however, must be interpreted against species in deference to their differing densities and extractives, which also influence electrical properties. Temperature and chemical treatments affect electrical properties as well.

Oven-Dry Method. A specimen taken from the wood representing the material to be measured is weighed using a balance sensitive to at least one tenth of a gram. The result is recorded as "original weight." The specimen is dried in an oven maintained at 215 to 220°F, and then reweighed. The difference in weight is the water content of the specimen. This weight of water, divided by the weight of the dried specimen and multiplied by 100, becomes the percent moisture content of the specimen, oven-dry basis. In formula form:

$$\text{Percent Moisture Content} = \frac{\text{Weight of Water}}{\text{Weight of Oven Dry Wood}} \times 100$$

$$\text{or} = \frac{\text{Original Weight} - \text{Dried Weight}}{\text{Dried Weight}} \times 100$$

Some procedural cautions help in assuring reliable results:

1. Use as large a specimen as is convenient to reduce weighing error.
2. Dry the specimen until three successive weighings show no change.
3. In determining the moisture content of lumber,

   (a) Cut the specimen so as to expose as much end grain as possible; e.g., the entire cross section of the board, and $\frac{1}{2}$ to 1 inch in the grain direction.

   (b) Cut the specimen back from the end at least twelve inches to eliminate possible gradients that may exist there.

(c) Use a sharp saw, since heat from a dull saw can drive out moisture before it can be weighed.

(d) Remove splinters that might fall off between weighings and be counted as moisture.

(e) Use a scale sensitive to 0.1 gram.

4. In determining moisture content of veneer, place veneer specimen in a rack to assure air flow to all sides.

5. Flakes and smaller elements should be dried in a tared container to avoid loss of bits and pieces during handling. (Equipment is available for drying particulate materials that incorporates a scale, a pan, and a heat source to provide a direct reading of moisture content. The heat source may be infrared, microwave, or radio frequency.)

6. In all cases, overheating may drive off water of constitution and other volatile constituents, thus overstating the moisture content.

Moisture Meters. Moisture meters, in relying on the relationship of electrical properties to moisture content, are subject to other variables that also influence electrical properties, such as density and extractives as mentioned earlier. The effect of these factors, however, is taken into account on a species basis by calculated adjustments to meter readings. Nevertheless, to the extent that these factors vary within a species, the readings will be less accurate. The resulting variability is partially compensated for the ease with which readings can be taken; more readings will be made and a workable average deduced.

Two main types of moisture meters for solid wood exist: resistance types and capacitance types. Resistance meters have needle-like probes or prongs that are driven into the wood. The meter senses the resistance to a direct electrical current imposed between the needles. It will reflect the path containing the most moisture—the least resistance. Resistance meters (Figure 10-2 (a)) are most suitable for lumber. Though not requiring specimens to be cut, these meters do leave pin holes in the wood that may be objectionable for some uses.

Capacitance meters do not make holes in the wood. The capacitance, being a function of the power loss or impedance to a radio frequency current, can be sensed through electrodes placed on the surface of the wood. These meters, though somewhat less accurate than resistance meters, are preferred for veneer moisture content measurement. One method of continuous monitoring of the output from a veneer dryer is shown in Fig-

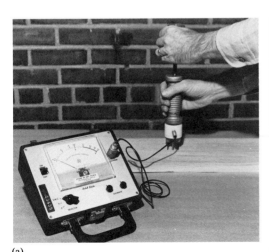
(a)

(b)

Figure 10-2 Moisture meters: (a) for lumber (USFPL), (b) for veneer. (Mann-Russell Electronics Co. brochure)

ure 10–2 (b). Spring wire feelers in contact with the moving veneer sense the moisture content on a 100 percent basis. These can be programmed to mark wayward veneer so it can be sorted out and recycled or reworked.

Methods other than oven drying for wood in particulate form are highly sophisticated in terms of their sensing mechanisms, but are still at the mercy of other wood variables, plus an additional one: bulk density. Generally these methods are intended to monitor moisture content of material on the move, either on belts or in air streams. They depend on the physical properties of water to absorb various kinds of energy, such as neutron, infrared, and microwave. Electrical resistance and capacitance properties are also invoked in special instruments that can produce a well-defined, though simulated, specimen. In one ingenious method the atmosphere surrounding the particles is sampled continuously and measured for humidity. In each of these methods any lack of accuracy is compensated for by uniformity, so that once the flow of material is stabilized and the instruments calibrated, readings become reliable. In continuous dryers, the outfeed air temperature is monitored as an indication of the amount of moisture being evaporated. When the temperature is higher than the calibrated temperature, too much water has been removed and the particles will be too dry. Too low a temperature indicates that insufficient drying is occurring. The rate of drying is then corrected by increasing the rate at which material is fed in if too much drying is occurring, and decreasing the rate of feed if too little drying is occurring.

Moisture Distribution. As important as the average moisture content within and between wood elements is the question of *distribution* of moisture within pieces. Uneven moisture distribution guarantees distortion before and after gluing, particularly in lumber and veneer as they attempt to equilibrate over time. In other products with smaller wood elements, uneven moisture distribution could cause local havoc on the glue line, but the multiplicity of elements ensures some leveling of the moisture distribution as a whole.

The main problem of uneven moisture distribution centers on lumber because of the large mass of wood involved with every glue line. Lumber has a significant probability of uneven moisture distribution both between the core and the surface, and between the center and the ends. Core-to-surface variation is due to inadequate drying, but center-to-end variation can be due both to improper drying and to improper storage. Even deadpiling cannot prevent center-to-end variation under prolonged storage when the relative humidity is different from that producing the original moisture content. In fact, deadpiling can promote variation, sometimes in only a few days of excessively humid or excessively dry conditions. A cartload of lumber parked near a radiator over a weekend in the winter time can become moisture imbalanced sufficiently to destroy its fitness for gluing.

In checking the core-to-surface variation, thin specimens are cut as previously described for measuring moisture content by the oven-dry method. However, before weighing and drying, the specimens are recut to separate the center from the shell or surface region (Figure 10–3 (a)). Shell and core are then weighed separately, and dried and reweighed to determine the amount of moisture in each region of the board. Several specimens should be taken at various intervals to determine the variability of the variation; i.e., whether there might be a local wet spot or whether it is general throughout the board. Further sampling would of course be needed to determine to what extent an entire batch of lumber is affected with this type of variability in moisture content.

Cutting up the boards to determine moisture distribution destroys the boards, unless they were to be cut up anyway. A procedure for measuring core-to-surface moisture distribution that does not require cutting is available in the resistance-type moisture meters. Normally one would expect the current to find the path of least resistance and indicate the maximum moisture without telling

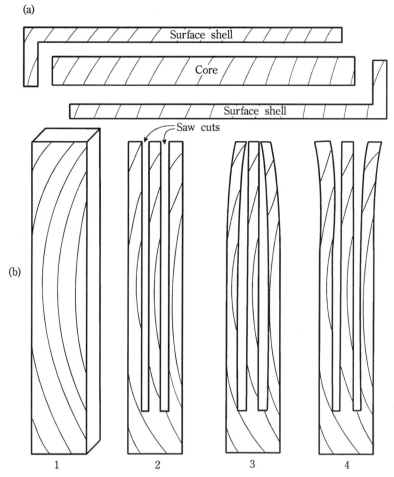

Figure 10–3   Specimen-cutting configuration for (a) measuring core-to-surface moisture distribution in lumber, and (b) observing the presence of casehardening. End-grain wafer from board ½″–1″ along grain (1). After saw cuts as shown: (2) no casehardening, (3) casehardened, (4) reverse casehardened.

where it is located thicknesswise. To provide a more positive sense of location, special prongs are provided that are active only at the tip. Thus the prongs will read the moisture at the particular depth to which they are driven.

Determining the moisture distribution from center to end involves similar techniques but applied at intervals from the end inward. As before, specimens may be cut and oven dried, or the resistance meter may be used.

A cautionary note: Moisture meters generally do not sense moisture accurately much below 5 percent and may be unreliable above the fiber saturation point. However, within the range of importance to gluing, a properly operated moisture meter can provide the information needed for monitoring an operation.

As a final note of caution, wood moisture content is not a fixed characteristic like porosity, but changes with ambient relative humidity. The changes can occur quite rapidly at end-grain surfaces. It is therefore prudent to store wood in a controlled atmosphere while awaiting gluing, especially if long delays are expected. Because end-grain surfaces are the location of rather steep moisture gradients, and hence sources of severe internal stresses on the bond, performance of the bond in service is improved if the end-grain surfaces are well sealed as soon after manufacture as possible.

Dimension Control

Dimensions that are important to bond formation are thickness uniformity in veneer and lumber to be face-glued (and to a less extent in flakes), and width uniformity in edge-glued lumber and veneer. In the smaller wood elements, particle dimensions relate more to product performance, though they impact gluing by affecting adhesive coverage and the temperature and moisture gradients in the mat during consolidation. Dimensions also affect *conformability* (the spatial relationships of particles to one another in the mat). In the case of scarf and finger joints, angles are the most important elements of fit, while in assembly joints, it is relative sizes of tenons and the holes into which they are glued.

Most of these are machine problems and deal with set-up procedures and drift of settings during operation. Consequently measurements should be emphasized at the beginning of a run, and periodically thereafter.

Thickness Measurement. In order to be useful, measurements need to have an accuracy to .001 inches. This accuracy is provided by the micrometer caliper (Figure 10–4(a)) and by the machinist micrometer (b). Direct-reading dial gauges (Figure 10–4(c)) are easier to use. A dial gauge permanently mounted on a stand as shown, though less portable, is more adaptable for small specimens. However, it harbors a possible source of random error. In using the mounted gauge, the post referencing the lower surface can induce misreadings if the wood is not held perfectly level. The error can be avoided altogether if the wood piece is pivoted slowly in all directions. The dial will go up and down, but the lowest reading will be the correct one.

Micrometers are also used to monitor the thickness of flakes. While less important per

se in influencing bond formation, it governs a number of other factors, notably binder coverage. Since binder is applied by weight percent of wood, and thickness determines surface area per unit weight of wood, the thinner the flake, the more area exists per pound of wood and the less adhesive per unit of area if the rate of addition remains the same.

For lumber, the greatest variability is likely to be between the ends and the middle because of possible moisture changes and because entering and leaving the planer sometimes provide different hold-down action. Any point-to-point disparity greater than plus or minus .005 inches should be cause for concern, as two minus areas coming opposite each other on a glue line can cause gaps of .010 or more inches. In face gluing, variation *within* board is of greater consequence than variation *between* boards because pressure cannot equalize when high and low spots exist. With between board variability the pressure is unaffected if one board is thicker or thinner than another.

For veneer, variability in thickness between ends and middle can also exist, but the greatest likelihood of excessive variability is across the grain. Unlike lumber, the variability between sheets of veneer is of equal, if not greater, importance than variation within sheets. This is because veneer pieces are frequently edge glued together. A thin piece next to a thick piece inevitably causes less pressure to be received by the thin piece. Variations in thickness also cause variations in the amount of glue applied by roll spreaders, thicker pieces receiving less glue than thin pieces. Besides hand-held measuring devices used for spot checking, a number of devices can automatically scan every piece of veneer for thickness. These can either be contact wheels registering through a transducer, or noncontact proximity gauges. In any case, a means of marking the wayward pieces is also part of the system. Tolerance limits depend on species and thickness. Low density species generally can tolerate a greater variability in thickness. Since conformity to adjacent surfaces is related to

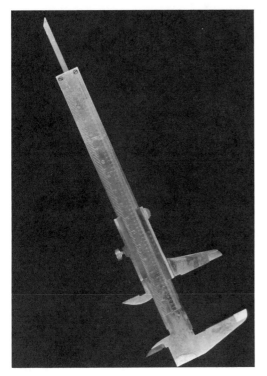

(a)

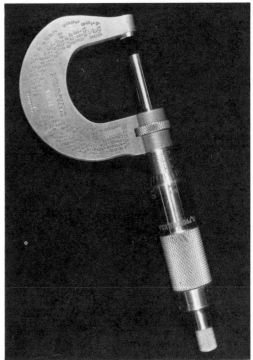

(b)

(c)

Figure 10–4 Thickness-measuring instruments: (a) micrometer caliper, (b) machinist micrometer, (c) mounted dial gauge.

stiffness, thicker pieces need to be more uniform.

Width Measurement. In edge gluing of both lumber and veneer, uniformity of width is the important dimension affecting the fit of the joints. Here the concern is within a piece rather than between pieces. While direct measurement with caliper or ruler may provide the desired data, a visual approach may be more indicative. Bringing the two mating edges together in front of a light source instantly reveals how good the fit is. Feeler gauges inserted where light comes through will provide a numerical reading for the gaps.

If bonding pressure is applied, without glue, a sense of how well the joint will close can also be measured. Experience will show how much gap can be tolerated with the gluing process being used. It is well to bear in mind that when there are gaps, the high spots are receiving more than their share of pressure, and may become starved during pressing.

Size measurements of the smaller wood elements are usually determined by screen analysis. A series of screens having mesh sizes incrementing over a suitable range (such as $\frac{1}{64}$ inch to $\frac{1}{2}$ inch) will separate material into fractions on the basis of passing thru or remaining on a certain mesh; for example, through $\frac{1}{4}$ inch, on $\frac{1}{8}$ inch would define a fraction that was less than $\frac{1}{4}$ inch and greater than $\frac{1}{8}$ inch. Weights of particles remaining on each mesh size compared to the total amount would quantitate the percent of each size and produce a particle-size distribution profile. This method is less satisfactory with linear particles due to the tendency to up-end and go through without regard to length. Air classification systems rely on the ratio of mass to size or shape, which pits gravity against aerodynamics to provide a separation. A horizontally blown mass of particles would produce a gradual size classification, beginning with the heaviest near the blower and the lightest farthest away. Bins placed in the path of the particles would catch their proportionate share of the different sizes.

In the case of fibrous materials, one means of direct measurement is to sprinkle a small quantity loosely on an overhead transparency prepared with a suitably dimensioned grid. Projecting onto a large screen provides an enlargement that can more readily be measured. Knowing the magnification permits reduction to actual dimensions.

Fibrous materials scheduled for wet process consolidation are mainly monitored by a draining test for "freeness." The fibers are suspended in water over a screen. A drain is opened and the time required for the water to drain through the mat of fibers collecting on the screen is noted. The test is standardized for the amount of water, concentration of fiber, and screen size. Other factors being equal, the coarser the fiber the faster the rate of drainage.

Wood Density

As an indicator of the quality of links 8 and 9, density is one of the more important characteristics of wood. This factor also has a strong bearing on bond performance through its relationship with the stress it can generate or deliver to the glue line.

Although Table 5–7 gives the density or its related equivalent, specific gravity, by species, it is sometimes necessary to measure the density of a particular piece of wood because densities at both ends of the range can give trouble. Density has parameters in weight per unit volume—pounds per cubic foot or grams per cubic centimeter. The determination therefore requires weighing the piece and measuring its dimensions to calculate its volume.

For example, a piece of lumber weighing 5 pounds and measuring 1 inch by 6 inches by 4 feet would have a volume of $\frac{1}{12} \times \frac{6}{12} \times 4$, which calculates to $\frac{1}{6}$ cubic feet. Five pounds per $\frac{1}{6}$ cubic foot equates to 30 pounds per cubic foot, the density of the board. The specific gravity equivalent relates to the weight of an equal volume of water. Water weighs 62.4 pounds per cubic foot. The weight of an equal volume of water, $\frac{1}{6} \times 62.4$, equals 10.4 pounds, about twice the weight of the wood, a specific gravity of 5 pounds divided by 10.4 pounds, or

0.5. This is a dimensionless quantity since "pounds" cancels out in the division. Adjustments for moisture content will change the values slightly.

For smaller or irregularly shaped pieces of wood where direct measurement is not feasible, a water-displacement method provides the measure of volume. By dipping the wood in hot wax, a coating is first applied to prevent the wood from absorbing water. Three approaches are available:

1. In a narrow graduated glass cyclinder partly filled with water, the specimen, impaled on a slender rod, is fully immersed. The change in elevation of the water level is noted, and if the cylinder is graduated in cubic centimeters, it becomes a direct reading for volume of the wood.

2. With a slender container of water filled just to overflowing, the wood is fully immersed as before and the overflow caught in another container. If this container is a cylinder graduated in cubic centimeters, it will read volume directly, otherwise the water can be weighed in grams to obtain its volume.

3. With a container of water balanced on a scale, thrusting the specimen into the water produces a force equal to the weight of the water displaced by the specimen. The force is read as weight on the scale and converted to volume as before.

Grain Direction

Along with moisture content and density, grain direction is a major source of variability in wood properties, in both bond formation and bond performance. The chief concern here is not ring direction but the direction of the main cell structure of which wood is composed, and which controls the so-called directional properties of wood; i.e., its anisotropy.

When wood has large rays, the grain direction on the tangential surface can easily be observed as it parallels the rays. Large pores allow grain direction to be observed on both the radial and the tangential surface. Surface checks, when present, also follow the grain. A problem exists when no indicators are visible. Visibility can be improved by placing on the surface a drop of red ink, such as is used in recording control instruments. The ink will spread up and down the grain in streaks that are plainly visible. In lieu of the red dye, a sharp stylus on a pivoted arm when drawn gently over the surface will follow the grain.

Grain direction is measured as the angle it makes with respect to the edges of the wood piece. It is usually reduced to a tangent function of inches per inch; i.e., inches along the edge per inch of deviation, such as 1:5, or 1:12, approaching zero angle as the grain becomes more parallel with the edge. When the grain deviates both across the thickness and across the face, a resultant angle must be calculated from the geometry of the other two angles.

Surface Quality

Determining the adhesion properties of a wood surface remains one of the weaker areas of gluing technology. There is yet no measure of "glueability" that can assure the quality of a bond. However, a number of measureable attributes can be defined. These begin with the quality of the surface that describes the grossest departures from smoothness or flatness, and end with the finest.

Warp. Defined as a deviation of the wood surface from a flat plane, warping (which includes cup, sweep, bow, and twist) comes about not as a result of machining error, but as a result of inherent properties of the particular piece of wood. If the warp is ignored, and conformity with another surface is forced by bonding pressure, the force will remain in the system after bonding, tending to restore the original warped condition. This is one of the internal destructive elements that tend to pull the bonded surfaces apart.

Warpage is easily observed and easily measured, though not easily interpreted in terms of bond effects. Direct measurement of the deviation from a flat surface provides the data. How much warp can be allowed cannot be stipulated as a generality because it depends upon species, thickness, and width. A trial-and-error approach is necessary with the materials and systems at hand. Clamping the assembly with warped boards included (without glue), the pressure is increased until closure is achieved, as judged by eye or feeler gauge. Pressure should not exceed 100 psi for low-density woods or 200 psi for high-density woods.

Trueness. Like warp, trueness implies a deviation from a true plane, but in this case as a result more of machining aberrations than an inherent wood response. The term mostly implies deviations from a straight line, and is more applicable to edge gluing. In face gluing, planer skips and dubbing are troublesome, but out-of-square and out-of-line edges are the main problems. The former can be checked with a carpenter square. Very little, if any, visible deviation should be permitted since an open or weak joint at one or the other surface is the risk.

Edge surfaces can be checked and measured for out-of-line by reference to a straight edge or to adjacent boards in the assembly. Allowable deviation can be determined by trial and error, as for warp, varying gaps and pressures.

Roughness. Defined as "within plane" deviations from the true surface, roughness is the result of local elevations or depressions of various magnitudes. Their effect is to produce small gaps or elevations scattered throughout the glue line. Gluing procedures respond to roughness generally in a negative manner but some more than others. Consequently there can be no standard of acceptability for this surface condition. Methods of measurement are nevertheless useful to allow a trial and error approach to the setting of limits for a particular situation. There are many methods, varying in sophistication. One of the simplest, for observation purposes only, is low-angle light, which tends to enhance visibility of surface deviations. If the low-angle light contains the shadow of a straight line, a point of reference is established from which to judge mag-

nitudes. More precise numerical information can be obtained by drawing a stylus gently across the surface, referencing from its own independent true surface. If the stylus is connected to a dial gauge, direct readings can be made, or "go," "no-go" limits can be set on the dial. Engaging the stylus to a transducer permits a trace of the surface to be made. Proximity gauges provide a trace without contact with the surface.

Interpretation of data is a problem in itself. Two factors interplay: magnitude of the deviations, and frequency of the deviations. The issue becomes a question of how to relate a few large deviations to many small ones. This can only be answered with data representing the species, construction, adhesive, and pressure system being used. A limit based on both magnitude and frequency should evolve in which, for example, certain magnitudes are permissible when they only occur so many times per inch.

Surface Inactivation. The obvious concern with surface inactivation is its interference with the formation of links 4 and 5, adhesion of adhesive to the wood substrate. However, it also affects wetting, penetration, solidification, and filtering. These are affected in large measure by the absorbent and adsorbent nature of the wood surface toward glue water. This applies to glues that harden by loss of water, by chemical reaction, at ambient temperatures or elevated temperatures in which the action is the same but the reactions are different.

The situation is a complex one involving free surface energies and their modification (reduction) by contaminants or chemical transformations. Rather than being *hydrophilic,* the surface has become to a certain degree *hydrophobic,* the terms implying affinity, or lack of it, for water. Hence a measure of inactivation involves some means of observing this affinity.

One of the simplest methods is to place a drop of water on the wood surface and observe what happens to it over a period of time. Two actions are immediately observable that represent the extremes of surface condition. If the drop remains in spherical shape on the surface for a long period of time (it may even evaporate) it indicates a highly hydrophobic surface. If, on the other hand, the drop flattens out to a thin film, spreading over the surface and disappearing quickly into the wood, it indicates a very hydrophilic surface. Intermediate shapes and disappearing times occur with intermediate inactivation. The observation becomes quantitative if the angle the drop makes with the surface is measured (Figure 3–7).

An alternative approach is to immerse the wood partly in water and observe the direction of the meniscus against the wood surface; down into the water is indicative of surface inactivation; up the surface of the wood is indicative of high affinity. If the wood piece is tilted until the meniscus is parallel with the surface of the water, the angle the wood surface makes with the water surface has the same connotations as the angle a drop makes with the surface.

While it would be most desirable from an adhesion standpoint to opt for a highly hydrophilic surface, in current practice it can cause as much trouble as the hydrophobic surface. Rapid drying of some glues (cold-setting ureas and resorcinolics, for example) is inimical to their ultimate chemical cure, and certainly to the assembly time some operations require. Nevertheless, it is important to reduce as much as possible the degree of inactivation that prevents essential adhesive actions.

In controlling surface inactivation, its causes suggest the remedies. Excessive heat in drying, prompted by the need for increased speed in production, is a major cause especially for veneer and flakes. Grease and oil dripping from machinery, and wax from markers, also contribute. Dull or improperly fitted knives in rotating cutters tend to burnish or burn the wood surface, as well as damage the subsurface. Worn or clogged sandpaper can also burnish the surface. A wood surface that is darker than normal should be suspected of inactivation by heat. Separating heartwood from sapwood in ele-vated-temperature drying helps reduce overexposure of sapwood to heat.

Subsurface Damage.   Considerable attention has been paid to subsurface damage because of its variable effect on performance of glue bonds. Despite its recognized effects, no means of measuring its presence or severity has been suggested. There are, to be sure, all degrees, from a few torn surface fibers to deep fissures into the wood. All the means of generating wood surfaces are capable of producing subsurface damage—knives, saws, sandpaper—even under the best operating conditions. The problem worsens through carelessness—not keeping knives and saws sharp and well fitted, using coarse sandpaper, or placing excessive pressure on feeding devices. In the case of sanding, the problem is not only one of weakening the surface but also of plugging up avenues through which the glue might penetrate to achieve some repair (links 6 and 7).

In observing subsurface damage, veneer lathe checks are the most visible, especially on the thicker sizes. Bending the veneer across the grain toward the tight side opens the checks for easier viewing. Severity of the checks is related not only to their depth and frequency, but also to their angle with the surface plane. As the angle becomes more inclined from the perpendicular (more parallel with the surface), it becomes more deleterious. Instances occur where the checks become parallel with the surface, join together, and form a continuous separation *within* the veneer, virtually destroying its integrity and placing it beyond the ability of the glue to repair. Checks in veneer are minimized by using the correct amount of nose-bar clearance. The setting of the nose bar is an operational technicality fine-tuned by trial and error to amend to the properties of the wood being cut.

Microscope observations at 100-power magnification are needed to see subsurface damages that involve cell wall fractures or cell cavity occlusions. Highly polished or microtomed sections of end-grain surfaces improve the observation.

The presence of loose surface fibers can be observed by pressing on a strip of adhesive tape, preferably duct tape, and peeling it off. If it is covered with fibers or splinters, the surface may be considered extensively damaged. Scoring alongside the tape with a razor blade before peeling increases the amount picked up with any one pull.

Another approach that has proved useful, but may have limited appeal because of its delayed results, is to bond a block of wood to the surface in question using a procedure that produces a shallow bond. A highly filled epoxy resin of pastelike consistency does not penetrate, bonding only to the very surface. When the block is pried off, it will lift the damaged surface with it. The extent of damage can then be appraised in terms of wood failure, noting the shallow character of the fracture surface.

Wood Temperature. Wood can vary in temperature from freezing cold to that encountered in drying processes, depending upon storage conditions and time prior to gluing. In an assembly of wood and glue, the wood far exceeds the mass of a thin glue line, and therefore controls the temperature of the glue until changed by heating while under pressure. In cold pressing where bundles of panels or lumber are held under pressure, the beginning temperature of the wood is particularly critical. Trying to compensate for cold wood by surrounding the press area with hot air may have little effect in raising the temperature at the interior regions of a charge. The effect of wood temperature on bond formation is exercised mostly through the hardening properties of the glue. Some of the possibilities include the chalky glue line condition that some vinyl glues suffer when used on wood below 60°F (usually cautioned on the label); the interrupted and possibly aborted curing of chemically hardening glues on wood below 60°F; and the rapid drying out of glues during assembly time on wood above 100°F. Veneer used too soon out of the dryer, and the smaller wood elements that have been force-dried and used with too short a dwell time in the bin, are most likely

to be too hot when given their coat of glue. On the other hand, glues that harden by cooling have a better chance at bond formation when placed on wood above room temperature.

In any case, it behooves quality control to keep a close watch on temperature of the wood *before* it is delivered to the glue applicators. A touch of the hand is all that is needed to alert the senses that something may be wrong. It can be confirmed with thermometers designed for gauging surface temperatures, or in the case of loose materials in a bin, an ordinary thermometer will do. Permissible temperatures depend on the glue, the assembly time, and the structure being glued. Communication with the glue manufacturer will help resolve questions of degree.

MONITORING GLUE PROPERTIES

From a users' standpoint, the greatest unknown may be what's in the glue pot. Most often it is a liquid, and it is supposed to hold wood together. Their primary concern tends naturally to whether it will harden and be strong enough. While this is a reasonable concern, it lies largely in the hands of chemists who have earlier established the strength potential and the hardening characteristics. Nevertheless, it would be prudent to check the hardening properties, especially with glues that require the addition of catalyst or have a restricted shelf life. Beyond this point, only the intervention of variables that arise *in the gluing operation* can affect the strength that is actually realized.

Of equal if not greater importance is the viscosity of the glue. It is one of the more fundamental properties of glues in bond formation, for it controls not only how it is applied but also how it flows and penetrates on the glue line, how it achieves proximity and hence adhesion to the wood, and how it might affect or be affected by other operational variables. Viscosity measurements are also used as indicators of a number of other adhesive properties that bear on bond formation, such as molecular size, solvent content, pot life, rate of hardening, storage life, and assembly time. It is well to recall at this point that viscosity is a transient property of glues, particularly on the glue line.

On their way to converting from a liquid to a solid, most adhesives pass through a well-defined stage, the *gel point,* beyond the reach of a conventional viscosity measurement, but signaling the onset of a stage in which rigid properties begin to develop. The time to reach gel point is used with some glues to indicate rate of hardening. It is a rather simple measurement in which the adhesive is observed in a test tube maintained at a constant temperature, or series of temperatures if time-temperature curves are desired. A coil of wire fitting closely the inside of the test tube, used like a stirrer, will signal the gel point when the adhesive, in curing, forms a plug in the coil.

Further beyond the gel point, the adhesive passes through a rubbery state before another sharply defined point marks the beginning of true solid properties, the *glass transition point.* This point varies with time and temperature and is reversible with thermoplastic glues but generally not with thermosetting glues. It often represents the peak bond strength, but in the case of thermosets not the peak durability. The lack of coincidence between strength and durability is partly because maximum durability comes with the chemical property of ultimate cure, whereas maximum strength is more related to mechanical properties that seem to appear at an earlier stage and may be due to rheological properties more favorable to stress distribution. Methods of measurement involve strength observations of the adhesive film against a timed series of temperatures to determine when solid properties develop. The information is more useful in understanding bond performance than bond formation because of its relationship with temperature and stress responses.

Viscosity Measurement

Viscosity is measured as a resistance to flow, and, as indicated above, can be interpreted to reveal many adhesive properties of a molecular and compositional nature. Although viscosity will be interpreted mostly for a sense of fluidity (mobility) of the glue as it might function on the glue line, it will also be used in establishing limits for time-dependent actions throughout the gluing operation. Some of the methods mentioned for measuring viscosity require instruments. These are more completely described in ASTM method D 2556. Two visual methods of judging fluid suitability are also given for quick estimations of viscosity suitability.

Brookfield Viscometer. The Brookfield viscometer measures the force necessary to turn a spindle or bobbin at constant speed while submersed in the liquid (Figure 10–5(a)). The higher the viscosity (i.e., the lower the fluidity), the higher the force necessary to turn the spindle. Spindles vary in size to provide different surface areas for developing measurable shear resistance over a wide range of fluidities. For example, water would require a larger spindle to develop measurable shear resistance than would a liquid with the fluidity of honey. The resistance is sensed as the power to turn a constant-speed motor, and this is translated into poises or centipoises and displayed on a scale. Speed of the motor can be varied to obtain evidence of thixotropic or dilatancy behavior.

In practice, a cup of the adhesive is placed so as to immerse the spindle to a definite depth, the motor is engaged, and a direct reading is obtained. Since viscosity is strongly affected by heat, readings are comparable only when taken at the same temperature.

Stormer Viscometer. The Stormer viscometer works on the same principle as the Brookfield except that the turning power is supplied by weights and pulleys. A string wound around the spindle stem and over a pulley allows gravity to provide a constant turning force. The force is controlled by the size of the weights. Since the force in this case is constant, the measurement is time for

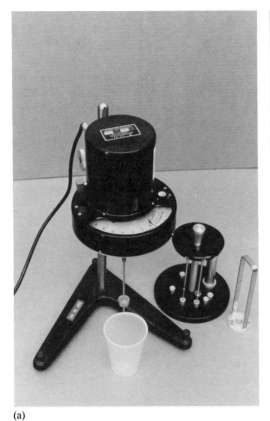

Figure 10-5 Instruments for measuring viscosity: (a) Brookfield viscometer, (b) bubble-rise tubes (shown sideways; top at right), (c) perforated cup.

the force to produce 100 turns of the spindle. The more fluid the adhesive, the less time it will take to allow 100 turns. The conversion of time (seconds) to viscosity in poises is done by tabled values, against the size of the spindle used.

The Bubble-Rise Method.

Simpler than the spindle methods, the bubble-rise method (Figure 10-5(b)) is restricted to a lower range of viscosities. A test tube is filled with the adhesive and corked, leaving a space between the top of the liquid and the bottom of the cork. This forms the bubble. When the test tube is inverted, the bubble will rise to the other end with a speed that is related to the viscostiy. Timing the rise between fixed points on the tube (in seconds) provides a reading that can be converted to centipoises by comparison with other tubes con-

taining liquids of known viscosity, a series of which is shown in Figure 10-5b.

Cup Method.

Perhaps the simplest of all, the cup method (Figure 10-5(c)), really seems to be measuring fluidity directly since the observation is the rate at which the liquid flows through a hole. The cup, with a hole at the bottom, is filled to the top and then allowed to drain through the hole. The time it takes to drain is converted to viscosity terms, taking into consideration the size of the hole, which can be varied to accommodate a range of viscosities.

Mixing Power.

When batches of glue are mixed in industrial sizes, they often are of exactly the same composition day after day. In such cases where size of the batch is constant, an ammeter on the mixer motor can provide a continuous indication of viscosity, once calibrated against one of the above methods of measurement.

Qualitative Methods. When no instruments are available, it may be necessary to resort to a spatula or even the fingers to obtain some indication of how the adhesive will flow. Dip the spatula in the glue, raise it about 10 inches above the surface of the glue in the pot, and allow the liquid to flow off the spatula back into the mix in a steady stream. As it strikes the surface, one of three actions will occur: (a) The adhesive stream immediately disappears into the mix creating a small crater at the point of entry (2) The falling stream creates a pile-up on the surface; (3) The falling stream disappears as fast as it meets the surface, creating neither

a crater nor a pile-up, but remaining level (Figure 10–6).

Since thixotropy and dilatancy affect flow properties, each of these three situations has significance only with respect to a particular glue, and only as a means of judging its suitability against a predetermined norm. With experience, this procedure can help track a glue mix that has been in use for some time. As a point of departure, the author has found that for veneer gluing with urea glues, the falling stream should form a slight crater; for edge gluing lumber, it should form a pile; and for face gluing lumber, it should remain level.

Further along the gluing operation, drawing the fingers through a spread film of wet glue on the wood surface also provides some clue as to its condition to receive pressure. The behavior of the displaced glue is indica-

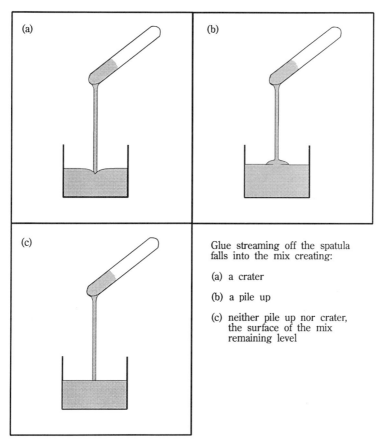

Glue streaming off the spatula falls into the mix creating:

(a) a crater

(b) a pile up

(c) neither pile up nor crater, the surface of the mix remaining level

Figure 10–6 Observing mix viscosity by a falling-stream method.

tive of its fluidity. Again, experience will sharpen the observation. Some starting points: If the glue on the surface immediately flows back to a uniform film, it probably is too fluid to be pressed, especially in hot pressing veneer and gluing of lumber, less so in cold pressing veneer. At the opposite extreme, if the glue barely moves under finger pressure, cold-press operations are at risk. A hot-press operation may succeed if wood moisture content is on the high side. It would have a better chance of succeeding if the glue were on both surfaces to be joined since the transfer action would be insured. In fact, if the glue were applied to both sides of the joint, even a glue that has dried tack-free has a chance if the glue line temperature is high enough to reach the fusibility point of the glue. Unextended phenolic glues can be formulated to be sufficiently fusible (with the help of moisture from the wood) that they can be applied in the dry state either in powder form or film form.

To repeat a lofty objective: The best glue-spread situation immediately prior to placing the assembly under pressure is one in which the glue has been applied to both surfaces of the joint with a crater-forming consistency, is tacky to the touch, but now deforms like butter as a result of assembly time actions. Skinning over or tackiness would not be a problem if the surfaces were in contact during the assembly time.

The matter of tackiness is a deceiving quality in some glues. One perceives tackiness by sensing that some force is necessary to pull the fingers off the surface (or to pull fingers apart after squeezing a drop of glue between them), and when separated, the glue forms a stringy mass in between. Many glues are not tacky in the pot or when first applied to a surface. In these cases, the glue develops tackiness as it advances toward the hardened state.

Storage Effects. Because many glues have a limited storage life to begin with, and are further limited by any severe storage conditions they may have encountered, the condition of the glue at time of use merits attention. Manufacturers always advise FIFO (First In, First Out) to reduce the chances of using overaged material. Any suspicion of material having exceeded its storage life should activate a checking procedure. This is done with a viscosity measurement. All resins or base components are manufactured to a narrow viscosity range. This range should persist to the time of use unless a change has been brought on during storage. Manufacturers usually record the shipped viscosity by lot number. Should the suspicious glue show an elevated viscosity, the manufacturer should be consulted to determine usability.

Pot Life. Glues that come ready to use are generally assumed to have an unlimited pot life almost by definition, since having survived storage they should be able to survive a few more hours in the pot waiting to be used. Glues that require mixing, especially those to which a catalyst must be added or is already incorporated as a dry powder, will have a limited pot life. Glues with a limited pot life experience a gradual increase in viscosity during its dwell in the pot or hopper as application proceeds. The increase in viscosity may be slow at first and then increase rapidly toward the end. Some glues—resorcinols, for example—may take an hour to double their original viscosity, and then solidify in the next ten minutes, due partly to the generation of heat by exothermic action.

In any case, one needs to know when the pot life has been exceeded. A viscosity limit can usually be set by the manufacturer of the glue. Alternatively, the method of application will provide the first indication that something is amiss. If the glue will no longer go through the nozzle, flow through the curtain coater, transfer off spreader rolls, or spread with a brush when it is supposed to, it should be discarded.

Will It Harden? As mentioned earlier, this is a universal concern, but one that is a property of the adhesive, built in by the manufacturer. Still, some verification would be reassuring. Making and breaking specimens is the sure way, applicable to all glues, but except for hot-press glues the answer may not come until the next day. Running a viscosity or gel-time check over time and temperature to plot a rate-of-hardening curve would give a more immediate answer, but again only for those glues that harden by chemical action. Glues that harden by loss of solvent are subject to the rate of loss of that solvent, and this, unfortunately, can only be checked by actually making a bond with the wood at hand.

For those glues that set cold by chemical reaction, some reassurance can be derived by saving a small cup of the mix and observing whether it has set up to a hardened state in the period of time specified in the instructions. The author has also found it informative to place a few drops of the glue (all those that set cold by any mechanism) on the surface of the wood being glued. After the hardening period, usually the next day, one can chisel off the drop and obtain a qualitative sense of its hardness. By careful chiseling, the drop can be lifted off the surface intact. A surprising observation will often ensue: The bottom of the drop will have wood adhered to it, wood failure. What is most surprising is that in the glue line, wood failure might not occur with the same glue. One is left with the speculation that something happens in the glue line that is inimical to bond formation—a disconcerting thought, but one that demands attention at the formulation level. The effect of the wood on the hardening properties of a thin glue line, or the penetration properties of the glue, are prime suspects in this case.

Another interesting observation is the shape of the outer edges of the drop. When first placed on the wood, the drop will often remain spherical (a high contact angle), indicative of rejection by a seemingly inactivated wood surface. During the time hardening is taking place, the drop may slowly spread, producing a minicus-type edge, and suggesting some affinitive action. This result confirms that wetting while partially spontaneous is time-dependent with wood adhesives, and therefore requires prompting by pressure in the glue line if it is to form the necessary attachments.

MONITORING THE GLUING OPERATION

The gluing operation begins with preparation of the glue, for those that require preparation, as prescribed by the glue manufacturer and verified by appropriate tests. Wood and glue then finally meet. The meeting is performed by some application system, and the main questions are how much is applied and how uniformly. The rest of the gluing operation involves questions of temperature, pressure, and time.

Mixing

All glues, come, or should come, with complete mixing instructions, either on the container or in separate bulletins that can be hung on the wall for continuous reference. Some glues come ready to use from the can, some require only the addition of water, others require only the addition of catalyst. Many industrial glues require the addition of several constituents. In addition to water and catalyst, these include fillers, extenders, solvents, and fortifiers. Some extenders require cooking or digesting before incorporating with glue solids. Thus the glue preparation procedure can get rather involved, particularly if order of addition, manner of addition (dumping, sifting, dripping), and time between additions are important. The best one can do is to follow instructions religiously, and set up the procedure so that each step occurs naturally and without failure. It is helpful in this connection to preweigh all the ingredients destined for mixing, place them in separate containers, and line them up in the order to be added. Liquid additives may be measured by volume in calibrated containers. Powdered additives, however, must always be weighed rather than measured by volume since it is impossible to judge how compacted they might be.

In the absence of specific instructions, some general cautions apply. In mixing a powder and a liquid, the liquid should not be added to the powder; instead, the liquid is placed in the mixer and the powder is added to it, preferrably sifting it in while stirring. This procedure avoids the lumps that inevitably form when a liquid is added to a powder. If the amount of liquid is large compared to the powder, it is better to begin with just enough liquid to make a stiff paste with all the powder, adding the remaining liquid after a good lump-free mix has been achieved. In general, the stiffer the mix, the better it will deal with lumps.

Catalysts are generally added last in order to ensure the longest mix life. Dyes added to the catalyst provide visual evidence that the catalyst has been added. Intensity of the color of the mix further confirms whether the catalyst might have been added in excess or in deficient amount.

Ironically, the addition of water, a seemingly simple step, sometimes becomes a source of error, especially in large-batch mixing. Two kinds of mistakes can occur, both leading to too much water. When the water is let into the mixer by gravity from a tank fitted with a sight gauge, a faulty valve or inattention in closing it can let in more water than called for. A solution: Vessels feeding directly to mixers should contain only the amount of material needed. Another mistake arises through the use of a dip stick. Dip sticks usually cannot reference from the bottom of the mixer because bottoms are sloped toward the let-down valve. They are therefore referenced from the top of the mixer after calibration with a measured amount of liquid. Unfortunately, the top of a mixer does not remain constant because of deposits collecting there from dumping in other ingredients. As the top edge rises so does the amount of water introduced.

Speed of rotation of mixers is often neglected but can be a source of problems in affecting the density of the glue by whipping air into it. Density of the glue becomes a problem when the glue is applied as it often is, by volume rather than by weight. Although adhesives operate volumetrically on the glue line, weight is a truer measure of the amount of glue that should be applied because the volume may include air in the form of bubbles. Bubbles do no good in the cured glue line, and may cause harm if not expressed out of the glue line in the pressing operation. Some glues are deliberately foamed to increase their volume for more uniform spreading, but these are specially formulated to collapse when heat and pressure are applied.

Finally, temperature of the mix has a profound influence on its pot life. Furthermore, as with all liquids, temperature affects apparent viscosity. The effect of temperature on pot life differs with the various glues. With proteinacious glues, high temperatures tend to break down the protein, reducing the viscosity as well as reducing strength of the hardened film. Catalyzed cold-setting glues are especially sensitive to heat, increasing the rate of hardening exponentially, and therefore reducing the length of time they remain viable in the pot. Resorcinolic glues generate their own heat by exothermic action and need watching closely when they are prepared in large batches and the heat is less able to dissipate. Mixers and containers are often water jacketed to keep the mixes cool, thereby prolonging their pot lives.

Keeping the above factors under control is essential if the glue is to meet wood in the best possible condition.

After assuring the stipulated addition of all the ingredients, quality control at the mixer is achieved with a viscosity measurement, along with a temperature measurement. The glue should come out of the mixer with a consistency that has previously been established with the manufacturer of the glue. When the size of the mix is constant day in and day out, the viscosity of the mix can be monitored with an ammeter on the motor of the mixer that has been previously calibrated against known viscosities and known glue volumes, as mentioned earlier.

Amount of Adhesive Applied

There are two distinct and sometimes confusing criteria for judging the amount of adhesive applied. They depend upon the ratio between the surface area of a wood element and its weight. When this ratio is low, as in lumber and even veneer, the criteria is weight of adhesive per unit area of glue line. In the U.S. it is pounds of adhesive per thousand square feet of glue line. Note that one unit area of glue line contains two unit areas of wood surfaces. In other words, 1000 square feet of glue line contains 2000 square feet of surface, 1000 on each side of the joint. It is the glue line that is important in this case, not the total surface.

When the ratio between the surface of the wood element and its weight is high, as it is in flakes, particles, and fibers, the criteria is weight of adhesive per unit weight of wood. In the U.S. it is pounds of adhesive per pound of wood, usually expressed as percent adhesive based upon weight of wood.

Another important distinction between the two criteria is that in the case of lumber and veneer, the adhesive is weighed in its liquid or as-applied form. In the case of flakes, particles, and fibers, the adhesive is calculated on a resin solids basis rather than the liquid basis. Moreover, the weight of the wood is the oven-dry rather than the as-is weight.

The most accurate method of measuring the amount of adhesive applied in the case of lumber and veneer is to weigh a piece of known dimensions before and after application of the glue. The difference in weight divided by the area covered gives the desired measurement. Figure 10–7 illustrates one of the more convenient methods of carrying out this observation. An appropriate piece of

wood—veneer in this case—is weighed on a diet scale. It is supported on a nail plate to avoid messing the scales with glue. Before applying the glue, the veneer and its nail support is tared off by turning the dial to zero. After applying the glue, the veneer is placed on the scale again and the new weight records only the actual amount of glue applied.

Because the observed numbers are so small, and the units so varied, some conventions have arisen to standardize the reporting. As mentioned above, all measurements are reported in pounds per thousand square feet of single glue line, abbreviated to lbs/MSGL. In metric units, it may be grams per square meter or some multiple thereof. Often the adhesive is weighed in grams, and the area is measured in square inches. Grams per square inch can be converted to pounds per thousand square feet by multiplying by the factor 317.2.

A procedural twist simplifies the conversion when mixed units are used in measuring. Veneer is often coated on both sides of one piece in order to make two glue lines with one pass through a spreader. A piece of veneer with an area 12 inches by 12 inches (one square foot) coated on both sides carries enough glue for two square feet of single glue line. By a quirk of arithmetic, the weight of glue this square foot of veneer carries, measured in grams, when multiplied by the factor 1.1 is numerically equal to pounds of adhesive per thousand square feet of single glue line. No calculations are needed; the procedure becomes direct reading, if the veneer size is increased by 1.1, or 12 inches by 13 inches.

An alternative to weight measurement is a film-thickness measurement. A wet film gauge used mostly for paint and varnish coatings measures the thickness of the liquid glue film. This thickness measurement produces a volumetric estimate in conjunction with the area covered, which can be converted to pounds, knowing the density of the glue. For example, a film thickness of .006 inches would have a volume of (.006/12) × 1000 or .5 cubic feet in a thousand square feet of glue line. If its density is 70 pounds

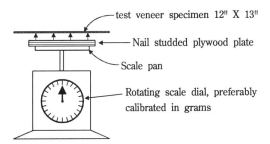

Figure 10–7   Diet-scale method for measuring amount of adhesive applied to veneer.

per cubic foot, the spread rate would be 70 × .5 or 35 lbs/MSGL. While relatively simple to use, this method is best used for films that are spread smoothly on a smooth surface.

Some applicators—particularly roller spreaders that have been rolling the same glue for a period of time, as they might over a coffee break or lunch hour—also tend to cut air into the glue, lowering its apparent density. Extended rolling of the same glue can also result in loss of solvent and increase in viscosity. In either case the amount of glue solids applied will vary from the setting. A weight measurement after spreading should catch both eventualities.

Each type of glue applicator has its own vagaries that must come under surveillance periodically. Sprayers and extruders can become plugged or, worse, partially plugged, because it may escape observation. Curtain coaters may be sensitive to viscosity changes. Roller spreaders may apply more on one surface than the other, or more on one end than the other, depending upon how well they are adjusted. Moreover, grooved rubber applicator rolls can become worn sometimes only at a certain point where the stock is fed the most. Consequently, the admonition is frequently given to feed the rollers over their full length to keep the wear uniform.

Roller spreaders have been used for decades to apply glues to lumber and veneer. They have accumulated their share of advantages and disadvantages, the latter leading to the more modern methods of application: spraying, extruding, and curtain coating. The most objectionable disadvantage is that roller spreaders do not deal well with thick and thin or rough and wavy stock, because the amount of glue applied depends upon two factors: the amount of glue doctored onto the rolls, and the pressure between the rolls and the surface of the wood. The rolls, being a fixed distance apart, vary in pressure with the variation in thickness or elevation of the wood. Consequently high or thick areas may receive an insufficient amount of glue because it is squeegeed off, and the low areas may have excess glue dumped onto

them. Although resilient roll spreaders have been developed that minimize these effects, the other methods of application apply the same amount of glue no matter what the condition of the wood, an apparent advantage.

However, there is a down side to this advantage. By applying the glue unevenly, the roll spreader is saying that there is something irregular about the wood surface. If the surface does not receive the glue well, it may not be suitable for incorporating into a glue line. In this sense the roll spreader is serving as a monitor for the quality of the wood to be bonded. But now it is late in the process, and the decision to discard includes both the wood and the glue on it.

The frequency of taking measurements on the amount of glue applied depends upon the control exercised prior to the spreading operation. Vigilance by those operating the spreader will spot the more drastic changes that occasionally occur. Drawing a practiced finger over a recently spread glue surface provides evidence that there is sufficient glue. However, a weight check should be made periodically to confirm and record the actual amount applied, certainly at the beginning of very day, or at changes in stock or glue. Weight measurements should also be taken several times during the day when no elected changes have been made, in order to make sure variability remains within established limits, and to forestall creeping changes due to machine wear or viscosity changes.

Density of the Mixed Glue

Because of variables introduced in mixing, the density of a glue may be different—probably less—than that indicated by the manufacturer. The density can easily be ascertained by weighing a known volume of the glue. The volume of a container is most easily measured in the metric system, where one gram of water is equal to one cubic centimeter of volume. Therefore, filling the container (previously tared by weighing it empty) with water and reweighing gives the volume directly in cc. Weighing the same container filled with glue shows how much glue the same volume holds. When reduced to weight per unit volume, at this point grams per cc, the result can be converted to pounds per cubic foot by multiplying by the factor 62.4. Example: A container is found to hold 250 grams of water. This is equivalent to 250 cubic centimeters of volume. The same container is found to hold 198 grams of glue. The density in metric is 198 divided by 250, or 0.792 grams per cubic centimeter. In English units it would be 0.792 multiplied by 62.4, or 49.4 pounds per cubic foot.

Assembly Time

As mentioned repeatedly, the assembly time is crucial in the life of an adhesive. It is lying on the surface exposed to air on one side and to wood on the other. Solvent is escaping in all directions. Viscosity is increasing. A skin may be forming on the glue surface. The important action of transfer may not yet have been executed. In time, it may become too dry to execute the other motions of forming a bond. Some glues need a period of time at this point to firm up for the pressure step. But most glues have a maximum time limit beyond which they cannot perform. Both the minimum and the maximum time depend not only upon the glue formulation but also upon the amount of glue applied, whether to one surface or to both surfaces, whether open or closed, the moisture content and temperature of the wood, the ambient temperature and relative humidity, and the pressing conditions of pressure and temperature. This all has to be established between the glue manufacturer and the operator, sometimes by trial and error.

Once established, the assembly time check is a simple task of watching a clock. When more than one assembly is to be charged into one press, such as plywood in a multiopening hot press, or laminations in a nest of beams, the time between applying glue on the first glue line to the time glue is applied to the last glue line may span minutes or hours. The critical glue lines are the first few through the glue spreader, and the last few, as these are the ones experiencing the extreme in assembly times. Keeping all glue lines within the assembly time limits becomes the major objective of a quality-control operator.

Ambient Temperature and Relative Humidity

In many gluing operations, where control is good and gluing factors are kept well within their limits, these two factors may have little effect on performance of the glue. Moreover, those glues that are particularly sensitive to them often are produced in two types, one to be used in winter conditions, and one to be used in summer conditions. Operations that are skirting along the edges of factor limits—for example, those needing every minute of assembly time to prepare a load for the press—court disaster with an increase in ambient temperature or decrease in relative humidity. If there is also rapid air movement, as there might be for the benefit of workers, and if long open assembly is required, the damaging effect is made more severe. In this particular case we would be worrying about possible drying out of the applied film of glue before pressure can be applied.

Instruments for measuring temperature and relative humidity are familiar accoutrements of many households, but strangely, not of glue rooms. For those doing serious glue work, a recording temperature and relative humidity indicator would provide knowledge needed to control the operation adequately, and the record would be useful in associating a cause if trouble did develop. If these devices also automatically actuated switches to correct an unfavorable atmosphere, there would be one less worry to occupy one's mind. This would be the optimum. But there are less sophisticated means of gauging the atmosphere.

A wet and a dry bulb thermometer with a fan to assure air movement over the wet bulb (Figure 10–8 (a)) provides all the information needed to observe ambient temperature

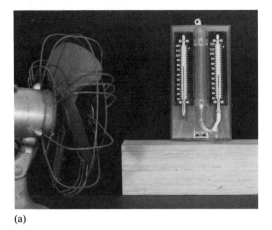

(a)

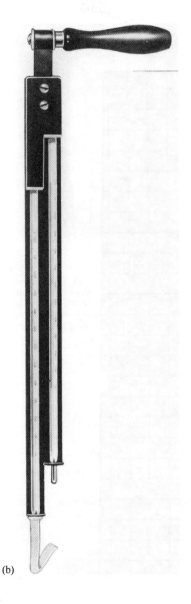

(b)

Figure 10–8 Measuring devices for relative humidity:
(a) wet and dry bulb with fan, (b) sling psychrometer.

and to calculate relative humidity. Figure 10–9 shows the relative humidity that pertains to various combinations of dry and wet bulb readings, and the resulting wood EMC.

In lieu of a fan, the wet and dry bulb assembly can be whirled through the air to provide the air movement. The sling psychrometer (Figure 10–8(b)) is portable and therefore can provide readings at various points throughout the gluing area.

The extreme limits of temperature and rel-ative humidity within which the gluing area should be maintained depend upon the particular glue being used; some can tolerate temperatures near freezing, others should not go below 70°F, and some may actually need higher temperatures. However, as a general rule, what is tolerable for people is tolerable for glues during the assembly time. It is frequently necessary to shorten or lengthen the assembly time when the temperature and rel-ative humidity demand it, or to abort the operation altogether if ambient conditions are out of line and cannot be controlled, particularly if the temperature is too low.

Pressure

When it comes time to apply pressure to the assembly of wood pieces, a number of different devices may be used, each with a different mechanical system, and therefore with a different means of monitoring what it is delivering. These may be reduced to just six basic mechanisms: live roll, dead weight, spring, screw, ram, and pneumatic or fluid pressure. One important distinction is that the screw differs from the others in a significant respect: The load-delivering head remains in a fixed position throughout the pressing cycle; the rest except for the roll, have a "follow-up" potential. There are movements that occur during pressing: moisture migrating; wood compressing, swelling, or shrinking; adhesive penetrating or flowing and reducing the volume of the glue line. Follow-up pressure devices can accommodate to these changes and maintain the pressure for the entire pressure period. Screws cannot adjust to the changing situation except by additional servomechanisms. It is not unusual for the system to experience a rise in pressure shortly after clamping, and to end up with zero pressure several hours later as the wood and glue make their adjustments.

In order to discuss the differences, some clarification of terms is necessary. Two terms in particular may be confusing: *load* (or force) and *pressure*. If a 200-pound rock were placed on a panel 12 inches by 12 inches, the load on the panel would be 200 pounds. The pressure on the panel would be 200 pounds divided by the area of the panel (12 × 12 = 144 square inches), or 1.39 pounds per square inch. (It is distressing to realize that a 200-pound stone, more than one person can handle, would produce such a pitifully small pressure on the glue line of so small a panel, when use instructions often call for a hundred times that pressure.) Conventionally, the load is always expressed in

pounds or other units of weight; pressure is always expressed in pounds per square inch (psi), or other units of load and area.

Another distressing thought is that the process goes blind at this point, and we do not really know specifically what is going on in the glue line, but only what is grossly happening. We know, for example, the average pressure, but not the pressure at any one point (even in a postmortem examination, except as there might have been too little or too much pressure).

It is well to recognize that two kinds of pressure can be applied: rigid and fluid. In the case of the rock applying pressure on a veneer panel, the rock being rigid will engage only a few points of contact, the high spots, or thickest parts of the panel. If the rock were replaced by a bag of sand of the same weight and covering the same area, since the bag is flexible and the sand relatively freeflowing, pressure would more likely be the same on both the high spots and the low spots—a more uniform pressure over the entire area of the panel. Because of this effect, fluid pressures can be much lower than rigid pressures in promoting adhesive motions. Placing a rubber pad between the stone and the panel would produce an effect similar to the sandbag.

With these points in mind, controlling the pressing operation becomes something more than reading a dial gauge. One has to sense what the loading system is generating, as well as what it is delivering to the glue line.

Dead-Weight System. In a dead-weight system, whether the load is produced directly or through pulleys, everything is direct reading. The weight of the loading object multiplied by any leverage applied is divided by the area of the assembly upon which it bears. For example, if the 200-pound rock on the 12-inch-square panel cited above were replaced with a 55-gallon drum of water, the load would be 55 × 8.3 (the weight of a gallon of water) = 456.5 pounds, and the pressure would be 456.5 divided by 144 or 3.17 pounds per square inch. Hanging the drum on a lever arm having a mechanical advan-

WET BULB DEPRESSION (°F.) (DIFFERENCE BETWEEN WET AND DRY BULB TEMPERATURES)

Relative humidity shown in bold face type in upper half of square — equilibrium moisture content in light face type in lower half of square.

DRY BULB TEMPERATURES (°F.) — each cell shows **relative humidity** (top) / equilibrium moisture content (bottom).

Dry Bulb	1°	2°	3°	4°	5°	6°	7°	8°	9°	10°	11°	12°	13°	14°	15°	16°	17°	18°
30°	89	78 / 15.9	67 / 12.9	57 / 10.8	46 / 9.0	36 / 7.4	27 / 5.7	17 / 3.9	6 / 1.6									
35°	90	81 / 16.8	72 / 13.9	63 / 11.9	54 / 10.3	45 / 8.8	37 / 7.4	28 / 6.0	19 / 4.5	11 / 2.9	3 / 0.8							
40°	92	83 / 17.6	75 / 14.8	68 / 12.9	60 / 11.2	52 / 9.9	45 / 8.6	37 / 7.4	29 / 6.2	22 / 5.0	15 / 3.5	8 / 1.9						
45°	93	85 / 18.3	78 / 15.6	72 / 13.7	64 / 12.0	58 / 10.7	51 / 9.5	44 / 8.5	37 / 7.5	31 / 6.5	25 / 5.3	19 / 4.2	12 / 2.9	6 / 1.5				
50°	93	86 / 19.0	80 / 16.3	74 / 14.4	68 / 12.7	62 / 11.5	56 / 10.3	50 / 9.4	44 / 8.5	38 / 7.6	32 / 6.7	27 / 5.7	21 / 4.8	16 / 3.9	10 / 2.8	5 / 1.5		
55°	94	88 / 19.5	82 / 16.9	76 / 15.1	70 / 13.4	65 / 12.2	60 / 11.0	54 / 10.1	49 / 9.3	44 / 8.4	39 / 7.6	34 / 6.8	28 / 6.0	24 / 5.3	19 / 4.5	14 / 3.6	9 / 2.5	5 / 1.3
60°	94	89 / 19.9	83 / 17.4	78 / 15.6	73 / 13.9	68 / 12.7	63 / 11.6	58 / 10.7	53 / 9.9	48 / 9.1	43 / 8.3	39 / 7.6	34 / 6.9	30 / 6.3	26 / 5.6	21 / 4.9	17 / 4.1	13 / 3.2
65°	95	90 / 20.3	84 / 17.8	80 / 16.1	75 / 14.4	70 / 13.3	66 / 12.1	61 / 11.2	56 / 10.4	52 / 9.7	48 / 8.9	44 / 8.3	39 / 7.7	36 / 7.1	32 / 6.5	27 / 5.8	24 / 5.2	20 / 4.5
70°	95	90 / 20.6	86 / 18.2	81 / 16.5	77 / 14.9	72 / 13.7	68 / 12.5	64 / 11.6	59 / 10.9	55 / 10.1	51 / 9.4	48 / 8.8	44 / 8.3	40 / 7.7	36 / 7.2	33 / 6.6	29 / 6.0	25 / 5.5
75°	95	91 / 20.9	86 / 18.5	82 / 16.8	78 / 15.2	74 / 14.0	70 / 12.9	66 / 12.0	62 / 11.2	58 / 10.5	54 / 9.8	51 / 9.3	47 / 8.7	44 / 8.2	41 / 7.7	37 / 7.2	34 / 6.7	31 / 6.2
80°	96	91 / 21.0	87 / 18.7	83 / 17.0	79 / 15.5	75 / 14.3	72 / 13.2	68 / 12.3	64 / 11.5	61 / 10.9	57 / 10.1	54 / 9.7	50 / 9.1	47 / 8.6	44 / 8.1	41 / 7.7	38 / 7.2	35 / 6.8
85°	96	92 / 21.2	88 / 18.8	84 / 17.2	80 / 15.7	76 / 14.5	73 / 13.5	70 / 12.5	66 / 11.8	63 / 11.2	59 / 10.5	56 / 10.0	53 / 9.5	50 / 9.0	47 / 8.5	44 / 8.1	41 / 7.6	38 / 7.2
90°	96	92 / 21.3	89 / 18.9	85 / 17.3	81 / 15.9	78 / 14.7	74 / 13.7	71 / 12.8	68 / 12.0	65 / 11.4	61 / 10.7	58 / 10.2	55 / 9.7	52 / 9.3	49 / 8.8	47 / 8.4	44 / 8.0	41 / 7.6
95°	96	92 / 21.3	89 / 19.0	85 / 17.4	82 / 16.1	79 / 14.9	75 / 13.9	72 / 12.9	69 / 12.2	66 / 11.6	63 / 11.0	60 / 10.5	57 / 10.0	55 / 9.5	52 / 9.1	49 / 8.7	46 / 8.2	44 / 7.9
100°	96	93 / 21.3	89 / 19.0	86 / 17.5	83 / 16.1	80 / 15.0	77 / 13.9	73 / 13.1	70 / 12.4	68 / 11.8	65 / 11.2	62 / 10.6	59 / 10.1	56 / 9.6	54 / 9.2	51 / 8.9	49 / 8.5	46 / 8.1
105°	96	93 / 21.4	90 / 19.0	87 / 17.5	83 / 16.2	80 / 15.1	77 / 14.0	74 / 13.2	71 / 12.6	69 / 11.9	66 / 11.3	63 / 10.8	60 / 10.3	58 / 9.8	55 / 9.4	53 / 9.0	50 / 8.7	48 / 8.3
110°	97	93 / 21.4	90 / 19.0	87 / 17.5	84 / 16.2	81 / 15.1	78 / 14.1	75 / 13.3	73 / 12.6	70 / 12.0	67 / 11.4	65 / 10.8	62 / 10.4	60 / 9.9	57 / 9.5	55 / 9.2	52 / 8.8	50 / 8.4
115°	97	93 / 21.4	90 / 19.0	88 / 17.5	85 / 16.2	82 / 15.1	79 / 14.1	76 / 13.4	74 / 12.7	71 / 12.1	68 / 11.5	66 / 10.9	63 / 10.4	61 / 10.0	58 / 9.6	56 / 9.3	54 / 8.9	52 / 8.6
120°	97	94 / 21.3	91 / 19.0	88 / 17.4	85 / 16.2	82 / 15.1	80 / 14.1	77 / 13.4	74 / 12.7	72 / 12.1	69 / 11.5	67 / 11.0	65 / 10.5	62 / 10.0	60 / 9.7	58 / 9.4	55 / 9.0	53 / 8.7
125°	97	94 / 21.2	91 / 18.9	88 / 17.3	86 / 16.1	83 / 15.0	80 / 14.0	77 / 13.4	75 / 12.7	73 / 12.1	70 / 11.5	68 / 11.0	65 / 10.5	63 / 10.0	61 / 9.7	59 / 9.4	57 / 9.0	55 / 8.7
130°	97	94 / 21.0	91 / 18.8	89 / 17.2	86 / 16.0	83 / 14.9	81 / 14.0	78 / 13.4	76 / 12.7	73 / 12.1	71 / 11.5	69 / 11.0	67 / 10.5	64 / 10.0	62 / 9.7	60 / 9.4	58 / 9.0	56 / 8.7
140°	97	95 / 20.7	92 / 18.6	89 / 16.9	87 / 15.8	84 / 14.8	82 / 13.8	79 / 13.2	77 / 12.5	75 / 11.9	73 / 11.4	70 / 10.9	68 / 10.4	66 / 10.0	64 / 9.6	62 / 9.4	60 / 9.0	58 / 8.7
150°	98	95 / 20.2	92 / 18.4	90 / 16.6	87 / 15.5	85 / 14.5	82 / 13.7	80 / 13.0	78 / 12.4	76 / 11.8	74 / 11.2	72 / 10.8	70 / 10.3	68 / 9.9	66 / 9.5	64 / 9.2	62 / 8.9	60 / 8.6
160°	98	95 / 19.8	93 / 18.1	90 / 16.2	88 / 15.2	86 / 14.2	83 / 13.4	81 / 12.7	79 / 12.1	77 / 11.5	75 / 11.0	73 / 10.6	71 / 10.1	69 / 9.7	67 / 9.4	65 / 9.1	64 / 8.8	62 / 8.5
170°	98	95 / 19.4	93 / 17.7	91 / 15.8	89 / 14.8	86 / 13.9	84 / 13.2	82 / 12.4	80 / 11.8	78 / 11.3	76 / 10.8	74 / 10.4	72 / 9.9	70 / 9.6	69 / 9.2	67 / 9.0	65 / 8.6	63 / 8.4
180°	98	96 / 18.9	94 / 17.3	91 / 15.5	89 / 14.5	87 / 13.7	85 / 12.9	83 / 12.2	81 / 11.6	79 / 11.1	77 / 10.6	75 / 10.1	73 / 9.7	72 / 9.4	70 / 9.0	68 / 8.8	67 / 8.4	65 / 8.1
190°	98	96 / 18.5	94 / 16.9	92 / 15.2	90 / 14.2	88 / 13.4	85 / 12.7	84 / 12.0	82 / 11.4	80 / 10.9	78 / 10.5	76 / 10.0	75 / 9.6	73 / 9.2	71 / 8.9	69 / 8.6	68 / 8.2	66 / 7.9
200°	98	96 / 18.1	94 / 16.4	92 / 14.9	90 / 14.0	88 / 13.2	86 / 12.4	84 / 11.8	82 / 11.2	80 / 10.8	79 / 10.3	77 / 9.8	75 / 9.4	74 / 9.1	72 / 8.8	70 / 8.4	69 / 8.1	67 / 7.7
210°	98	96 / 17.7	94 / 16.0	92 / 14.6	90 / 13.8	88 / 13.0	86 / 12.2	85 / 11.7	83 / 11.1	81 / 10.6	79 / 10.0	78 / 9.7	76 / 9.2	75 / 9.0	73 / 8.7	71 / 8.3	70 / 8.0	68 / 7.6

WET BULB DEPRESSION (°F.) (DIFFERENCE BETWEEN WET AND DRY BULB TEMPERATURES)

Figure 10–9   Wood-moisture content and relative humidity from wet and dry bulb readings. (Compiled by Irvington Moore Co.)

Directions: Find dry bulb temperature on side margin and wet bulb depression at top or bottom of chart. Follow these columns to square at point of intersection. Relative humidity values are in bold face type in upper half of square, equilibrium moisture content values in light face type in lower half of square.

Combined R. H. — E. M. C. Chart

This chart is a valuable drying tool. It combines on the same chart percentage of Relative Humidity (R. H.) and Equilibrium Moisture Content (E. M. C.) values for a given dry bulb temperature and wet bulb depression.

Relative Humidity is the amount of water vapor actually in the air, expressed as a percentage of the amount it would hold at saturation. E. M. C. is the moisture content at which lumber will finally equalize when exposed to air of a given dry bulb temperature and wet bulb depression.

As an example: for 130° dry bulb temperature and 120° wet bulb temperature (wet bulb depression of 10°), follow 130° line across to square under 10° depression column — this will show R. H. of 73% and E. M. C. of 12.1%.

WET BULB DEPRESSION (°F.) — diagonal scale labels: 25°, 26°, 27°, 28°, 30°, 32°, 34°, 36°, 38°, 40°, 45°, 50°

Compiled from chart data furnished by Forest Products Laboratory

Each cell shows **R. H.** (upper, bold) over E. M. C. (lower, light) as "R.H. / E.M.C."

Dry Bulb (°F.)	19°	20°	21°	22°	23°	24°	25°	26°	27°	28°	30°	32°	34°	36°	38°	40°	45°	50°
30°																		
35°																		
40°																		
45°																		
50°																		
55°																		
60°	9 / 2.3	5 / 1.3	1 / 0.2															
65°	16 / 3.8	13 / 3.0	8 / 2.3	6 / 1.4	2 / 0.4													
70°	22 / 4.9	19 / 4.3	15 / 3.7	12 / 2.9	9 / 2.3	6 / 1.5	3 / 0.7											
75°	28 / 5.6	24 / 5.1	21 / 4.7	18 / 4.1	15 / 3.5	12 / 2.9	10 / 2.3	7 / 1.7	4 / 0.9	1 / 0.2								
80°	32 / 6.3	29 / 5.8	26 / 5.4	23 / 5.0	20 / 4.5	18 / 4.0	15 / 3.5	12 / 3.0	10 / 2.4	7 / 1.8	3 / 0.3							
85°	36 / 6.7	33 / 6.3	30 / 6.0	28 / 5.6	25 / 5.2	23 / 4.8	20 / 4.3	18 / 3.9	15 / 3.4	13 / 3.0	9 / 1.7	4 / 0.9						
90°	39 / 7.2	36 / 6.8	34 / 6.5	31 / 6.1	29 / 5.7	26 / 5.3	24 / 4.9	22 / 4.6	19 / 4.2	17 / 3.8	13 / 2.8	9 / 2.1	5 / 1.3	1 / 0.4				
95°	42 / 7.5	39 / 7.1	37 / 6.8	34 / 6.4	32 / 6.1	30 / 5.7	28 / 5.3	26 / 5.1	23 / 4.8	22 / 4.4	17 / 3.6	14 / 3.0	10 / 2.3	6 / 1.5	2 / 0.6			
100°	44 / 7.8	41 / 7.4	39 / 7.0	37 / 6.7	35 / 6.4	33 / 6.1	30 / 5.7	28 / 5.4	26 / 5.2	24 / 4.9	21 / 4.2	17 / 3.6	13 / 3.1	10 / 2.4	7 / 1.6	4 / 0.7		
105°	46 / 7.9	44 / 7.6	42 / 7.3	40 / 6.9	37 / 6.7	35 / 6.4	34 / 6.1	31 / 5.7	29 / 5.4	28 / 5.2	24 / 4.6	20 / 4.2	17 / 3.6	14 / 3.1	11 / 2.4	8 / 1.8		
110°	48 / 8.1	46 / 7.7	44 / 7.5	42 / 7.2	40 / 6.8	38 / 6.6	36 / 6.3	34 / 6.0	32 / 5.7	30 / 5.4	26 / 4.8	23 / 4.5	20 / 4.0	17 / 3.5	14 / 3.0	11 / 2.5	4 / 1.1	
115°	50 / 8.2	48 / 7.8	45 / 7.6	43 / 7.3	41 / 7.0	40 / 6.7	38 / 6.5	36 / 6.2	34 / 5.9	32 / 5.6	29 / 5.2	26 / 4.7	23 / 4.3	20 / 3.9	17 / 3.4	14 / 2.9	8 / 1.7	2 / 0.4
120°	51 / 8.3	49 / 7.9	47 / 7.7	45 / 7.4	43 / 7.2	41 / 6.8	40 / 6.6	38 / 6.3	36 / 6.1	34 / 5.8	31 / 5.4	28 / 5.0	25 / 4.6	22 / 4.2	19 / 3.7	17 / 3.3	10 / 2.3	5 / 1.1
125°	53 / 8.3	51 / 8.0	48 / 7.7	47 / 7.5	45 / 7.2	43 / 7.0	41 / 6.7	39 / 6.5	38 / 6.2	36 / 6.0	33 / 5.5	30 / 5.2	27 / 4.8	24 / 4.4	22 / 4.0	19 / 3.6	13 / 2.7	8 / 1.6
130°	54 / 8.3	52 / 8.0	50 / 7.8	48 / 7.6	46 / 7.3	45 / 7.0	43 / 6.8	41 / 6.6	40 / 6.4	38 / 6.1	35 / 5.6	32 / 5.3	29 / 4.9	26 / 4.6	24 / 4.2	21 / 3.8	15 / 3.0	10 / 2.0
140°	56 / 8.4	54 / 8.0	53 / 7.8	51 / 7.6	49 / 7.3	47 / 7.1	46 / 6.9	44 / 6.6	43 / 6.4	41 / 6.2	38 / 5.8	35 / 5.5	32 / 5.1	30 / 4.8	27 / 4.4	25 / 4.1	19 / 3.4	14 / 2.6
150°	58 / 8.3	57 / 8.0	55 / 7.8	53 / 7.5	51 / 7.3	49 / 7.1	48 / 6.9	46 / 6.7	45 / 6.4	43 / 6.2	41 / 5.8	38 / 5.5	36 / 5.2	33 / 4.9	30 / 4.5	28 / 4.2	23 / 3.6	18 / 2.9
160°	60 / 8.2	58 / 7.9	57 / 7.7	55 / 7.4	53 / 7.2	52 / 7.0	50 / 6.8	49 / 6.7	47 / 6.4	46 / 6.2	43 / 5.8	41 / 5.5	38 / 5.2	35 / 4.9	33 / 4.6	31 / 4.3	25 / 3.7	21 / 3.2
170°	62 / 8.0	60 / 7.8	59 / 7.6	57 / 7.3	55 / 7.2	53 / 6.9	52 / 6.7	51 / 6.6	49 / 6.4	48 / 6.2	45 / 5.7	43 / 5.5	40 / 5.2	38 / 4.9	35 / 4.6	33 / 4.4	28 / 3.7	24 / 3.2
180°	63 / 7.8	62 / 7.6	60 / 7.4	58 / 7.2	57 / 7.0	55 / 6.8	54 / 6.5	52 / 6.4	51 / 6.2	50 / 6.0	47 / 5.7	45 / 5.4	42 / 5.2	40 / 4.8	38 / 4.6	35 / 4.4	30 / 3.8	26 / 3.3
190°	65 / 7.7	63 / 7.4	62 / 7.2	60 / 7.0	58 / 6.8	57 / 6.6	56 / 6.4	54 / 6.2	53 / 6.0	51 / 5.9	49 / 5.5	46 / 5.3	44 / 5.0	42 / 4.8	39 / 4.5	37 / 4.4	32 / 3.8	28 / 3.3
200°	66 / 7.5	64 / 7.2	63 / 7.0	61 / 6.9	60 / 6.6	58 / 6.4	57 / 6.2	55 / 6.0	54 / 5.9	53 / 5.7	51 / 5.4	48 / 5.2	46 / 4.9	43 / 4.7	41 / 4.5	39 / 4.3	34 / 3.8	30 / 3.3
210°	67 / 7.4	65 / 7.1	64 / 6.9	63 / 6.8	61 / 6.5	60 / 6.3	59 / 6.1	57 / 5.9	56 / 5.8	54 / 5.5	52 / 5.3	50 / 5.1	47 / 4.8	45 / 4.6	43 / 4.4	41 / 4.2	36 / 3.7	32 / 3.2

Right margin: DRY BULB TEMPERATURES (°F.) — 110, 115°, 120°, 125°, 130°, 140°, 150°, 160°, 170°, 180°, 190°, 200°, 210

Bottom scale: WET BULB DEPRESSION (°F.) — 19° 20° 21° 22° 23° 24°   25° 26° 27° 28° 30° 32°   34° 36° 38° 40° 45° 50°

tage of 4:1 would jump the load to 1826 pounds and the pressure to 12.68 psi, still not very much. Dead weight systems are not very effective.

Spring-Loading Systems. Springs are convenient and appropriate methods of applying loads to gluing assemblies, especially if pressure must be maintained overnight. Since they apply a relatively constant pressure despite small changes in dimensions of the assembly, they are of the follow-up variety. In order to know how much load a spring is delivering, it is necessary to know its spring constant, the pounds required to produce a unit deflection. If not available, the spring constant can be determined by applying a series of known loads to the spring and measuring the deflection. Plotting this information produces a relationship that may be interpolated to estimate what load is being produced when the spring is compressed a certain amount. The plot may produce a straight line or a curved line depending upon the nature of the spring.

In use, the springs are compressed and restrained by a rod that then delivers the load to a platen. A system of four springs, each requiring 400 pounds per inch of deflection and each deflected $1\frac{1}{2}$ inches, would produce a load of 600 pounds per spring and 2400 pounds for the system. Delivered to a 12-inch-square panel, this would produce a pressure of 16.67 psi. In order to develop 100 psi, the number of such springs similarly loaded would have to be increased to 24.

The toggle clamp is a form of spring that is often used to hold parts in position until glue hardens. While the load it delivers is relatively small, they can be ganged to provide a quick and easy pressing system for small parts with accurately machined surfaces. The load they can deliver is a function of their size, and is usually given on the package. However, the load they actually deliver depends upon the thickness of the assembly they engage. Obviously if the assembly is too thin, no load will be developed. On the other hand, if the load is too thick, it will be im-

possible to snap the clamp shut. There is an optimum thickness for maximum pressure and this can be discovered by trial and error.

Screw Presses. The majority of small shop gluing, and a good share of industrial gluing, is done on screw pressing devices. The loads generated are a function of the pitch of the threads on the screw, the diameter of the screw, the torque applied to turn the screw, and the friction in the system. Bypassing the engineering calculations necessary to figure out the relationships in mathematical detail, the information important to a user is what torque or force on one end of the screw will produce what load at the other end. Because screws differ widely in design as well as in condition, each one, even if they are similar, should be calibrated. Calibration involves the use of two pieces of equipment, a torque wrench, and a load cell. A load cell is a device that can be inserted into a pressing system and it reads the load it receives. There are many versions of these using strain gauges, calibrated beams, or hydraulics. In any case, the procedure is to place the load cell in the press as if it were the assembly being glued. The screw is tightened as it normally would be and the load is read on a dial. More useful information can be obtained if a torque wrench is used to tighten the screw. This would provide data of torque versus load, which can be plotted. From the plot it is possible to specify how much torque would be needed to obtain a particular load as demanded by the size of the assembly being pressed.

Some industrial screw presses are driven by pneumatic wrenches. In these cases, the calibration follows the procedure as with the torque wrenches except that air pressure to the wrench rather than torque becomes the operating variable. The plot would then show air pressure versus load delivered by the screw.

When screws are driven by electric motors, the force they deliver to torque the screw is more difficult to measure and control. A clutch-type device can be attached to the shaft of the motor that reads torque. However, it might be nearer to actuality if a load cell were used directly in the bed of the press to measure the true load delivered. An ammeter on the motor then can be coordinated with the load produced, and so plotted.

A number of hand screw pressing devices are frequently used, singly or in multiples to create pressure. These include C-clamps, wood clamps, bar clamps, and toggle clamps mentioned earlier (Figure 10–10). The loads they can deliver, hand torquing by an average person, are shown in Table 10–1. These

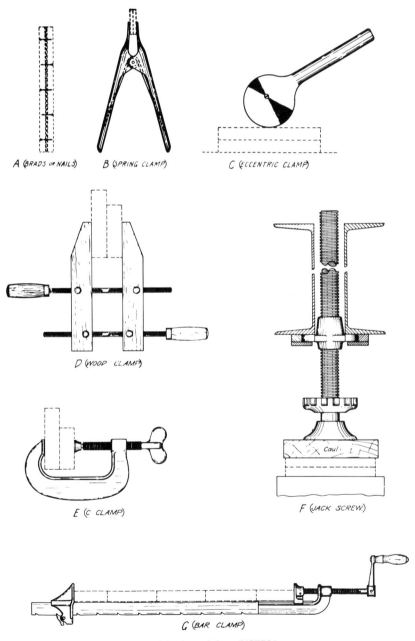

Figure 10–10   Typical clamps used in shop gluing. (USFPL)

Table 10-1. Loads That Can Be Delivered by Various Clamping Devices. (USFPL)

CLAMP	POUNDS TOTAL LOAD	INCHES SCREW DIAMETER	COMMENTS
A	no data	——	Depends upon driving force and withdrawal resistance
B	5–30	——	Larger spring and shorter lever arm gives higher load
C	no data	——	Longer handle and smaller diameter produces greater force, but depends also on thickness
D	1585 1370 2700	$\frac{9}{16}$ $\frac{11}{16}$ $\frac{9}{16}$	Lever arm $1\frac{3}{4}''$ Lever arm $2''$ Lever arm $3''$
E	750–1000 1500–2000	$\frac{3}{8}$ $\frac{3}{8}$	With work at tip as shown With work close to screws
F	34,000 16,000 17,000 7500	$1\frac{5}{16}$ $1\frac{5}{16}$ $2\frac{5}{16}$ $\frac{7}{8}$	Lever arm $37''$ Lever arm $18''$ Lever arm $31''$ Lever arm $7''$
G	2800 1100	$\frac{1}{2}$ $\frac{5}{8}$	Lever arm $2\frac{1}{2}''$ Lever arm $1\frac{3}{4}''$

loads, divided by the area of the glue line being pressed, give the approximate pressure in psi delivered to the assembly.

Hydraulic Presses. Operating on the principle of the hydraulic jack, a piston or ram is pressured by a liquid medium, usually oil. The pressure of the liquid acting on the bottom of the ram is multiplied by the cross-sectional area of the ram to produce the force at the top, a consequence of liquid and mechanical properties. No matter what the size of the pipe carrying the liquid from a pump to the chamber at the bottom, the multiplication at the top is the same. It is the pressure of the liquid and the area of the ram that is important; the two produce the force that will be delivered to the assembly. A simple formula connects the ram with the assembly being glued:

$$\text{AREA of the RAM} \times \text{PSI on the RAM} = \text{AREA of the ASSEMBLY} \times \text{PSI on the ASSEMBLY}$$

This equation can be reordered to determine any factor when any three are known, for example:

$$\text{PSI on the RAM} = \frac{\text{AREA of the ASSEMBLY} \times \text{PSI on the ASSEMBLY}}{\text{AREA of the RAM}}$$

For those presses that have more than one ram, the sum of all the ram areas must be used in the calculations. However, for a press having multiple openings in a stack, the area of the assembly on only one opening need be considered. This is because the same load is being transmitted from the assembly in one opening to that in the next. Also, the area of the platens in the press has no bearing on the pressure delivered to the assembly unless they are totally covered by the assembly; i.e., equal in area. For example, a fifteen-opening press with platens 52 inches by 102 inches, and powered by eight rams of ten-inch diameter, is to be loaded with 15 panels 36 inches by 100 inches, and given a pressure of 150 psi. What should the line pressure to the rams be?

The total ram area is
$$3.1417 \times 10 \times 10 \times 8 / 4 = 628.34 \text{ sq in}$$
The total area of the panels is
$$36 \times 100 = 3600 \text{ sq in}$$
Load needed on panels is
$$3600 \times 150 = 540,000 \text{ lbs}$$

Line pressure on rams is

$$540,000 \ / \ 628.34 \ = \ 859 \ \text{psi}$$

With a given press, it is often desirable to graph the relationship between panel area and gauge pressure on the rams. For the above press an applicable graph is shown in Figure 10–11. Such a graph should be placed near the press for easy reference when panel sizes are frequently changed.

Pressure Period. Press times are usually suggested by the adhesive supplier for the formulation and the product being glued, and the temperature existing, whether room or elevated. Some fine-tuning is usually needed to account for local variations. After

that, monitoring is only a question of clock-work. If everything is manual, an eye must be engaged with the clock. Timers that control elapsed time and ring a bell are better; and better still are timers that control both time and the pressure valves. For many room-setting glues, time may be several hours or overnight, and the exact time is not critical. Assembly glues are usually not timed directly but are part of a total cycle that includes applying glue, assembling parts, placing in the press, and removing from the press.

In the production of comminuted panel materials, pressure is applied as needed to compress the mat to the desired thickness, usually to stops, in a prescribed time period.

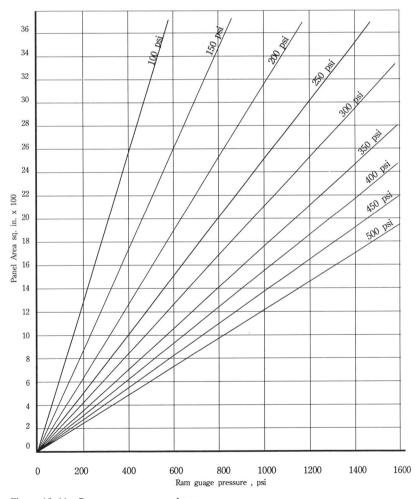

Figure 10–11   Ram gauge pressure chart.

After stops are reached, pressure is reduced to that needed for holding at the stops. At definite points during the pressure period, pressure is reduced in order to allow steam pressures inside the panel to dissipate. "Breathing" the press in this manner is necessary because of the build-up of moisture at the center of the panel and its conversion to steam as the temperature there rises above 210°F. If this steam pressure is not allowed to escape, blisters may form, ruining the panel. Since these involve closely timed intervals the entire press cycle is controlled by timing devices tied directly to the press.

Glue Line Temperature

Monitoring the glue-line temperature is more a research objective than a production-control tool. It is usually done by inserting thermocouples directly in the glue line at various distances from the heat source and following the temperature rise with a calibrated potentiometer. The glue-line temperature can also be calculated from the platen temperature and the physical laws governing heat transfer and diffusion. However, direct measurement provides confirmation against other variables that may affect a temperature gradient, such as density and moisture. The innermost glue line is the one requiring close attention because it is farthest from the heat source and most at risk of being undercured.

Reconditioning

Many panel products come out of the press too hot or too dry and must be restored to ambient conditions. Panels made with urea resins in hot-press operations are in need of cooling in order to prevent overcuring and degradation of the bond. For this purpose they are placed on edge in a fixture that can hold a press load, and air is allowed to circulate for cooling (Figure 9–22). Panels bonded with phenolic resins, on the other hand, are placed in a stack immediately out of the press to retain the heat of the press as long as possible and complete the cure. This hot stacking permits shorter press times and

speeds production. When panels emerge too dry, they are sprayed with a calculated amount of water before stacking. Reconditioning is also done in kilns where they can be subjected to a calculated amount of humidity and temperature for a period of time to effect equilibrium. Time, temperature, and humidity are thus the controlling parameters.

MONITORING THE GLUED PRODUCT

After the glued product comes out of the pressure system, all of the bond-forming actions have taken place, except for some final molecular adjustments to achieve ultimate strength and durability, or for equalizing moisture or temperature. Handling the product at this stage and during trimming and sanding provides an opportunity to inspect for off-grade products. Decisions to reject and repair or discard have to be made quickly as the products are passing by. Hence only visual characteristics can be used. To aid in inspection, various kinds and angles of lighting are used. For example, for observing surface defects low-angle lighting reveals differences in point-to-point elevation. Ultraviolet light reveals bleed-through of urea glues.

A preliminary look at bond quality is often obtained by examining trim ends and edges. Although these locations can contain the poorest parts of the glue line because of undercuring, low pressure, or edge dry-out, it can be reassuring when good results appear. For plywood, a knife is used to pry up the end-grain edges of the face veneer. The knife should preferably have a raised shoulder at the handle end of the blade to protect the hand from scraping against the veneer should the knife slip. It is important that both ends of the same panel be checked in this manner, because grain direction through the thickness of the face veneer can create an entirely different impression of the glue bond; from zero to 100 percent wood failure.

For glued lumber, a chisel is used to break open the joints. The chisel should be about two inches wide and fairly sharp. With the edge of the chisel exactly over the glue line at the end-grain edge, a blow of a heavy hammer should cleave the joint apart. If a second blow is needed, the chisel should be repositioned over the glue line. If the chisel is not on the glue line, it is probable that it will cleave entirely in the wood, giving a false impression, unless the bond is so bad that it cleaves in the glue line despite starting in the wood. If possible, duplicate samples should be obtained from the same location, and the chisel driven from opposite directions to ensure grain effects are accounted for.

Once into the glue line, one can apply all the appearance characteristics of Chapter 3 to obtain some estimate of how the glue is working.

In addition to these glimpses in the glue line, observations of the entire glued product should spot

- Open joints
- Loose surface elements
- Splits
- Blisters
- Darkened surface areas (indicating possible blows)
- Other discolorations
- Warpages
- Misassembly
- Miscuts

Since at this point the equivalent of 100 percent inspection exists, a powerful additional quality-control tool could be activated: records. Stamping each item with a serial number relating to date and time of manufacture provides a log not only of the amount of production but also of the number of defects. When defects are tied to a time line, they are much easier to trace to their origin. Analysis of the data over time will aid in deciding whether corrective action would be cost effective or whether to accept the rate of failure as inevitable, and therefore perhaps normal. Because of the knowledge such a history produces about the conduct of an operation, the most important element in quality control is the record.

Assurance that the product will perform as specified requires physical and mechanical tests. This normally falls into the province of a separate entity, a testing laboratory. The laboratory employs standard procedures to determine conformance to specifications using criteria promulgated; in the case of commercial products, by the U.S. Department of Commerce in cooperation with trade associations, producers, consumers, and general interest parties. The resulting product standards are usually administered by various associations whose stamp on the product certifies that it has been made according to specifications and that it will perform as stipulated. This is the ultimate achievement of a materials science of wood. Some representative test procedures that underlie these standards for several products are given in Chapter 11.

11

Bond Performance: Testing and Evaluation

After an adhesive has performed its five motions, and solidified in one or more of its nine states, a bond has been formed between two surfaces. At this stage, the bond is considered to be composed of nine links, each fashioned to various degrees depending upon how all the bond-forming factors have operated. Frozen in the completed bond is the history of what transpired during its formation. Some of this history is obvious to the naked eye, particularly when the bond is a bad one. When the bond is a very good one, the history is locked in the woodwork, as it were, out of sight and out of reach of instruments. Qualities of bond in between these extremes are generally revealed by various kinds of invasive or destructive procedures, some yielding numerical values and some yielding purely qualitative indications. Noninvasive or nondestructive test methods are useful mostly for revealing major flaws in the glue line.

This chapter is concerned with methods of determining the quality of bonds for purposes primarily of appraising potential performance. However, by proper manipulation of procedures and experimental designs, these methods are also useful for observing effects of most of the factors in the equation of performance.

Interpreting observations is as important as making them, and of course these will vary with the experience and the objectives of the tester. As a first priority, however, it is necessary to have accurate and relevant data.

In approaching the subject of measurement and its attendant questions of quality, the distinction between *bond formation* and *bond performance* takes on added significance. Often tests for the latter are used to observe factors operating in the former. The resulting data can be ambiguous and misleading. Some separation of effects should be experimentally possible, however, since each part invokes different properties of the adhesive, or at least invokes them at different times. Adhesive bond formation is essentially a two-step process of first establishing adhesion and then achieving strength. These are dependent upon fluid properties and hardening properties and should be observable by physiochemical methods. Adhesive bond performance, on the other hand, is a consequence of the solid phase that emerged from the fluid phase, reacting with the solid properties of the wood. It is properly studied by the methods of material science in which strength and permanence are the critical properties.

On the assumption that the physiochemical properties have already been established by the adhesive manufacturer, interest in this chapter centers on the completed bond, its status, and how it might perform, given its composition, the structure it is holding together, and the process by which it was formed; in other words, getting the summative answer to the equation of performance.

As a backdrop for a discussion on testing and evaluation, it is instructive to review some sources of misleading results by way of warning against hasty conclusions. Every test has limitations, not all obvious, and frequently overstepped.

PITFALLS IN TESTING

Some extreme examples illustrate the problems that may occur between the test tube and the glue line. Grease is not ordinarily considered to have adhesive properties on the basis of test-tube tests. It is nonpolar in nature and has hydrophobic properties. Yet when used like an adhesive between two solid blocks such as glass, a considerable force is needed to pull them apart in tension perpendicular to the surfaces. We could conclude that grease is an adhesive. If the test were repeated, but this time the glued blocks received a shear force, they would easily slide apart, and we would say there were no adhesive properties. Similarly, water, at the opposite end of a chemical scale, is both highly polar and hydrophilic. When used in a similar manner, it too exhibits adhesive properties against a tension force but not against a shear force. If frozen (i.e., hardened), the water bond would resist a shear force but the grease bond, having very low adhesion, would not provide much resistance to separation. Neither grease nor water if observed in bulk form would be considered an adhesive. Yet under certain conditions, some type of bond can be formed. Therefore, to characterize fully the adhesive qualities of two unknown materials, represented in this case by grease and water, both kinds of tests are necessary.

Coming at the question from a different

direction, it is known that phenol-formaldehyde resins produce the strongest and most durable bonds in the manufacture of plywood. In liquid form, the resins show no adhesive properties; they are like water. When cured to a solid state, they are hard and brittle like glass. If a phenolic resin is tested by making bonded specimens, a bond may or may not form depending upon the conditions chosen for making the bond. Hence if conditions were ill chosen, the resulting failure would confirm the unfavorable bulk observations. The determination of adhesive properties thus can produce either positive or negative conclusions depending upon the observation system used.

Other pitfalls can arise even if an apparently good bond forms. Some adhesives can show sufficiently high bond strength in test specimens, and fail in structures. Two insidious factors are at play here. One is long-time loading, and the other is stress concentrations. Different adhesives respond obscurely to these two factors and often may not register in a test, or if it does will do so capriciously. For example, polyvinylacetate gives bonds that appear to be stronger than wood when tested by the block shear method (described later). Yet the same adhesive could fail if used to bond the same wood in a laminated beam. The reason lies in the plastic deformation (rheological) properties of the resin, which result in yielding under the long-time loading that occurs in a beam. A resorcinol resin might show a lower strength in block shear tests but perform satisfactorily in a beam. The delusion is furthered by stress concentrations that are higher in a block shear specimen than in a beam. The polyvinyl acetate, because of its more plastic (versus brittle) nature, tends to distribute stress more uniformly over a bond area, and thus is better able than resorcinol to deal with the stress concentrations in a specimen. For this same reason, the polyvinyl acetate could actually have an intrinsically lower shear strength; its ability to distribute stress more uniformly being actually an expression of greater efficiency in utilizing bond area.

The question of how stress interacts on the glue line raises many issues regarding the geometry of a specimen, the type of load it can deliver to the bond, and how the load resembles that in the actual product. Superimposed on geometry are the environmental conditions the bonded assembly must resist in service. The geometric relationship of specimen to product, and the environmental relationship of test conditions to natural atmospheric conditions, pose problems of interpretation that have not been fully resolved. Figure 11-1 shows how blind or capricious a bond test can be in accounting for its true quality.

It is fairly obvious, therefore, that in order to sort out or isolate effects that affect performance, it is necessary to recognize and account for all the pertinent factors integrating themselves positively and negatively into an observation. The following section is a general overview of different approaches to testing that allow for the registration of various groups of factors in the gluing operation.

SPECIMEN CONSIDERATIONS

Depending upon the knowledge objective sought, several test options exist that incorporate an increasing number of variables that may have an influence on the quality of the bond or the specificity of the observation. In the following series of test specimens, observations become less specific with respect to any single factor, but more specific with respect to the glued structure being tested. The final test includes so many variables (all of them in all their interactions) that the observation can apply with confidence only to the one product. The main dif-

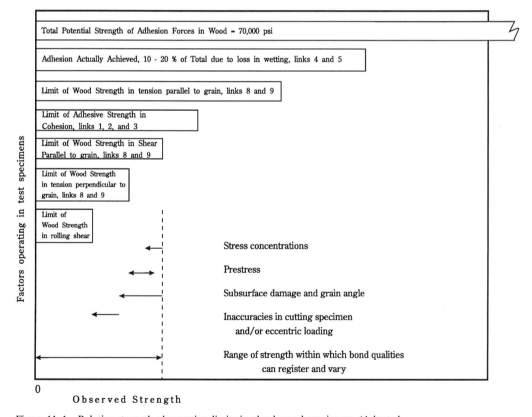

Figure 11-1 Relative strength observation limits in glued wood specimens. (Adapted from Frank Reinhart J. Chem. Ed. 128, 31, 1954)

ferences between the tests are in the geometry of the specimens and the procedure by which they are produced. The series begins with the simplest in terms of structural purity, and proceeds to the most complex in terms of number of factors operating. The specimens are more fully described in a later section.

Adhesive Tests of Maximum Specificity

In making observations that represent only intrinsic properties of the adhesive, it is desirable to impose a relatively pure stress on the bond and minimize the influence of the substrate on the reading. The development of a pure stress is difficult to accomplish in wood because of its anisotropic and quasi-plastic nature. The button specimen for tension, and the cylinder torsion specimen for shear (described later), however, come the closest to single-stress systems. Most other specimens introduce combined stresses. Although the real world of glued wood products involves mostly combined stresses, at the research level it is often desirable to know how shear and tension properties develop in an adhesive in cases where it is desired to improve one or the other.

Standard Specimens

When history plays a part in interpretation of results or when results must be repeatable, reliable, and comparable, for purposes of describing an adhesive to be bought or sold on specification, specimens must be standardized for species, surfaces, and sizes, and the gluing procedure also stipulated within a narrow range of handling properties. A number of specimens have been standardized by the American Society of Testing and Materials for testing adhesives for lumber and plywood. For the other smaller wood elements, a standard specimen for determining adhesive properties is still evolving, mainly because it is difficult to standardize sizes, surfaces, and compositing procedure. Consequently the bond must be tested as part of an assemblage of many elements, and usu-

ally compared with another adhesive of known performance in an identical procedure. Because of varying environments and orientations within the assembly, one can never know with reasonable certainty what factor or factors were responsible for contrary results when differences occur.

Standard Specimens Simulating the Product

Since geometry has such a profound effect on the performance of a glue bond, every specimen that has a different geometry will produce a different estimate of bond quality. Consequently the predictive value of a specimen resides in how close the geometry of the specimen resembles that of the ultimate glued product it purports to represent. For example, the button specimen, the torsion specimen, and the cross-lap specimen bear very little resemblance to any glued product. On the other hand, the block shear and the plywood shear specimens contain grain orientation and wood element thicknesses found in beams and plywood. These later specimens, even though made as samples, have a better chance of simulating what the glue would experience in an actual construction. But not quite.

Standard Specimens Cut from Glued Products

Specimens prepared as specimens (as above) are made under relatively ideal laboratory conditions and are appropriate for the precise comparison of adhesives. However, they do not contain the vagaries of the mill floor, the effects on bond formation that shape and size of wood elements introduce, the effects of wood preparation factors, and especially the effects of looser controls. To obtain a fuller accounting of performance and provide some evidence of tolerance to mill conditions, the informational output of standard specimens can be increased by cutting them out of products that have come off a production line. When the data from such specimens is compared with that of laboratory-made specimens simulating the product, the difference can be attributed to effects of workmanship.

Portions of Glued Product as Specimens

Portions of glued products may contain, in addition to workmanship, factors of shape and size both of the product and of the specimen that have affected bond formation and can now register to a better degree their effects on bond performance. Being somewhat discretionary, these specimens may vary from product to product, and from point to point in the same product, and can no longer be standardized. The evaluation now more approaches the performance of the product, while the performance of the bond itself is buried in the overall results. Moreover, strength test and exposure test results cannot be freely extrapolated to other products. Thus, although the test becomes more revealing for a particular product, it also becomes less general.

The Entire Product as Specimen

The ultimate test of bonding qualities is made of the whole glued product where all the pertinent factors in the equation of performance have had a chance to operate and assert themselves. Again, the results cannot be extrapolated beyond the structure tested, because part of the observed performance is due to the size of the wood elements and the total shape of the product, as well as any special techniques entailed by these factors.

EXPOSURE CONSIDERATIONS

The above sequence of specimen sources can yield information on the strength of bonds made in various ways, that is, the strength at the moment of test and under the prevailing conditions. Of greater interest in some circumstances is information on whether strength will be maintained over the expected life of the product in the environment in which it will be used. This invokes the durability aspects of the bond, and raises the concern of workmanship in achieving the

maximum resistance to deterioration promised by the composition of the adhesive. In essence, the question becomes: Did bond formation capture all the qualities built into the adhesive, and did it produce the ultimate in bond performance?

In assessing the durability of a bond, information is needed that will predict not only how it will perform in the environment where it will be placed in service, but also for how long. The time element is the problem; except for a few cases of long experience, it is really not known how to expose a specimen in a laboratory so that results obtained today can be extrapolated into the future.

However, all may not be lost. Several courses of action exist to allow workable estimates. First, a base line durability is measured on the hardened glue alone against the expected degradative factors that might exist in the service environment, heat and water being the most likely. This will provide a yes or no answer at either extreme: no effect, or complete destruction. It can then be said that heat at some level of temperature has no effect, or that immersion in water has no effect. The converse could also provide a definitive statement: Heat above a certain temperature and immersion in water cause complete destruction of the bond. Time is a variable in this case, leading to an open question. What, for example, would be the consequence in performance of a bond between an adhesive that required four hours to destruct, compared to one that remained intact for six hours? And would that consequence hold the same for lumber as for veneer, or other wood elements?

Cases that fall in between, with some effect but not complete destruction, introduce further questions. Now it is necessary to estimate how much strength loss is acceptable, and worse, how much heat and/or water will result in strength loss, and no more, *over time;* again versus the configurations of product. At this point it would be helpful if there were a quantum of heat, water, and time so that a measure could be struck on how many quantums of this effect a glue can sustain indefinitely.

In a bonded structure complications multiply because water has six possible effects, three on the glue: dissolving, softening, hydrolyzing; and three on the wood: dimension change, strength change, and decay, plus a number of interactions. The effects on the wood are further complicated by differences among species and sizes of the wood elements, which affect the amount of stress that would be generated with gain and loss of water. Species and sizes also affect the rate with which the water enters or leaves the specimen, which destroys the comparability of results where timed exposures are part of the test. Heat has three effects on the glue: softening, further curing, and breakdown; and four effects on wood: temporary strength loss, permanent strength loss, dimension change, and drying (which produces a strength gain). This is elaborated more fully in the section on specimens. In most cases a further loss in performance is generated by the amount of stress imposed on the glued specimen while it is being exposed. This stress is effective whether internal or external. All this is more than a computer can handle with the knowledge we currently have to feed it.

One recourse is to establish standard exposure procedures that with experience will allow prediction of performance with a degree of confidence. ASTM and other associations have done this for a number of products by way of assuring proper adhesive selection and assured product performance. The standards define how a test shall be conducted, including preparation of the wood, application of the adhesive, method of strength measurement, design of the specimen, number of specimens, exposure treatment, and reporting of results. Species is also specified, but can be selected to suit a particular objective.

It should be noted that standardization produces rigidity in the method of testing in order to achieve uniformity, precision, and repeatability of results. The results are intended to provide information for product specifications where it is important to know whether stipulated criteria are met or not met, pass or do not pass; the judgment must be in black or white terms.

When doing research, on the other hand, it is desirable to have hue and color as well as black and white in order to discern trends and make extrapolations, trace the effects of certain factors, observe interactions, and in general obtain a broader picture of the adhesive. Hue and color is obtained by an ordered series of several factors, providing opportunities for the adhesive to register strengths and sensitivities that are inevitably missed in following a standard procedure. In the one-shot picture provided by a standard procedure, the risk is that the combination of conditions specified can reflect the best or the worst performance of a particular glue (or anything in between) depending upon interactions with its bond-forming characteristics. Later in this chapter, a procedure is outlined that provides not single values of performance but a context of values that describe in useful detail the performance profile of an adhesive. The context of values can then be integrated statistically into a tolerance index (TI) that in a sense quantifies the ability of the adhesive to overcome vagaries in the gluing operation. When an adhesive has a low TI, it can be expected to perform more reliably in day-to-day operations.

As a means of sensing degrees of severity imposed by various test-exposure atmospheres, the following list suggests a sequence of conditions increasing in destructive capacity on a given bonded wood specimen.

1. Controlled room atmospheric conditions, usually used to condition wood before gluing, recondition specimens after gluing, or maintain control specimens at constant moisture and temperature

2. Controlled room atmospheric conditions plus applied stress to observe creep behavior under reproducible conditions

3. Continuous exposure to heat: a. tested hot to determine temporary effects;

b. tested at room temperature for permanent effects

4. Continuous exposure to heat plus external stress to accentuate the effects of heat

5. Continuous exposure to moisture in vapor form: a. tested at high moisture to observe temporary effects; b. tested reconditioned to observed permanent effects

6. Continuous exposure to moisture in liquid form, immersion: a. tested wet to determine temporary effects; b. tested reconditioned to determine permanent effects

7. Continuous exposure to moisture, plus external stress to accentuate the effect of moisture

8. Continuous exposure to both heat and moisture (steaming or boiling): a. tested wet and hot to determine temporary effects; b. tested wet and cooled to distinguish effects; c. tested redried and cooled for permanent effects

9. Continuous exposure to heat and moisture plus external stress to accentuate effects

10. Cyclic exposure to heat, imposing heat and deformations due to gradients in conjunction with breakdown and softening of the adhesive: a. tested at hot end of cycle; b. tested at cool end of cycle

11. Cyclic exposure to moisture (vapor or liquid), imposing deformations and stresses due to gradients in conjunction with adhesive breakdown or softening: a. tested at high end of moisture cycle; b. tested at reconditioned end of cycle

12. Cyclic exposure to heat and moisture: a. tested hot and wet; b. tested reconditioned

13. Combinations of cyclic heat, cyclic moisture, and cyclic external stress providing both in-phase maxima as well as out-of-phase maxima

14. Same as 13 plus the inclusion of a freezing cycle

15. Uncontrolled atmospheric conditions introducing reality into long-time exposures to determine actual life expectancy, and to provide a basis for judging reliability of the above exposures (usually unprotected outdoor exposure at specified locations in the country)

The above list is far from suggesting quantums of exposure, but it provides a starting point in assessing the durability level of an adhesive. The remaining unknown factor is time. Because it is desired to accelerate the test but predict for the long term, it is necessary to consider how much extra severity must be imposed in the test to account for the shorter time of exposure. This is an age-old and ongoing problem. Fortunately, long-term exposures have been, and are continuing to be conducted to correlate with short-term accelerated tests. Some surprising results have emerged, causing reconsideration of certain exposures. For example, the introduction of heat in a test procedure can complete the cure of an otherwise undercured hot-press glue and give it a durability rating it might not exhibit in long-time natural exposure.

The lack of quantums of exposure means that a universal test does not yet exist with which to screen all adhesives and put them in some well-defined order of resistance with any more rigor than that provided by the SHMO intuitive ranking of Figure 4-2. However, a few rather simple immersion tests can separate most glue bonds into three rather distinct resistance categories:

1. Those that survive boiling water (waterproof)
2. Those that fail boiling water but survive cold soaking (water-resistant)
3. Those that fail cold soaking (non-water-resistant)

Within categories, glues can be further separated by noting the time to failure in each immersion or noting the percent loss in strength with a given immersion. This classification, of course, would only hold if all observations were carried out with the same specimen size, configuration, and species.

GENERAL CONSIDERATIONS
IN TESTING

In the previous section, it was pointed out that testing for strength and permanence involves considerations of specimen size, organization, and configuration, and the imposition of a controlled environment indicative or reflective of service exposure. Hundreds of test methods have been developed over the years, and a large number are in use throughout the glued wood products field. As one attempts to read the quality of a bond from tests, three different kinds of data are considered:

1. Strength—as load per unit area of glue line, psi
2. Wood failure—as percent of total fracture area
3. Delamination—as percent of open joint edges

Strength and delamination are directly measurable quantities and their analysis would seem to be relatively straight forward. Wood failure, on the other hand, is more difficult to assess and is subject to a great deal of interpretation.

Considerations in Observations of Strength

Although strength would seem to be an unambiguous reading of a dial, it suffers from ignorance of the stress distribution in a bond area under load, and this introduces the question of generality of the observed result. When strength is reduced to pounds per square inch of bond area, it is presumed that each square inch carries the same load. This is seldom the case, for many reasons. Visualize that the load on the specimen is produced by a machine with a gauge on it. The load is delivered to the specimen through some type of grips or jig that engages the wood. The wood then transmits the load to the bond. Because wood is anisotropic and elastoplastic in nature, it cannot distribute stress equally in all directions. At the point of contact with the grips, the wood receives all the load. From there to the bond area, the load fans out and distribution responds to all the mechnical variables of wood. In most cases, depending upon configuration of the specimen, stress is delivered most intensely at the edges of the bond, usually at end-grain edges, creating stress concentrations at those points.

A consequence of the stress concentration is that the maximum load will first be sensed at the edges, causing failure there and then progressively to the rest of the bond. Although the rupture may appear in some cases to be catastrophic, actually few parts of the bond area are at maximum stress at the same time. The actual bond area supporting the maximum load therefore is indeterminate. However, for lack of a stress/area summing system, dividing the total bond area into the load at failure provides the best available estimate of unit bond strength.

It is for this reason that specimen size must be stipulated in standardized test procedures. Smaller specimens tend to give higher unit strengths. This can be demonstrated by shortening the length of the bond in simple lap specimens while keeping the width constant. A similar effect can be noted with the block shear specimen by shortening the height between the notches. Alternatively, the size of the specimen can be left unchanged, but the area of the bond reduced by applying wax or graphite incrementally to the bond area, blocking bond formation in those areas. Interesting variations can be imposed. If the blocking is applied to portions of the joint area near the edges receiving the test load, the load-carrying capacity will be reduced accordingly, and the calculated unit bond stress will remain fairly constant. However, if the blocking is applied to the central portions of the bond, very little, if any, difference in load to rupture will occur, and the calculated unit bond stress will rise proportionately.

Adhesives differ in response to the absence of bond in central portions of a joint. Low-modulus adhesives, thermoplastic or creep prone, tend to be more sensitive to unbonded central areas, increasing in load capacity as bond area increases. High-modulus, brittle, glassy adhesives do not recognize a central bond area loss. The difference is due to the plastic adhesives being able, because of their slight deformability, to transmit stress more efficiently to all areas of the joint. Hence when a bond area is absent, the load it would otherwise carry is also absent—a strength loss. The brittle adhesives, being too rigid to undergo the deformation necessary to transmit stress, do not use the inner areas of a bond, and fail catastrophically from the outside where the stress remains concentrated.

This uncertainty in actual strength of a bond as read from the numbers on a dial is partially assuaged by standardization of procedure. However, the specimen size and bond area interaction can become a source of information when used as controlled variables in an experiment. A change in observed strength measured against a change in internal bond area provides information on plastic properties of the bond. This would aid in the formulation of adhesives with properties more in harmony with the properties of the wood, optimizing brittleness versus plasticity to produce the same stress reactions as wood, and thereby reducing both creep and stress, concentrating effects for maximum performance.

Another useful objective of observing the response of an adhesive to change in central bond area is to calculate the maximum, or intrinsic, bond strength. Observations of bond strength against a series of central bond area changes provide data that can be plotted as strength versus percent of area bonded (Figure 11-2). Extrapolation to zero bond area, the intercept on the strength axis, wipes out the effect of size of the bond area, and yields a figure that is free of stress-distribution differences.

Considerations in Observations of Wood Failure

Observations on the fractured surface of a bond should provide some indication of which of the nine links failed, and with more intense analysis, the reason(s) for failure. Often more than one link has failed in a given fracture, and therefore the assumption must be that the mechanisms of bond formation have encountered differing conditions from point to point on the glue line or differing wood strengths, e.g., springwood or summerwood. Hence while the focus for many purposes is on the amount of wood failure observed, the analysis can be broadened to include observations on many facets of the gluing operation. For example, failure in links 8 or 9—wood failure—in the same bond line with failure of link 4 on the unspread side of a joint—transfer—may be an indication of uneven machining of the surface.

In general the assessment of wood failure is made by eye, although optical scanning devices have been developed to measure this feature more accurately. It has been shown by round robin observations of the same set of specimens that different viewers may differ on the amount of wood failure on a given specimen, but the average reported on a sample of ten specimens is fairly uniform. Various kinds of visual aids can be used, one of the most useful being a grid the size of the fracture area drawn on a clear plastic sheet. The grid can be scaled to divide the area into halves, quarters, and eighths, or into decimal equivalents. By placing the grid over the fracture surface, it is possible to observe the wood failure under each subunit and then to add these together. Practiced wood-failure readers can estimate within plus or minus 5 percent simply by visualizing a grid of various configurations depending upon the pattern of wood failure. The literature will sometimes contain wood-failure values reported to some decimal place. This is usually the result of averaging a number of observations and is of limited value except in precise experiments.

Interpretations of wood failure can be improved by characterizing it in various ways. Two distinctions are of importance: shallow wood failure and deep wood failure. Deep wood failure may be taken as 10 or more cells below the surface, characteristics of

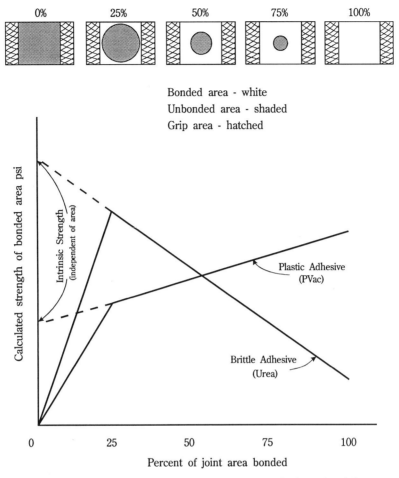

Figure 11–2 Variation in calculated psi strength for two adhesives when joint area remains constant and bond area is varied by wax blocking in the cross-lap specimen. (Marra; FPJ)

good penetration by the adhesive and repair of subsurface damage, links 6 and 7. Shallow wood failure, less than 10 cells deep, is sometimes referred to as fuzzy fiber pull. It suggests inadequate penetration and repair of subsurface damage, and is often not counted as wood failure.

It is at times useful to note which surface provided the wood failure as well as which was spread with glue. This is especially true in plywood constructions where grain is crossed and the glue is applied only to one surface. When wood is pulled from the element sustaining shear along the grain, the face ply, it indicates good repair of subsurface damage in the cross-grain element.

Wood-failure estimates of 0 percent and 100 percent should be reasonably clear-cut indications of bond quality: The bond is weaker than the wood, or the bond is stronger than the wood. The values between 0 and 100 are more difficult to interpret because while the balance of the bond may be termed adhesion failure or adhesive failure, there are many reasons for the lack of wood failure at any particular spot on the glue line. Compared to the relatively narrow range of the factors that contribute to a good bond, the factors that contribute to a poor bond are unlimited in range. It is generally true that bond line areas that have no wood failure contain more information regarding the conditions that existed during pressing than wood failure areas.

Considerations in Observations of Delamination

The tendency of an adhesively bonded assembly to delaminate is usually observed after subjecting the specimen to some type of exposure. This means the glue line and the wood surrounding it may be exposed to moisture increases and decreases, as well as temperature increases and decreases. Hence most of the factors that degrade a bond may be operating, including stress that generates in the wood and crosses the glue line. The amount of stress generated is dependent upon the species of wood (density), thickness of the wood, moisture gradient, and grain direction on either side of the bond. The same adhesive will therefore show a different degree of delamination if any of these factors are varied. Therefore comparisons can only be valid if made with identical procedures. Some of the tests that follow are designed to impose a source of degrading effects coupled with gradients to develop internal stress.

TEST SPECIMENS

Glues for each wood element, and the composites they produce, tend to have their own special test specimens and procedures for evaluating bond performance. Following is a sampling representative of the major products. Typical geometries and procedures are given together with a brief analysis of mechanisms and cautions that apply.

Specimens for Testing Adhesive Properties

In trying to assess the strength properties of adhesives, it is desirable to disencumber the results from the interactions of the geometry and size of the specimen and produce data that reflect only a property of the adhesive. While this is difficult to do with wood as the substrate, several methods approach the

ideal sufficiently to enable useful approximations. The following specimens are used mainly for observing the bond properties of adhesives mostly of the self-curing types, although some have been adapted for heat curing. Because of their greater sophistication, they are not widely used by adhesive users, but by researchers bent on elucidating particular characteristics. The cross-lap specimen, of considerably simpler construction and production than the others, is more amenable for spot testing and comparison testing.

The Button Specimen. The name derives from the configuration of the specimen, resembling two buttons glued together (Figure 11–3(a)). Its design is intended to minimize distortion while being loaded, and to impose a relatively pure tensile stress on the adhesive bond. If rupture occurs within the adhesive (i.e., link 1), the observed value is indicative of the tensile strength of the hardened adhesive.

The specimen is made by first gluing together two lumber-size wood elements, grain parallel, to produce a blank. After reconditioning, two-inch-square blocks are cut and these are then reduced to the final configuration on a lathe. The test is carried out by inserting the specimen in a stirrup that allows tensile load to be applied perpendicular to the bond area. (See ASTM Method D-897 for details.)

The notch at the glue line tends to insure that the load is delivered in somewhat concentrated form to the glue line, thereby lessening the tendency for rupture to occur in the wood. Nevertheless, the limit of strength that can be measured is the tensile strength of wood perpendicular to the grain.

In theory it is presumed that the stress is uniform over the area of the joint and that there are no edge effects. While possibly true for metal, it is less true for wood, due to its anisotropic properties. However, if the specimen were modified so end-grain surfaces were being bonded, a more uniform stress could be delivered, though at the risk of poorer anchorage to the wood surface. This specimen is not suitable for exposure testing

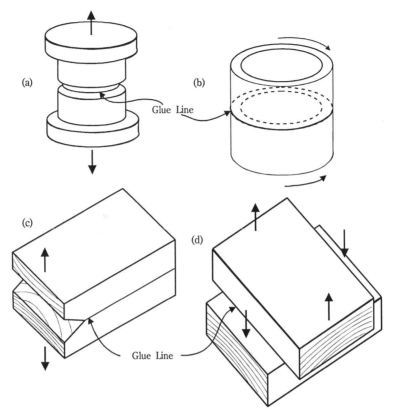

Figure 11-3 Specimens primarily for measuring strength properties of adhesives per se: (a) button specimen, (b) torsion specimen, (c) cleavage specimen, (d) cross-lap specimen.

because of possible distortions that would affect seating in the test jig.

The Torsion Shear Specimen. Similar in aspect to the bonding of two pieces of pipe endwise, the torsion specimen is designed to deliver a pure shear force to the glue line, and without edge effects since the bond area is a continuous circular band (Figure 11-3(b)). Its use with wood, however, is limited by the difficulty of providing wood in tubular form, and gripping it to deliver the shear force in torsion. Thin veneer, coiled and glued around a mandrel, offers one means of producing elements with which to make specimens. After preparation, they can be slipped onto the same mandrel to make the test bond. As with all wood specimens, some means must be provided to separate deformations of the wood from deforma-

tions of the adhesive if a shear modulus is being determined.

Cleavage Specimen. There are many configurations of cleavage test specimens. All, however, introduce a tensile force at one end-grain edge of a bond area with the objective of determining the force necessary to pry it open. The area of bond resisting the force at any one time is largely unknown. However, the width of the bond line at the point of application of the force is related to the strength and is directly measurable. Consequently results are reported in pounds per inch of bond width rather than pounds per square inch of bond area as is normally done. Rupture can occur progressively as well as catastrophically, depending upon the configuration of the specimen, its materials, and the nature of the bond. In progressive rupture, the load rises and falls as separation increases, suggesting successive build-up and release of stresses. Hence it is useful to record both maximum and minimum loads observed during the test.

Figure 11-3(c) is only a schematic representation of the specimen. Many modifications in length, thickness, and reinforcements, and methods of applying the cleavage force are in the literature, and should be consulted before undertaking cleavage tests. (J. M. Yavorski et al., Forest Products Journal, 1955, 5 [5] and 5 [1]). Ebewele Koutsky and River, Wood and Fiber Science, 1979, 11 [3], have refined the test to allow measurement of fracture toughness that then can be used to observe the rheological properties both during cure and during degradative exposures.

Like other specimens the cleavage specimen is limited by the presence of end grain at the edge of the bond area coinciding with the point where the force is applied. An effective means of avoiding the consequences of this juncture is the insertion of very thin tape for some distance in from the edge of the bond before it is formed. Alternatively, the bond can be prevented from forming at the edge by coating the surface of the wood with wax before bonding.

The Cross-Lap Specimen.   Combining the properties of the three previous specimens, the cross-lap specimen (Figure 11-3(d)) imposes tension, cleavage, and a minor amount of shear stress, and therefore produces data that is too ambiguous to use in engineering. However, its configuration, which removes end grain from the edges of the bond area, reduces the tendency to wood failure and makes the data more reflective of the actual condition of the adhesive layer. Also, since the specimens are easy to make and no machining is required after bond formation, the test is most useful for observing the rate of strength development. The ability to vary the tension/cleavage interaction is also of interest. This is done by varying the thickness of the blocks either before gluing or, preferably, after gluing by cutting away increments

of thickness, or by bonding additional blocks to the backs of the original specimen.

Specimens for Testing Glued Lumber Elements

Throughout this century, the workhorses of adhesive testing have been the so-called block shear specimen for lumber and the plywood shear specimen (described later) for veneer. These two specimens have been involved in the development of virtually every wood adhesive on the market today. They are therefore well standardized and entrenched in adhesive technology, providing a long history of data against which to compare new adhesives and bonding techniques. Along with block shear for strength assessment, lumber gluing is backed by a delamination specimen for measuring permanence.

The Block Shear Specimen.   Designed to test adhesive performance in lumber-size wood elements, the block shear specimen offers the opportunity to pit the bond against one of the stronger properties of wood, shear along the grain. It thus is able to provide the registration of high strength values before wood failure occurs to obscure the results.

When used to evaluate or compare adhesives, the specimen is produced as described in standardized detail in ASTM Method D-905. The standard specimen (Figure 11-4(a)) has a glue line 1.5 inches along the grain by 2 inches across the grain, providing 3 square inches of glue line area. The load at failure divided by 3 thus produces the strength of the joint in psi. Conventionally, it is seldom made one at a time, but in multiples of five, cut out of a glued assembly (billet), as shown in Figure 11-4(b).

Standardization of the procedure is necessary in this and other tests because of the variables that can cause unreliable, inaccurate, or unrepeatable results. Nevertheless, observations cannot be considered absolute but only relative since they are still dependent on the procedure used. A different procedure—for example, a different rate of

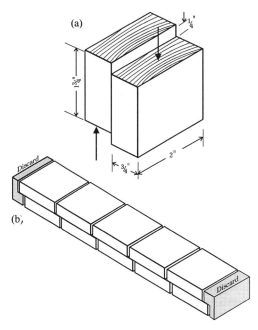

Figure 11–4   (a) The block shear specimen: (b) method of cutting from a glued billet.

loading, a different specimen size, or a different jig for holding the specimen during testing—can produce a different result. Cutting inaccuracies are a major cause of varying results. Okkonen and River, 1989, Forest Products Journal, vol. 1, no. 1, have studied some of the factors affecting the strength of block shear specimens and confirmed several sources of variability that included radial versus tangential plane of cut, species, shear tool, nature of notching, and specimen size. Differences as much as 60 percent were observed with the variables well controlled, and the procedure well executed.

Of equal importance are variables that are uncontrolled and unknown playing capriciously in a procedure. One of these is accuracy of notching. The specimen has a natural tendency to cleave due to the way the load is applied, eccentrically, and the way the specimen is held while loading. Out-of-square notching can accentuate this effect, increasing or decreasing the observed result and altering the wood failure. Figure 11–5(a, b, c) illustrates three ways the specimen can be misnotched.

Another variable that is often not sufficiently controlled is cellular-grain direction (as opposed to ring-grain direction in radial and tangential orientation). In Figure 11–5(d, e) the grain is at an angle with respect to the plane of the bond, identical in both specimens as they might be cut out of the same billet. However specimen (e), if by chance notched as shown, will invariably register a higher load at rupture than specimen (d), also so notched by chance. This has nothing to do with quality of the bond, the same in both cases, but with the location or orientation of the notches. In (d) the direction of load vectors with respect to the grain produces a component in tension perpendicular to the glue line, whereas in (e) the component is in compression perpendicular to the glue line, a situation more favorable to resist the load. In (f) and (g) the herringbone orientation of grain on either side of the glue line would produce the same result in both notch orientations, since the weaker one would determine the result in each case.

As a means of eliminating this source of variation, or a least observing it, a procedure for cutting specimens provides for both possibilities to register when or if the conditions exist. It is suggested that the blank from which the specimens are cut be lengthened to provide for six rather than five specimens, three to be cut with one notch orientation and three to be cut in the opposite orientation, Figure 11–5(h). If the specimens are numbered 1 through 6 before cutting, results can be correlated with their notching, then averaged to obtain the value for the sample.

When it is desired to observe the shear quality of a bond in a glued assembly such as an edge-glued lumber panel or a face-glued lumber beam, it is necessary to contrive a means of cutting the specimen from the product. In an edge-glued lumber panel, the resulting specimen can be of the normal length along the grain, but no wider than the thickness of the panel, Figure 11–6(a). A suggested cutting pattern would be to cut two strips each two inches along the grain from across the panel as shown. These are marked A and B in congruent corners so as

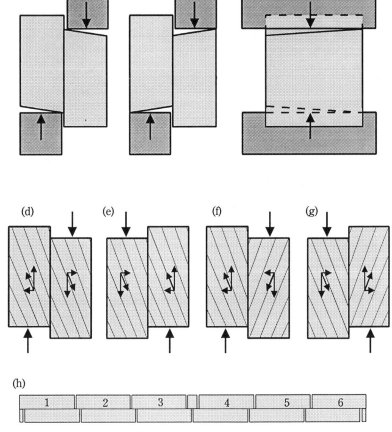

Figure 11-5   Some sources of variability in test results with the block shear specimen: (a, b, c) inaccuracies in cutting block shear specimen, (d, e, f, g) grain direction possibilities that can become incorporated unknowingly into the block shear specimen, (h) cutting pattern to equalize grain and other differences.

to insure their original spatial relationship with each other. An outline of a specimen is accurately drawn at each glue line, notching those in A opposite to those in B, and identifying each as A1, A2, A3, B1, B2, B3, and so on. This assures that any grain-deviation effects will be accounted in the observations. The specimens are then cut from the panel with the same cautions as in cutting the standard specimen.

Cutting block shear specimens from a laminated beam requires greater skill in laying out the cutting pattern and in sawing out the specimen. The specimens for each glue line in this case remain attached to each other, forming the so-called stair-step shear

specimen (Figure 11-6(b)). A section of the beam (previously side-dressed), 6+ inches long, is cut off beyond the normal end trim and resawn to a width of 2 inches. Again the specimen is sketched accurately around the glue lines in succession across the beam, offsetting each notch $\frac{1}{4}$ inch as in the standard specimen. Dotted lines from the surface to the glue lines establish where the saw cuts are to be made for each notch. Making two matching sets of step shear specimens, with each notched oppositely, helps wash out differences due to grain direction.

Delamination Test Specimen.   The delimination test for lumber, glued face to face, is

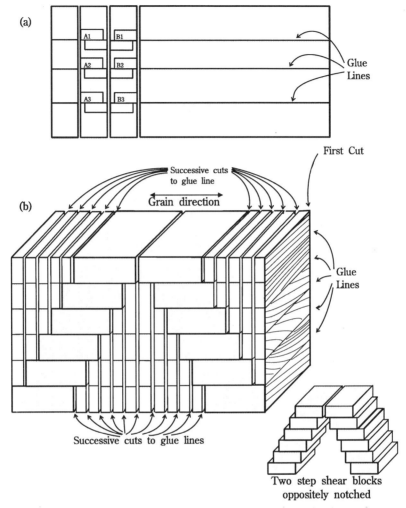

Figure 11-6 Method of cutting bock shear specimen from (a) edge-glued panel, and (b) face-glued lumber beam.

intended to produce the greatest force on the glue line that can be generated by the wood itself as it goes through a maximum moisture-content change. Because water immersion is involved as well as elevated temperatures, the bond is also being subjected to these degradative factors at the same time as the forces are being exerted.

The standard specimen is cut from a sample beam comprised of six $\frac{3}{4}$-inch boards glued with parallel grain, using the adhesive and procedure in question. The beam section is 6 inches wide by 40 inches long, to provide a representative bond area for the adhesive to function and to yield enough of both types of specimens for statistical analysis. The specimen itself (Figure 11-7(a)) measures 3 inches along the grain and 5 inches across after side dressing.

Standardized in ASTM Specification D-2559 (Testing Adhesives for Structural Laminated Wood Products for Use Under Exterior [wet use] Exposure Conditions), the procedure is based on producing a steep moisture gradient in the test piece. This is done by first loading the specimen with water to at least 50 percent using vacuum/pressure methods as in wood impregnation, followed by rapid drying in an oven at 150°F with forced circulation. This cycle is re-

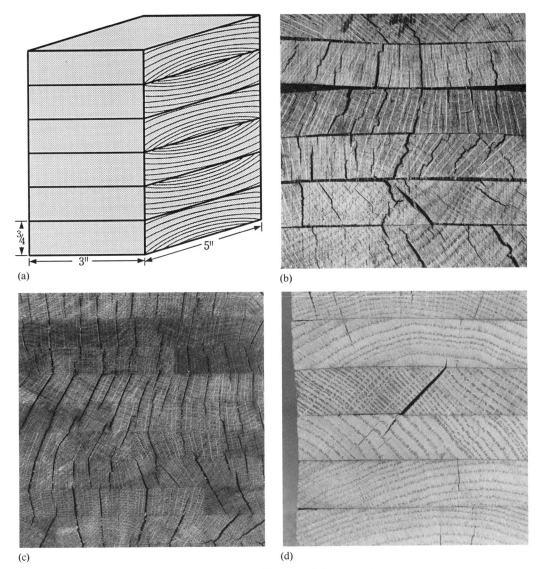

Figure 11-7   The delamination specimen for face-glued lumber: (a) the standard specimen, (b) open joints after cyclic exposure (USFPL), (c) no open joints after cyclic
exposure (USFPL), (d) edge distortions due to differential instability of individual laminations following cyclic exposure.

peated two or three times, sometimes with
steaming as part of the immersion cycle. The
test ends after a drying part of a cycle, which
is intended to leave the specimen with a high-
moisture core and a near-zero moisture at
the end-grain faces. Thus the core will be
fully swollen, while the faces will be fully
shrunken. The result is the development of a
tension force perpendicular to the plane of
the glue lines, the force being maximum at

the end-grain face, diminishing inward
toward the core, Figure 11-8(a). Separations
(delaminations) show up in the bond line
where weaknesses exist. Compare Figure
11-7(b) with 11-7(c) which sustained no delamination despite severe checking along
rays. These separations are measured for
length, added together, and reduced to percent based on the total length of glue lines
exposed at the end grain faces. Separations

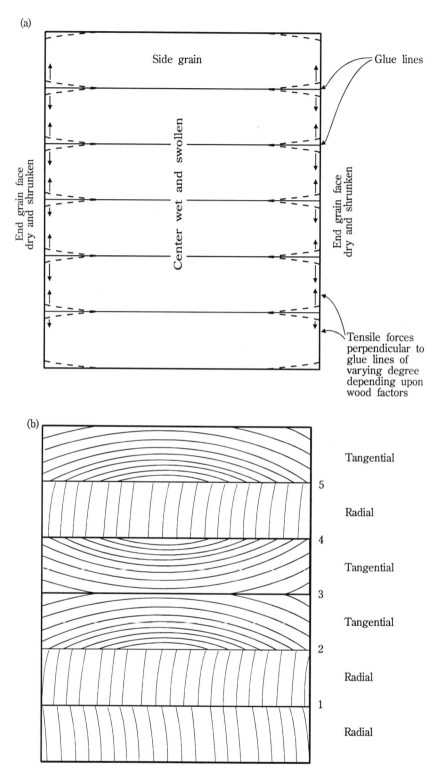

Figure 11-8 Forces acting in the lumber delamination specimen: (a) due to moisture gradient during cycling, (b) due to dimensional change differential between radial and tangential cut boards.

along the side grain are usually not considered in the calculation. In the standard size specimen, there are 50 inches of end-grain glue lines. Thus the calculation would be total inches of delamination divided by 50, multiplied by 100 to yield the percent delamination.

Because this is a particularly severe test even for solid wood, all kinds and degrees of separation can occur, some of which may be discounted. For example, the wood, particularly oak, may become extensively checked along rays, or the wood may split next to the glue line. These are not counted since they do not represent bond failure. Separations associated with knots are not counted. Sporadic, short, isolated separations in the glue line are also not counted since they could be due to a small local aberration of some sort. A probe, machinist feeler gauge, .003 to .005 inches thick, is sometimes used to verify a failure.

A question that is open to speculation regards hairline separations that may be visible to the naked eye or made visible with a magnifying glass. While these may be regarded as incipient failures, they are generally disregarded on the supposition that since the test is so severe, they would pose no threat to performance under natural exposure conditions.

It should be noted that there is an optimum time when the separations are maximum. If drying is continued to complete equilibrium—i.e., when the face and core are at the same moisture content and therefore at the same dimensions—glue line breaks that were once open may close. It is proposed, as a check to assure against this possibility, that the specimens be weighed before exposure so the drying can be stopped when they reach not less than 1.15 nor more than 1.4 times their original weight.

The method has interesting implications from the standpoint of wood properties. Since the wood is the source of the forces creating the destructive effects (as well as a participant in creating the bond), different results can ensue out of variables arising in the wood even though the bond quality may be the same. For example, quarter-sawn and flat-sawn lumber can have differing effects alone and particularly in mixtures. Since the development of tensile stress on the glue line is due to shrinkage of the wood perpendicular to the plane of the bond, quarter-sawn lumber should provide a more severe effect than flat-sawn lumber due to its greater shrinkage in the thickness direction. Glue line 1 in Figure 11–8(b) would be more severely stressed than glue in line 3. A mixture of quarter- and flat-sawn lumber introduces an additional stress component, shear, as adjacent boards shrink and swell across the grain different amounts with the moisture change, glue lines 2, 4, and 5. Note the edge distortions produced by dissimilar shrinkage of the laminations in Figure 11–7(d).

An even greater apparent difference in performance would be observed between, for example, white pine lumber and oak lumber since the latter is capable of developing a higher shrinkage force. In either case, however, it should be noted that the amount of stress development is limited to the lowest strength property of wood: tension perpendicular to the grain. When shrinkage produces stress above this amount, the wood begins to fail, relieving stresses. It is logical therefore that an adhesive failing on oak might pass on pine, an effect that is more an expression of density than it is of species. Specifications recognize this difference by allowing, for example, a 10 percent delamination for an adhesive to be used on hardwoods, but only 5 percent if used on a softwood.

The relationship between delamination and how it is affected by species, density, and plane of cut must be tempered by the actual strength of the wood, radial and tangential, and by its actual dimension change with moisture change, radial and tangential. The ray structure of oak in effect represents a line of weakness in the plane of the rays that can negate the greater shrinkage in the tangential direction that might otherwise produce greater stress and more delamination.

Standardization of this test seeks to reduce

the effects of these types of variables and make the comparison and specification of adhesives more reliable.

Testing End-Grain Joints

In end-grain joints where the scarf joint and finger joint are the primary objects of test, grain direction, joint design, and joint preparation have such a strong influence on both bond formation and bond performance, they are very specific to the construction and the method of production. Consequently use of test results is most appropriately limited to comparisons within the experiment, or for observing day-to-day qualities of the same product.

Both the scarf joint and the finger joint can be tested in a beam mode or in direct tension along the grain. In the beam mode, the joint is placed at center span and the load applied at two points as shown in Figure 11–9(a). In the figure, the load is applied perpendicular to the plane of the glue lines, which places the potentially weakest part of the joint, the feather edge, at the highest stress point, the extreme fiber, potentially biasing the results downward. A more uniform reaction of the joint to the imposed load is obtained by turning the specimen 90 degrees

so that the glue line is parallel with the direction of loading. Calculations are made as for solid wood in determining modulus of rupture. Comparison with solid wood can therefore be made under the same conditions of test. In the direct-tension mode, the specimen is simply gripped at the ends and an endwise tensile force applied to failure. In order to insure breakage at the joint and avoid slippage at the grips, the width of the specimen is reduced gradually at the joint, Figure 11–9 (c and d). This is also similar to the method of obtaining the tensile strength of solid wood parallel to the grain, and comparisons can be made as a percentage of the strength of solid wood. In calculating strength of the joint, therefore, it is necessary to use the cross-sectional area of the necked-down portion of the specimen rather than the area of the bond.

Alternatively, in the case of the scarf joint, block shear specimens can be cut that are smaller than normal size in order to include the glue line, Figure 11–9(b). In this situation, grain direction, as discussed earlier, is particularly important because of the relatively steep angles that are inherent in scarves. Hence it is necessary to insure that any group of specimens representing a single joint contain an equal number notched both ways as shown.

TESTS FOR GLUED VENEER PRODUCTS

As mentioned earlier, one of the most used specimens in the development of adhesives is the plywood tensile shear specimen (Figure 11–10(a)) so called because the shear load is applied as a tension force (as opposed to the block shear specimen where the shear load is applied as a compressive force). It is a typical three-ply plywood assembly with face and back veneers and a cross-grained core veneer. Several advantages attend the configuration of this specimen:

1. The thinness of veneer makes it possible to introduce platen heat and therefore suitable for evaluating both hot-press and cold-press adhesives.

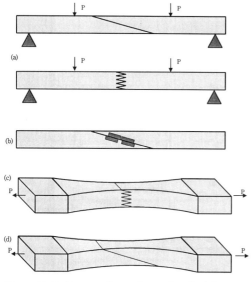

Figure 11–9 Test specimens for end-grain joints: (a) flexural test, (b) shear test, (c and d) tensile test.

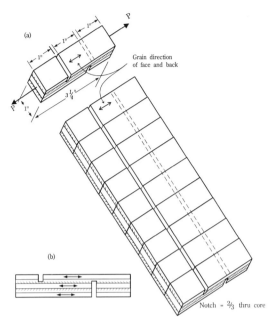

Figure 11-10 The plywood shear specimen and method of cutting from a sample panel of (a) three plies, and (b) notching for a five-ply panel.

2. The small size of the specimen amends it to accelerated exposures, with rapid introduction of heat and moisture.
3. Its ease of fabrication and testing means many specimens can be included in a test to allow more controlled variables and to smooth out the effects of uncontrolled variables.
4. Its cross-grained construction imposes stress on the glue line during moisture changes.

There are also several disadvantages:

1. The cross-grain center ply introduces a very weak wood property in the shear zone, shear in the plane of the fibers or "rolling shear," often limiting the maximum strength that can register.
2. The small size of the specimen magnifies edge effects.
3. Lathe checks and grain angles play capriciously.

Like the block shear specimen, the plywood shear specimen is made in multiples for easier control of bonding factors, and to provide a sample size sufficient to analyze statistically. In a typical procedure, three plies of veneer 4 inches by 14 inches, $\frac{1}{16}$ inch in thickness, are assembled with the grain crossed, face plies with grain in the short direction, and core with grain in the long direction. After gluing, the sample panel is trimmed to $3\frac{1}{4}$ inches, and two saw kerfs are cut the length of the panel on opposite sides, equally spaced from both edges, and two-thirds the distance through the center ply (Figure 11-10(a)). This produces a shear area 1 inch long, which will become 1 square inch when the specimen is cut from the panel in 1-inch widths. When five-ply panels are to be tested, notching would be as shown in Figure 11-10(b).

With very thin face veneer, which cannot carry the load to the glue line without breaking, the shear area is reduced by shortening the distance between notches to $\frac{1}{2}$ inch. The resulting breaking load is then multiplied by two to obtain strength in psi. The calculated strength should be decreased by 10 percent to make it comparable to what it would have been if the specimen had been cut normally; a consequence of the smaller bond area mentioned earlier.

Factors that influence the results obtained in testing the plywood shear specimen include lathe checks and grain direction. As discussed in the chapter on wood preparation, lathe checks are fissures along the grain of veneer produced by the action of the knife and nose bar as the veneer is cut from the log. These checks can be severe or mild depending upon the species, its preparation for cutting, and the relationship of knife to nose bar. Normally the lathe checks are at an angle with the surface and go part way through the thickness of the veneer, always on the "loose" side.

In the plywood shear specimen these lathe checks, besides influencing bond formation (penetration of the adhesive), affect the rolling shear effect and, unfortunately, more in one direction than the other. Both the observed strength and the observed wood failure thus can vary independently of the quality of the bond.

When veneer is assembled into construction plywood, the loose side of face veneers is always toward the glue line. This assures that the tight side, which has a greater degree of integrity, will be the outside surface where finish treatments require the tightest surface. The tight side also weathers better. Turning the loose side in inevitably casts a difference between any two glue lines in a three-ply construction with respect to bonding surfaces, one being loose side to loose side, and the other loose side to tight side. Tight side to tight side can also occur, particularly in random assemblies or assemblies of more than three plies. In a test situation, the weakest of the two glue lines is the one that will determine the observed result.

When the plywood specimen is under load in a testing machine, the inclination of the lathe checks interacts with the shear force, affecting the observed strength and wood failure. Two situations are at play, one in which the shear action tends to close the checks (Figure 11–11(a)), and one in which the action is to open the checks (Figure 11–11(b)). The "close" orientation can result in higher strength readings and lower wood failures, while the "open" orientation can result in lower strength and higher wood failure, with the bond identical in both cases.

Results may differ as much as 100 percent through the arbitrary choice of procedure in cutting the notches. However, it is possible to observe the inclination of the lathe checks

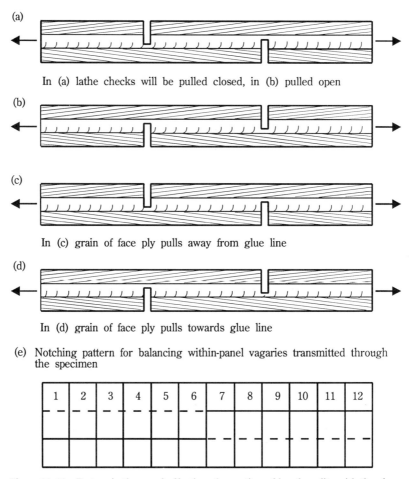

(a)

In (a) lathe checks will be pulled closed, in (b) pulled open

(b)

(c)

In (c) grain of face ply pulls away from glue line

(d)

In (d) grain of face ply pulls towards glue line

(e) Notching pattern for balancing within-panel vagaries transmitted through the specimen

1	2	3	4	5	6	7	8	9	10	11	12

Figure 11–11 Factors in the wood affecting observation of bond quality with the plywood shear specimen: (a) and (c) notched oppositely from (b) and (d); (e) recommended notching pattern to equalize these wood factors.

and thus to choose which way to cut them. The visibility of the lathe checks can be enhanced by sanding the edge and applying a red dye. This introduces the question of what notch orientation to choose. If one is interested in adhesive properties, there is more to be learned with the orientation that allows the greatest strength to register, i.e., "closed." On the other hand, if interest is in recording the maximum amount of wood failure, the "open" situation would be the one chosen.

Since there is useful information to be deduced from both orientations, it is desirable that test reports include observations of both. Each batch of specimens cut from any one panel should therefore contain an equal number of each. This is accomplished by cutting the test panel in half before notching and reversing one of the halves end to end as it is fed through the saw to notch it. Correct notching can be assured by penciling them on the faces of the panel before cutting in half.

It is helpful in revealing intrapanel conditions if the specimens are numbered consecutively as they would have appeared in the original panel (Figure 11–11(e)). This permits dividing the specimens into two subgroups on the basis of notching, and averaging them separately before pooling them into the value for the panel. The amount of difference is significant. A wide difference may signify, for example, that the adhesive has bonded near the limit of its assembly time, providing a bond at the very surface but insufficient mobility to penetrate and rebond the lathe checks. On the other hand, no difference between the two halves is indicative that the adhesive has performed optimally. Overpenetration could also result in no difference but at a lower strength or wood failure.

It should be noted that if it is desired to compare post-gluing treatments with original strength as in durability testing, it is impossible to obtain two equal samples of specimens from the cutting schedule shown in Figure 11–10(a). Choosing the even-numbered specimens for one treatment and the

odd-numbered for another, as one is tempted to do, would bias each sample, one with more evens and the other with more odds, introducing a difference even without the effect of the treatment. In the interest of equality, it is therefore necessary either to discard two specimens from each sample panel or to make a larger panel to yield 12 specimens, six in each group, as suggested in Figure 11–11(e).

Reversing the notching also partially circumvents the effects of another wood factor, grain direction through the thickness of the face veneers. This angle of grain interacts with the force it is carrying, limiting its strength. It also interacts with the bond in how the force is delivered (Figure 11–11(c, d)). The effect is similar to that in the block shear specimen. In (c) the grain tends to pull away from the bond, while in (d) the grain tends to push into the bond. The differences may be small, but at critical limits they could be decisive. In any case there is no need for this bias either, since by proper notching both orientations can appear in the same results.

A steep diagonal grain angle (with respect to an edge rather than the surface) in the core veneer can also cause havoc in the strength readings. In cutting the notch two-thirds through the core veneer, it is assumed that the other third has zero strength in helping to carry the shear load. However, if the angle allows grain to extend across the notch, in a sense bridging the gap, it will carry some of the load. Unusually high strength readings should raise suspicions that this factor is operating.

The need to notch the specimen from opposite sides leads to eccentricity in the specimen, which produces distortion during loading and a peeling effect on the bond. If the testing grips or the machine also have eccentricities, the effect can be additive or subtractive with the eccentricity of the specimen. Hence one of the standard cautions in testing is always to insert the specimens the same way in the grips, e.g., notch always up and to the left, or vice versa.

Because of the low strength that can be

read, and because soaking and other accelerated exposure tests further weaken the wood, especially when the specimens are tested wet, a burden of interpretation falls on the user of test information. Absolute values of strength tend to become less meaningful. Observations take on greater significance when they are made as part of a controlled sequence of variables, or when they are compared with some standard.

Single-Lap Specimens. The low strength readings that plywood tensile shear specimens deliver restrict their use in exploring effects of factors that influence the bond at levels of strength above that of cross-grain constructions. In order to raise the strength reading and widen the range of observations that can register, the weak core ply can be omitted, leaving the two face plies with grain parallel glued together (Figure 11–12(a)). This restores the strong direction of wood on either side of the glue line but removes the cross-grain element, which provides restraint and stress in making accelerated tests more effective. Thus results from these specimens can be considered more indicative of adhesive performance per se than the plywood specimen. The latter, however, is more representative of actual performance in the cross-grain plywood construction.

The single-lap specimen shown in Figure 11–12(a) is widely used because it can be made from most materials that are glued. Its simplicity, however, is marred by its eccentricity, which induces distortion under load and peeling effect on the bond. The peeling effect can be reduced by easing the end-grain edge of the lap as shown in Figure 11–12(b). Distortion can be further reduced somewhat by filling in the offset part of the specimen with veneer of the same thickness (Figure 11–12(c)). This configuration is easier to glue since it begins with two full-size pieces of veneer. However, the saw cut to make the notches must be precise to reach just to the glue line. If not deep enough, some grain will be continuous and raise the observed reading; if too deep, greater wood failure may develop. In any case, reverse notching as in the plywood specimen will produce more reliable data for comparison purposes.

As an alternative to notching by saw, it may be preferrable to have two sizes of veneer, one to make the lap and one to make the fill-ins. With exact fit in a jig, or by taping them together, the pieces can be made to stay in place while being glued. The butt joints at end grain play little or no part in the strength reading since they break at early load.

In order to remove eccentricity further, the double-lap specimen of Figure 11–12(d) provides for greater alignment of forces. It is made by adding one more ply to the specimen in (c) such that it mirrors the ply on the opposite face.

All of these lap specimens can also be used in accelerated aging exposures, but again will reflect primarily adhesive bond durability free of the stress effects of cross grain.

Delamination Tests of Glued Veneer. In measuring durability of less durable adhesives, it is sometimes more definitive and also easier to observe delamination resistance against less rigorous exposures than it is to measure strength or strength loss. Two specimens have been standardized for size. One, measuring 2 inches by 5 inches (with face grain in the long direction), is used to

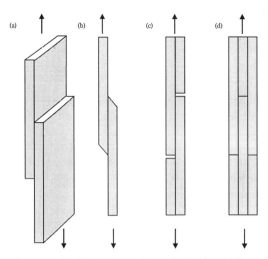

Figure 11–12   Single-lap specimens (a, b, c), and double-lap specimen (d).

observe delamination produced by soaking at room temperature, followed by forced drying at elevated temperature to a low moisture content. The other, measuring 6 inches by 6 inches, is soaked at room temperature and dried at room temperature. Soaking and drying can proceed through one or more cycles. The resulting delamination is measured and totaled or reduced to a percentage. Sometimes a determination is made on the basis of how many cycles can be withstood before a certain amount of delamination occurs.

Tests for Comminuted Wood Products

Once past solid wood into the world of comminuted wood, geometry factors necessitate a different approach to bond evaluation. Gone are the continuous, discrete glue lines and the constant wood element sizes. Surfaces and orientations are variable. Individual wood pieces are small, too small to form isolated glued specimens or to grip them in a testing machine. The glue line is now a pattern of spots. Nevertheless, it must still perform its five functions, though more obscurely. All seven links are still to be formed, though less obvious, and links 8 and 9 still represent the wood just as dominantly in total though not as much individually. No single piece of wood nor spot of glue controls the process nor determine the product. However, each sums its smaller contribution into a total performance. It is out of measurements of this total performance that some estimate of bond quality must be extracted.

Unfortunately, performance is not uniform in any given panel. One can never be sure what factor or factors are playing at any given point and are responsible for a reading of strength or other attribute. Almost all reconstituted wood products are consolidated with heat using heat-curing adhesives. The nature of heat input and distribution and the effects it has on moisture distribution create a wave of transient conditions that sweeps through and across panels during consolidation (Figure 9-18). Nowhere in a panel are the conditions the same; intensities and durations vary from point to point vertically as well as horizontally. Consequently, *a panel is not a uniform entity;* local differences exist edge to edge and face to face.

The importance of this situation is that while panels may function satisfactorily as a whole, test specimens by nature may at times reflect only a local status. In a laboratory where test panels are small, local differences are reduced due to the size effect. The shorter distances in the plane of the panel result in faster moisture movement and egress, with less build-up. Thus the sweep of conditions has a shorter period and therefore less of a differential effect on bond formation. Mat formation, however, can be less uniform in the laboratory, producing variations in density and orientation that can easily overshadow differences in the environment for bond formation. Because of the effect of panel size on bond formation, it is sometimes difficult to extrapolate laboratory results to full-size mill panels.

Since bond quality is difficult to measure directly as it is in glued solid wood products, and unknown variations pervade test panels, it is necessary to introduce into the test panel a means of inferring the condition or contribution of the bond. One approach is to include in the experiment a sequenced variable that primarily influences the bond. Any one of the following factors may be varied: amount of adhesive, assembly time, temperature, pressure cycle, and total moisture content.

The following five tests are commonly used to evaluate properties of particulate products with strong inference to bond quality.

Tensile Strength Perpendicular to the Surface. The specimen is made by cutting a rectangular piece out of the test panel, 2 inches by 2 inches by the thickness of the panel. To facilitate the uniform loading in tension perpendicular to the surface, metal blocks are glued to both surfaces, using a hot melt adhesive or some other adhesive that carries no water. The metal blocks are grooved to receive grips to the testing ma-

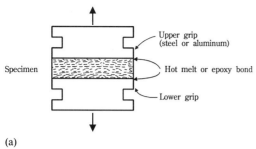

(a)

Figure 11–13   Internal bond specimen set-up (a), and testing (b). (USFPL)

(b)

chine (Figure 11–13). At rupture, the force necessary to pull the specimen apart is recorded and reduced to pounds per square inch of area, in this case 4 square inches. Although this is one of the weaker strength properties of comminuted products, the test is one of the most indicative of bond quality because all the nine links in the bond chain are in sequence and parallel to the direction of the applied load. For this reason it is also called the "internal bond test" (IB).

It is noteworthy that the weakest link is usually located at or near the midpoint through the thickness. Often it is link 1 that has malformed due to accumulation of moisture and its effect on curing of the adhesive. This is also the region where the greatest amount of springback often occurs, an action that can destroy bonds before testing can be performed. Sometimes the weakest link is near the surface of the panel. The same link 1 may then be involved but for a different reason, precure before pressure was fully applied. In order to measure the stronger links, and to assess the quality at

other levels in the thickness it is necessary to cut the specimen into layers and test each one separately.

Linear Expansion.   The stability of a reconstituted panel also provides a strong indication of bond quality. Consider a well-bonded flakeboard, and assume that it is well felted and randomly oriented, in which case each flake can be presumed to have another flake bonded to it with grain crossed at 90 degrees *on average* throughout the thickness. Each flake is therefore restrained from swelling across the grain by the flake to which it is attached, but only if it is well bonded. The amount of expansion a panel undergoes as moisture content increases can therefore be considered a reflection of the degree of bonding.

Since the amount of expansion may be small, extreme accuracy of measurement is necessary. The standard specimen is 3 inches wide by 12 inches long by the thickness of the panel. Measurements are made using a dial gauge reading to .001 inches appropriately mounted to accept the 12-inch span. A test consists of conditioning the specimen to a relative humidity of 50 percent, measuring the length, and then equilibrating to a relative humidity of 90 percent, after which the specimen is remeasured to determine the change in length. The change divided by the length and multiplied by 100 produces a percent change. A wider range of 30 to 90 percent relative humidity can also be included if desired. During exposure, the ends of the specimen may become uneven due to differential movement of individual flakes at the edge surface. In order to prevent these irregularities from biasing the readings, a flat bar is interposed between the ends of the specimen and the point of the dial gauge. Greater precision in measurement is obtainable by driving two small brads or tacks about 10 inches apart in the face of the panel, scribing a line on the heads, and measuring the distance between lines. An eyepiece with cross hairs traveling on a calibrated bar allows measurement without engaging the ends.

The accuracy of this method depends upon establishing an accurate relative humidity range, and achieving equilibrium at both ends of the range. Relative humidities are best controlled over aqueous salt solutions. However, conditioning chambers in which moisture is removed or added in response to sensors are the more common. It is essential, of course, that any condensation that may develop inside the chamber be prevented from dripping onto the specimen.

Knowing when equilibrium has been achieved requires repeated weighing of the specimen. When no change in weight occurs during a 24-hour period, the specimen can be considered to have reached equilibrium. By plotting weight gain against time, it can be easily observed when the plot is becoming asymptotic and the end of the exposure is approaching.

Since panels often contain directional differences due to felting mechanics, it is important to obtain specimens that represent perpendicular directions in the same panel. During exposure to high humidity, some specimens may incur warp. When measur-

ing, it is necessary that the specimen be flattened in the jig before reading the expansion.

Thickness Swelling. The swelling of reconstituted products in the thickness direction (i.e., in the direction of compression) is not in itself a consistent indicator of bond quality, though it does reflect strongly on product performance. This is because the swelling movement is perpendicular to the glue lines, and the bond may not be highly stressed by the deformations occurring. Visualize, for example, plywood, where swelling in the thickness direction is more obviously independent of how good the bonds are. The veneers are free to swell in thickness whether bonded or not. In fact, a well-bonded plywood specimen may swell more in thickness than a poorly made one. The reason in this case is that in being well bonded, the veneers are restrained from swelling in width; they then express a portion of their swelling in the thickness direction. In the case of panels made of flakes, the elements are not in precise layers but in an overlapped and interwoven structure. Moreover, the amount of swelling is not uniform but varies from point to point due to nonuniform felting and consolidation conditions. The overlapping and the interweaving coupled with the nonuniform swelling creates cleavage situations throughout the panel. The actual dimensional swelling that occurs is assumed to be related to how well the bonds hold. The observation that comminuted wood specimens become weaker and less dense after swelling and reconditioning suggests that bonds have been broken and voids created. Certainly the performance of the product is therefore related to the amount of swelling that occurs on exposure to moisture especially since a portion of this swelling is not reversible as in solid wood but remains permanently as "springback."

Swelling tests involve the simple measurement of thickness with a micrometer, comparing the original thickness with the change in thickness resulting from moisture intake. Since specimen size, rate of infusion, method of exposure (vapor or liquid), length of exposure period, and location of measurement with respect to edges have a bearing on the readings taken, only a standardized procedure can produce comparable data (ASTM D-1037). Conventionally, the thickness of the 12-inch by 12-inch specimen is measured one inch in from the edges at four different points. A typical exposure would be 24 hours immersion in water at room temperature. Many variables intrude on the interpretation of results. One of the most worrisome is the question of whether the measurement reflects equilibrium conditions. In practice, equilibrium may not be reached. Consequently conclusions usually are only estimates. The ultimate swelling potential of the specimen is deduced from the amount of swelling that occurs within the scope of the procedure. The deduction is further compromised by the amount of wax that is normally incorporated in comminuted wood products for the purpose of reducing the rate of water uptake. Wax only affects the rate of intake but not the amount nor the effects.

Flexural Strength. Although it is axiomatic that strength in bending is totally dependent upon degree of bonding, the relationship is more obscure in this case. One reason for the obscurity is that bending strength is dependent primarily upon the development of shear resistance between elements, and this can be accomplished with a minimum of bonding. Differences in bond quality therefore have less opportunity to register through a bending test. Nevertheless, it provides a useful tool when used in conjunction with accelerated exposures. Resistance to deterioration—calculated as "strength retained," a percentage of original strength—can be taken as a measure of how well the panel has been consolidated; bond quality is then inferred.

The test specimen (ASTM D-1037) is 2 inches wide if the panel is less than $\frac{1}{4}$ inch thick and 3 inches wide if the panel is greater than $\frac{1}{4}$ inch thick. The length is taken as 24 times the nominal thickness plus 2 inches. This ratio of length to thickness is standardized in order to minimize the influence of ho-

rizontal shear deformation on the determination of a true flexural modulus. A true modulus is important for engineering purposes and for comparing panels of different thicknesses. When interest is primarily on quality of the bond, it may be more informative to reduce the length of the span in order to increase the shear effect.

In carrying out the test the specimen is supported at the ends and loaded at midspan, as is normally done for solid wood (Figure 11-14). Rate of loading is controlled to provide a consistent rate of fiber strain, which is related to the thickness of the specimen. The rate of cross-head travel of the testing machine has been calculated to be .12 inches per minute per each $\frac{1}{4}$ inch of panel thickness. Slower speeds will produce observations that are lower than the standard, and faster speeds will result in higher observed strengths. As in dimensional stability tests, the possible effects of directional properties should be accounted for by sampling in both directions of a panel.

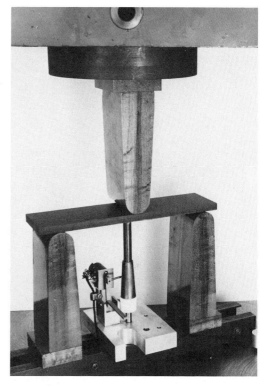

Figure 11-14   Flexural testing. (USFPL)

Interlaminar Shear.   Observations of shear strength in the plane of the panel should be directly related to the quality of the bond. However, if specimens similar to those used to test glue bonds in solid wood are cut from a panel, the shear plane would be located near the center, and the test would reflect only that portion. In the standard interlaminar shear test of ASTM D-1037, steel plates are glued to the faces of the specimen and the shear load is applied through these plates. Shear planes throughout the whole thickness of the panel thus come under load, and failure is free to occur at the weakest point. Such information is useful in determining performance as structural members in sandwich constructions, gusset plates, or flanges in box beams. When information is desired about specific regions in a panel, it is necessary to plane away or otherwise remove all other regions before gluing to the steel plates.

This shear test interacts with a felting characteristic in the same manner as the block shear test interacts with grain direction in solid-wood gluing. As a mat is formed progressively on the moving caul, a small inclination between successive layers of wood elements is inevitably built into the mat. This inclination persists in the consolidated panel and produces what amounts to a slope of grain. The situation probably does not exist in laboratory-made panels, but it should be presumed to occur in commercial panels. Account may be taken, as for block shear, by making specimens in pairs, reversing the direction of loading so both the strong and the weak grain direction are represented in the results.

ACCELERATED EXPOSURES FOR TESTING BOND PERFORMANCE

A graded series of exposure conditions progressing from mild to severe was presented earlier. When these conditions are appropriately vested with numbers defining temperature, moisture, time, and cycles, they become useful as predictors of bond performance against expected service conditions. Experience with these exposures enables their use in specifications for quality assurance involving particular glues and structures. Some representative exposure schedules and their uses follow.

Accelerated Aging for Performance in Extreme Exposures

Three cyclic tests involving heat and moisture are intended to create a maximum of adhesive degradation coupled with a maximum of internally generated stress to loosen the bond.

Face-Glued Lumber Products.   This exposure test is used to evaluate bond performance in laminated lumber structures such as beams for exterior locations. The specimen is that shown in Figures 11-7 and 11-8.

1. Submerge in water 65 to 80° F, weighted down.
2. Draw vacuum, 25 inches mercury, 15 minutes.
3. Release vacuum and apply pressure 75 psi, 1 hour.
4. Repeat steps 2 and 3.
5. Dry at 150°F, air speed 500 ft/min over end grain 21 to 22 hours.
6. Repeat entire procedure two times.
7. Observe percent end-grain delamination.

The procedure is often modified to increase severity by introducing steam after the drying cycle, or to shorten the time by increasing the drying temperature.

All Veneer Plywood Panels.   More drastic with respect to adhesive degradation but lower generation of internal stresses, this test, PS 1-83, uses the plywood tensile shear specimen as the instrument of observation (Figures 11-10, 11-11, 11-12). The specimen is subjected to the following:

1. Boil in water for 4 hours.
2. Dry at 145°F for 20 hours.
3. Boil again for 4 hours.
4. Cool in water and test wet.

One of the oldest accelerated aging tests, this procedure reveals the potential performance of veneer bonds intended for the most severe exposure conditions. It can be used to separate bonds made with the most durable adhesives—such as phenol, resorcinol, and melamine, which are expected to produce "waterproof" bonds—from the less-durable adhesives, such as urea and casein, which are only expected to produce "water-resistant" bonds. It is also used to reveal *quality* of bond made with the most-durable adhesives, under the presumption that a lower-quality bond will degrade to a greater degree than a high-quality bond. In practice, however, amount of degrade is seldom measured, except in laboratory adhesive tests where it is important to compare adhesives on this count.

Instead the actual strength, or wood failure exhibited after exposure is compared to that required in specifications for the intended use. In the case of softwood plywood, specimens are required to have both a high average wood failure (e.g., 90 percent) and a minimum for the sample (e.g., 70 percent). In the case of hardwood plywood, the wood-failure requirement is related to the observed strength; for example, 50 percent wood failure if the strength is under 250 psi, and 15 percent if the strength is over 350 psi, with minimums of 25 and 10 percent respectively for individual specimens. The test defines type I hardwood plywood as well as establishing performance for exterior softwood plywood.

The plywood tensile shear specimen, as mentioned earlier, is easily made for liberal replication. In small batches for spot testing, specimens are simply boiled in an open container and then dried on screens in an oven controlled for temperature and air flow. (The procedure is so simple that many a pot has boiled dry, burning the specimens, while the operator thought other things could be done at the same time.) When large quantities of specimens are to be tested, the process is automated with the specimens strung on wires, boiled, and then swung over in position for drying on a timer-controlled schedule. Some cautions with regard to the testing of this specimen were given in an earlier section, pages 396–399.

Comminuted Wood Panels. Materials made of comminuted wood elements present a special challenge because it is impossible to target a particular glue line. The structure must be approached as a whole, with the best outcome being the identification of the weakest part. The following exposure procedure originally developed by the National Bureau of Standards and refined into ASTM D-1037 imposes a high degree of adhesive degradation as well as internal stress due to recovery from compression. Called the 6-cycle test it is used to simulate the most severe exposure a building material might encounter.

1. Submerge in water at 120°F for 1 hour.
2. Steam and water vapor at 200°F for 3 hours.
3. Freeze at 10°F for 20 hours.
4. Heat at 210°F, dry air, for 3 hours.
5. Repeat step 2.
6. Repeat step 4 for 18 hours.
7. Repeat entire procedure 5 more times.
8. Recondition 48 hours at 68°F and 65 percent RH.
9 Test for strength retained and/or observe for disintegration or delamination.

This aging test is intended to predict how a material might withstand unprotected outdoor weathering conditions. It is usually applied to specimens prepared for testing strength or other properties in order to obtain a "before" and "after" comparison and calculate the "strength retained" for use in engineering structures. When an entire panel is exposed and subsequently cut into specimens, edge effects that may extend varying distances inward can introduce variability that must receive some accounting in analyzing the results.

Accelerated Aging for Intermediate-Durability Bonds

At this level of durability testing, the severity of adhesive degradation is reduced, but the generation of internal stresses remains fairly high. Three exposure schedules are given, each with a slightly different combination of internal stress and degradation. They are used mainly on plywood-type panels.

The first exposure is used primarily for hardwood plywood and is generally referred to as the "three-cycle soak test." The specimen in this case is a small piece of plywood 2 inches by 5 inches, with the face grain in the long direction (NBS PS 51-71).

1. Submerge in water at room temperature for 4 hours.
2. Dry at 120 to 125°F for 19 hours with air flow.
3. Repeat two times.

The specimen is considered to have failed if 2 inches of continuous delamination occurs along the edge in any glue line. The test defines type II bond quality when 8 of 10 specimens pass the test.

For softwood interior–type plywood made with interior glues, the exposure is intensified somewhat by increasing the water temperature, the amount of water uptake, and the severity of drying as in the following schedule (PS 1-83):

1. Submerge in water at 110°F.
2. Draw vacuum of 15 inches of mercury for 30 minutes.
3. Soak for 4.5 hours, no additional heating.
4. Dry at 150°F for 15 hours with forced air.

The same 2-by-5-inch specimen is used as for hardwood plywood, and the same criteria for failure, 2 inches of continuous delamination.

For softwood interior–type plywood made with glues of higher durability, the exposure is intensified further by increasing water temperature and soaking time (PS 1-83):

1. Submerge in water at 120°F.
2. Draw a vacuum of 15 inches of mercury for 30 minutes.
3. Soak for 15 hours at not less than 75°F.

In this case the specimen is the plywood tensile shear, which insures rather complete saturation. It is tested while still wet and judged for percent wood failure, which should average not less than 45 percent.

The durable hot-press glues that may be undercured out of the press and perform poorly in service are sometimes further cured by the heat and used in accelerated exposures. To guard against such an occasion, a special cold soaking procedure is used (PS 1-83):

1. Submerge in cold tap water.
2. Draw a vacuum of 25 inches mercury for 30 minutes.
3. Follow with 65 to 70 psi pressure for 30 minutes.

The specimens, plywood tensile shear, are then tested wet and the wood failure read after drying.

Accelerated Aging for Low-Durability Glues

A specimen 6 inches by 6 inches is used to define bond quality for type III hardwood plywood in which only a minimum amount of moisture resistance is required. The procedure is known as the "two-cycle soak test" and consists of these steps (NBS PS 51-71):

1. Submerge in water at 75°F for 4 hours.
2. Dry at 75°F, open room, 20 hours.
3. Repeat once.

Failure is deemed to have occurred when 2 inches of continuous delamination appear in any glue line having a depth of $\frac{1}{4}$ inch and a separation of 0.003 inches, sensed with a feeler gauge.

ESTABLISHING A PERFORMANCE PROFILE

The above tests and procedures define performance as a single value derived from a fixed set of parameters. The result is like determining if something fits into a mold. They can only provide a yes-or-no point in the total performance potential of an adhesive. A much broader picture of potential could be observed if there were a series of molds representing different bond-forming conditions to provide a series of fit observations. One can then determine what conditions result in the best fit, as well as what conditions result in misfits. Both strengths and weaknesses are revealed, and with the proper choice of conditions, both ends of an operational range can be established for a given glue.

Such a series of molds is fashioned by providing a set of bonding parameters, incrementing over a practical range. The most revealing set is comprised of an adhesive variable, a wood variable, and an operational variable, organized to allow a three-factor analysis. In one organization that has been shown to be especially sensitive, the adhesive variable is viscosity expressed as age of the mixed glue; the wood variable is moisture content, and the operational variable is assembly time. These three variables, each introduced at three levels, all interact to affect bond formation, compounding both positive and negative effects and summing them into separate observations. In doing so, each becomes a sort of micrometer gauging the effects of the others. The three variables, each at three levels, provide 27 different combinations (molds) of bonding conditions for the adhesive to react to and yield a reading of its response.

The complete scenario is shown in Figure 11-15. It can be seen that when data from tests is entered, the rows and columns can be added and compared to determine individual factor effects, as well as combined factor effects. It can also be sensed that two corners of the table, lower left and upper right, represent compounding negative actions and therefore extreme effects on *mobility,* while the other two tend to represent optimizing or cancelling effects. One diagonal may then contain the worst results, and the other the better results, with the best results clustered around the middle.

The conditions that produced the best results can be determined by casual inspection of the Table. Conditions that produced the poor results can also be identified without further analysis. If the best results are not at the center of the table but are clustered in one corner, it indicates that the level of factors chosen needs to be shifted further in the direction suggested by the location in the Table.

In Figure 11-16, plywood shear specimens that had been glued with a phenolic resin in a hot press and sheared after the cycle boil exposure are arranged in the same location in the Table into which their data would have been entered. Four specimens are shown in each box, two pulled with, and two pulled against, the angle of the lathe checks. They

Veneer Moisture Content	Mix Age									Moisture Content Avgs.
	Short			Mid			Long			
	Assembly Time									
	0	10	20	0	10	20	0	10	20	
4										
8										
12										
Assembly Time Avgs.										
Age Avgs.									Grand Avg.

Figure 11-15 An experimental matrix for determining the performance profile of an adhesive.

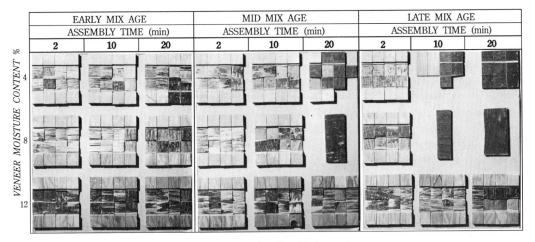

Figure 11-16   Interactions of a fast-setting, hot-press phenolic adhesive to three operational variables, each at three levels. Shown are plywood shear specimens after boil-dry-boil exposure and tested wet. They are arranged in groups of four, two of which have been pulled in the "open" lathe check direction, and two in the "closed" direction. Each group is located in a cell represented by the row and column intersection. In this matrix of variables, the entire range of the *mobility parameter* has expressed itself by the condition of the glue lines and the amount of wood failure. (Marra, John Wiley & Sons, revised)

are shown with the broken halves opposite each other. The white wood failures show sharply against the dark glue line and provides a visual estimate of bond quality. Scanning the specimens, it can be seen by the absence of wood failure that the proper bonds appear on the inclining diagonal of each adhesive variable box as well as on the inclining diagonal of the total display, being poorest at each end. Conversely, the better bonds are on the declining diagonals.

At the lower right end of the inclining diagonal the bonds are poor by reason of excessive mobility occasioned by the compounded effect of low viscosity, short assembly time, and high moisture content. At the upper right end of the same diagonal, the compounding effect of high viscosity, long assembly time, and low moisture content has reduced mobility to the point where none of the bond-forming adhesive actions could occur; the adhesive has not moved from the spread pattern in which it was applied.

Also to be seen, with reference to the declining diagonal, is some indication of factors counteracting each other. At the upper left, dry veneer seems to have partially compensated for both the high mobility of the glue and the short assembly time. However, there is room for further improvement that assembly time can give. At the lower right corner, high-moisture veneer seems to have partially compensated for low-mobility glue and long assembly times, though not quite optimally.

This organization of variables can be duplicated with other factors that have a similar effect on bond formation, such as speed of cure or catalyst level, platen temperature, wood temperature, extender or filler content, pressure, amount of adhesive, or surface inactivation—any factor that can be quantified and serialized and has an effect on mobility of the adhesive.

In addition to the visual interpretation described above, numerical data when inserted into the system can be analyzed statistically to yield confirmation regarding the significance of the differences observed, and therefore the importance of each factor or combination of factors can be determined. When this is done, the total variation produced by the system of factors can be quantified and reduced to a *coefficient of variation*. Since this variation is a consequence of the effects of the factors playing in the system, it also represents the sensitivity of the adhesive to these particular factors. To put it differently, it represents a *tolerance index,* the ability of the adhesive to tolerate these variables when ranged in this manner. The lower the coefficient of variation, the more tolerance the adhesive has and the less likely it is to succumb to chance variation in the operation. If, for example, every box in the table showed high wood failure, one could conclude that the adhesive is very tolerant and would produce a minimum of defects under production conditions. On the other hand, a mixture of low and high values throughout the box would indicate low tolerance and the likelihood of frequent or sporadic defects. Uniformly low values would also give a numeric value indicative of good tolerance. In this case, it would be necessary to consider the actual values in arriving at an estimate of potential quality.

12

Guide to Troubleshooting

Defective or substandard products are the natural enemies of the manufacturing world. They occur wherever men, machines, and materials interact in producing a product. Each has its own independent system of quality variables, any one of which could wipe out the best qualities of the other. The general situation in wood gluing, as in other processes, is that small wayward variations in one sector, which may not be serious in themselves, can and often do add to others in another sector, creating an unfavorable condition for the glue.

Troubleshooting is detective work—in this case, looking for glue clues. Much of the search is based simply upon knowledge of how a glue works, or failed to work in a particular case. It is not an exact science, but a process of making observations and retrospectively sensing the operation of factors that had a bearing on them. The concern then centers on discovering which factors may have been out of line when the glue was forming the bond. But even if the precise culprit cannot be identified, it is possible in many cases to institute corrective action by manipulating factors to alter *mobility* at one or more stages in the bonding operation. For example, the effects of high-moisture veneer in hot-press gluing can sometimes be offset by such procedures as increasing the assembly time or lowering the temperature.

The equation of performance shows the variables in plus and minus terms, suggesting their influence to add or subtract from the quality of the bond. In troubleshooting, we know that a subtractive effect has occurred. We would like to counter it with an additive effect. For this purpose, we need to know in what direction mobility aid is needed, and what factors will drive in that direction. A sense of direction must therefore be generated in order to undertake corrective action. This sense of direction, once developed, is as easy as knowing which way is downhill. Because we are dealing in the glue line with predominantly fluid properties of the glue, we only need to know what has interfered with the fluidity at the time bond formation was going on.

The mobility parameter that operates throughout the equation of performance reduces observations and corrective actions to either up or down on the mobility scale. Even with no numbers as yet on the scale, it is possible, with cut-and-try methods, to nudge the mobility in the desired direction. The state of mobility achieved by the glue after placement on the glue line can be read by physical characteristics visible on the broken joint surface, as indicated in Chapter 3. They are the same for all glues. Factors that affect mobility up and down are summarized in a later section (see page 429).

Because some clues or symptoms have multiple causes, it always improves the anal-

411

ysis to have as many parameters as possible documented by direct measurement. The more important include mix proportions, age of the mixed glue, viscosity, amount applied, assembly time, wood-moisture content and temperature, and curing pressure and temperature. These nine factors comprise the basis of a good quality-control program, discussed in Chapter 10. Unfortunately, those who have problems are those who are unlikely to have a regime of regular measurement. As a general rule, the further in time the defective glue bond is observed beyond the pressing operation, the more such information is lost or becomes nebulous, and the more difficult it is to assay the problem and identify the specific cause. Most of the clues that may be available for observation downstream are those frozen in the glue line by the hardening of the glue between the two surfaces of the joint.

It is important to realize that bonds can be represented by a continuum of quality, beginning with no bond, or complete failure. The next increment of quality would be one barely holding together for the product to be handled but probably not to survive finishing or fabrication operations. Further along the quality scale would be a poor bond strong enough to get into service but due to fail shortly thereafter. Another quality would see the bonded product performing for a period of time before beginning to fail. Finally, a quality is reached that performs indefinitely in the intended manner. Two separate but variously associated factors are operating in the above scenario: strength and durability. When looking at a failed joint, it may sometimes be necessary to decide whether the bond lacked strength to begin with, or lost it due to lack of durability against the elements, or was simply overstressed.

Trouble in a gluing operation comes in various guises, some obvious, some subtle, some latent, the latter spreading over a time span of minutes to years. The best trouble, if there is such a thing, is that which occurs immediately out of the press with the joint falling apart. In this case, corrective action

can be taken at once and losses confined to one press load. The worst trouble is that which occurs in the customer's hands, after many units have been made, and many customers have registered their complaints.

Two broad categories of trouble thus arise on the basis of time alone: (1) *immediate*, out of the press, or during subsequent trimming or sanding, and (2) *delayed*, in process or in service after stressing by loading, moisture, or temperature.

The immediate trouble is one of no bond, and it normally can be assayed with the use of visual characteristics. This trouble is usually evidenced in a gluing operation by broken joints, joints open along ends or edges, or blisters. Mechanical or operational factors are the first suspects in this case.

In the delayed-failure case, we are dealing primarily with a poor bond, one that has some strength but that has not achieved full strength or durability. This trouble, the more insidious and less obvious, also has operational components but is more enmeshed in chemical or physical irregularities, both in the wood and the glue. In this case some testing such as accelerated aging, Chapter 10, may be needed to supplement visual observations and make a determination about the quality of the bond and the reasons for that quality.

In approaching a given problem, an overall plan of attack is proposed in which the schedule of inquiry is designed to categorize the cause of the defect and thus to localize the search. A sequence of observations further narrows the search until the possible sources of the problem are found.

COMMONALITY RESTATED

A premise that underlies much of the rationale in this book now comes to enlighten the entire process of troubleshooting: Glues have certain actions or motions that are common to all. This means at once that no matter what glue was used, in what product, and by what process, we can expect certain behaviors from the glue. More importantly, departures from expected behavior also produce glue line symptoms that are common to all.

The power of this premise is that it reduces the thousands of failure situations to a few observations that can be made on the glue line. A failed joint always provides two surfaces that should have been bonded together. These two surfaces often contain information needed to trace the difficulty to a probable cause. For example, a prehardened glue line is always due to delay in closing and pressurizing the joint, no matter what the glue, what construction it is in, or what kind of pressure system—hot, cold, screw, hydraulic.

Although all glue lines share common traits that relate to the circumstances existing at the time of their solidification, some differentiations can nevertheless be made. The most obvious is whether or not a definite glue line exists. Generally speaking, there is no glue line in highly comminuted products such as fiberboard, particleboard, and flakeboard. Being applied in droplet or powder form, the glue does not have the mass to exhibit mobility effects with obvious characteristics. In lumber and veneer products where the glue is applied as a film, mobility effects can manifest themselves with observable characteristics. For this reason the troubleshooting guide will deal primarily with lumber and veneer products. However, it should be noted that the glue in the more comminuted products responds in exactly the same way to factors affecting mobility on film-forming glue lines; the main difference is that the consequences are largely invisible.

DISCOVERY PROCEDURE

In approaching a defect problem, several possibilities can be addressed without even looking at the failed joints:

1. The frequency of the defect
2. The location of the defect
3. The time of occurrence

Each carries evidence that can direct attention to possible causes, or at least help narrow the field.

Frequency of the Defect

How often does the defect occur? What percentage of the glue lines were defective in a given lot? The following reasoning can be applied as a first order appraisal.

More Than 50 Percent Defective. There is something radically wrong here. When more bad than good products are being made, a major factor is out of line. This is one of the few instances where the glue itself comes under strong suspicion, either its composition, its age, its preparation for use (catalysts, water, extenders), or its condition when applied.

Other factors that can wreak major havoc arise primarily in equipment failure, producing:

1. *Inadequate pressure* due to error in calculation or inaccurate gauges. This cause can be confirmed by glue line observation and corrected by recalculation or checking gauges. Two characteristics related to the prehardened condition may be observed: thick glue line, and two levels in the glue line (areas in which the glue line is thinner and did not make contact). Shallow wood failure may also occur.

2. *Low stock or press temperature,* confirmed by chalky glue line in PVAc glues, prehardened in heat-loss glues, and undercured glue lines in chemically cured resins. Checked by thermometer, and corrected by preconditioning stock before gluing or increasing press temperature if needed.

3. *Excessively high moisture content* in the wood, which can occur in batches if lumber dry kilns or veneer dryers drift out of control. Check moisture content and restore control. This factor produces different effects with different glue types. For example, water-loss glues will not harden adequately; heat-loss glues will not achieve adequate anchoring; chemically cured glues will experience undercure, starved or filtered in hot press gluing.

4. *Excessively low moisture content* in the wood causes mobility effects at the opposite extreme from those due to high moisture.
5. *Poor machining:* in lumber, out-of-square or out-of-line edges due to saw or fence maladjustment, or loose feeding mechanisms; in veneer, rough surfaces due to dull knife, vibrating knife, or inadequate pretreatment of logs.
6. *Wood treatments* for fire, decay, or stability can interfere with bond formation in a wholesale manner.

Less Than 50 Percent Defective. Here the problem may lie more in operational factors and less in the glue (except for any fine tuning that may be needed), and more in operators than in machines. We are looking for situations that produce mostly good parts, rather than mostly bad as in the previous section; factors that *vary* rather than factors that are relatively *fixed*. These include:

1. Wood widely varying in moisture content around the target average
2. Wood with rough, loose, or malfitting surfaces due to warping or machining aberrations against grain or moisture variations
3. Temperature of the wood, the shop, or the press, varying up and down throughout the day
4. Heartwood/sapwood conflicts with adhesive that has been tuned to one or the other, and moisture effects associated with either
5. Inadequate inspection and control to overcome any of the above three conditions

Less Than 5 Percent Defective. The problem at this level is usually due to factors that fall within the range of normal variation in men, machines, and materials but outside the limits of acceptability, i.e., in the tail ends of the distribution curve. In many cases it may not be economically feasible to search for or correct them unless high-value prod-

ucts are involved. It is also probable that several factors, each varying within acceptable levels, have by chance occurred when they were all at an extreme where their effects on mobility were in the same direction. For example, a fresh batch of glue combined with assembly time on the short side and moisture content on the high side could result in too much mobility in a hot-press operation.

Defects Occurring at Regular Intervals. Gluing defects that occur or repeat at definite intervals are responding to factors that also change at the same time. The most common of these are:

Change of Seasons. Adhesive manufacturers often note a rash of complaints during seasonal changes from hot to cold weather or cold to hot. Wood that has been equilibrated to summer atmospheres is poised to change during processing in the drier conditions of the following winter. Conversely, wood that is equilibrated to winter conditions is set to change in summer. The changes in wood are due to humidity changes in the atmosphere. These produce moisture changes in the wood, which produce dimension changes. Because dimension changes are seldom uniform within or between pieces of wood, joint surfaces tend to move or deform during machining, after machining, before gluing, and after gluing. The result is poor-fitting joints before gluing, or stress on the joint after gluing.

Time of Day. Defects that seem to occur at the same time of day may have their causes in operator fatigue, which leads to carelessness, reduced attention, or slower speed, especially toward the end of the day. Knife wear also begins to affect surface quality toward the end of a working period as personnel try to stretch a run without stoppage.

Spreading equipment slowly becomes fouled with accumulated debris or thickened glue after running all day. With the resultant changes in glue viscosity, the amount of glue spread changes up or down depending upon the type of applicator—down if spray, up if roller spreader.

Defects are sometimes associated with coffee breaks and lunch breaks. When these breaks occur in the course of building a charge for the press, the assembled portion will bear an additional assembly time, sometimes carrying it beyond the specified time limit. These breaks can also allow glue on spreading equipment to thicken and override the settings.

Work Shift.   Quality differences sometimes appear in the production from different shifts. This could be due to personnel or attitudes responding differently to established procedures. However, it could also be due to faulty workmanship of one shift carrying over into another shift. For example, high-moisture veneer created in one shift could end up in the press of another. Moreover, batches started in one shift may be finished in the next with unaccounted delays in assembly time and dwell time of glue on a spreader.

Start-Ups.   When equipment is cold or needs adjusting while operating, the first materials through the operation can be defective. Production from start-ups should always receive extra inspection.

Change in Procedure.   When routines are interrupted by changes in procedure, time settings and pressure settings may not get changed at all, get changed enough, or get changed soon enough to avoid unfavorable bond-forming conditions. Sometimes just an interruption in the rhythm of the crew can cause delays in time-dependent processes that could ripple throughout an operation.

Change in Materials.   A change in materials is often upsetting, and a spate of defects can develop before the crew adjusts to any differences that might accompany them. A change in wood species may in certain circumstances require some change in procedure or in glue composition. A change in

glue often brings defects at first, sometimes for no apparent reason. Even with no differences, something as simple as a change in brand name has been known to disrupt an operation.

Change in Construction.   Defects often occur during the first few moments after changes have been made in constructions being glued, even with the same species and the same glue. Changes in the sizes, thicknesses, number of plies, or number of assemblies per press load all require adjustmenta and can have a disruptive effect on the conduct of the gluing operation. They invite little irregularities that accumulate to the disadvantage of the bond.

Change in Machinery.   When new or different machines are introduced into a gluing operation, effects occur that are similar to those at start-up, plus effects resulting from new learning that may be needed, as well as any fine-tuning to adjust to characteristics special to the operation.

Beginning or End of a Run.   In making a run of identical pieces, adjustments are often necessary at the beginning that are usually fine-tuned after the operation has started. During this time it is possible for some critical factor to be overlooked; the most likely relates to close time requirements. Pressure and temperature settings may also need adjustment.

At the end of a run, monotony may dull the alertness of operators, the glue mix may be getting too old, scheduled knife sharpening or periodic adjustments may get ignored, or quitting time may shorten attention.

Rush Order.   Time constraints decrease operator ability to correct errors; they may not, for example, inspect the underside of spread veneer to check for uniformity of glue application. There may be an increased inclination to cut corners, speed up assembly times and press times, and generally allow completion to supersede vigilance.

Defects Occurring Sporadically. Defects that occur independently of any definite occasion can be assumed to be due to factors that vary around a norm but at times may slip beyond specified limits or, as indicated earlier, may in their random variation come together at their extremes and produce an excessively high or low mobility. Examples include wood-moisture content variation, heartwood, temperature, and carelessness.

Location Factors of the Defect

There are two main locational factors that have an evidential bearing on the causes of a defect. These are location within the glued product (as well as between products), and location in the press. The latter type of information requires some means of tagging products to reflect position in the press both vertically and horizontally. If the defect can be associated with location in the press, the problem is mechanical in nature and may only require maintenance.

Tracing a defect with respect to location in the product primarily involves record keeping to establish whether the defects always occur in the same place or may occur in all areas of the piece; also whether the defects appear in all types of products or are confined to certain ones. When defects appear in only one construction of several being produced, the cause is associated with what is different about that construction.

If defects always occur in the same spot in a construction, the problem is either in the pressure or heating system, or in dimensional variation in the wood. The latter has two sources: moisture change after machining and the machining operation itself, particularly the feeding and hold down mechanisms. If defects are random in a construction, the cause may lie in uneven surfaces or uneven glue application. When defects occur among all types of products, the problem encompasses the entire process.

This information, when coupled with observations of the broken joint surfaces, provides a means of focusing on the problem. Causes as well as remedies emerge from the resulting analysis. An important caveat: Analysis must be confined to a single adhesive mixture. Different mixtures demand separation of products into different populations on this basis. Defects can then be properly attributed to one glue or all of them, with significant implications. If defects are associated with one glue, it may be cause for suspicion; if all glues are implicated, the problem is more general and may involve some aspect of the entire operation.

Defects Within a Construction. Defects that occur only in a particular construction engender two considerations: whether they are localized and always in the same place, or whether they are scattered throughout the piece.

Same Location. When defects always occur at the same location, the causes can be assumed to arise from or be associated with that location.

I. Edge-glued lumber
    A. Open joints at ends
        1. Moisture change
        2. Dubbing in jointing
        3. Skip in spreading
    B. Open joints on one side
        1. Edges cut out of square
        2. Clamps not balanced both sides of panel
        3. Moisture content change in bias-cut lumber
II. Face-glued lumber
    A. Open joints at ends
        1. End clamps not sufficiently drawn
        2. Dubbing in surfacing
        3. Glue wiped off in handling piece
        4. Skip in spreading
    B. Open joints along one side
        1. Clamps not seated properly
        2. Assembly not centered in press
    C. Defect in first glue line assembled
        1. Too long an assembly time
    D. Defect in last glue line assembled
        1. Too short an assembly time

III. End-glued lumber
  A. Outside fingers open
    1. Insufficient lateral pressure
    2. Excessive end pressure
    3. Skip in glue application
  B. Inside fingers open
    1. Miscut
    2. Insufficient end pressure
  C. Open scarf joint
    1. Endwise slippage in pressing
    2. Inadequate adhesive application
    3. Variation in thickness of joined boards
  D. Open at one end of scarf
    1. Inaccurate angle
IV. Plywood hot press
  A. Glue line farthest from heat source: filtered, undercured, or blistered
    1. High-moisture veneer
    2. Too much glue applied
    3. Assembly time too short
    4. Inactivated veneer surfaces
    5. Heartwood (possible)
  B. Glue line nearest heat source: prehardened, unanchored, delayed blisters
    1. Slow closing press
    2. Slow loading
    3. Excessive assembly time
    4. Glue too fast setting, over catalyzed, old mix
    5. Veneer too dry
    6. Sapwood (possible)
    7. Low spread rate
  C. All glue lines along one edge, others okay
    1. Exposed to heat of press while waiting to be loaded
  D. All glue lines along all edges, centers okay
    2. Curling or wavy veneer allowing open assembly effects
  E. All glue lines one end
    1. Press rams not fully activated
    2. Warped platens
  F. All glue lines one location
    1. Cold spot in platen
    2. Depression in platen
  G. Glue lines on high-density composites

    1. Inactivated surfaces
    2. Glue water unable to dissipate
V. Plywood cold press
  A. Always same glue line, failure at surface of one ply
    1. Species effect
    2. Surface inactivation
    3. Impervious ply (with adhesives needing water loss)
  B. One edge or one end
    1. Stacked assemblies not in vertical alignment
  C. Alternate glue lines
    1. Top and bottom spreader rolls not applying equally

Random Locations. When defects occur randomly, two situations exist: within a product and between products.

When defects occur randomly *within a product,* the causes include:

1. Heartwood to heartwood areas
2. Summerwood to summerwood areas
3. Local steep grain areas
4. Wet pockets or dry pockets; hot spots or cold spots
5. Surface deviations: roughness, skips, thickness variation
6. Inactivated surfaces
7. Uneven glue application
8. Uneven pressure

When defects occur randomly among different products, it usually signifies a general looseness in control of the gluing process. Unless a study of the glue lines reveals a specific condition, all aspects of the operation need to be scrutinized.

Defects that have a relationship with *location in the press* often point directly to the cause. Because some defects do not show until after the product leaves the press room, it is necessary to find some means of maintaining location identity on each piece as long as possible. A mark that can be placed by platens or clamps would insure location information. Otherwise, hand or mechanical marking as pieces come out of the press provides a good alternative. The mark should remain at least through inspection. If the mark can be placed in an inconspicuous place on the piece so that it stays throughout its service life, so much the better.

Defects that can be associated with location in the press may have the following causes:

Lumber Gluing

1. Clamps: undertightened, overtightened, slipped, spaced too far apart
2. Insufficient assembly time on last glue line assembled
3. Too much assembly on first glue line assembled

Veneer Gluing

1. Hot pressing

- Defects that come consistently from the top platens or bottom platens of multiopening hot presses arise from the method of loading and closing the press. The problem is due to the extended time panels dwell on hot platens before the press closes. During long dwell times the glue begins to dry or cure, predisposing the glue line to prehardened conditions.
- Defects coming consistently from one platen or area of platen indicate a cold spot or unevenness in surface of platen.
- Defects coming consistently from one end of the press indicate a problem with rams.
- Occasional failures when loading more than one small panel side by side in same opening may be due to varying thicknesses that then vary the pressure, the thinnest receiving less pressure and the thickest receiving the most pressure.

2. Cold pressing

- Since constructions are assembled and stacked one over another in bundles to fit in the press opening, the first assembled panels receive a longer assembly time than the last, thereby producing a mobility difference from top to bottom of the stack. Problems are more likely to arise from the bottom than from the top in this case.

Time Defect Occurred (or Was Observed)

Strictly speaking, all glue line defects are cast at the time of pressing. However, many may not become evident until some time later. The life of a bond may be divided into three observation periods:

1. While in the factory of origin
2. While being reprocessed or fabricated
3. In service

When defects appear while the glued product is in the plant of origin, it is still possible to associate it with most of the circumstances of its production, including all of the factors mentioned in the sections "Same Location" and "Random Location," above.

After the glued product has left the plant, transit and warehouse conditions, as well as subsequent remanufacturing conditions, have an opportunity to aggravate a preexisting deficiency in the bond. This complicates the analysis as far as pinpointing the causes. However, study of the glue lines and their contiguous wood can often establish a list of possible causes from which responsibility can be deduced.

When glue line defects appear after the product has been in service, more complications enter, such as environmental effects, installation effects, and possible abuse. These must be factored in to determine a cause. Defects appearing in service are most often associated with a moisture change, particularly in lumber constructions, and more particularly with denser species of wood and in constructions containing crossed grain, exotic woods of high extractive content, and treated wood.

Moisture change creates internal stress due to differential dimension changes that can rupture bonds containing some weakness. Even well-made bonds can fail under the forces of internal stress. Well-made bonds can also fail in service due to degradative effects of heat, water, and organisms. At this late date, reliance must be placed on knowledge of the adhesive type and on the appearance of the glue line.

Glue lines that are starved, unanchored, or prehardened trace back to the gluing operation. The gluing operation can be suspected also if both ruptured surfaces have glue on them; i.e., the break is within link 1 and the appearance is frothy, grainy, or filtered. Undercure can also be a factor at this juncture but would need to be gauged against the possibility that the choice of glue was in error—lacking strength or durability for the use to which it was exposed.

READING THE GLUE LINES

Without all the information obtainable in the previous section, and left with only a broken joint or delaminated piece of panel, there remains the glue line. In failing, it has said that something went wrong. What went wrong, as far as bond-forming actions is concerned, is written in the appearance of the adhesive in the failed area. The script was produced by adhesive actions that were underachieved, overachieved, or aborted altogether. It is also written in the links that were malformed, overformed, or absent altogether. A reading of the actions and the links as described in Chapter 3 yields a history that can be interpreted in mobility terms to implicate causes stemming from many sources. In addition, it can, even without pointing to a specific cause, suggest approaches to remedies. A number of approaches usually exist, some that can be applied immediately (at the next press load), or at various points before, as circumstances dictate. Following is a summary of what the glue line can say.

I. Starved: link 1 missing
  A. General causes
    1. Insufficient glue, or no glue, applied (ordinarily this condition would be avoided by continuous monitoring of the spreading operation, thus allowing focus on other causes of starving.)
    2. Sufficient glue applied, but lost due to high mobility
    3. Sufficient glue applied, but wiped away
  B. Appearance
    1. If none applied, bare wood both surfaces
    2. If insufficient applied
      a) Open or unfilled cell structure
      b) Glue remains on the spread surface, little or no transfer
      c) Often subject to predrying because of its low mass
      d) May develop a glossy or low sheen surface
    3. Sufficient applied but lost
      a) Open or unfilled cell structure
      b) Glue on both surfaces
      c) Often associated with blisters in hot press operations
      d) Also associated with excessive squeeze-out or bleed-through, observed on faces or edges
  C. Specific causes for insufficient application
    1. Applicators out of adjustment
    2. Wood varying in dimensions, which interacts with applicator, e.g., wood thicker than normal will receive less adhesive from a double roll spreader
    3. Plugged ports in extruder-type spreaders or curtain coaters
    4. Viscosity increase in spray-type applicators
    5. Wipes by careless hands, or dragging one piece over another during lay-up
    6. Out of square, out of line, or machine-dubbed edges
  D. Specific causes for lost glue: too much mobility
    1. Excessive squeeze-out
      a) Too much pressure
      b) Too-short assembly time
      c) Too-low viscosity
      d) Especially in edge gluing where glue lines are narrow
      e) More particularly with high-density species
    2. Excessive bleed-through

a) Too-short assembly time
b) Too-low viscosity
c) Thin veneer of porous decorative species
d) Steep grain angles
e) More likely to occur on hot press than cold press

3. Excessive penetration
a) Low-viscosity glue
b) Viscosity further lowered by heat in pressing
c) Viscosity lowered or prolonged by moisture in system
d) Wood grain inclined with gluing surface
e) Lathe checks
f) Too much pressure
g) Slow-hardening glue
h) Inactivated surfaces: slow dewatering—higher mobility
i) Heartwood: inactivation effects—slow dewatering
j) Springwood: greater porosity—permits greater mobility
k) High-moisture content wood: permits greater mobility—slower dewatering, slower curing in hot press

II. Bonded: Ordinarily the bonded condition is the desired one. In this case all links are presumably well formed, but a failure occurred.
A. General causes of failure
1. Wrong glue
2. Deficiency in permanence, link 1 (also 2 and 3)
3. Deficiency in strength, link 1 (also 2 and 3)
B. Appearance
1. Remains of glue film on both surfaces
2. Fractured wood on both surfaces
C. Specific causes for latent failure
1. Glue ill-chosen for intended service
2. Glue exposed beyond its durability potential
3. Joint overloaded

III. Unanchored glue line: links 4 or 5 missing

A. General causes
1. Inactivated wood surfaces
2. High-viscosity glue
3. Contaminated surfaces
B. Appearance
1. Imprint of opposite surface in glue line
2. Occasional loose fiber embedded in glue film
3. Replica of vessel or fiber walls sometimes visible in glue film
C. Specific causes
1. Poor wetting due to inactivated wood surfaces: overheating during drying, contaminated
2. Too-long assembly time, especially open, reducing mobility
3. Hot or warm wood, reducing mobility of water loss or chemically reactive glues during assembly time
4. Cold wood, reducing mobility of heat loss adhesives during assembly time
5. Exposure of prepared assemblies to heated air flow reducing mobility of spread glue, especially at edges
6. Using mixed glue beyond its pot life, reducing mobility
7. Speed of hardening too fast, reducing mobility
8. Low moisture content wood: dewatering of glue line too fast during assembly time, reducing mobility
9. Sapwood: too-fast dewatering, reducing mobility
10. Heartwood: surface inactivation

IV. Prehardened glue line: all links one side missing
A. General causes
1. Glue line hardened before pressure applied
2. Pressure never arrived
B. Appearance
1. Glue line appears as applied: spreader pattern, brush marks, extruder ribbons
2. Often more than one elevation in

glue line, as if glue tried to flow at high spots

3. Glue film shiny if neat resin; grainy if heavily extended
4. No transfer to unspread surface
5. Unspread surface may be dented by being pressed against hardened glue

C. Specific causes
1. Virtual total lack of glue mobility when pressed
2. Pressure arrived too late
3. Pressure inadequate or never arrived
4. Same factors as for unanchored, except more severe
5. Glue too fast in hardening speed
6. Wood very dry: dewatering excessively during assembly time, increasing viscosity and reducing mobility

In addition to the glue line conditions observed above, the glue line can reveal several states of matter that primarily influence the quality of link 1, and are also associated with bond-forming factors.

Undercured Glue Line. As the name suggests, this state of the glue line is immature in the sense that it has not yet reached full solidification. It does not have any distinguishing appearance characteristics since it is the product of an otherwise favorable bond-forming situation. However, delaminated areas will usually show glue adhering to both surfaces as a result of failure within link 1. Undercure is especially insidious since it may respond well to most short-term tests and not show up until sometime after being placed in service. Its weakness is mostly in durability against wet exposure. While most often associated with hot-press operations where insufficient heat or press time has occurred, the phenomenon can also be found in cold-press operations, where wood is too cold to allow curing to take place, or the wood is too dry, causing a glue to dry out before it has a chance to cure (for glues that cure by chemical reaction).

Overcured Glue Line. Again the name suggests the problem. This is a defect associated with such glues as the urea-based adhesives, which have a tendency to break down with heat after they have been cured. Sometimes this can occur during consolidation or at subsequent fabrication procedures where heat is used. In practice, the problem is avoided by cooling after hot pressing, and by restricting the amount of heating in secondary gluing operations as well as in service environments. Like undercure, it too lacks characteristic appearance features in the glue line and failure usually occurs within link 1. This type of failure is often associated with particleboard overlayed in a hot-press operation. In this case, the failure is seen as a layer of particleboard elements attached to the overlay after being pulled away from the particleboard surface.

Filtered Glue Line. Glues that are heavily loaded with fillers and extenders sometimes lose a portion of their resin fraction to the porous structure of the wood, leaving link 1 with inadequate fortification. The bond may retain strength enough to reach an end use and fail after some exposure to moisture or stress. In tests it will have low strength, low moisture resistance, and low wood failure. In appearance the failure zone looks like any other weak glue line with glue attached to both surfaces, sometimes grainy due to the preponderance of filler.

Filtered glue lines are associated with the same mobility factors that cause starved conditions, in this case starved of resin because the mobility of the resin exceeded the ability of filler action to control it. High moisture content and slow temperature rise contribute to filtering actions, and for this reason filtered glue lines are often found in blisters, particularly in plywood. A remedy therefore is to manage bond-forming factors in a manner to reduce mobility. (See later section on manipulating mobility.)

Grainy Glue Line. Although not a defect in itself, a grainy glue line is symptomatic of other problems. While partaking of some of the same appearance characteristics of the filtered glue line that are associated with hot pressing, the grainy glue line is also likely to occur with filler-loaded resin formulations in cold-press operations. In cold pressing it is to be found under the same circumstances that produce the prehardened condition, especially on wood of low moisture content where the adhesive dried out before it had a chance to cure. In the latter case, the remedy is not to shorten the assembly time as would be useful with other prehardened situations, since the same rapid drying would occur after pressure was applied and the glue line would remain undercured. A more appropriate approach, if low moisture is incorrectible, would be to speed the rate of cure so it beats the drying process.

Frothy Glue Line. A frothy glue line is a weak glue line full of bubbles in a meringue-like structure. It is relatively easy to observe on a broken joint surface, most often seen in glued lumber assemblies cured with dielectric heat. Besides being weak in itself, the condition usually indicates the presence of a rather large gap between the surfaces, and therefore also a lack of pressure at that location. The frothy appearance can be due to the action of a mixer beating air into the glue, or a long dwell time on a roller spreader that can also fold air into the glue. The more serious conditions are produced by dielectric heating, which heats the glue to the boiling point so fast that steam erupts into bubbles before the glue cures. Sometimes arcing accompanies bubble formation. Remedies include monitoring the mixing and spreading operations, and reducing the amperage of the dielectric current to slow down the heat input.

Chalky Glue Line. As the name implies, this failed glue line has a chalky appearance. It is associated mainly with vinyl-based glues and is caused by low-temperature wood that destroys the emulsion before it has a chance to coalesce. The condition is irreversible and can only be avoided by assuring wood tem-

peratures above the minimum recommended for that particular adhesive.

Double-Level Glue Line. When a broken glue line shows the glue film at two levels, one level apparently having touched the opposite surface and the other level not, it is a sure sign that insufficient pressure was applied to close the joint.

Stippled, Dentate, or Splotchy Glue Line. When tendrils of glue project from both surfaces of a broken joint, it suggests that the joint was once closed but for some reason opened just before the glue lost fluid properties. In pulling apart, the surfaces drew the adhesive into peaks of various configurations depending upon the rheological properties existing at the time.

In situations where a prior gap existed in the joint, assembly glues, especially those that incur shrinkage as a result of hardening, may sometimes respond in a slightly different manner. As shrinkage takes place, due either to water leaving the glue film or to molecular contraction, the loss in volume forces some adhesive adjustments. Some glues, such as the vinyls, tend to form into pillars connecting the two surfaces, leaving voids in between. Other glues may simply contract to the edges of the gap, forming a fillet-type structure and leaving the gap entirely open.

Crazed Glue Line. Some glues, such as ureas, that produce a highly cross-linked and brittle structure in curing may at times develop a maze of hairline cracks throughout the film as it shrinks. This is crazing, and it tends to occur where gaps exist or in glue lines thicker than the glue can tolerate, generally .015 inches or more in thickness.

Solid Glue Line. In the above list of glue line states some operational misfunction has altered the solidified integrity of link 1. The "solid" state may be considered one in which all factors have operated to an optimum degree, all ingredients are properly incorporated and distributed, and all chemical and physical actions have taken place—the best situation that can be achieved.

The Pressure Factor

Pressure performs several functions besides the obvious one of bringing the surfaces together. It helps express from the bond surface the air that becomes trapped in small surface interstices as the glue is applied. If not driven out, it can obstruct the wetting action of the glue, links 4 and 5. Pressure is also needed to overcome bridging across surface irregularities, and to encourage penetration into the pore structure of the wood as well as into the damaged subsurface regions to produce links 6 and 7. Finally, pressure helps strengthen link 1 by reducing bubbles sometimes occurring in glues that have gone through a mixing operation. Glues that carry a high load of fillers or extenders also benefit by the compaction effect of pressure.

Many types of pressure devices are used in bonding procedures—hydraulic, screw, spring, pneumatic, dead weights, live rollers. Aside from delivering the loads needed to close the joint and assist in bond formation, these devices differ in one important respect: whether or not they can maintain a constant or follow-up pressure. One can visualize that a screw clamp, once tightened against the assemblage, can only maintain a fixed closure, no matter what is going on in the assembly that might change the pressure. Dead weights, on the other hand, do not maintain a fixed closure, but are free to advance in response to internal changes.

Several events occur in the glue line during bond formation that affect dimensions of the assembly reacting to the pressure system:

1. Most glues contract as they harden, due either to solvent (water) loss or to molecular rearrangements.

2. The water in some adhesives diffuses into the wood, swelling it with a gradient away from the glue line, a swelling wave. With a fixed clamp, the swelling has nowhere to go. A chain of events involving swelling, shrinking, and pressure change thus reacts against the glue film.

3. As glue penetrates into the structure of the wood, its volume on the surface decreases, with the same effect as a contraction.

4. In hot-press operations, considerable moisture loss from the wood occurs. The pressure device should adjust to the resulting shrinkage as well as to the compression of the wood.

Two commonplace observations confirm that events such as noted above occur. In face-gluing lumber pieces into billets for block shear testing, they are sometimes made in multiples, clamped in a nut and bolt system, and left overnight to cure. The next day, the nuts may be found to be loose, a consequence of swelling, shrinking, penetrating, contracting, and perhaps some compression set. In another situation, bonding comminuted wood, the mats are compressed by platens pressurized until they meet stops that are intended to control thickness. Panels emerge from the press, however, at a thickness slightly less than the thickness of the stops. This is due to shrinkage accompanying loss of moisture from the panel. Since shrinkage could not occur in a panel unless each piece of wood were well bonded to the next throughout the thickness, this is taken as an indication of good consolidation.

A conclusion to be drawn from the above is that the most effective pressure-delivery system would be one that responds to the varying internal situation and maintains an effective pressure on the assembly while the adhesive is solidifying; in other words, some form of follow-up pressure.

Another aspect of pressure has to do with observations of the glue line condition where insufficient movement of the glue has occurred—various degrees of the prehardened state. Lack of movement is always related to pressure. Between the pressure device and the glue line a series of losses can be visualized:

1. Pressure generated by the system
2. Pressure lost in friction of screws, weight of rams, misfitting molds

3. Pressure lost in bending the wood to shape
4. Pressure lost in compressing high spots
5. Pressure delivered to glue on the glue line

Except at the high spots and bending points, only the last of these participates in movement of the glue on the rest of the glue line. The following questions may be raised:

1. *Did the system produce pressure?* When a pressure factor is being questioned, one must not only recheck the calculations and confirm their registration on dials or recording gauges, but also whether the dials and gauges are reflecting the pressure accurately.

2. *Did the pressure arrive on all the glue lines?* Since glue lines are usually in series across a pressure field, if one receives pressure then all should receive pressure. When some are bereft of pressure, fit of joints or surface irregularities may be suspected. In edge gluing lumber in batches, sometimes one panel, due to uneven lengths or varying widths, may interfere with pressure delivery to an adjacent one. Also in producing small panels by placing several in a single opening of the press, any that are slightly thinner than the rest will not be fully pressurized.

3. *Did pressure arrive, but too late?* This produces the classic example of prehardened glue line. The evidence for this situation is to be found on the unspread surface. Indentations of the glue pattern will be embossed in the wood surface, sometimes with a slight amount of the glue.

4. *Did pressure arrive at all?* This also produces the classic prehardened glue line but without the indentations in the opposing surface.

5. *Did pressure arrive soon enough, but insufficiently?* In this case, there will be a mixture of glue line effects, some flow, some transfer, very little penetration, and a shallow bond if any, perhaps with fiber pull. A thick glue line may also be present, as well as lack of squeeze-out. Sometimes, looking at the broken surface that contains the glue line, the adhesive layer appears to have two thickness levels, one that made contact with the other surface and one that remains close to the surface on which it was applied.

6. *Was the pressure received but then relieved?* Slippage of clamps or readjustment of the wood in response to pressure can cause this situation. Evidence is a splotchy, stippled, or dentate glue line with tendrils of glue extending from *both* surfaces.

7. *Was the pressure uniformly distributed in individual glue lines?* When prehardened and bonded conditions appear in the same glue line, pressure was not uniform. Two sources of causes apply. One, that the pressure was not delivered uniformly due to clamping errors; and two, that the wood was not prepared with well-mating surfaces. Both of these conditions are often associated with edge-glued lumber when edges are cut out of square, out of line, dubbed at the ends, or have undergone a rapid moisture change between machining and gluing, with accompanying warping or dimension change at ends.

The Squeeze-Out Factor

Questions of pressure and its uniformity can often be answered by observing squeeze-out. This is especially indicative in edge- and face-gluing lumber elements. While an assembly is under pressure, observations of squeeze-out can reveal how the pressure factor is operating. The normal expectation is that there will be a slight bead of squeeze-out around all ends and edges of the assembly. Areas where there is no squeeze-out are indicative, though not conclusively, of insufficient pressure. Conversely, areas with copious squeeze-out may be assumed to have received sufficient pressure, and perhaps too much for the viscosity of the glue, with possible starving of the joint.

Several situations can occur. If there is squeeze-out along all glue lines on one side of an assembly but not on the other side, there is probable cause for suspicion of uneven application of pressure. If ample squeeze-out occurs at one end of an assembly but not at the other, clamping errors are a certainty. If ample squeeze-out occurs with glue lines at the top of an assembly but not at the bottom, there is an assembly-time problem since glue would have been applied to the bottom course of laminates first and been the oldest before pressure was applied. If there is no squeeze-out anywhere, pressure was too little or too late. If glue lines from top to bottom show varying amounts of squeeze-out, pressure cannot be a problem, and assembly time cannot be a problem. The amount of glue spread could be irregular.

The analysis of squeeze-out is intertwined with questions of the amount of glue applied. However, the latter is subject to prior control, and if well monitored should not enter the pressure or assembly-time issues.

Blows, Blisters, and Delaminations

These defects, primarily in panel products, can be defined simply as large separations between layers in an assembly. They normally are formed during the consolidation stage but may or may not manifest themselves immediately. Some may not be evident until the panel is cut into smaller pieces and a separation is revealed along the new edge. Although the consequences may be similar, the causes vary. Blows are caused by internal vapor pressures produced by high heat and moisture in the panel. When the hot press is opened, this pressure, unable to relieve itself while under pressure of the press, escapes suddenly and noisily, and provides audible evidence of the conditions in the press. Blows may or may not indicate poor bonding, but they do signal that conditions are marginal and need watching.

Blisters are basically a manifestation of separations in the interior of panels away from the edges. Typically they are rounded, raised areas on the panel surface caused by internal pressure that had no opportunity to escape. They tend to form where the bond is weakest, or near the center of the panel. The rupture may be either in the glue line or in the wood if the bond is well formed, which sometimes happens despite poor bonding conditions. Sometimes they are so massive as

to wedge themselves between press platens and disrupt the entire press cycling procedure. Ordinarily they are much smaller, though easily visible. Generally they do not enter the marketplace. However, there are also latent blisters that remain flat until a moisture increase swells the surface layers and reveals their presence. These are difficult to perceive by eye, but scratching the surface with a fingernail or a coin will create a distinctive sound that is easily detected. Also, sonic scanners and other instruments are able to pick up the separation or discontinuity that lies underneath.

Delaminations are basically separations caused by poor bonding between various layers in a panel. Usually the poor bonding is caused by factors that cause starving, filtering, undercure, or precure and the remedies include those that correct these conditions.

The remedies for blows and blisters include lowering the press temperature and concomitantly increasing the press time, lowering the total moisture content of the press charge by either lowering the moisture in the wood or lowering the water in the glue, or applying less glue. Means for allowing the vapor to escape before the press is opened is also practiced. This includes "breathing the press"—reducing the pressure near to, but not at, zero at some time during the press cycle, usually toward the end, to allow vapor to escape through the faces of the panel without disrupting the bond. Sometimes it is advantageous to breathe the press early in the cycle to eliminate moisture before it has a chance to affect bond formation. Escape through the edges can be promoted by increasing the press time without a pressure reduction. This might save a press load at the risk of disrupting the flow of operations upstream.

Manipulating Mobility

The final objective of troubleshooting should be a determination of what to do to correct the problem. In the previous sections, suggestions were made that define the problem and perhaps trace it to a source, which then often signals the correction needed. When the problems are of a mechanical origin, they generally require rather obvious mechanical adjustment. However, when the problem has to do with mobility effects that may have been thwarted by some operational factor, the correction need only involve making a change, almost any change, that will adjust the mobility in the right direction during bond formation. The changes that are made are usually dictated by urgency, sometimes as soon as the next press load, or back into the wood preparation operations.

The first determination that must be made is whether the mobility needs to be increased or decreased. If the problem is a starved condition, the mobility should be reduced; if unanchored or prehardened, mobility must be increased. The second determination to be made is what factors to change. Usually they will be those that are less costly and most immediate. The problem reduces to the effects and consequences portrayed in Figure 12–1. In the display, the vertical axis represents an increasing mobility, and the horizontal axis represents an increasing factor level affecting mobility. Some factors cause a rise in mobility as they increase (represented by solid lines). Another group of factors causes a decrease in mobility as they increase (represented by dashed lines). The effects are plotted to show only direction and separation; the curvature has no meaning other than to suggest they vary in some fashion.

The mobility scale has no numbers, first because its physical reality is difficult to measure on the glue line, and second because it is composed of other parameters besides viscosity—pressure, porosity, fillers, molecular sizes, thixotropy, and so on. The absence of numbers does not present a major problem, however. It can be dealt with by making purely qualitative observations of the four conditions glue lines develop during bond formation. Once it is determined what direction mobility must be shifted, an operational change followed by another glue line observation will show whether a sufficient

shift has occurred. If not, another incremental change is made until the mobility level lands in the *bonded* region as evidenced by complete wood failure.

Thus one can visualize ascending and descending mobility effects of different factors varying in rates and magnitudes, most of which are operating to some degree in every glue line. While operating somewhat independently, they all interact negatively and positively, and compound in various ways to produce a combined effect. During solidification of the glue on the glue line, all the downward effects on mobility may be thought of as integrating or summing themselves into curve 1; all the upward effects on mobility are integrated into curve 2. The intersection of the two compounded curves, one ascending and the other descending, suggests an equilibrium of effects and therefore the final combined effect on mobility that becomes frozen in the glue by the solidification action.

Depending upon magnitudes of the resultants, the intersection can fall in any of the four regions that produce each of the distinctive glue line conditions: starved, bonded, unanchored, or prehardened. These zones of glue line conditions are delineated tentatively by horizontal bands across the factor playing field in Figure 12–1. Actually, they are not sharply defined but grade into each other. Nevertheless, a given mobility, however it may be generated, has a definite impact on bond formation. This is an important premise, for it suggests that if the existing mobility is not favorable to bond formation, it can be moved up or down by the proper selection of factors, increasing or decreasing them as the situation demands.

Referring again to Figure 12–1, the intersection at A suggests that all the effects equilibrated at a level that produced an optimum mobility, creating all seven links and forming the desired *bonded* condition. With some factors out of control that cause mobility to increase, the intersection might rise to B and produce a *starved* condition. In this case, maximum flow, transfer, penetration,

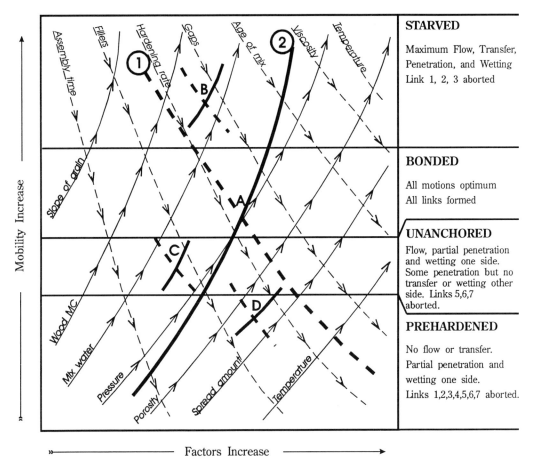

STARVED

Maximum Flow, Transfer, Penetration, and Wetting Link 1, 2, 3 aborted

BONDED

All motions optimum
All links formed

UNANCHORED

Flow, partial penetration and wetting one side. Some penetration but no transfer or wetting other side. Links 5,6,7 aborted.

PREHARDENED

No flow or transfer.
Partial penetration and wetting one side.
Links 1,2,3,4,5,6,7 aborted.

Mobility Increase

Factors Increase

Figure 12–1   Interactions in the glue line. Factors that depress mobility as they increase (dashed lines) sum themselves into curve 1; those that increase mobility as they increase (solid lines) sum themselves into curve 2. The intersection at A is the resultant of all actions, in this case producing a *bonded* condition. Had the intersection fallen at B, the resultant would have been a *starved* glue line. Similarly, intersections at C and D produce *unanchored* and *prehardened* glue lines respectively. Moving B, C, and D toward the *bonded* zone involves increasing (or decreasing) factor levels to slide up or down one or more of the curves. (Marra, John Wiley & Sons, revised)

and wetting would occur and links 2 through 7 would be well formed, but link 1 would be aborted due to loss of glue. If factors integrate such that mobility moves in the opposite direction, the intersection could occur at point C, suggesting an equilibrium in the *unanchored* zone. Here flow could occur along with some spontaneous penetration and wetting on the spread surface, while on the unspread surface some forced penetration into the coarser cavities would occur, but no wetting and no transfer. Links 1, 2, 3, and 4

would be formed, but links 5, 6, and 7 would be left unformed. A further compounding of factors that decrease mobility might lower the intersection to point D, the *prehardened* zone, where none of the essential motions can occur, not even flow, although on the surface to which glue was applied some penetration and wetting may have developed without help of pressure.

An example will illustrate how the information in Figure 12–1 can be used in practice. Assuming that it is known from inspection of the broken joint that a prehardened condition exists, the question becomes how to move the intersection point up to the bonded zone. A number of options present themselves. From the figure it is seen that there are two ways up: one using the factors represented by curve 2, increasing them, and one using factors represented by curve 1, decreasing them. If this is a high-production situation, a quick solution is needed to avoid producing more reject material. The most immediate option—a tempting one—is to increase pressure on the current load in the press. However, because of the negative consequences of excessive pressure (crushing wood and building in stresses), using this factor to overcome prehardening effects has limits and should not become a common procedure. Moreover, it is unlikely that such a procedure would raise mobility above the unanchored zone. Another immediate option in the case of hot pressing would be to increase the temperature in the hope of inducing a plasticizing effect to increase mobility. This could save the next load, but it too has limitations. Because higher temperature also accelerates curing, it could further preharden glue lines close to the platens. The likelihood of inducing blisters would also be increased. Hence the opportunities for alleviating the effects of prehardening at the press are unreliable. It is already too late.

One step back from the press lies the most appropriate factor to consider: decreasing assembly time. However, time is not the only factor operating between the spreader and the press. Air flow and elevated ambient temperatures also intrude, as well as assemblies left "open" instead of the prescribed "closed," and curled or wrinkled veneer in the case of plywood manufacture. Prepassing—applying pressure during the assembly period—eliminates these undesirable effects while at the same time accomplishing the transfer motion earlier in the bond-forming process.

Looking back further from the press to the next previous operation, the most convenient factor to intervene with a change is at the glue spreader. An increase in the amount of glue applied would help increase mobility at the press. The main objections to this option are the amount of water that would also be added to the assembly, and the extra cost. Alternatively, applying half the glue (plus 10 percent) to each surface being joined eliminates the transfer problem, making one less action to be performed in the press.

At the mixer three choices present themselves: reduce the rate of hardening (applicable only to those glues to which a catalyst is being added), decrease filler or extender content, or increase water content of the mix (a limited option because of dilution effects, and because more water would be added to the assembly). As a general admonition, adding more water than the recipe calls for should not be done without consulting the manufacturer of the adhesive, since the need for water may be signaling other problems such as exceeding the pot life or storage life.

Further back to wood preparation, the moisture content may have been too low, and therefore may have drawn too much water from the glue during the assembly time. It may need to be increased, but not without checking it first because if it is on target, increasing it will disturb other factors and lead to another cause of bond failure or product malfunction. Gaps in the joint are the equivalent of zero pressure leading to zero mobility and the appearance of prehardening. Since gaps are caused in preparation of the wood by warping, moisture change, and miscutting (as well as by low pressure in the press), corrective action can only follow further observation to determine what specifically needs correcting.

There are, of course, many factors in the bonding process that arrive at the glue-application stage immutably fixed and out of control at that point. These include such factors as slope of grain, porosity, wood density, glue line dimensions, and degree of surface inactivation. The adverse effects on mobility of factors such as these still read as glue line conditions, but cannot be directly addressed and instead must be combated by adjustments of other factors that move mobility in an opposite direction.

Several interesting side observations can be drawn from the interactions depicted in Figure 12-1. Little deviations that are not serious in themselves can compound with others to create an inoperable mobility. For example, a few more minutes longer assembly time can usually be tolerated. But if in addition the wood is on the dry side, as well as the warm side, and the amount of glue applied on the light side, the stage may be set for prehardening or unanchoring.

Given the potential for intense interactions of the many factors operating in a glue line, it is not surprising that both extremes of the mobility scale can sometimes be seen in the same glue line. Wood-failure values that vary specimen to specimen cut from the same sample are further confirmation that

critical variables are at work in the glue line, adding and subtracting as the glue line environment changes from point to point.

Finally, Figure 12-1 illustrates what may be a disturbing thought: There seem to be more opportunities for producing a poor bond than for producing a good one. If so, it is because good bonds require a collusion of factors that produce optimums—factors at or near prescribed levels. Poor bonds partake of every little deviation that drives the mobility toward either extreme. Sometimes the variation of a single factor, such as pressure, can alone drive mobility from one extreme to the other.

It is nevertheless reassuring to realize that with some understanding of how factors in the glue, in the wood, and in the gluing operation function, they can be controlled and manipulated to promote good bond formation in a most efficient manner.

Bibliography

GENERAL

Books

Bikerman, J. J. 1960. *The Science of Adhesive Joints.* New York: Academic Press.

Clark C. Heritage Memorial Series on Wood. University Park, Pa.: Pennsylvania State University Press, vols. 3 and 4. Blomquist, R. F., Christiansen, A. W., Gillespie, R. H., and Myers, G. E., eds. 1983. *Adhesive Bonding of Wood and Other Structural Materials,* vol. 3. Freas, A. D., Moody, R. C., and Soltis, L. A., eds. 1984. *Wood: Engineering Design Concepts,* vol. 4.

Guy, A. G. 1976. *Essentials of Materials Science.* New York: McGraw-Hill Book Co.

Haygreen, J. G., and Bowyer, J. L. 1982. *Forest Products and Wood Science.* Ames, Iowa: Iowa State University Press.

Kaeble, D. H. 1971. *Physical Chemistry of Adhesion.* New York: John Wiley and Sons.

Van Vlack, L. H. 1982. *Materials for Engineering.* Reading, Mass.: Addison-Wesley Publishing Co.

Journals

Adhesive Age
Forest Products Journal
Holz als Roh und Werkstoff
Industrial and Engineering Chemistry
Industrial Forestry
Journal of Adhesion
Journal of Applied Polymer Science
Journal of Polymer Science
Journal of the Japan Wood Research Society
Wood and Fiber Science
Wood and Wood Products

Associations

American Hardboard Association
American Institute of Timber Construction
American Plywood Association
Hardwood Plywood Manufacturers Association
National Particleboard Association
Technical Association of the Pulp and Paper Industry
Wood Machinery Manufacturers of America

Tapes

1. A series of slide tapes on wood, adhesives, composites, and related technologies is available for sale or loan from the Forestry Media Center, Oregon State University, Corvallis, OR 97331.
2. Woodworking Video Journal is available from Nicholas Weedhaas, North Carolina State University, Raleigh NC 27695.

ADHESIVES

Alfrey, T., and Gurney, E. F., 1967. "The Organic Chemistry of Polymers." Chap. 8 in *Organic Polymers.* Englewood Cliffs, N.J.: Prentice Hall.

Houwink, R., and Salomon, G. 1965. *Adhesion and Adhesives.* New York: Elsevier Publishing.

May, C. A., and Yoshio, T. 1973. *Epoxy Resins: Chemistry and Technology.* New York: Marcel Dekker.

Meyer, Beat. 1979. *Urea Formaldehyde Resins.* New York: Addison-Wesley Publishing Co.

Patrick, R. L., ed. 1966–91. *Treatise on Adhesion and Adhesives.* New York: Marcel Dekker.

Pizzi, A., ed. *Wood Adhesives: Chemistry and Technology* New York: Marcel Dekker.

Skeist, Irving. 1990. *Handbook of Adhesives.* 3d ed. New York: Van Nostrand Reinhold.

Wake, W. C. 1976. *Adhesion and the Formulation of Adhesives.* London: Applied Science Publishers.

Wood Adhesives. 1990. Proceedings of the Forest Products Research Society. Madison, Wis.

WOOD

Browning, B. L., ed. 1963. *The Chemistry of Wood.* New York: Interscience Publishers.

Clark C. Heritage Memorial Series on Wood. University Park, Pa.: Pennsylvania State University Press., vols. 1 and 2. Wangaard, F. F., ed. 1981. *Wood: Its Structure and Properties,* vol. 1. Dietz, A. G. H., Schaffer, E. L., and Gromala, D. S., eds. 1982. *Wood as a Structural Material,* vol. 2.

Core, H. A., Cote, W. A., and Day, A. C. 1976. *Wood Structure and Identification.* Syracuse, N.Y.: Syracuse University Press.

Goldstein, I. S., ed. 1977. *Wood Technology: Chemical Aspects.* American Chemical Society.

Hoadley, B. 1980. *Understanding Wood.* Newtown, Conn.: Tauton Press.

——. 1990. *Identifying Wood.* Newtown, Conn.: Tauton Press.

Koch, Peter. 1972. *Utilization of Southern Pines.* 2 vols. Agriculture Handbook no. 420. Washington, D.C.: U.S. Department of Agriculture.

——. 1985. *Utilization of Hardwoods Growing on Southern Pine Sites.* 3 vols. Agriculture Handbook no. 605. Washington, D.C.: U.S. Department of Agriculture.

Kollman, F. P., Kuenzi, L. W., and Stamm, A. J. 1975. *Principles of Wood Science and Technology.* Vol 2, *Wood-Based Materials.* New York: Springer-Verlag.

Panshin, A. J. and DeZeeuw, C. 1980. *Textbook of Wood Technology.* New York: McGraw-Hill.

Snogen, R. C. 1974. *Handbook of Surface Preparation.* New York: Palmerton Publishing Co.

Stamm, A. J. 1964. *Wood and Cellulose Science.* New York: Ronald Press.

U.S.D.A. Forest Service. *Wood Handbook: Wood as an Engineering Material.* Latest revision. Agriculture Handbook no. 72. Washington D.C.: U.S. Government Printing Office.

PROCESSES

Baldwin, R. F. 1975. *Plywood Manufacturing Practices.* San Francisco: Miller Freeman Publications.

Cagle, C. V. 1973. *Handbook of Adhesive Bonding.* New York: McGraw-Hill.

Eckelman, Carl A. 1989. *Design, Construction, and Testing of Adhesive-Based Furniture Assembly Joints.* Department of Forestry. West Lafayette, Ind.: Purdue University.

Feirer, John L. 1970. *Cabinetmaking and Millwork.* Peoria, Ill.: Charles A. Bennett Co.

Feirer, John L., and Hutchings, Gilbert R. 1972. *Advanced Woodwork and Furniture Making.* 4th ed. Peoria, Ill.: Charles A. Bennett Co.

Gillespie, R. H., Countryman, D., and Blomquist, R. F. 1978. *Adhesives in Building Construction.* Agriculture Handbook no. 516. Washington, D.C.: U.S. Department of Agriculture.

Kelly, M. W. 1977. *Relationships Between Processing and Properties of Particleboards.* U.S.D.A. Forest Service General Technical Report no. 10. Madison, Wis.: U.S. Department of Agriculture.

Maloney, T. M. 1977. *Modern Particleboard and Dry Process Fiberboard Manufacturing.* San Francisco: Miller Freeman Publications.

Modern Plywood Techniques. Proceedings of the First and Second Plywood Clinics. San Francisco: Miller Freeman Publications.

Selbo, M. L. 1975. *Adhesive Bonding of Wood.* Forest Service Technical Bulletin no. 1512. Madison, Wis.: U.S. Department of Agriculture.

Sellers, T., Jr. 1988. *Plywood and Adhesive Technology.* New York: Marcel Dekker.

Suchland, O. and Woodson, G. E. *Fiberboard Manufacturing Practices in the United States.* Agriculture Handbook no. 640. U.S. Department of Agriculture.

TESTING AND STANDARDS

American Society for Testing and Materials. *Standards: Wood Adhesives.* Latest revision. 1916 Race St. Philadelphia.

Anderson, G. P., Bennett, S. J., and DeVries, K. L. 1977. *Analysis and Testing of Adhesive Bonds.* New York: Academic Press.

Adhesives: Methods of Testing. 1975. Federal Test Methods Standard no. 175a. Washington, D.C.: General Services Administration.

Gillespie, R. H., and Lewis, W. C. 1972. *Evaluating Adhesives for Building Construction.* Forest Service Research Paper no. 172. Madison Wis.: U.S. Department of Agriculture.

U.S. Department of Commerce. Product Standards. Latest issues. *Construction and Industrial Plywood, Structural Glued Laminated Timber, Hardwood Plywood.* Washington, D.C.: U.S. Government Printing Office.

Glossary

Abhesive A material that exhibits little or no attraction for other materials: a release agent.

Abrasion The rubbing away or wearing off of a surface by grinding or wearing action.

Absorption The act or process by which liquids or gases are taken up in the voids or interstices of a solid, as in a sponge. See also ADSORPTION.

Accelerator A material that triggers or speeds up a reaction and becomes part of the resulting compound, as opposed to catalyst, which has similar action but does not become part of the resultant. See also CATALYST and HARDENER.

Accelerated Aging A test used to obtain a measure of the inherent ability of a material to withstand long-term service by using severe exposure conditions in testing.

Acoustical Board A low density structural fiberboard suitably textured for use as sound absorption.

Additive Any material incorporated into an adhesive or composite to impart special properties.

Adhere The act of attachment to surfaces.

Adherend The material being held together by an adhesive.

Adhesion The state in which two surfaces are held together by molecular forces.

Adhesion, Mechanical Adhesion between surfaces in which the adhesive holds the parts together by interlocking action.

Adhesion, Specific Adhesion between surfaces held together by valence forces of the same type as those that hold all substances together.

Adhesive A substance capable of holding materials together by surface attachment. Synonymous with GLUE. See also BINDER.

Adhesive, Assembly An adhesive used for bonding parts or components together in the fabrication of consumer products such as furniture. The term distinguishes so-called "joint glues" from those used in consolidating wood elements such as lumber, veneer, and comminuted wood.

Adhesive, Cold-setting An adhesive that sets at temperatures below 68°F (20°C). (Also generally refers to adhesives that set at room temperature.)

Adhesive, Contact An adhesive that is applied as a liquid and may be dried to the touch but will adhere to itself instantaneously upon contact; also called dry-bond adhesive.

Adhesive, Gap-filling Adhesive suitable for use where the surfaces to be joined may not be in close contact but separated some distance. This distance has not been defined but may be considered to be anything greater than .010 inches.

Adhesive, Heat-activated An adhesive the curing and/or fluid properties of which are dependent on heat.

Adhesive, Hot-melt An adhesive applied in a molten state that forms a bond on cooling to a solid state.

Adhesive, Hot-setting An adhesive that requires a temperature at or above 212°F (100°C).

Adhesive, Intermediate-setting An adhesive that sets in the temperature range 87 to 211°F (31 to 99°C).

Adhesive, Room-temperature-setting An adhesive that sets in the temperature range 68 to 86°F (20 to 30°C).

Adhesive, Separate Application An adhesive consisting of two parts, one part of which is applied to one surface of a joint, and the other part to the other surface. When the surfaces are brought together the two parts interdiffuse and instigate the curing reaction.

Adhesive, Solvent An adhesive dissolved or suspended in a volatile organic liquid, as opposed to one having water as the vehicle.

Adhesive, Solvent-activated A dry adhesive film rendered tacky just prior to use by application of a solvent.

Adsorption The process by which the surface of a solid holds, by molecular attraction, an extremely thin layer of a gas, liquid, or dissolved substance, as in an activated carbon filter. See also ABSORPTION.

Air-classifier Equipment that separates particles into size classes by floating them in a stream of air from which gravity separates them on the basis of weight and surface area.

Air-felting The process of forming a mat of comminuted wood by deposition from a suspension in air, as opposed to a suspension in water.

Anisotropic A material whose values for its properties differ in the three axial directions. Wood is highly anisotropic. An isotropic material would have similar properties in all directions. See also ORTHOTROPIC.

Assembly Time A term used to designate the period after the adhesive has been applied until the pressing stage in the process. This period is divided into two parts: "open," in which the parts remain unassembled and the adhesive is exposed to the atmosphere, and "closed," in which the parts coated with adhesive are assembled and awaiting pressure.

Axial A directional term with reference to an axis.

Balanced Construction A construction in which conjugate layers on either side of a center plane are of similar properties. Such a construction minimizes warping that is due to dimensional changes.

Binder A term used synonymously with glue or adhesive, but which refers mainly to the bonding agent for materials in particulate form.

Bleed-through Glue or components of glue that have penetrated through the outer layer or ply of a veneered product and that show as a blemish or discoloration on the surface. The cured glue on the surface also interferes with finishing operations and with subsequent gluing.

Blending The process of applying binder to materials in particulate form.

Blisters Bulges on the surface of panel products due to local bond failures.

Blows Accumulation of gases in the interior of panels during hot pressing that escape with pops or hisses and sometimes produce blisters.

Bond The attachment of materials by means of an adhesive. When properly done, it can exceed the strength of wood adherends.

Bond Line See GLUE LINE.

Bond Strength The load required to rupture an adhesive bond, reduced to unity with respect to the area involved.

Buffer An additive that regulates or stabilizes pH.

Buffering Capacity The ability of a material to resist change in pH as acids or alkali are added.

Bulk Density The weight of uncompacted particle or fiber material; usually expressed as weight per cubic foot or other volumetric measure.

Caliper A reference to the thickness of a panel (for example, its caliper is 0.375 inches). Also an instrument used to measure thickness.

Casehardening A condition of stress and set in dry lumber characterized by compressive stress in the outer layers and tensile stress in the center or core. A cause of warping when surfacing unequally on either side.

Catalyst A substance that, added in minor amounts, changes the rate of reaction of a chemical system without becoming part of it; usually an additive to speed the cure of resins. See also ACCELERATOR and HARDENER.

Caul A sheet of material, usually made of aluminum, stainless steel, or hardboard, used under and/or over an assembly to aid in transporting it to and into the press, or to protect its faces or the surfaces of the platens.

Caulless System A manufacturing process in which particle mats are formed and conveyed on moving belts or other mat carriers and then pressed directly between the press platens without the use of caul plates.

Check A separation along the grain of wood.

Chemical Bond A molecular attachment involving covalent forces and forming a new chemical entity across the interface.

Chips Small chunks of wood, generally half a thumb in size, although hand-size are also produced for special purposes. One of the primary breakdown products from logs that will be further refined into flakes or fibers.

Chord Either of the two outside members of a truss, connected and braced by the web members.

Classifier Equipment for separating particles according to size or weight.

Closed Assembly See ASSEMBLY TIME.

Closing Time The time interval to compress a mat from initial to final thickness in a press.

Cobwebbing A phenomenon observed during application of an adhesive characterized by formation of weblike threads between applicator and the adherend surface.

Coefficient of Thermal Expansion The change in unit volume per degree temperature increase over a specified initial temperature; commonly stated as the average coefficient over a given temperature range.

Coefficient of Variation The amount of deviation from an average value expressed as a percent of the average.

Cohesion The state in which the particles of a single substance are held together by primary or secondary valence forces. As used in the adhesive field, the state in which the particles of the adhesive or the adherend are held together.

Cold Pressing A bonding operation in which an assembly is subjected to pressure without the application of heat.

Comminuted Wood Wood reduced to small particles.

Composite A combination of two or more materials bonded together and performing as a single unit.

Composition Board A board usually incorporating comminuted, materials.

Compression The act of applying stress in a direction such as to cause compaction or densification.

Compression Wood Abnormal wood formed on the lower side of branches and inclined trunks of softwood trees. Compression wood is identified by its relatively wide annual rings, usually eccentric; its relatively large amount of summerwood, sometimes more than 50 percent of the width of annual rings in which it occurs; and a lack of sharp demarcation between springwood and summerwood in the same annual ring. Compression wood shrinks excessively lengthwise as compared with normal wood, and is much weaker.

Compressometer A device for measuring force or pressure.

Condensation   1. A physical process by which vapors change state to liquids. 2. A chemical reaction in which two or more molecules combine with the separation of water or some other simple substance.

Conditioning (Pre- and Post-)   The exposure of a material to the influence of a prescribed atmosphere for a stipulated period of time or until a stipulated relation is reached between material and atmosphere.

Consistency   That property of a liquid adhesive by virtue of which it tends to resist deformation such as flowing. It is not a fundamental property, but is comprised of viscosity, plasticity, and other phenomena. See also VISCOSITY.

Consolidation   The operation in which the parts of an assembly with or without coating of adhesive are made into a unified whole.

Construction   Normally a term used in reference to the erection of a building or other shaped object. In reconstituted products the term is used to denote the organization of a mat or assembly of wood elements, their disposition in layers, orientation, or densities.

Construction Adhesive   Any adhesive used to assemble building materials into components or structures; generally applies to elastomer or mastic type adhesives.

Contact Cement   See ADHESIVE, CONTACT.

Core   The layer of a panel located centrally in the thickness direction.

Crazing   The breakdown of an adhesive layer into fine cracks that may extend in a network throughout the glue line.

Creep   The deformation of a material under long-time loading, following the instantaneous elastic or rapid deformation; sometimes called cold flow.

Crossbands   Layers of wood placed with the grain at right angles to that of face plies in order to minimize shrinking and swelling, particularly in plywood of five or more plies.

Cup   A distortion of a board in which there is a deviation from a straight line across the width of the board.

Cure   To harden or otherwise change the physical properties of an adhesive by chemical reaction, which may be vulcanization, condensation, or addition polymerization, usually accomplished by the action of heat and catalyst, alone or in combination, with or without pressure.

Curtain Coating   Applying a liquid film of adhesive or other material to a surface by passing under a thin falling stream of the liquid.

Daylight   The distance between platens in a press into which a charge or assembly is inserted to receive pressure. The size of this opening determines the maximum thickness of mat that can be consolidated.

Decompression   The lowering of pressure on a mat during the hot pressing cycle in order to release vapor pressure trapped in the mat or to stop further compaction.

Decorative Laminate   A multilayered panel made by consolidating sheets of resin-impregnated paper, the surface layer of which carries a decorative design.

Defect   Any irregularity in a part or product that lowers its performance beyond acceptability.

Defiberize   The process of converting chips or other relatively large particles into fibers.

Delamination   The separation of layers in a product due to failure of the bond, either in adhesion at the interfaces or in cohesion within the adhesive layer or in the adherend.

Density   The mass of a material in a unit volume, generally expressed as pounds per cubic foot, kilograms per cubic meter, or grams per cubic centimeter. Note that in the case of wood, weight and volume vary with the moisture content.

Density Profile   A diagrammatic representation or sectional elevation of the variation in density across a board in any of the three directions.

Deposition   The act by which materials to

be consolidated are caused to fall onto a caul or receptacle.

Desorption The act by which liquids or gases leave a solid, the opposite of adsorption. The term *sorption* encompasses both.

Deviant Grain The collective term for any grain direction varying from the true edges or faces of solid wood pieces due to growth or cutting processes, not that due to assembly processes.

Dielectric Substance Any of a group of materials that are nonconductors of electricity.

Dielectric Constant The property of a material measured as the ratio of the capacitance of a given configuration of electrodes with the material as the dielectric between them, to the capacitance of the same electrode configuration with a vacuum as the dielectric.

Diffusion The spontaneous process by which physical states or concentrations equalize, such as differences in heat, gas pressures, and dissolved substances.

Diluent An additive that reduces the effect or concentration of a mixture.

Dimensional Stability The degree to which a material retains its dimensions when exposed to varying conditions of temperature and humidity. See also SHAPE STABILITY.

Disc Flaker A flaking machine in which the knives are mounted radially on the face of a rotating disc.

Doctor Bar or **Blade** A scraper mechanism that regulates the amount of adhesive on the spreader rolls or on the surface being coated.

Doctor Roll A roller mechanism operating in close conjunction with a roller applicator and with a controlled space between; regulates the amount of adhesive carried on the applicator and thus controls the amount of adhesive that can be applied.

Dosing Metering of materials as in application of binders.

Double Spreading Application of adhesive to both surfaces of a joint.

Dressed Size The dimensions of lumber after being surfaced with a planing machine. The dressed size may be $\frac{1}{2}$ inch less than the nominal or rough size. A 2- by 4-inch stud, for example, actually measures $1\frac{1}{2}$ by $3\frac{1}{2}$ inches.

Drum Flaker A flaking machine with knives mounted on the outside surface of a rotating cylinder.

Dry Kiln A chamber having controlled airflow, temperature, and relative humidity for drying lumber and other wood products.

Dry Out Loss of moisture or solvent in an adhesive film during the assembly time, causing all adhesive motions to cease.

Dry Solids The weight of oven-dry solid material in a mixture or solution. A term used with liquid adhesives to specify the amount of actual adhesive material that is present. The term *resin solids* goes further and refers only to the resin fraction of the dry solids content.

Durability The life expectancy of adhesive bonds under the anticipated service conditions of the product.

Dwell Time That period during which a material or assembly remains in a machine or stage of the manufacturing process.

Early Wood The portion of the annual growth ring that is formed during the first part of the growing season. It is usually less dense and mechanically weaker than wood formed later in the season.

Edge Banding A thin, flat strip of material bonded to edges of panels as a decorative and protective adjunct.

Elastic Limit The greatest stress a material is capable of sustaining without any permanent strain remaining upon complete release of the stress. See also PROPORTIONAL LIMIT.

Elastomer A macromolecular material that, at room temperature, is capable of recovering substantially in size and shape after removal of a deforming force.

Element See WOOD ELEMENT.

Embossing A process by which the surface

of a product is given a relief effect, usually by using a patterned caul or die, and heat and pressure.

Endothermic   A chemical or physical reaction that produces a cooling effect. See also EXOTHERMIC.

Engineered Board   A product, usually structural, the properties of which are the result of design, and the performance of which can be dependably specified.

Equilibrium Moisture Content   The moisture content at which wood neither gains nor loses moisture when surrounded by air at a given relative humidity and temperature.

Exothermic   A chemical or physical reaction accompanied by the evolution of heat. See also ENDOTHERMIC.

Extender   A low-cost additive such as cereal flour used with an adhesive, reducing its cost while retaining its bulk for uniform spreading. Extenders also generally reduce durability of a given resin.

Exterior Exposure   Service conditions for a product where all atmospheric forces are free to impinge on performance.

Exterior Panel   A product designed to withstand exterior exposure.

Extractives   Substances in wood, not an integral part of the cellular structure, that can be removed by solution in hot or cold water: ether, benzene, or other solvents that do not react chemically with wood components.

Extruded Particleboard   A particleboard made by endwise pressure through a die. The particles lie with their longer dimensions perpendicular to the direction of pressure. See also MAT-FORMED PARTICLEBOARD.

Face   That surface of a panel on which the grade or quality is judged. When both surfaces are of the same quality, both are regarded as faces.

Failure, Adherend   Rupture of an adhesive bond such that the separation appears to be within the adherend. See also WOOD FAILURE.

Failure, Adhesion   Rupture of an adhesive bond such that the plane of separation appears to be at the adhesive-adherend interface—unanchored.

Failure, Adhesive   Rupture of an adhesive bond such that the plane of separation appears to be in the adhesive layer. Glue is to be seen on both of the ruptured surfaces. See also COHESION.

Fiber   A structural cell element of wood or lignocellulosic materials. Thick-walled and linear, its primary function is to contribute strength in the tree. In producing paper, it is delignified and further refined to its fibrillar structure.

Fiber Bundle   Groups of fibers still held together with lignin, retaining a fibrous aspect and used at that stage of comminution to make fiberboards.

Fiber-saturation Point   The stage in the drying or wetting of wood at which the cell walls are saturated and the cell cavities are free from water. It is usually taken as approximately 30 percent moisture content, based on oven-dry weight.

Fiberboard   A broad generic term referring to panels of widely varying densities manufactured of refined or partially refined wood (or other vegetable) fibers. Bonding agents and other materials may be added to improve some property such as strength, or resistance to moisture, fire, or decay. The furnish may also be used to produce molded objects.

Filler   1. An additive used with adhesives to control penetration into the wood and to improve characteristics of the hardened film. Finely ground minerals and wood and nut shell flours are commonly used (compare with EXTENDER). 2. A material used in finishing a wood-based panel to fill surface pores and provide a more uniform surface for the finish coats.

Fillet   A portion of a cornered structure that fills and reinforces the angle between members. The squeeze-out from glue lines sometimes performs this function especially when bonding to thin edges such as those in honeycomb structures.

Fines   Small, dustlike wood particles, usually discarded but sometimes used in the face layer of particleboard to provide a smoother surface, or added to adhesives as a filler or to make a paste.

Finishing The final sanding, filling, and coating to prepare a product for its end use.

Fire Retardant A chemical or preparation of chemicals used to reduce flammability, to retard spread of a fire over the surface, and to reduce smoke.

Flat-platen-pressed Particleboard A particleboard manufactured by pressing a mat between parallel platens with pressure applied perpendicular to the face. The wood elements in this case tend to lie with their long axis parallel to the faces of the panel. See also EXTRUDED PARTICLEBOARD.

Former The equipment that produces a mat of flakes or particles. Also called a *felter* in fiber mat forming.

Forming The process of converting a mass of wood elements to a mat in preparation for consolidation. Also referred to as *felting* in the case of fibers.

Furnish Wood elements en masse, with or without the binder, before consolidation.

Glue Originally a hard, proteinaceous gelatin obtained from hides, tendons, cartilage, and bones of animals and processed to provide adhesive properties. Although the term has come to be used synonymously with ADHESIVE, there lingers a tendency to use it for adhesives of natural origin.

Glue Line The layer of adhesive that attaches two adherends; also called bond line.

Graduated-layer Board A board made of comminuted wood in which elements grade imperceptibly from coarse in the core to fine on the surface, a characteristic of wind-sifter formers.

Grain 1. Orientation of fibers in the wood piece with respect to the surface planes. 2. Orientation and structure of the annual rings.

The following terms are associated with the first meaning:

Long grain A plane parallel to the long axis of the fibers.

Cross grain A plane perpendicular to the long axis of the fibers.

Deviant grain A general term.

Short grain A plane in which the deviant grain is inclined with respect to the face of a piece.

Diagonal grain A plane in which the deviant grain is inclined with respect to the edges of the piece.

Spiral grain Sloped grain in a tree, resulting in diagonal grain in boards or veneer.

Interlocked grain Sloped grain direction that reverses in successive annual rings creating an interlocking effect.

The following terms are associated with the second meaning:

Coarse grain Having large, open pores or lumens as in red oak and redwood. Also called *open grain*.

Fine grain Having small pores or lumens, as in maple and cedar. Also called *close grain*.

Hard grain Wood in which a high percentage of the annual ring is composed of summerwood.

Soft grain Wood in which a high percentage of the annual ring is composed of springwood.

Fancy grain Wood in which grain patterns produce decorative values.

Green The moisture condition of wood, freshly derived from the log and having a water content above the fiber-saturation point.

Green Strength The strength of wood when it is freshly cut and above the fiber-saturation point. The term is also used in reference to bonded assemblies when the adhesive has not yet reached full cure.

Growth Ring The layer of wood growth a tree produces during a single growing season. Different species produce distinctive ring structures. In individual species, ring structure varies with conditions of growth and serves as an indicator of wood quality.

Gusset A flat metal, plywood, or similar material used to reinforce connections at intersections of wood members, as in the joints of trusses.

Hammermill A machine having pivoted hammers revolving at high speed, used for reducing chips or other particles to a finer stage of comminution, the size being controlled by the openings in a grate or screen adjacent to the path of the hammers.

Hardboard A generic term for a panel manufactured primarily from fibrous elements and consolidated to a density of 50 pounds per cubic foot or more.

Hardener A substance or mixture of substances added to an adhesive to promote or control the curing reaction by taking part in it. The term is also used to designate a substance added to control the degree of hardness of the cured film.

Heartwood The wood extending from the sapwood to the center of the tree, the cells of which are dead and no longer participate in life processes. Heartwood may contain phenolic compounds, gums, resins, and other materials that usually make it darker and more decay-resistant than sapwood.

Heat Treating The process of subjecting a wood-based panel material, usually hardboard, to a special heat treatment after hot pressing to increase some strength properties and water resistance.

Honeycomb A sandwich core material constructed of thin sheet materials or ribbons formed to a porous structure resembling a honeycomb.

Hot Pressing The process of consolidation in which heat and pressure are used.

Hot Stacking A post-conditioning operation in which panels, after removal from a hot press, are tightly dead piled to prolong their temperature and assure complete cure of the adhesive.

Hygroscopic The property of a material that allows it to readily take up moisture from the atmosphere. Wood is hygroscopic.

I-beam A configuration of materials in the form of an "I," designed to enhance bending strength and stiffness by the strategic placement of material at points of maximum stress.

Improved Wood A general term for wood that has been specially treated in various ways to reduce hygroscopicity, reduce absorbing capacity, or to increase strength and surface qualities. Treatments include heating, impregnating with different resins or esterifying agents, and densification. Some specific terms include:

Compreg Wood that has been impregnated with synthetic resins such as phenol-formaldehyde, and densified under heat and pressure.

Flapreg A form of improved flakeboard, similar to compreg.

Impreg Wood treated as for compreg but not compressed.

Staypak Wood that has been compressed to a high density under heat and pressure but not impregnated with resins.

Internal Bond (IB) Tensile strength perpendicular to the surface of a reconstituted product; taken to be a measure of the bond strength between elements, but influenced by the amount of interfelting that develops in mat formation.

Joist One of a series of parallel beams used to support floor and ceiling loads and supported in turn by larger beams, girders, or bearing walls.

Laminate 1. A product made by bonding together two or more layers of material or materials. 2. The process of bonding such a product.

Laminate, Cross A laminate in which some of the layers of material are oriented at right angles to the remaining layers with respect to the grain or strongest direction in tension.

Laminate, Parallel A laminate in which all the layers of material are oriented with grain parallel to each other.

Laminated Wood An assembly made by bonding layers of veneer or lumber with the grain parallel.

Lamination Any layer in a laminate.

Latewood See SUMMERWOOD.

Lathe Checks Fissures formed along the grain of veneer during cutting due to splitting ahead of the knife. The checks form on the veneer side of the knife, producing the "loose side." The log side of the knife remains relatively intact, and will form the "tight side" of the next layer of veneer to be cut.

Linear Expansion Change in length with change in moisture content, an important property in reconstituted materials.

Longitudinal A direction in wood that is parallel to the direction of the fibers.

Loose Side See LATHE CHECKS.

Lumen The void space or cavity within plant cells, the total volume of which determines the density of a piece of wood; its size has a bearing on the porosity of wood.

Mat An assembly of comminuted wood formed loosely in prescribed composition, structure, and size of the product prior to consolidation.

Mat-formed Particleboard A board that is formed into a mat before consolidation, as opposed to a board that is consolidated as it is formed. At the present stage of technology, the distinction separates face-pressed boards from edge-pressed or extruded boards.

Modifier Any ingredient added to an adhesive formulation that changes its properties.

Modulus of Elasticity (MOE) The ratio of stress to corresponding strain below the proportional limit, reduced to unity (psi in USA) by accounting for dimensional parameters; a measure of stiffness.

Modulus of Rupture (MOR) The value of maximum tensile or compressive stress (whichever causes failure) in the extreme fiber of a beam loaded to failure in bending, reduced to unity (psi in USA) by accounting for dimensional parameters (a measure of strength).

Moisture Content The amount of water contained in the wood, usually expressed as a percentage of the weight of the oven-dry wood.

Nail Gluing The use of nails to deliver bonding pressure.

Nail Popping Protrusion of nailheads above the surface after they have been driven, due to shrinking and swelling of wood.

Nominal Size As applied to timber or lumber, the size by which it is known and sold on the market; often differs from the actual size because shrinkage and surfacing losses are included.

Nondestructive Testing A means of measuring the strength of materials without breaking them, involving such techniques as proof loading, sonics, and piezoelectric properties.

On-site Bonding Gluing carried on at a construction site, often under outdoor—and therefore uncontrolled—conditions.

Open Assembly See ASSEMBLY TIME.

Orthotropic Having unique and independent properties in three mutually orthogonal (perpendicular) planes of symmetry. A special case of ANISOTROPY.

Oven-dry Wood Wood dried to a relatively constant weight in a ventilated oven at 214 to 221°F (101 to 105°C).

Overlay A thin layer of paper, plastic film, metal foil, or other material bonded to one or both faces of panel products or to lumber so as to provide a protective or decorative face or a base for painting.

Particle A generic term representing a distinct fraction of comminuted wood or other lignocellulosic material produced mechanically. Types of particles include:

Flake Thin, flat, leaflike element, essentially a small piece of veneer, rectangular in shape, sometimes square and sometimes longer along the grain than across.

Granule A small element the dimensions of which are approximately equal, such as coarse sawdust.

Shaving A small wood element varying in dimensions having a peculiar structure because of its method of production, being a by-product of surfacing operations involving rotary cutterheads. The cutting action produces a thin chip varying in thickness from one edge to the other, usually curled, and with short grain throughout.

Sliver An element having small but equal cross-section dimensions and a length parallel to the grain of more than four times the thickness.

Splinter An alternative term for sliver.

Strand A narrow flake.

Wafer A thick flake.

Wood wool (excelsior) Long, slender, thin elements of wood, usually curled, made by scoring and slashing knives along the grain.

Particleboard A generic term for boards made from any of the comminuted wood elements. In practice the term refers to those boards composed of the more granular or sliver-type elements.

Particleboard Corestock A common term given to a particleboard manufactured specifically for use as a core (smooth to surface) to be overlayed with veneer or film.

Particleboard Underlayment A grade of particleboard made or sanded to close thickness tolerances for use as a leveling course, and to provide a smooth surface under floor-covering materials.

Permeability The ease with which liquids or gases pass into or through a material.

Performance-rated Panels Panels that have been given a use classification based upon a series of tests, composition, and quality of workmanship. This avoids the publication of values that must then be interpreted by the engineer or contractor to determine suitability for a given purpose.

Permanence See DURABILITY.

Piezoelectric Effect An electrical polarization produced in a crystal by the application of mechanical stress. Because portions of the cellulose in wood is crystalline, it too exhibits this effect. Bearing a relationship to density and modulus of elasticity, the effect can be used in nondestructive tests of wood.

Pith The small, soft portion of a tree trunk or limb produced during the early stages of growth.

Plank A broad board, usually 2 inches or more in thickness, laid with its wide dimension horizontal and used as a bearing surface to span supporting members.

Plastic Laminate See DECORATIVE LAMINATE.

Plasticizer A material incorporated in an adhesive to increase its flexibility, or distensibility. The addition of the plasticizer may cause a reduction in melt viscosity, lower the temperature of the second-order transition, or lower the elastic modulus of the solidified adhesive.

Platens Thick metal plates usually heated with steam or hot oil, between which panel assemblies are pressed.

Plywood A panel composed of an assembly of veneer or veneer in combination with different core materials. Except for special constructions, the grain of alternate plies is always approximately at right angles.

Polymer A compound formed by the reaction of simple molecules having functional groups that permit their combination to proceed to high molecular weights under suitable conditions. When two or more different monomers are involved, the product is called a copolymer. The reaction is called *polymerization.*

Pore Anatomically, a pore is a vessel element such as those in hardwoods. The opening in a wood surface created by such an element.

Porosity A term that describes the openness of a wood piece or its penetrability by liquids. Thus porosity can vary in the same piece of wood depending upon which surface is addressed. Although porosity may be defined as the ratio of the void volume to the solid volume of a piece of wood in terms useful to gluing or treating, it must be observed by direct testing.

Postcure Additional cure after pressing.

Pot Life The period of time during which an adhesive, after mixing with catalyst or other ingredients, remains suitable for use.

Power Factor Also known as the *dielectric loss factor,* a measure of the power per unit volume that will be converted to heat in a nonconducting material by an external electric field that is oscillating with a given amplitude and frequency.

Precure The curing of a resin after application to the adherend and before pressure is applied, resulting in no bond-forming actions. The term DRY-OUT (which see), has similar connotations and similar effects but its cause is not curing too soon, but loss of solvent. The term *Prehardened* covers both causes.

Prepressing A cold-press operation preceding hot pressing designed, in the case of plywood, to achieve early transfer of adhesive to the unspread surface, and to develop enough tack in the bond line to hold veneers together, keep them from curling, and permit easier loading into the narrow openings of the hot press. In the case of mat-formed products, prepressing compacts the mat and provides increased integrity of the mat for transporting through cut-off operations, weigh stations, and press loaders. It also reduces the height of a mat so that less opening is needed between platens, and less time is needed to close the press.

Preservative A substance impregnated into wood for the purpose of reducing or preventing decay, mildew, borer, or insect attack. A *water-repellant preservative* is one that includes a substance to reduce water absorption.

Pressing to Caliper Applying pressure to a mat until its final thickness has been reached.

Pressing to Stops Applying pressure to a mat until the platens seat themselves on stops. See also STOPS.

Press-cycle Time Total time of the pressing operation, including the time necessary to load and close the press on the assembly.

Press Time The time during which the assembly is actually under pressure.

Profile See DENSITY PROFILE.

Proportional Limit The point in the stress/strain curve where the relationship ceases to be constant, but instead produces greater strain for the same increment of stress. See also ELASTIC LIMIT.

Psychrometer An instrument for measuring the amount of water vapor in the atmosphere. It has two thermometers, one dry and the other wet, kept moist by a knitted sock dipping into a water tube. Since evaporation is a cooling process and related to the dryness of the air, the wet

bulb reading is always lower, and the difference between the two thermometers is used, by reference to charts, to obtain the relative humidity of the atmosphere.

Pyrometer An instrument for measuring temperature, usually employing bimetallic elements that either produce a measurable current or a differential expansion when heated.

Racking The deformation of a rectangular structure subjected to a lateral force when anchored at one edge but with the other edges free to move.

Radio Frequency (RF) Heating A method of generating heat within a dielectric material (dry wood being one), by use of relatively low electric current but high voltage at high (radio) frequencies. Three aspects of the process are practiced: 1. in which the RF energy is directed parallel to the glue lines (*parallel heating*), and heating primarily the glue lines because they are more conductive (e.g., edge gluing lumber); 2. in which the RF energy is directed perpendicular to the glue lines (*perpendicular heating*), in this case heating both the glue lines *and* the intervening wood (e.g., plywood); 3. a combination of 1 and 2 in which positive and negative electrodes are at the same face of the assembly, and the current seeks the more conductive path, the glue line (*stray field heating*), used primarily in assembly gluing of parts or edge banding.

Relative Humidity The ratio of the amount of water vapor present in the air to that which the air would hold at saturation at the same temperature, reduced to a percentage.

Resin A solid or semisolid organic substance of natural or synthetic origin of high molecular weight or convertible to high molecular weight.

Resin Content A term used in glued wood products referring to the amount of dry resin solids present expressed as a percentage of the dry weight of the wood present.

Rheology The science of matter that deals with flow and deformation in response to stress.

Ring Flaker   A flaking machine in which flaking is accomplished by knives mounted axially on the inside surface of a rotating cylinder. A concentric impeller rotating at high speed in the opposite direction forces chips or other small wood pieces against the knife edges, producing flakes. See also DRUM FLAKER.

Sander Dust   Very fine particles produced during sanding operations; usually burned or used as an additive in plastics or adhesive formulations.

Sandwich Panel   A layered assembly comprised of relatively thin facing materials bonded to a low density core material. A *structural sandwich panel* would be one in which the facing materials were of high strength and the adhesive of high durability.

Sapwood   The wood, usually pale in color, near the outside of a tree, still participating in the life processes of the tree. In general sapwood is more permeable and more susceptible to decay than heartwood.

Sawdust   Small, granular wood particles produced by the action of saw teeth.

Scarf Joint   A method of joining board or panel members in which the edges of each are oppositely sloped through the thickness to provide greater bonding area.

Set   The conversion of an adhesive from a fluid to a solid state by chemical or physical action or both. Also used synonymously with cure, although the latter often connotes conversion by chemical action, while set implies physical action such as loss of solvent. The term *harden* is less specific. See also CURE.

Shape Stability   The resistance of a material or structure to distortion by atmospheric conditions, such as warping. See also DIMENSIONAL STABILITY.

Shaving   See PARTICLE.

Shear   A condition of stress or strain in which parallel planes slide relative to one another.

Shear Strength   The load a material is able to withstand when subjected to a shear effect. It is measured as the load at rupture

reduced to unity by dividing by the area carrying the load.

Sheathing   The covering, usually structural, such as fiberboards flakeboards, or plywood, placed over exterior studding or rafters of a building. It serves many purposes including strength and closure.

Showthrough   The appearance of blemishes on the surface of a panel due to defects in the interior. See also TELEGRAPHING.

Siding   The finish covering of the outside wall of a frame building, such as clapboards, shingles, and so on.

Simultaneous Closing   The closing of all the openings in a multiopening press at the same time rather than sequentially as they do in some presses.

Single Opening Press   A press that has only one opening for materials to be pressed.

Size   1. To cut to a specified dimension. 2. An additive such as alum, wax, asphalt, resin, or other material incorporated into a product primarily to improve water resistance or surface properties.

Sliver   See PARTICLE.

Solids Content   The percentage by weight of the nonvolatile matter in an adhesive.

Specific Gravity   The ratio of the weight of a material to the weight of an equal volume of water. In the case of wood, the weight is usually that of oven dry, and the volume is either in the green condition (standard) or as measured.

Spread   The quantity of adhesive per unit joint area applied to an adherend, for example, in pounds of adhesive per thousand square feet of glue line area. This quantity may be applied only to one of the two joint surfaces, *single spread,* or to both of the joint surfaces, *double spread.*

Springback   The tendency of compressed wood to recover some or all of its uncompressed state.

Springwood   See EARLY WOOD.

Squeeze-out   The glue expressed from a joint when pressure is applied.

Starved Glue Line   Insufficient adhesive on

the glue line to form all links of a bond, due either to loss by various mechanisms of the amount applied or to inadequate amount applied.

Static Bending Bending under a constant or slowly applied load.

Stickering The use of strips of material (stickers), usually wood, between courses of lumber or panels in a pile to facilitate air circulation and promote even cooling, heating, or drying.

Stops Long, narrow strips or pieces of metal machined to precise thicknesses, used as spacers along the edges of platens to assure a desired thickness during consolidation of comminuted materials.

Storage Life The period of time during which a packaged adhesive can be stored under specific temperature conditions and remain suitable for use. Sometimes called *shelf life*.

Strain The unit change in the size or shape of a body due to the application of a force compared to its original dimension. The calculation theoretically results in a dimensionless quantity, but in practice the dimensions are retained as inches per inch or centimeters per centimeter.

Strand See PARTICLE.

Strength Usually assumed to be the load or stress a member can carry without breaking. It is measured, however, as the maximum load or stress sustained when loaded to failure.

Strength, Wet The strength of an adhesive joint or material measured after immersion in a liquid for a specified time, temperature, and pressure.

Stress The external force per unit area developed in resistance to loading, or under certain conditions, self-generated internally as a result of variations in moisture, temperature, or both.

Stress-graded Lumber Lumber that has been judged by visual characteristics or by nondestructive testing to have a specified strength or performance class.

Stressed Skin Construction A special sandwich-type construction in which panels are glued to a framed core as in a stud wall, producing an I-beam effect where the panels act as the flanges carrying the applied loads.

Structural Adhesive A bonding agent used in materials destined to be used in buildings—i.e., high durability—or an adhesive to be used at a building site.

Structural Board (Timber) A panel or laminated beam of high strength and durability suitable for use in load-bearing structures.

Stud One of a series of slender wood structural members used as supporting elements in walls and partitions, traditionally 2 × 4's.

Subfloors Boards or panels laid on joists over which a finish floor is to be laid.

Substrate A material upon the surface of which an adhesive-containing substance is spread for any purpose, such as bonding or coating. See also ADHEREND.

Summerwood The portion of the annual growth ring that is formed later in the growing season (also called *latewood*). It is denser and stronger than *springwood*.

Surfaced Lumber Lumber that has been dressed from its rough-sawn condition, usually by planing, to provide a smoother surface for subsequent operations.

Tack The property of an adhesive that enables it to form a bond of measurable strength immediately after adhesive and adherend are brought into contact under low pressure, i.e., stickiness.

Tack, Dry The property of certain adhesives to adhere to themselves on contact even though they seem dry to the touch.

Tangential With respect to a log, a plane that is tangent to the annual rings. Flat sawn is lumber that came from such a plane.

Tannin Wood or bark extractive that has phenolic properties. Like phenol, it is capable of reacting with formaldehyde to form resins.

Telegraphing A condition in a laminate or other type of composite construction in which irregularities, imperfections, or patterns of an inner layer are visibly transmitted to the surface. This condition is also referred to as *photographing*. See SHOWTHROUGH.

Temperature, Curing The temperature needed by an adhesive to achieve cure. (Time is also a factor in curing; time at the specified temperature and time to reach the specified temperature).

Tempered Hardboard Hardboard that has been specially treated with a drying oil and heat to greatly improve its properties, particularly strength and moisture resistance.

Tension Wood A form of wood found in leaning trees and branches of some hardwood trees, and characterized anatomically by the presence of gelatinous fibers. Tension wood exhibits excessive longitudinal shrinkage and fuzzy sawn surfaces and is inclined to severe warpage. See COMPRESSION WOOD.

Thermal Conductivity The quantity of heat that flows in unit time across unit area of a substance of unit thickness when the temperature of the faces differs by one degree.

Thermal Expansion A reversible increase in dimensions due to an increase in temperature.

Thermocouple Two dissimilar elements, usually wires twisted and soldered together, which produce a definite electromotive force at different temperatures. Calibrated and used to measure temperature.

Thermoplastic A material that softens when heated and hardens when cooled, i.e., remains fusible. It also remains soluble.

Thermoset A material that has undergone a chemical reaction that produces an insoluble and infusible state.

Thinner A volatile liquid added to a mixture to modify consistency or other properties.

Thixotropy A property of liquid mixtures to appear thinner under agitation and to thicken at rest.

Three-layer Board A reconstituted board in which face material differs from core material; one of the important means by which board properties can be engineered for specific purposes.

Torque A force applied through a lever arm that tends to twist or rotate a body, such as a wrench on a nut. It is numerically expressed as the product of the force and the length of the lever arm. With the force in pounds and the lever arm in feet the units would be in foot-pounds.

Torque Wrench A wrench equipped with a device to measure torque.

Torsion A turning or twisting action at one end of a body while the other end is held fast.

Truss A special case of an I-beam in which the web is formed as a series of short members arranged and connected in triangles.

Twist A distortion in a product such that the four corners of any face are no longer in the same plane.

Underlayment A sheet material, part of a floor system, placed under finish coverings to provide a smooth surface.

Vapor Barrier A relatively impermeable sheet such as aluminum foil, plastic film, or specially coated paper, used to control moisture movement in or out of a material or building.

Vehicle The liquid portion of an adhesive or a finishing material.

Veneer A thin layer or sheet of wood generally produced by knife action in lathes (*rotary veneer*), or sliced (*sliced veneer*). Very thick veneer, or species that are difficult to knife cut, are sawn (*sawn veneer*).

Viscosity A quality of the fluid properties of a liquid that reflects its ability to flow; measured as a resistance to flow as through a hole in a cup, or a resistance to shear as in stirring.

Volatile Solvent Any nonaqueous liquid that has the distinctive property of evaporating readily at room temperature and atmospheric pressure.

Warp Any variation from a true plane, including bow, cup, twist, or any combination.

Water Absorption A property measured

as the amount of water taken up during prescribed conditions of time, temperature, size of specimen, and depth of immersion in water.

Waterproof A term applied to materials bonded with highly resistant adhesives, capable of withstanding prolonged exposure to severe service conditions without failure in the glue bonds. A typical criteria is ability to withstand boiling water.

Water Resistant A term applied to materials bonded with adhesives capable of withstanding limited immersion in water at room temperature.

Wax Emulsion A sizing agent incorporated into reconstituted products to reduce water absorption.

Weathering The effects, mechanical and chemical (but not decay), upon wood or wood products, caused by exposure to atmospheric forces such as sunlight, wind, dust, heat, and moisture changes.

Webbing Filaments or threads that may form when adhesives are transferred from one surface to another while in the fluid state.

Wettability A condition of a surface that determines to what extent a liquid will be attracted to it, affecting absorption, adsorption, penetration, and spreading.

Wet Felting Forming a mat of fibrous elements from a suspension in water.

Wood Element Any subdivision of a log that may be assembled into a structure by fastening or gluing. In the context of this book they begin with lumber sizes and end with fiber sizes.

Wood Failure The rupturing of a bond in the wood part of the joint; expressed quantitatively as a percentage of the joint surface that has failed in the wood.

Wood Flour Wood reduced to finely divided particles approximating those of cereal flours in size, appearance, and texture.

Wood, Glued-laminated A bonded assembly of veneer or lumber with the grain of all laminations parallel to each other.

Working Life The period of time during which an adhesive, after mixing with catalyst, solvent, or other compounding ingredients, remains suitable for use. See also POT LIFE and STORAGE LIFE.

Wood Wool See PARTICLE.

Working Properties The properties of an adhesive that affect or dictate the manner of application to the adherends to be bonded and the assembly of the joint before pressure application, such as viscosity, working life, permissible assembly time, curing time, curing temperature, and so on.

Yield The ratio of product output to raw material input expressed as a percentage.

Yield Value The level of stress in a material that causes increased strain without an increase in load, suggesting that molecular bonds are being broken rather than merely deformed.

Some of the definitions in this glossary have been drawn from the following sources, and in many cases modified as deemed appropriate:

American Society for Testing and Materials
Borden Chemical Division of Borden, Inc.
Heritage Series, Volume III, Pennsylvania State University Press
Modern Particleboard, Maloney, Miller Freeman Publications

Index